Lecture Notes in Computer Science   12931

More information about this subseries at https://link.springer.com/bookseries/7407

Dimitris E. Simos · Panos M. Pardalos ·
Ilias S. Kotsireas (Eds.)

Learning and Intelligent Optimization

15th International Conference, LION 15
Athens, Greece, June 20–25, 2021
Revised Selected Papers

 Springer

Editors
Dimitris E. Simos 🅙
SBA Research
Vienna, Austria

Panos M. Pardalos 🅙
University of Florida
Gainesville, FL, USA

Ilias S. Kotsireas 🅙
Wilfrid Laurier University
Waterloo, ON, Canada

ISSN 0302-9743 ISSN 1611-3349 (electronic)
Lecture Notes in Computer Science
ISBN 978-3-030-92120-0 ISBN 978-3-030-92121-7 (eBook)
https://doi.org/10.1007/978-3-030-92121-7

LNCS Sublibrary: SL1 – Theoretical Computer Science and General Issues

This Springer imprint is published by the registered company Springer Nature Switzerland AG
The registered company address is: Gewerbestrasse 11, 6330 Cham, Switzerland

Guest Editorial

The fifteenth installment of the conference series "Learning and Intelligent Optimization" (LION 15) was scheduled to be held in Athens, Greece, during June 20–25, 2021, but regrettably it was canceled, due to travel restrictions imposed world-wide by the COVID-19 pandemic. However, we were fully prepared to convert the event into an all-digital conference experience. Moreover, we felt it was important to publish the proceedings of the conference, in order to minimize the disruption to the participant's careers and especially the potentially devastating negative effects in the careers of PhD students, post-doctoral fellows, and young scholars. An additional reason for us to undertake the publication of these LNCS proceedings, was to ensure the continuity of the LION conference series.

LION 15 featured five invited speakers:

- "$B_k - VPG$ Graphs – the String Graphs of Paths on a Grid", plenary talk given by Martin Charles Golumbic (University of Haifa, Israel)
- "Communication and Mobility in Optimization for Infrastructure Resilience", plenary talk given by Evangelos Kranakis (Carleton University, Canada)
- "Temporal Networks and the Impact of Availability Patterns", plenary talk given by Paul Spirakis (University of Liverpool, UK, and University of Patras, Greece)
- "Combinatorial Difference Methods in AI", tutorial talk given by Rick Kuhn (NIST, USA)
- "On the Use of Ontologies for Automated Test Suite Generation", tutorial talk given by Franz Wotawa (Graz University of Technology, Austria)

We would like to thank the authors for contributing their work and the reviewers whose tireless efforts resulted in keeping the quality of the contributions at the highest standards. The volume contains 30 refereed papers carefully selected out of 48 total submissions, thus LION 15 bears an overall acceptance rate of 62%.

The editors express their gratitude to the organizers and sponsors of the LION 15 international conference:

- MATRIS Research Group, SBA Research, Austria
- Laboratory of Algorithms and Technologies for Networks Analysis (LATNA), Higher School of Economics (HSE), Niznhy Novgorod, Russia
- CARGO Lab, Wilfrid Laurier University, Canada,
- APM Institute for the Advancement of Physics and Mathematics.

A special thank you goes to the Strategic Innovation and Communication Team at SBA Research (Nicolas Petri and Yvonne Poul) for sponsoring the virtual infrastructure of the conference, as well as the LION 15 volunteers (junior and senior researchers of the MATRIS Research Group at SBA Research) who made sure that the virtual technical sessions could be carried out flawlessly.

Even though organization of all physical conferences is still on hiatus, we are very pleased to be able to deliver this LNCS proceedings volume for LION 15, in keeping with the tradition of the most recent LION conferences [1, 2] and [3]. We sincerely hope we will be able to reconnect with the members of the vibrant LION community next year.

October 2021

Dimitris E. Simos
Panos M. Pardalos
Ilias S. Kotsireas

References

1. Roberto Battiti, Mauro Brunato, Ilias S. Kotsireas, Panos M. Pardalos: Learning and Intelligent Optimization - 12th International Conference, LION 12, Kalamata, Greece, June 10–15, 2018, Revised Selected Papers, Lecture Notes in Computer Science, LNCS 11353, Springer, (2019).
2. Nikolaos F. Matsatsinis, Yannis Marinakis, Panos M. Pardalos: Learning and Intelligent Optimization - 13th International Conference, LION 13, Chania, Crete, Greece, May 27–31, 2019, Revised Selected Papers, Lecture Notes in Computer Science, LNCS 11968, Springer, (2020).
3. Ilias S. Kotsireas, Panos M. Pardalos: Learning and Intelligent Optimization - 14th International Conference, LION 14, Athens, Greece, May 24–28, 2020, Revised Selected Papers, Lecture Notes in Computer Science, LNCS 12096, Springer, (2020).

Organization

General Chair

Panos M. Pardalos   Higher School of Economics, Niznhy Novgorod, Russia/University of Florida, USA

Technical Program Committee Chair

Dimitris E. Simos   SBA Research and Graz University of Technology, Austria/NIST, USA

Local Organizing Committee Chair

Ilias S. Kotsireas   Wilfrid Laurier University, Canada

Program Committee

Francesco Archetti	Consorzio Milano Ricerche, Italy
Annabella Astorino	ICAR-CNR, Italy
Amir Atiya	Cairo University, Egypt
Rodolfo Baggio	Bocconi University, Italy
Roberto Battiti	University of Trento, Italy
Christian Blum	Spanish National Research Council (CSIC), Spain
Juergen Branke	University of Warwick, UK
Mauro Brunato	University of Trento, Italy
Dimitrios Buhalis	Bournemouth University, UK
Sonia Cafieri	Ecole Nationale de l'Aviation Civile, France
Antonio Candelieri	University of Milano-Bicocca, Italy
Andre de Carvalho	University of São Paulo, Brazil
John Chinneck	Carleton University, Canada
Kostas Chrisagis	City University London, UK
Andre Augusto Cire	University of Toronto, Canada
Patrick De Causmaecker	Katholieke Universiteit Leuven, Belgium
Renato De Leone	University of Camerino, Italy
Luca Di Gaspero	University of Udine, Italy
Clarisse Dhaenens	Université de Lille, France
Ciprian Dobre	University Politehnica of Bucharest, Romania
Adil Erzin	Sobolev Institute of Mathematics, Russia
Giovanni Fasano	University Ca'Foscari of Venice, Italy
Paola Festa	University of Napoli Federico II, Italy

Antonio Fuduli	Università della Calabria, Italy
Martin Golumbic	University of Haifa, Israel
Vladimir Grishagin	Nizhni Novgorod State University, Russia
Mario Guarracino	ICAR-CNR, Italy
Youssef Hamadi	Uber AI, France
Cindy Heo	Ecole hôtelière de Lausanne, Switzerland
Laetitia Jourdan	Université de Lille, France
Valeriy Kalyagin	Higher School of Economics, Russia
Alexander Kelmanov	Sobolev Institute of Mathematics, Russia
Marie-Eleonore Kessaci	Université de Lille, France
Michael Khachay	Krasovsky Institute of Mathematics and Mechanics, Russia
Oleg Khamisov	Melentiev Institute of Energy Systems, Russia
Zeynep Kiziltan	University of Bologna, Italy
Yury Kochetov	Sobolev Institute of Mathematics, Russia
Ilias Kotsireas	Wilfrid Laurier University, Canada
Dmitri Kvasov	University of Calabria, Italy
Dario Landa-Silva	University of Nottingham, UK
Hoai An Le Thi	Université de Lorraine, France
Daniela Lera	University of Cagliari, Italy
Vittorio Maniezzo	University of Bologna, Italy
Silvano Martello	University of Bologna, Italy
Francesco Masulli	University of Genova, Italy
Nikolaos Matsatsinis	Technical University of Crete, Greece
Kaisa Miettinen	University of Jyväskylä, Finland
Laurent Moalic	University of Haute-Alsace, France
Hossein Moosaei	Charles University, Czech Republic
Serafeim Moustakidis	AiDEAS OU, Greece
Evgeni Nurminski	FEFU, Russia
Panos M. Pardalos	Higher School of Economics, Niznhy Novgorod, Russia/University of Florida, USA
Konstantinos Parsopoulos	University of Ioannina, Greece
Marcello Pelillo	University of Venice, Italy
Ioannis Pitas	Aristotle University of Thessaloniki, Greece
Vincenzo Piuri	Università degli Studi di Milano, Italy
Mikhail Posypkin	Dorodnicyn Computing Centre, FRC CSC RAS, Russia
Oleg Prokopyev	University of Pittsburgh, USA
Helena Ramalhinho	Universitat Pompeu Fabra, Spain
Mauricio Resende	Amazon, USA
Andrea Roli	University of Bologna, Italy
Massimo Roma	Sapienza Università di Roma, Italy
Valeria Ruggiero	University of Ferrara, Italy
Frédéric Saubion	University of Angers, France
Andrea Schaerf	University of Udine, Italy
Marc Schoenauer	Inria, France
Meinolf Sellmann	GE Research, USA

Contents

An Optimization for Convolutional Network Layers Using the Viola-Jones Framework and Ternary Weight Networks

Rhys Agombar$^{(\boxtimes)}$, Christian Bauckhage, Max Luebbering, and Rafet Sifa

Fraunhofer Institute for Intelligent Analysis and Information Systems, Schloss Birlinghoven, 53757 Sankt Augustin, Germany

Abstract. Neural networks have the potential to be extremely powerful for computer vision related tasks, but can be computationally expensive. Classical methods, by comparison, tend to be relatively light weight, albeit not as powerful. In this paper, we propose a method of combining parts from a classical system, called the Viola-Jones Object Detection Framework, with a modern ternary neural network to improve the efficiency of a convolutional neural net by replacing convolutional filters with a set of custom ones inspired by the framework. This reduces the number of operations needed for computing feature values with negligible effects on overall accuracy, allowing for a more optimized network.

Keywords: Ternary Weight Networks · Viola-Jones Object Detection Framework · Neural net optimization · Haar-like features

1 Introduction

With the advent of convolutional neural networks, the field of computer vision was revolutionized. Suddenly we had access to a system that produced incredibly powerful and generalizable classifiers. While modern neural networks perform extremely well, they do have the downside of requiring large amounts of memory and processing power. As applications like self-driving cars or augmented reality have emerged, real-time execution has become an increasingly desirable property for a system to have - something that complex neural networks struggle with, due to their computational complexity.

Our paper is an attempt to solve this problem. One of our key insights is that, while neural networks are powerful but slow, classical methods of computer vision tend to be relatively fast and light weight. This indicates that there may be speed-ups to be gained if we found a way to integrate some useful properties or components from these classical methods into a modern neural network architecture. Of particular interest to us is the Viola-Jones Object Detection Framework

Supported by the Competence Center for Machine Learning Rhine Ruhr (ML2R) which is funded by the Federal Ministry of Education and Research of Germany (grant no. 01—S18038B). The OrcidIDs for R. Agombar and Max Lueberring are resp. 0000-0001-5574-9754 and 0000-0001-6291-9459.

D. E. Simos et al. (Eds.): LION 2021, LNCS 12931, pp. 1–6, 2021.
https://doi.org/10.1007/978-3-030-92121-7_1

[7] from 2001, which is capable of detecting objects like faces extremely quickly, even on weak hardware. While this framework is quick at detecting faces in an image, however, it does run into the same problems that classical methods tend to have, namely, a difficulty generalizing or handling things like noise. Some of its components have the potential to be easily transferred to convolutional networks, though, which could increase performance, giving us an ideal starting point for our work.

Our contribution in this paper is a combination of components from the Viola-Jones Object Detection Framework with a modern, convolutional neural net. By training a ternary convolutional network, and then modifying the first layer's weights to match the structure of the features used by Viola-Jones, we are able to accelerate the first layer of a network with little to no cost to the network's overall accuracy.

2   Background

The Viola-Jones Object Detection Framework was originally proposed in the paper 'Rapid object detection using a boosted cascade of simple features' [7]. It was designed to detect faces or other objects by sliding a small detection window across a larger image, classifying areas as containing a face (or object) or not as it goes. These classifications were made using Haar-like features within the sliding window, which were inspired by Haar basis functions.

Haar-like features are made of a single large rectangle that itself is composed of multiple, smaller, sub-rectangles representing positive or negative regions. These features can be positioned anywhere in an image or detection window, and they compute their feature values by summing the pixel values in the positive regions and subtracting the sums of the negative regions from them. The main benefit of these features is that, when using an integral image as an input, because of their rectangular shape, the value of any sub-rectangle can be computed using only 3 math operations and 4 array lookup operations, regardless of size. In the paper [7], the number of sub-rectangles was limited to two, three, or four, depending on the type of feature, meaning that, when using an integral image, they would require at most, 14, 21, or 28 total operations to compute their values, allowing for extremely fast execution.

The integral images used for this are data structures that are optimized towards the specific task of summing array values within a rectangular area. They are created such that the value at every (x, y) coordinate in them is equal to the sum of all the original image intensities above, to the left of, and on the coordinates of the given point.

Moving to the neural network part of the background, as previously mentioned, modern networks, while powerful, tend to require a lot of memory to run and can be slow to process. Ternary Weight Networks [4] were designed as a solution to this. The main idea is that the weights of a network's neurons can be approximated with ternary values. By restricting the values to -1, $+1$, or 0, instead of requiring 32-bit floating point representations in memory, the weight values require only two bits to be stored, thus producing a $16\times$ reduction in memory use. Additionally, network operations on ternary weights can be estimated using only addition and subtraction, without computationally expensive multiplication operations. This means that an approximately $2\times$ increase in speed can be achieved while maintaining similar accuracy to a standard network [4].

Now, the problem of selecting which ternary value to use for a given weight in a network is non-trivial, as is the question of how to use gradient descent on a network made of discrete values. The paper 'Ternary Weight Networks' [4] details methods to solve both of these using things like threshold-based discrete value optimizations, and alterations to the training process, but these are out of the scope of this paper, so we will not discuss them in depth.

3   Methodology

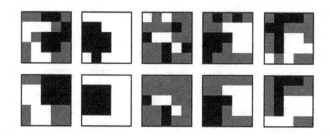

Fig. 1. Ternary weights from our LeNet-5 MNIST experiment. The top row is the unmodified ternary weights, the bottom row is filled with our corresponding custom ternary weights (grey represents 0 values, white represents -1, and black represents 1).

We've mentioned before that Haar-like features are composed of positive and negative rectangles, and that they can be arranged in any position in a given image, possibly ignoring large parts of it. These positive, negative, and uncovered regions can be represented by the values of $+1$, -1, and 0, respectively, which correspond to the same values present in a ternary weight network. This means that we can represent Haar-like features, or other similar ones, directly in a ternary convolutional network's convolutional filters (Fig. 1).

Now, limiting our Haar-like features to the arrangements defined in [7] is restrictive, and would hamper a network's ability to learn. Thankfully, so long as they are composed of rectangles, any arbitrary arrangement should still produce a speed-up by reducing the number of operations. For example, a 5 convolutional filter requires $5 \times 5 - 1 = 24$ math operations and $5 \times 5 = 25$ lookup operations, equaling a total of 49 operations. For contrast, when using integral images and these custom Haar-like convolutional filters, a custom filter would require only 4 lookup operations and 3 math operations, per rectangle. If we assume our filters will always require four rectangles (the same as the most complex Haar-like features in [7]), then regardless of the filter's size, its value can always be computed with a total of $4(4+3) = 28$ operations, which is a significant reduction for larger filter sizes.

The way we put all of this together is by first training a ternary network using the processes detailed in [4], then visualizing the weights for transformation. To transform the network's weights into Haar-like ones, we modify the ternary weights of the first layer by hand, such that the new ones are composed of up to four rectangles. We do this by trying to best match the set of rectangles to the patterns observed in the visualized weights, minimizing the number of missed or wrong patches, while maximizing the expressive abilities of the modified filters. Once these custom filters have been defined, we have to retrain the network, but first, we remove the top convolutional layer and replace it with a function that can leverage the rectangular filter structures and integral image inputs to get the speedup we want. Without this, the convolutional filters would behave exactly like ordinary ones, with no efficiency gains.

Table 1. A comparison of number of computations required for a single filter to process a 28×28 greyscale image.

	3×3	5×5	7×7
Integral image operations	3025	3025	3025
Custom filter operations	21952	21952	21952
Total operations	24977	24977	24977
Ternary filter operations	13328	38416	76048
(Total custom/ternary)	**1.84**	**0.65**	**0.33**

Table 1 shows the theoretical performance for this method, compared to using standard convolutional kernels. Note that while our method performs worse on 3×3 filters, as the size of the filters increases, so to does our method's efficiency. This means that it still has potential for use with networks that use large filters.

4   Experiments

While our calculations show improved computational efficiency, this is only useful if the network's accuracy remains at an acceptable level. To determine if our solution meets this requirement, we conducted a set of experiments on the MNIST [3] and CIFAR-10 [1] datasets, using modified LeNet-5 [2] and VGG7 [4] architectures, respectively.

For LeNet-5, we used the implementation given in the Ternary Weight Networks paper [4], which is summarized as "32-C5+MP2+64-C5+MP2+512-FC+ SVM". The one modification that we made was that we replaced the SVM with a Softmax classification layer. For parameters, the network used the Adam algorithm for optimization with a weight decay of 1e−5, and a learning rate of

0.001. For inputs, the network's batch size was set to 100, with the data being normalized.

VGG7 is a modified version of VGG16 [6]. It was also published in the afore-mentioned paper [4], and can be summarized as "2 × (128-C3) + MP2 + 2 × (256-C3) + MP2 + 2 × (512-C3) + MP2 + 1024-FC + Softmax". Again, we used a similar implementation, though, since our method only gives useful efficiency gains when using larger feature sizes, we replaced the first two 3 × 3 convolutional layers with a pair of 5 × 5 ones. This network used Stochastic Gradient Descent (SGD) as its optimization algorithm, with a learning rate of 0.1, a momentum value of 0.9, and a weight decay of 1e−4. For inputs, it used a batch size of 100, as well as normalized batch data augmented by random crops and horizontal flips. The last parameter, the learning rate decay, was set to divide the given rate by 10 after 80 and 120 epochs, in order to fine-tune the network.

For each of these architectures, five different networks are trained: a full precision network, a full ternary network, an N1T network, a custom full ternary network, and a custom N1T network. The 'full' and 'custom' networks are self explanatory, being entirely implemented in their chosen style or using our custom filters for the first layer, respectively. The N1T and custom N1T networks are full precision networks, except for the first layer, which has been replaced by a ternary convolutional layer. N1T uses a set of learned ternary weights for this, while the custom N1T network uses our custom filters.

5   Results

The results of our experiments can be seen in Table 2. In these, the ternary net-works routinely make less accurate classifications than full precision networks, as expected, but the networks using our custom filters did not deviate significantly from the levels given by standard ternary layers. Even when the networks were very shallow (such as with LeNet-5), where our custom filters would represent a greater proportion of the network, they did not hamper performance. Indeed, in many instances, the networks actually performed slightly better than their ternary counterparts. This indicates that our method is capable of speeding up the first layer of a network with negligible losses in accuracy.

Table 2. Validation accuracies (%) from the different networks and datasets.

	LeNet-5 on MNIST	VGG7 on CIFAR-10
Full precision	99.36	92.41
Full ternary	95.11	91.59
N1T	99.31	91.87
Custom full ternary	95.49	91.30
Custom N1T	99.48	91.90

6   Discussion and Future Work

Our method works well for what it set out to do, but it does have a significant drawback. The modifications needed for this method are done by hand, which is time consuming and limits our optimization to the first layer of the network (deeper layers require too many custom filters to be practical). While admittedly this is not ideal, in many architectures, the first layer contains the largest filters, making it one of the more expensive parts of the network. The original YOLO [5] network is an example of this. Its first convolutional layer uses 7×7 filters, which our method could improve the speed of.

This problem is something we would like to solve in future work by creating an algorithm that can automatically construct optimal custom filters given a ternary weight filter. With this, we could extend our method to work with more complex networks, and more importantly, extend our optimizations to the deeper layers of the network, allowing us to bring significant performance gains to larger architectures. Unfortunately, devising such an algorithm is a complex task, and is beyond the scope of this paper.

7   Conclusion

In conclusion, we have demonstrated a method for combining the Haar-like features and integral images from the Viola-Jones Framework [7] with a modern convolutional neural network to create optimized, custom filters. This optimization reduces the number of operations needed to compute the first layer's filter values, with negligible effects on the network's overall accuracy, theoretically improving performance. Unfortunately, the custom filters we use have to be crafted by hand, which is impractical for larger networks. A task for future work will be to create an algorithm to construct the filters automatically, instead of relying on hand-crafted ones.

References

1. Krizhevsky, A., Hinton, G., et al.: Learning multiple layers of features from tiny images (2009)
2. LeCun, Y., Bottou, L., Bengio, Y., Haffner, P.: Gradient-based learning applied to document recognition. Proc. IEEE **86**(11), 2278–2324 (1998)
3. LeCun, Y., Cortes, C., Burges, C.: MNIST handwritten digit database. ATT Labs [Online] 2 (2010). http://yann.lecun.com/exdb/mnist
4. Li, F., Zhang, B., Liu, B.: Ternary weight networks. arXiv preprint arXiv:1605.04711 (2016)
5. Redmon, J., Divvala, S., Girshick, R., Farhadi, A.: You only look once: unified, real-time object detection (2015)
6. Simonyan, K., Zisserman, A.: Very deep convolutional networks for large-scale image recognition. arXiv preprint arXiv:1409.1556 (2014)
7. Viola, P., Jones, M.: Rapid object detection using a boosted cascade of simple features. In: Proceedings of the 2001 IEEE Computer Society Conference on Computer Vision and Pattern Recognition, CVPR 2001, vol. 1, p. I (2001)

Learning to Optimize Black-Box Functions with Extreme Limits on the Number of Function Evaluations

Carlos Ansótegui[1], Meinolf Sellmann[4], Tapan Shah[2], and Kevin Tierney[3(✉)]

[1] DIEI, Universitat de Lleida, Lleida, Spain
`carlos@diei.udl.ca`
[2] General Electric Global Research Center, Niskayuna, USA
`tapan.shah@ge.com`
[3] Bielefeld University, Bielefeld, Germany
`kevin.tierney@uni-bielefeld.de`
[4] Shopify, Ottowa, Canada
`meinolf@ge.com`

Abstract. We consider black-box optimization in which only an extremely limited number of function evaluations, on the order of around 100, are affordable and the function evaluations must be performed in even fewer batches of a limited number of parallel trials. This is a typical scenario when optimizing variable settings that are very costly to evaluate, for example in the context of simulation-based optimization or machine learning hyperparameterization. We propose an original method that uses established approaches to propose a set of points for each batch and then down-selects from these candidate points to the number of trials that can be run in parallel. The key novelty of our approach lies in the introduction of a hyperparameterized method for down-selecting the number of candidates to the allowed batch-size, which is optimized offline using automated algorithm configuration. We tune this method for black box optimization and then evaluate on classical black box optimization benchmarks. Our results show that it is possible to learn how to combine evaluation points suggested by highly diverse black box optimization methods conditioned on the progress of the optimization. Compared with the state of the art in black box minimization and various other methods specifically geared towards few-shot minimization, we achieve an average reduction of 50% of normalized cost, which is a highly significant improvement in performance.

1 Introduction

We consider the situation where we need to optimize the input variables for unconstrained objective functions that are very expensive to evaluate. Furthermore, the objective functions we consider are *black boxes*, meaning we can use no knowledge from the computation of the objective function (such as a derivative) during search. Such objective functions arise when optimizing over simulations,

© Springer Nature Switzerland AG 2021
D. E. Simos et al. (Eds.): LION 2021, LNCS 12931, pp. 7–24, 2021.
https://doi.org/10.1007/978-3-030-92121-7_2

engineering prototypes, and when conducting hyperparameter optimization for machine learning algorithms and optimization solvers.

Consider the situation where we wish to optimize the geometry of a turbine blade whose effectiveness will be evaluated by running a number of computational fluid dynamics (CFD) simulations over various different scenarios simulating different environmental conditions. Each evaluation of a geometry design requires running the CFD simulations for all scenarios and may take hours or even days to complete. In such a setting, we may evaluate multiple scenarios and geometries at the same time to reduce the elapsed time until a high quality design is found. We could, for example, evaluate 5 designs in parallel and run 20 such batches sequentially. In this case, whenever we set up the new trials for the next batch, we carefully choose 5 new geometries based on the designs and results that were obtained in earlier batches. In total, we would evaluate 100 designs, and the total elapsed time to obtain the optimized geometries would take 20 times the time it takes to evaluate one design for one scenario (provided we are able to run all environmental conditions for all 5 candidate designs in parallel).

From an optimization perspective, this setup is extremely challenging. We are to optimize a black box function in which there is only time enough to try a mere 100 input settings, and, on top of that, we have to try 5 inputs in parallel, which implies we cannot learn from the results obtained from the inputs being evaluated in parallel before choosing these inputs. We refer to this optimization setup as parallel few-shot optimization, as an analogy to the idea of parallel few-shot learning.

The objective of the work presented in this paper is to develop a hyper-parameterized heuristic that can be tuned offline to perform highly effective parallel few-shot optimization. Our approach involves a Monte-Carlo simulation in which we try out different combinations of candidate points from a larger pool of options. The hyperparameters of our approach control a scoring function that guides the creation of batches of candidates, and we select our final batch of candidates based on which candidates appear most often in the sampled batches. It turns out that setting the hyperparameters of our heuristic is itself a type of black-box optimization (BBO) problem that we solve using the algorithm configurator GGA [3] as in [6,8]. In this way, our heuristic can be customized to a specific domain, although, in this work, we try to make it work generally for BBO functions. A further key advancement of our approach is that our heuristic adjusts the scoring of new candidates to be added to a sampled batch based on the candidates already in the batch, ensuring a balanced exploration and exploitation of the search space.

In the following, we review the literature on BBO and few-shot optimization, introduce our new method, and, finally, provide numerical results on standard BBO benchmarks.

2   Related Work

We split our discussion of the related work into two groups of approaches: candidate generators and candidate selectors. Given a BBO problem, candidate generators propose points that hopefully offer good performance. This group of approaches can be further divided into evolutionary algorithms, grid sampling, and model-based approaches. Candidate selectors choose from a set of candidate points provided by candidate generators, or select which candidate generators should be used to suggest points, following the idea of the no free lunch theorem [2] that no single approach (in this case a candidate generator) dominates all others. These approaches include multi-armed bandit algorithms and various forms of ensembles.

2.1 Candidate Generators

Generating candidates has been the focus of the research community for a long time, especially in the field of engineering, where grid sampling approaches and fractional factorial design [21] have long been used to suggest designs (points) to be realized in experiments. Fractional factorial design suggests a subset of the cross-product of discrete design choices that has desirably statistical properties. Latin hypercube sampling [29] extends this notion to create candidate suggestions from continuous variables. These techniques are widely used when no information is available about the space, i.e., it is not yet possible inference where good points may be located.

Evolutionary algorithms (EAs) offer simple, non-model-based mechanisms for optimizing black-box functions and have long been used for BBO [11]. EAs initialize a *population* of multiple solutions (candidate points) and *evolve* the population over multiple generations, meaning iterations, by recombining and resampling solutions in population i to form population $i + 1$. The search continues until a termination criterion is reached, for example, the total number of evaluations or when the average quality of the population stops improving.

Standard algorithms in the area of EAs include the $1 + 1$ EA [16,33] and differential evolution (DE) [35]. These approaches have seen great research advancements over the past years, such as including self-adaptive parameters [31,37] (SADE, L-SHADE), using memory mechanisms [13] (iL-SHADE), specialized mutation strategies [14] (jSO), and covariance matrix learning [10] (L-SHADE-cnEpSin). The covariance matrix can also be used outside of a DE framework. The covariance matrix adaptation evolutionary strategy (CMA-ES) [24] is one of the most successful EAs, and offers a way of sampling solutions from the search space according to a multivariate normal distribution that is iteratively updated according to a maximum likelihood principle.

EAs have also been used for BBO in the area of algorithm configuration (AC). In the AC setting the black-box is a parameterized solver or algorithm whose performance must be optimized over a dataset of representative problem instances, e.g., when solving a delivery problem each instance could represent the set of deliveries each day. The GGA method [5] is a "gender-based" genetic algorithm that partitions its population in two. One half is evaluated with a racing mechanism and is the winners are recombined with members from the other half.

Model-based approaches are techniques that build an internal predictive model based on the performance of the candidates that were chosen in the past. These methods can offer advantages over EA and grid-sampling approaches in their ability to find high quality solutions after a learning phase. However, this comes at the expense of higher computation time. In the area of AC, the GGA++ [3] technique combines EAs with a random forest surrogate that evaluates the quality of multiple candidate recombinations, returning the one that ought to perform the best.

Bayesian optimization (BO) has become a widespread model-based method for selecting hyperparameters in black-box settings (see, e.g., [19]) and for the AutoML setting [18,34]. BO models a surrogate function, typically by using a Gaussian process model, which estimates the quality of a given parameter selection as well as the uncertainty of that selection. A key ingredient for BO is the choice of an *acquisition function*, which determines how the optimizer selects the next point to explore. There are numerous BO variants, thus we only point out the ones most relevant to this work. For few-shot learning, [38] proposes a deep kernel learning approach to allow for transfer learning. In [9], a BO approach for few-shot multi-task learning is proposed. Recently, the NeurIPS 2020 conference hosted a challenge for tuning the hyperparameters of machine learning algorithms [1], with the HEBO approach [15], emerging as the victor. HEBO is based on BO and selects points from a multi-objective Pareto frontier, as opposed to most BO methods which only consider a single criterion.

2.2 Candidate Selectors

The line between candidate generator and candidate selector is not clear cut, indeed even the fractional factorial design method not only suggests candidates (the cross product of all options), but also provides a mechanism for down-selecting. Thus, by candidate selector, we mean methods that can be applied generally, i.e., the candidates input to the method can come from any variety of candidate generators, or the selector could accept/choose candidate generation algorithms, such as in the setting of algorithm selection [12]. In [27], which also competed in the challenge described above, an ensemble generation approach for BBO is presented using GPUs. The resulting ensemble uses the Turbo optimizer [17] (itself a candidate selector using a bandit algorithm) and scikit-optimize [25]. In [39], an ensemble of three approaches is created and a hierarchy is formed to decide which to use to select points.

Multi-armed bandit approaches are a well-known class of candidate selectors. As we consider the case where multiple candidates in each iteration should be selected, combinatorial bandits with semi-bandit feedback (e.g., [26]) are most relevant. These approaches generally assume the order of observations (between batches) is irrelevant, however we note that, in our case, this is not true. For example, some approaches may work better at selecting points in the first few rounds, while others may excel later on or once particular structures are discovered. Contextual bandits [28] allow for the integration of extra information, such as the current iteration, to be included in arm selection. The CPPL approach of [30] uses a Placket-Luce model to choose the top-k arms in a contextual setting, but is meant for situations with a richer context vector, such as algorithm selection, rather than BBO candidate selection.

3   Hyperparameterized Parallel Few-Shot Optimization (HPFSO)

Having reviewed the dominant methodologies for BBO, we now introduce our new hyperparameterized approach for parallel few-shot optimization (**HPFSO**). The idea for our approach is simple: When determining the next batch of inputs to be evaluated in parallel, we employ multiple different existing methodologies to first produce a larger set of candidate inputs. The core function we introduce strategically selects inputs from the superset of candidates until the batch-size is reached. The mechanism for performing this reduction is a hyper-configurable heuristic that is learned using an AC algorithm offline. The selected candidates are evaluated in parallel and the results of all of these trials are communicated to all the candidate-generating methods. That is to say, every point generator is also informed about the true function values of candidates it itself did not propose for evaluation. This process is repeated until the total number of iterations is exhausted. In the end, we return the input that resulted in the best overall evaluation.

3.1   Candidate Generators

We first require several methods for generating a superset of candidate inputs from which we will select the final batch that will be evaluated in parallel. To generate candidates, we use:

- Latin Hypercube Sampling (**LHS**): For a requested number of points k, partition the domain of each variable into k equal (or, if needed, almost equal) sized intervals. For each variable, permute the k partitions randomly and independently of the other variables. Create the i-th point by picking a random value from partition i (in the respective permuted ordering) for each variable.
- Bayesian Optimization: Bayesian optimization updates a surrogate model with new point(s) x_n and their evaluated values. Using the surrogate model as response surface, an acquisition function $\alpha(x)$ is minimized to derive the next query point(s)

$$x_{n+1} = \mathrm{argmin}_x \alpha(x)$$

We run the above minimization k times to generate k points, each time with a different seed.
  • Gradient Boosting Tree – Lower Confidence Bound (**GBM-LCB**): We use a GBM as a surrogate model followed by the LCB acquisition function

$$\alpha(x) = \mu(x) - \kappa\sigma(x),$$

where $\mu(x)$ is the posterior mean, $\sigma(x)$ is the posterior standard deviation and $\kappa > 0$ is the exploration constant. The parameter κ adjusts the bias regarding exploration vs. exploitation. In our experiments, we set κ to 2.

- Modified Random Forest (**GGA++**): We use the surrogate proposed in [3], which directly identifies areas of interest instead of forecasting black-box function values. We use local search to generate local minima over this surrogate without using any form of uncertainty estimates to generate points.

- Covariance Matrix Adaptation: This is an evolutionary approach that samples k points from a Gaussian distribution that is evolved from epoch to epoch. The covariance matrix is adjusted based on the black box evaluations conducted so far [24]. We use the covariance matrix in two ways to generate points:
  - For sampling using the current best point as mean of the distribution (**CMA**).
  - For sampling using the mean of the best point suggested by the GGA++ surrogate (**CMA-N**).

- Turbo (**TUR**): This method employs surrogate models for local optimization followed by an implicit multi-armed bandit strategy to allocate the samples among the different local optimization problems [17].

- Recombinations of the best evaluated points and points that were suggested but not selected in earlier epochs: We create a *diversity store* of all the points recommended in the previous iterations by BO, covariance matrix adaptation and Turbo, but not evaluated after the down-selection. We create recombinations of these points using two methods:
  - **REP**: We select a random point in the diversity store and use path relinking [20] to connect it with the best point found so far. We choose the recombination on the path with the minimum value as priced by the GGA++-surrogate.
  - **RER**: In a variant of the above method, we use random crossover 1000 times for randomly chosen points from the diversity store. We perform pricing again using the GGA++ surrogate, and suggest the best k points to be evaluated.

Summarizing, we use 8 candidate point selectors: LHS, GBM-LCB, GGA++, CMA, CMA-N, TUR, REP and RER. We will use these acronyms going forward.

3.2 Sub-selection of Candidates

Having generated a set of candidate points, we next require a method to down-select the number of points to the desired batch size of function evaluations that can be conducted in parallel. This function represents the core of our new methodology and is the primary target for our automated parameter tuning, our goal in this process.

The selection of candidates is shown in Algorithm 1 and works as follows. In each iteration of the main while loop (line 4), we select a candidate for our final

---

Algorithm 1. Candidate sub-selection

1: **Input:** F: set of feature vectors for C candidates; w: feature weights; B: # of final candidates, N: # simulations
2: **Output:** S: Indices of selected candidates
3: **Initialize** $S \leftarrow \{\}$, $R \leftarrow \{1, 2, \ldots, C\}$
4: **while** $|S| < B$ **do**
5: $\quad$ $Q \leftarrow$ vector of length C of zeros
6: $\quad$ **for** j in $1, 2, \ldots, N$ **do**
7: $\quad\quad$ $S_b \leftarrow S$
8: $\quad\quad$ **while** $|S_b| < B$ **do**
9: $\quad\quad\quad$ Update and normalize diversity features in F w.r.t. S_b
10: $\quad\quad\quad$ **for** $c \in R \setminus S_b$ **do**
11: $\quad\quad\quad\quad$ $f^c \leftarrow F(c)$; $\quad s_w(f^c) \leftarrow \frac{1}{1+e^{w^T f^c}}$
12: $\quad\quad\quad$ **end for**
13: $\quad\quad\quad$ $s_c \leftarrow s_w(f^c) / \sum_{c=1}^{C} s_w(f^c)$ $\forall c \in R \setminus S_b$
14: $\quad\quad\quad$ Sample k from $R \setminus S_b$ with distribution $\{s_c\}$.
15: $\quad\quad\quad$ $S_b \leftarrow S_b \cup \{k\}$.
16: $\quad\quad\quad$ $Q[k] \leftarrow Q[k] + 1$
17: $\quad\quad$ **end while**
18: $\quad$ **end for**
19: $\quad$ $S \leftarrow S \cup \{\text{argmax}\{Q\}\}$ $\qquad \triangleright$ Choose only one value; break ties uniformly at random.
20: **end while**

---

selection S. We next simulate N completions of S. For each candidate that we could add to the batch, we compute its features apply a logistic regression using a weight vector w that is tuned offline (line 11), providing us with a score as to how good (or bad) candidate c is with respect to the current batch. Note that this is a key part of our contribution; we do not simply take the best B candidates, rather, we ensure that the candidates complement one another according to "diversity" features. For example, solutions that are too similar to the solutions in S_b can be penalized through the diversity features to encourage exploration, and will receive a lower score than other candidates, even if they otherwise look promising. Alternatively, the hyperparameter tuner can also decide to favor points that are close to each other to enhance intensification in a region. In any case, given the scores for each candidate, we form a probability distribution from the scores and sample a new candidate for S_b (line 14).

Figure 1 shows the selection of the next candidate for S_b graphically. The blue squares represent the candidates in S, which are fixed in the current simulation. The subsequent two orange cells were selected in the previous two iterations of the current simulation. Now, given three categories of features, which are explained in more detail later, we compute the scores for each of the remaining candidates and choose one at random according to the probability distribution determined by the scores.

Once a candidate has been chosen, we increment a counter for the chosen candidate (line 15). Having simulated batches of candidates of the desired size, we then use the frequencies with which the respective candidates appear in the sample batches. The candidate that appears most often gets added to the batch S that we will send to the black box for parallel evaluation, with ties broken uniformly at random.

Figure 2 provides a graphical view of the frequency selection of a candidate for S. In Fig. 2(a), we depict again in blue the candidates that are already

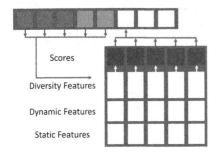

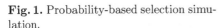

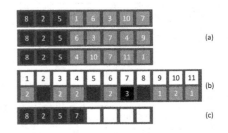

Fig. 1. Probability-based selection simulation.

Fig. 2. Frequency-based determination of next candidate. (a) Three simulated selections of batches. (b) Frequencies of candidates in simulations. (c) Augmented set of candidates. (Color figure online)

determined to be part of the final batch of eight. The orange cells show completions of the batch from three simulations. In Fig. 2(b), we calculate the frequencies (given as Q in the algorithm) with which the different candidates appear in the three sample batches. In green, we show candidate 7, which was selected most often. Finally, in Fig. 2(c), we provide the augmented, partially completed batch with candidate 7. The algorithm will then zero out its frequency table and begin a new round of simulations to fill in the remaining cells with candidate points.

3.3   Hyperparameterized Scoring Function

The final piece missing from our approach is the scoring function of candidates that is called during the randomized construction of sample batches. The score for each candidate depends on two pieces of information, the definition of candidate features and the determination of the feature weights w. We start by listing the features used to characterize each remaining candidate. The features fall into three different categories:

1. *Diversity features*, which rate the candidate in relation to candidates already selected to be part of the sample batch under construction,
2. *Dynamic features*, which characterize the point with respect to its expected performance, and
3. *Static features*, which capture the current state of the BBO as a whole.

Diversity Features: The first set of features considers how different the candidate is with respect to three different sets of other candidates. These three sets are 1. the set of points that were already evaluated by the black box function in earlier epochs, 2. the set of points already included in the sample batch under construction (denoted with the blue and orange colors above), and 3. the subset of candidates that are already part of the current sample batch and that were generated by the same point generator as the point whose features we are computing. For each of these three sets, we compute the vector of distances of the candidate to each point in the respective set. To turn this vector into a fixed number of features, we then compute some summary statistics over these three vectors: the mean, the minimum, the maximum, and the variance.

Dynamic Features: The second set of features characterizes each candidate in terms of its origin and its expected performance. In particular, we assess the following. What percentage of points evaluated so far were generated by the same point generator as the candidate (if there are none we set the all following values to 0)? Then, based on the vector of function values of these points that were generated by the same generator earlier, what was their average objective value? What was the minimum value? What was the standard deviation? Then, also for these related points evaluated earlier, what was the average deviation of the anticipated objective value from the true objective value? Next, based on the GBM used as a surrogate by some of the point generators, what is the expected objective value of the candidate? What is the probability that this candidate will improve over the current minimum? And, finally, what is the uncertainty of that probability?

Static Features: The last set of features considers the general state of the optimization in relation to the respective candidate. We track the following. What method was used to generate the candidate? Note that the method is one-hot encoded, so there are as many of these features as point generator methods. And, finally, what is the ratio of epochs remaining in the optimization?

After computing all features above for all remaining candidates in each step of building a sample batch, before applying the logistic scoring function, we normalize the diversity and the dynamic features such that the range of each feature is 0 to 1 over the set of all candidates. That is to say, after normalization, for each distance and each dynamic feature, there exists a candidate for which the feature is 0 and another candidate for which the feature is 1 (unless all feature values are identical, in which case they are all set to 0), and all feature values are in $[0, 1]$.

Hyperparameters: Finally, we need to determine the feature weights $w \in \mathbb{R}^n$ to fully define the scoring function. We use an algorithm configurator to determine the hyperparameters, as was previously proposed in [6,8], in which AC is used for determining weights of linear regressions in a reactive search. We tune on a set of 43 black box optimization problems from the 2160 problems in [22]. As is the practice in machine learning, the functions we tune the method for are different from the functions we evaluate the performance of the resulting method on in the following section. Our test set consists of 157 additional problems sampled at random from the same benchmark.

4   Numerical Results

We study the performance and scaling behavior of the approach developed in the prior sections by applying it to the established standard benchmarks from the black-box optimization community.

4.1   Experimental Setup

Benchmark: We use the Python API provided by the Comparing Continuous Optimizers (COCO) platform [22] to generate BBO training and evaluation instances. The framework provides both single objective and multi-objective BBO functions. We use the noiseless BBO functions in our experiments. The suite has 2160 optimization problems (using the standard dimensions from COCO), each specified by a fixed value of identifier, dimension and instance number. We randomly select 2% (43) problems for training the hyperparameters and 7.5% (157) problems for testing.

As each test instance works on its own scale, we normalize the solution values obtained by applying a linear transformation such that the best algorithm's solution value is zero and the worst is one. The algorithms considered for this normalization are all individual point generators as well as the state-of-the-art black box optimizers, but not sub-optimal HFPSO parameterizations, or solutions obtained when conducting different numbers of epochs. Note that, for the latter, values lower than zero or greater than one are therefore possible.

Configuration of HPFSO's Hyperparameters: We tune HPFSO using PYDGGA [4], which is an enhanced version of the GGA++ [3] algorithm configurator [7] written in the Python language. We tune for 50 wall-clock hours on 80 cores.

Contenders: To assess how the novel approach compares to the state of the art we compare with the following approaches: CMA-ES, HEBO [15], and multiple differential evolutionary methods, in particular DE [35], SADE [31], SHADE [36], L-SHADE [37] iL-SHADE [13], jSO [14] and L-SHADE-cnEpSin [10]. We use open source Python implementations for CMA [23], HEBO [15] and the DE methods [32].

Compute Environment: All the algorithms were run on a cluster of 80 Intel (R) Xeon CPU E5-2698, 2.20 GHz servers with 2 threads per core, an x86_64 architecture and the Ubuntu 4.4.0-142 operating system. All the solvers are executed in Python 3.7.

Table 1. Comparison of randomly chosen hyperparameters with the tuned hyperparameters after different generation of tuning by PYDGGA

	Random		At generation			
	A	B	5	10	20	40
Mean	0.198	0.191	0.108	0.088	0.085	**0.067**
Std	0.287	0.199	0.188	0.176	0.146	**0.115**
Mean/Gen 40	2.955	2.851	1.612	1.313	1.269	**1.000**
Std/Gen 40	2.496	1.730	1.635	1.530	1.270	**1.000**

4.2   Effectiveness of Hyperparameter Tuning

We begin our study by conducting experiments designed to assess the effectiveness of the hyperparameter tuning. In Table 1, we show the normalized (see Benchmarks) quality of solutions on the test set.

We provide the aggregate performance as measured by the arithmetic mean over the normalized values over all test instances. We compare versions of our novel approach that only differ in the hyperparameters used. The first two versions apply two different random parameterizations (Random A and B). PYDGGA is based on a genetic algorithm, thus, after each generation, it provides the best performing parameters for HPFSO in that generation. We provide the test performance of the parameterizations found in generation 5, 10, 20 and 40, respectively. Note that the parameters at generation 40 are the last parameters output by PYDGGA.

We observe that tuning is indeed effective for this method. A priori it was not certain that stochastically tying the static, dynamic, and diversity features to the decision as to which points are selected would have a significant effect on performance at all. Moreover, nor was it certain that we would be able to learn effectively how to skew these stochastic decisions so as to improve algorithm performance. As we can see, however, both of these were possible and this leads to an improvement of about a factor of three in normalized performance over random parameters. At the same time, we also observe a significant drop in variability. As the method gets better at optimizing the functions, the standard deviation also drops, which tells us that the tuning did not lead to outstanding performance on just some instances at the cost of doing a very poor job on others.

4.3   Importance of the Selection Procedure

Next, we quantify the impact of the main contribution of our approach, namely the sub-selection procedure. To this end, we compare with each of the 8 point generators included in HPFSO, in isolation, as well as the performance of an approach that employs all point generators and then randomly selects points from the pool that was generated (RAND), and another approach that also uses

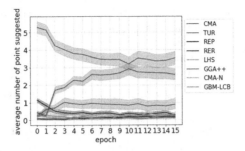

Fig. 3. Average number of points per epoch for each candidate method over test instances.

all 8 point generators and compiles a set of 8 candidate points to be evaluated by selecting the best point (as evaluated by the GBM surrogate) from each point generator (best per method - BPM).

In Table 2, we show the aggregate performances, again as measured by the means of the normalized solution qualities. We observe that the best individual point generator is CMA-ES, followed by BO based on a GBM as surrogate using LCB as acquisition function. Given that CMA-ES is the best point generator in isolation, it may not be surprising that the method also beats a random sub-selection of points over all points generated (RAND). What may be less expected is that the best point generator (CMA) in isolation also performs about two times better than choosing the batch consisting of the best points provided by each point generator in each epoch (BPM). And, in fact, this shows the difficulty of the challenge our new method must overcome, as combining the respective strengths of the different point generators appears all but straight forward.

However, when comparing HPFSO with BPM, we nevertheless see that the new method introduced in this paper does manage to orchestrate the individual point generators effectively. On average, HPFSO leads to solutions that incur less than half the normalized cost than any of the point-generation methods employed internally. Moreover, we also observe that the down-selection method works with much greater robustness. The standard deviation is a factor 1.8 lower than that of any competing method.

In Fig. 3, we show the average number of points selected from each point generator as a function over epochs. We observe that points generated by

Table 2. Comparison of HPFSO with individual point generators, the best point per method (BPM), and randomly sub-selected points (RAND)

	LHS	CMA-N	RER	REP	TUR	GGA++	GBM-LCB	CMA	RAND	BPM	HPFSO
Mean	0.606	0.612	0.549	0.543	0.534	0.521	0.341	0.142	0.266	0.269	**0.067**
Std	0.304	0.357	0.336	0.311	0.304	0.307	0.317	0.207	0.228	0.252	**0.115**
Mean/HPFSO	9.182	9.273	8.318	8.227	8.091	7.894	5.167	2.152	4.030	4.076	**1.000**
Std/HPFSO	2.667	3.132	2.947	2.728	2.667	2.693	2.781	1.816	2.000	2.211	**1.000**

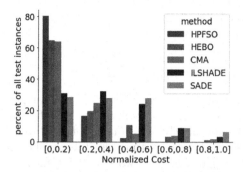

Fig. 4. Distribution of normalized function values for HPFSO and nearest competitors

CMA-ES (CMA) are generally favored by our method, which fills between 40% and 60% of the batches with points generated by that method, where the ratio starts at around 60% and drops gradually down to 40% in later epochs. Curiously, we also see that CMA-ES is favored in the very first epoch when clearly no learning of the covariant influence of the input variables could have taken place yet. Note that our method would have had the option to employ latin hypercube sampling instead, but did not. What this tells us is that it appears to be better, even in the very beginning, to gain deeper information in one region of the search space rather than distributing the initial points. We suspect that the superiority of this strategy is the result of the very limited number of black box function evaluations that can be afforded.

Most of the remaining points selected are generated by BO using the GGA++ surrogate, which ramps up quickly to 20% after two epochs and then steadily grows to about 35%. This behavior makes intuitive sense, as the GGA++ surrogate is designed to identify favored regions quickly, but does require at least some training data to become effective. The points generated by the Turbo method, which also employs BO internally, show a similar dynamic. After two epochs, about one in eight points is selected from this method, and this ratio stays very steady from then on.

Less intuitive is the strategy to choose a point generated by recombination via path relinking (REP) in the first epoch. We assume that the tuner "learned" to add a random point to the batch so as to gauge whether CMA-ES is not searching in a completely hopeless region. However, after two epochs, the influence of REP dwindles down to the same level of influence that all other point generators are

Table 3. Comparison of HPFSO with state-of-art methods

	HPFSO	HEBO	CMA	iL-SHADE	SADE	SHADE	L-SHADE	jSO	LSHADEcnEpSin	DE
Mean	0.067	0.135	0.142	0.303	0.343	0.365	0.379	0.384	0.410	0.410
Std	0.115	0.199	0.207	0.259	0.274	0.302	0.325	0.321	0.313	0.326
Mean/HPFSO	1.000	2.045	2.152	4.591	5.197	5.530	5.742	5.818	6.212	6.212
Std/HPFSO	1.000	1.746	1.816	2.272	2.404	2.649	2.851	2.816	2.746	2.860

afforded by the method. In all epochs, at most 0.3 points from all other point generators are selected on average.

4.4 Comparison with the State of the Art

Finally, we provide a comparison with some of the best performing black-box optimization algorithms to date as well as the recent winner from the 2020 NeurIPS Black Box Challenge, HEBO. Note that the established black-box optimizers (such as CMA-ES) have been tuned for the very benchmark we consider, but were not specifically designed to work well for such an extreme limit on the number of function evaluations. HEBO, on the other hand, was developed for the BBO challenge where the objective was to optimize hyperparameters of machine learning algorithms by testing sets of hyperparameters in 16 epochs of 8 samples per epochs.

In Table 3, we show the aggregate normalized performance of comparing HPFSO with other competitors. In Fig. 4, we also show the histogram of the normalized function values evaluated by HPFSO and its nearest competitors.

We observe that CMA-ES and HEBO are the closest contenders, but nonetheless produce solutions that have normalized costs over two times the quality produced by HPFSO. We use a Wilcoxon signed rank test against our data and find that the p-value for the hypothesis that HEBO outperforms HPFSO is less than 0.55%, which allows us to refute this hypothesis with statistical significance.

In Fig. 4, we see that, on 80% of our test instances, HPFSO yields a solution that is close to the best performing method for that instance: the normalized function value is between 0 and 0.2. CMA-ES and HEBO also perform relatively well, but still notably worse than HPFSO. We also see that on none of the test instances, HPFSO fails completely. Only on less than 3% of the test instances, the normalized function value exceeds 0.4, and it is never over 0.6. This again shows the robustness of HPFSO. As we see in Table 3, HEBO is the closest contender in terms of variability, but even it exhibits a standard deviation that is over 1.7 times larger than that of HPFSO.

We also investigate how different setups regarding the number of epochs would affect these results. In Table 4, we vary the number of epochs between 4 and 24. We use the same batch size of 8 points per parallel trial as before, with the exception of the DE methods which do not allow us to specify parallel trials, so we give these methods a competitive advantage by allowing them to conduct 8 times the number of epochs with one trial per epoch.

Note that HPFSO is trained exclusively using runs with 16 epochs. Nevertheless, the method produces the best results across the board, from optimizing the black box when only 32 function evaluations are allowed in 4 epochs of 8, to an optimization where 192 function evaluations can be afforded in 24 epochs with 8 parallel trials each. This shows that, while the training is targeting a specific number of function evaluations overall, the parameters learned do generalize to a range of other setting as well.

Table 4. Mean and standard deviation of normalized performances of different methods when different numbers of epochs are affordable. The batch size is held constant at 8, except for DE methods whose implementations do not allow for parallel trials. For these methods we allow 8 times the number of epochs, with one trial per epoch.

Epochs	4		8		12		16		24	
	Mean	Std	Mean	Std	Mean	Std	Mean	Std	Mean	Std
HPFSO	**0.419**	**0.350**	**0.217**	0.245	**0.137**	0.201	**0.067**	**0.115**	−0.068	1.123
CMA	0.477	0.408	0.265	**0.219**	0.177	**0.189**	0.142	0.207	−0.031	1.080
HEBO	0.825	2.434	0.345	0.570	0.200	0.252	0.135	0.199	0.073	**0.183**
ILSHADE	1.844	11.54	0.512	0.328	0.349	0.262	0.303	0.259	0.230	0.212
GBM-LCB	0.670	0.481	0.450	0.351	0.365	0.320	0.341	0.317	0.251	0.255
LSHADE	30.32	366.5	1.555	12.19	0.532	1.120	0.379	0.325	0.299	0.278
SADE	0.811	1.385	0.672	2.061	0.478	0.478	0.343	0.274	0.303	0.288
jSO	19.08	222.9	9.327	108.6	9.683	115.9	0.384	0.321	0.322	0.323
LSAHDECNEP*	39.80	487.3	1.530	12.197	0.521	0.893	0.410	0.313	0.325	0.280
SHADE	23.30	281.6	0.635	1.560	0.460	0.414	0.365	0.302	0.334	0.288
DE	2.227	16.92	5.708	61.93	1.173	8.582	0.410	0.326	0.370	0.300
REP	1.246	1.709	0.818	0.807	0.691	0.748	0.543	0.311	0.483	0.311
RER	8.014	85.32	0.710	0.543	0.600	0.389	0.549	0.336	0.494	0.322
GGA++	1.177	1.776	0.761	0.503	0.648	0.425	0.521	0.307	0.510	0.322
TUR	1.119	1.506	0.753	0.545	0.633	0.354	0.534	0.304	0.518	0.331
LHS	1.272	3.378	0.783	0.598	0.661	0.417	0.606	0.304	0.520	0.312

5   Conclusion

We considered the problem of optimizing a black box function when only a very limited number of function evaluations is permitted, and these have to be conducted in a given number of epochs with a specified number of parallel evaluations in each epoch. For this setting, we introduced the idea of using a portfolio of candidate point generators and employed a hyperparameterized method to effectively down-select the set of suggested points to the desired batch size. Our experiments showed that our method can be configured effectively by the PYDGGA algorithm configurator, and that the primary strength of the method is derived from the parameterized, dynamically self-adapting down-selection procedure. Furthermore, we saw that the resulting method significantly outperforms established black box optimization approaches, as well as a recently introduced method particularly designed for black box optimization with extreme limits on the number of function evaluations.

References

1. Blackbox BBO Challenge. https://bbochallenge.com/. Accessed 12 Mar 2020
2. Adam, S.P., Alexandropoulos, S.-A.N., Pardalos, P.M., Vrahatis, M.N.: No free lunch theorem: a review. In: Demetriou, I.C., Pardalos, P.M. (eds.) Approximation and Optimization. SOIA, vol. 145, pp. 57–82. Springer, Cham (2019). https://doi.org/10.1007/978-3-030-12767-1_5

3. Ansotegui, C., Malitsky, Y., Samulowitz, H., Sellmann, M., Tierney, K.: Model-based genetic algorithms for algorithm configuration. In: IJCAI, pp. 733–739 (2015)
4. Ansotegui, C., Pon, J.: PyDGGA. https://ulog.udl.cat/
5. Ansótegui, C., Sellmann, M., Tierney, K.: A gender-based genetic algorithm for the automatic configuration of algorithms. In: Gent, I.P. (ed.) CP 2009. LNCS, vol. 5732, pp. 142–157. Springer, Heidelberg (2009). https://doi.org/10.1007/978-3-642-04244-7_14
6. Ansótegui, C., Heymann, B., Pon, J., Sellmann, M., Tierney, K.: Hyper-reactive tabu search for MaxSAT. In: Battiti, R., Brunato, M., Kotsireas, I., Pardalos, P.M. (eds.) LION 12 2018. LNCS, vol. 11353, pp. 309–325. Springer, Cham (2019). https://doi.org/10.1007/978-3-030-05348-2_27
7. Ansótegui, C., Pon, J., Sellmann, M.: Boosting evolutionary algorithm configuration. Ann. Math. Artif. Intell. (2021). https://doi.org/10.1007/s10472-020-09726-y
8. Ansótegui, C., Pon, J., Sellmann, M., Tierney, K.: Reactive dialectic search portfolios for MaxSAT. In: Proceedings of the AAAI Conference on Artificial Intelligence, vol. 31 (2017)
9. Atkinson, S., Ghosh, S., Chennimalai Kumar, N., Khan, G., Wang, L.: Bayesian task embedding for few-shot Bayesian optimization. In: AIAA Scitech 2020 Forum, p. 1145 (2020)
10. Awad, N.H., Ali, M.Z., Suganthan, P.N.: Ensemble sinusoidal differential covariance matrix adaptation with Euclidean neighborhood for solving CEC2017 benchmark problems. In: 2017 IEEE Congress on Evolutionary Computation, pp. 372–379, June 2017
11. Bäck, T., Schwefel, H.P.: An overview of evolutionary algorithms for parameter optimization. Evol. Comput. **1**(1), 1–23 (1993)
12. Bischl, B., et al.: ASlib: a benchmark library for algorithm selection. Artif. Intell. **237**, 41–58 (2016)
13. Brest, J., Maučec, M.S., Bošković, B.: iL-SHADE: improved L-SHADE algorithm for single objective real-parameter optimization. In: 2016 IEEE Congress on Evolutionary Computation (CEC), pp. 1188–1195, July 2016
14. Brest, J., Maučec, M.S., Bošković, B.: Single objective real-parameter optimization: algorithm jSO. In: 2017 IEEE Congress on Evolutionary Computation, pp. 1311–1318 (2017)
15. Cowen-Rivers, A.I., et al.: HEBO: heteroscedastic evolutionary Bayesian optimisation. arXiv preprint arXiv:2012.03826 (2020)
16. Droste, S., Jansen, T., Wegener, I.: On the analysis of the $(1 + 1)$ evolutionary algorithm. Theor. Comput. Sci. **276**(1–2), 51–81 (2002). https://doi.org/10.1016/S0304-3975(01)00182-7
17. Eriksson, D., Pearce, M., Gardner, J., Turner, R.D., Poloczek, M.: Scalable global optimization via local Bayesian optimization. In: Advances in Neural Information Processing Systems, pp. 5496–5507 (2019)
18. Feurer, M., Hutter, F.: Hyperparameter optimization. In: Hutter, F., Kotthoff, L., Vanschoren, J. (eds.) Automated Machine Learning. TSSCML, pp. 3–33. Springer, Cham (2019). https://doi.org/10.1007/978-3-030-05318-5_1
19. Frazier, P.I.: A tutorial on Bayesian optimization. arXiv preprint arXiv:1807.02811 (2018)

20. Glover, F.: A template for scatter search and path relinking. In: Hao, J.-K., Lutton, E., Ronald, E., Schoenauer, M., Snyers, D. (eds.) AE 1997. LNCS, vol. 1363, pp. 1–51. Springer, Heidelberg (1998). https://doi.org/10.1007/BFb0026589

21. Gunst, R.F., Mason, R.L.: Fractional factorial design. Wiley Interdisc. Rev. Comput. Stat. **1**(2), 234–244 (2009)

22. Hansen, N., Auger, A., Ros, R., Mersmann, O., Tušar, T., Brockhoff, D.: COCO: a platform for comparing continuous optimizers in a black-box setting. Optim. Methods Softw. **36**(1), 114–144 (2020)

23. Hansen, N., Akimoto, Y., Baudis, P.: CMA-ES/pycma on Github, February 2019. https://doi.org/10.5281/zenodo.2559634

24. Hansen, N., Müller, S.D., Koumoutsakos, P.: Reducing the time complexity of the derandomized evolution strategy with covariance matrix adaptation (CMA-ES). Evol. Comput. **11**(1), 1–18 (2003)

25. Head, T., Kumar, M., Nahrstaedt, H., Louppe, G., Shcherbatyi, I.: scikit-optimize/scikit-optimize, September 2020. https://doi.org/10.5281/zenodo.4014775

26. Lattimore, T., Kveton, B., Li, S., Szepesvari, C.: TopRank: a practical algorithm for online stochastic ranking. arXiv preprint arXiv:1806.02248 (2018)

27. Liu, J., Tunguz, B., Titericz, G.: GPU accelerated exhaustive search for optimal ensemble of black-box optimization algorithms. arXiv preprint arXiv:2012.04201 (2020)

28. Lu, T., Pál, D., Pál, M.: Contextual multi-armed bandits. In: Proceedings of the Thirteenth International Conference on Artificial Intelligence and Statistics, pp. 485–492 (2010)

29. McKay, M.D., Beckman, R.J., Conover, W.J.: A comparison of three methods for selecting values of input variables in the analysis of output from a computer code. Technometrics **42**(1), 55–61 (2000)

30. Mesaoudi-Paul, A.E., Bengs, V., Hüllermeier, E.: Online preselection with context information under the plackett-luce model (2020)

31. Qin, A.K., Suganthan, P.N.: Self-adaptive differential evolution algorithm for numerical optimization. In: 2005 IEEE Congress on Evolutionary Computation, vol. 2, pp. 1785–1791, September 2005

32. Ramón, D.C.: xKuZz/pyade, March 2021. https://github.com/xKuZz/pyade. Accessed 05 Oct 2017

33. Rechenberg, I.: Evolutionsstrategien. In: Schneider, B., Ranft, U. (eds.) Simulationsmethoden in der Medizin und Biologie. MEDINFO, vol. 8, pp. 83–114. Springer, Heidelberg (1978). https://doi.org/10.1007/978-3-642-81283-5_8

34. Snoek, J., Larochelle, H., Adams, R.P.: Practical Bayesian optimization of machine learning algorithms. In: Advances in Neural Information Processing Systems, pp. 2960–2968 (2012)

35. Storn, R., Price, K.: Differential evolution - a simple and efficient heuristic for global optimization over continuous spaces. J. Glob. Optim. **11**(4), 341–359 (1997). https://doi.org/10.1023/A:1008202821328

36. Tanabe, R., Fukunaga, A.: Success-history based parameter adaptation for differential evolution. In: 2013 IEEE Congress on Evolutionary Computation, Cancun, Mexico, pp. 71–78. IEEE, June 2013

37. Tanabe, R., Fukunaga, A.S.: Improving the search performance of SHADE using linear population size reduction. In: 2014 IEEE Congress on Evolutionary Computation (CEC), Beijing, China, pp. 1658–1665. IEEE, July 2014

38. Wistuba, M., Grabocka, J.: Few-shot Bayesian optimization with deep kernel surrogates. In: International Conference on Learning Representations (2021). https://openreview.net/forum?id=bJxgv5C3sYc

39. Ye, P., Pan, G., Dong, Z.: Ensemble of surrogate based global optimization methods using hierarchical design space reduction. Struct. Multidiscip. Optim. **58**(2), 537–554 (2018). https://doi.org/10.1007/s00158-018-1906-6

Graph Diffusion & PCA Framework for Semi-supervised Learning

Konstantin Avrachenkov[1], Aurélie Boisbunon[2], and Mikhail Kamalov[1(✉)]

[1] INRIA, Sophia Antipolis, Valbonne, France
{konstantin.avratchenkov,mikhail.kamalov}@inria.fr
[2] MyDataModels, Sophia Antipolis, Valbonne, France
abb@mydatamodels.com

Abstract. A novel framework called Graph diffusion & PCA (GDPCA) is proposed in the context of semi-supervised learning on graph structured data. It combines a modified Principal Component Analysis with the classical supervised loss and Laplacian regularization, thus handling the case where the adjacency matrix is *Sparse* and avoiding the *Curse of dimensionality*. Our framework can be applied to non-graph datasets as well, such as images by constructing similarity graph. GDPCA improves node classification by enriching the local graph structure by node covariance. We demonstrate the performance of GDPCA in experiments on citation networks and images, and we show that GDPCA compares favourably with the best state-of-the-art algorithms and has significantly lower computational complexity.

Keywords: Semi-supervised learning · Principal Component Analysis · Citation networks

1 Introduction

The area of graph-based semi-supervised learning (GB-SSL) focuses on the classification of nodes in a graph where there is an extremely low number of labeled nodes. It is useful in applications such as paper classification to help researchers find articles in a topic, where the data is represented through a citation network, and it is especially beneficial for the classification of medical studies, where collecting labeled nodes is an expensive procedure. In particular, we prepared a real dataset for our experiments which consists of paper abstracts with clinical trials[1] regarding the coronavirus (COVID) topic. Also, GB-SSL is applicable for post labelling in social networks and for detecting protein functions in different biological protein-protein interactions [7].

In GB-SSL, the data consists of the feature matrix $X = [X_i]_{i=1}^n$, where $X_i = (X_{i,j})_{j=1}^d$ lies in a d-dimensional feature space (e.g. from bag-of-words

---

[1] https://clinicaltrials.gov.

Supported by MyDataModels company.

D. E. Simos et al. (Eds.): LION 2021, LNCS 12931, pp. 25–39, 2021.
https://doi.org/10.1007/978-3-030-92121-7_3

[15]), and of the label matrix $Y = [Y_{i,j}]_{i,j=1}^{n,k}$ such that $Y_{i,j} = 1$ if $X_i \in \mathcal{C}_j$ and $Y_{i,j} = 0$ otherwise, $\{\mathcal{C}_1, \ldots, \mathcal{C}_k\}$ being a set of k classes. The aim of semi-supervised learning is to estimate Y by a classification result $Z = [Z_{i,j}]_{i,j=1}^{n,k}$ when there is a low number of labels available, while X contains information for both labeled and unlabeled observations. We also assume that the dataset (X, Y) can be represented through the undirected graph $\mathcal{G} = (\mathcal{V}, \mathcal{E})$, with $n = |\mathcal{V}|$ the number of nodes with features (e.g. papers) and $e = |\mathcal{E}|$ is the number of edges (e.g. citations). Let $A = [A_{i,j}]_{i,j=1}^{n,n}$ denote the adjacency matrix associated with the $\mathcal{G}$, and $D = \text{diag}(D_{i,i})$ be a diagonal matrix with $D_{i,i} = \sum_{j=1}^{n} A_{i,j}$.

Many works in GB-SSL [1,31,32] consider the following minimization problem:

$$\min_{Z \in \mathbb{R}^{n \times k}} \left\{ \sum_{i=1}^{n} \sum_{j=1}^{n} A_{i,j} \|Z_i - Z_j\|_2^2 + \mu \sum_{i=1}^{n} \|Z_i - Y_i\|_2^2 \right\}, \tag{1}$$

where μ is a Lagrangian multiplier, n is the number of nodes, $A = [A_{i,j}]_{i,j=1}^{n,n}$ is an adjacency matrix, $Z = [Z_i]_{i=1}^{n}$ is a classification result and $Y = [Y_i]_{i=1}^{n}$ is a matrix that represents labels. The first part of the objective in (1) is a Laplacian regularization, which penalizes nodes connected from different classes, while the second part is a supervised loss. For the specific problem of paper classification in citation graphs, Problem (1) has the following particular issues:

1. *Sparse A*: the Laplacian regularization cannot estimate classification results Z for graphs with an extremely small number of edges [18] (e.g. citations). Moreover, binary weights ($A_{i,j} = 0$ or $A_{i,j} = 1$) are a poor reflection of node similarity which can lead to a weak estimation of the Laplacian regularization;
2. *Curse of dimensionality*: it arises when A is replaced by a similarity matrix $W = [h(X_i, X_j)]_{i,j=1}^{n} \in \mathbb{R}^{n \times n}$ with a positive definite kernel $h(\cdot)$ and $d \to \infty$ where $X = [X_i]_{i=1}^{n}$ is a matrix of node features and d is the node features space. This replacement is made to avoid the sparsity of A. This issue is especially noticeable in the case of paper classification, for example, based on the Heaps law [2], the d-space of features (bag-of-words [15]) is increasing with respect to the number and length of papers.

The first issue is resolved by Graph Convolution Network (GCN) [19] and Plane-toid [32] by prediction of edges, however these solutions are limited in their ability to generalize predicted edge structure. The second issue is treated in GCN as well as in Semi-supervised embedding (SemiEmb) [31], and Least-squares kernel PCA (LS-KPCA) [30], but they require high computation complexity.

In this work, we propose to add a reorganized principal component analysis (PCA) loss to Problem (1) and denote our framework as Graph diffusion & PCA (GDPCA). Not only does it address the aforementioned issues, but we also prove that there exists an explicit solution to the corresponding problem. We apply it to real datasets, and show that GDPCA is the best among GB-SSL state-of-the-art linear algorithms, and that it has comparable performance with GB-SSL neural network algorithms with significantly lower computational complexity.

Finally, we show that GDPCA can also be applied to datasets with no explicit graph structure such as images, and that it outperforms both linear and neural network algorithms for GB-SSL on this type of datasets.

2   Graph-Based Semi-supervised Learning

The recent advances in GB-SSL can be classified into the following rapidly growing directions: **(1) classical linear graph diffusion** algorithms which apply the graph structure for spreading the information of labelled nodes through it, such as Label Propagation (LP) [33], PageRank SSL (PRSSL) [1], or manifold regularization (ManiReg) [4]; and **(2) graph-convolution based neural network** algorithms. The latter category can be further seperated into *(i) nonlinear graph diffusion* algorithms, which apply convolution on the graph's adjacency matrix A with node features, such as Graph Convolution Network (GCN) [19], approximated Personalized graph neural network (APPNP) [20], Planetoid [32], or DeepWalk [23]; and *(ii) graph convolution deep generative* models, focusing on the application of nonlinear graph convolution algorithms with respect to the latent representation of nodes/edges: GenPR [18], Graphite [13].

Linear graph diffusion models are interesting because of their simplicity, but they suffer greatly from the curse of dimensionality. On the other contrary, graph-convolution based neural networks outperform classical linear graph diffusion algorithms and solve the *Curse of dimensionality* issue [19,32]. However, they are oriented only on computations on small, sparse graphs, leading to the *Sparse A* issue. Furthermore, they do not provide a transparent solution of the classification result Z.

In this work, we present the novel **Graph diffusion & PCA** (GDPCA) framework aiming at solving both the *Curse of dimensionality* and *Sparse A* issues while maintaining a low computational complexity. Moreover, our framework provides an explicit solution of the combination of (1) with a reorganized PCA loss. Also we show that GDPCA outperforms the main state-of-the-art GB-SSL classical linear algorithms on various datasets. Our framework also has comparable performance with state-of-the-art GB-SSL neural network algorithms and significantly lower computational complexity.

3   Graph Diffusion with Reorganized PCA Loss

This work is motivated by the idea that principal component analysis (PCA) can solve at least the *Curse of dimensionality* issue. Different works [3,17,25,27,30] consider a transformation of X by principal components $XU^T = Z$ to the classification results, where $U \in \mathbb{R}^{d \times k}$ is a matrix of principal component vectors from PCA. Instead, we consider principal components which are straightforwardly related to the classification result ($U \in \mathbb{R}^{k \times n}$, $U^T = Z$), as explained in the sequel.

One of the main ideas of this work is that the nodes from different classes have high covariance. This idea lies under the hood of Linear Discriminant Analysis (LDA) [10], which was developed for supervised learning. We extend this idea so that it can also be applied in both unsupervised (PCA-BC) and semi-supervised learning (GDPCA).

3.1 PCA for Binary Clustering (PCA-BC)

In this section, we restrict the setting to the case where no labels are available, and where the nodes come from two clusters. Let us assume that the feature matrix X is sampled from the Gaussian distribution:

$$X_1, \ldots, X_{\frac{n}{2}} \sim \mathcal{N}(\mu_1, C) \text{ and } X_{\frac{n}{2}+1}, \ldots, X_n \sim \mathcal{N}(\mu_2, C), \tag{2}$$

where C is the covariance matrix and μ_1, μ_2 are the expectations of classes $\mathcal{C}_1$ and $\mathcal{C}_2$ respectively. Furthermore, let $||C||_2 = O(1)$, $||\mu_1 - \mu_2||_2 = O(1)$, and the ratio $c_0 = n/d$ be bounded away from zero for large d.

Remark 1. The assumptions $||C||_2 = O(1)$ and $||\mu_1 - \mu_2||_2 = O(1)$ are needed to save the essential variations in d linearly independent directions and define a non-trivial classification case for extremely large d. In particular, this assumption allows us to work with bag-of-words [15] where the d-space is increasing with respect to the number and the length of papers, which leads to the *Curse of dimensionality* issue.

Based on the proof of Theorem 2.2 in [9] and the above restrictions on X, there exists a connection between the binary clustering problem and the PCA maximization objective given by:

$$\max_{U \in \mathbb{R}^{k \times n}} ||\bar{X}U^T||_2^2, \; s.\,t.\; U^T U = 1 \tag{3}$$

where $\bar{X} = [\bar{X}_i^T]_{i=1}^d \in \mathbb{R}^{d \times n}$ with $\bar{X}_i^T = X_i^T - \frac{1}{d}\sum_{j=1}^d X_j^T$; $U = [U_i]_{i=1}^k \in \mathbb{R}^{k \times n}$ is a matrix of principal component vectors. Moreover, $U_{i=1} = U_1 = (U_{1,j})_{j=1}^n$ is the direction of maximum variance, and it can be considered as clustering results in the following way: if $U_{1,j} \geqslant median(U_1)$ then $X_j \in \mathcal{C}_1$ otherwise $X_j \in \mathcal{C}_2$. Figure 1 illustrates the idea that the covariance between nodes from different classes is high. We further demonstrate the applicability of PCA on the binary clustering task with a small numerical experiment. We generated several synthetic datasets (2) with various ratios c_0 and fixed values for expectation ($\mu_1 = (0.5, \ldots, 0)$; $\mu_2 = (0.1, \ldots, 0)$;) and covariance matrix ($C = diag(0.1)$) with $\frac{n}{2}$ the number of nodes in each class: $n = 100$, $d = 1000$, $c_0 = 0.1$; $n = 1000$, $d = 100$, $c_0 = 10$. The code of these experiments is publicly available through a GitHub repository[2]. Figure 2 shows examples of how U_1 discriminates the two classes, even for large d-spaces.

---

[2] https://github.com/KamalovMikhail/GDPCA.

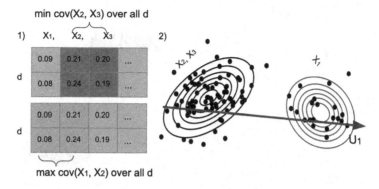

Fig. 1. The intuition behind PCA-BC: 1) Transpose X and visualise the nodes with the maximum and minimum covariance ($cov(\cdot)$) in between; 2) Normalize transposed X and find the direction of maximum covariance by PCA.

3.2 Generalization of PCA-BC for GB-SSL

We propose to modify Problem (1) by adding the reorganized PCA loss (the minus sign being necessary to account for the maximization of the covariance between classes). The optimization problem thus consists in:

$$
\min_{Z \in \mathbb{R}^{n \times k}} \left\{ \sum_{i=1}^{n} \sum_{j=1}^{n} A_{i,j} \| D_{ii}^{\sigma-1} Z_i - D_{jj}^{\sigma-1} Z_j \|_2^2 \\
+ \mu \sum_{i=1}^{n} D_{ii}^{2\sigma-1} \| Z_i - Y_i \|_2^2 - 2\delta \| \bar{X} Z \|_2^2 \right\}
\tag{4}
$$

where δ is a penalty multiplier and σ is the parameter controlling the contribution of node degree. We control the contribution of a node degree through the diagonal matrix D to the power in Problem (4) based on the work in [1]. It should be noticed that in Problem (4) we do not require the orthogonality condition $Z^T Z = 1$ as in (3). An interesting feature of Problem (4) is that there exists an explicit solution given by the following proposition.

Proposition 1. *When Problem (4) is convex, the explicit solution is given by:*

$$
Z = \left(I - \alpha \left(D^{\sigma-1} A D^{-\sigma} + \delta S D^{-2\sigma+1} \right) \right)^{-1} (1 - \alpha) Y,
\tag{5}
$$

where $\alpha = 2/(2 + \mu)$, $I \in \mathbb{R}^{n \times n}$ is the identity matrix and $S = \frac{\bar{X}^T \bar{X}}{(d-1)} \in \mathbb{R}^{n \times n}$ is the sample covariance matrix.

Proof. See Appendix A.

Remark 2. Proposition 1 provides the global minimum of Problem (4) in cases where it is convex, which occurs when the matrix

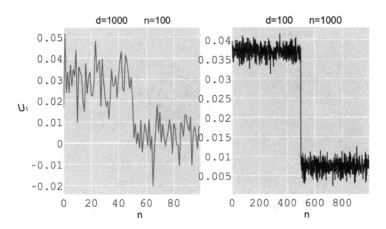

Fig. 2. Mean value of U_1 (the direction of maximum variance in the PCA) on 100 sets of random synthetic data.

$I - \alpha(D^{\sigma-1}AD^{-\sigma} + \delta SD^{-2\sigma+1})$ has positive eigenvalues (Theorem 1 in [12]). This condition can be achieved by values of δ such that the sum in brackets will not be upper then 1 and α always less than 1.

Direct matrix inversion in Eq. (5) can be avoided thanks to efficient iterative methods such as the PowerIteration (PI) or the Generalized minimal residual (GMRES) [28] methods. PI consists in iterative matrix multiplications[3] and can be applied when the spectral radius verifies $\rho(\alpha(D^{\sigma-1}AD^{-\sigma} + \delta SD^{-2\sigma+1})) < 1$. GMRES consists in approximating the vectors' solution in Krylov subspace instead of explicit matrix inversion. In practice, PI is more convenient for the computation of Eq. (5) as it converges faster to the best classification accuracy and it can be computed in a distributed regime over nodes [6, p. 135]. The accuracy is computed by comparing maximum values per row between label matrix Y and classification results Z. Furthermore, instead of explicitly computing the spectral radius mentioned above, we can use the following proposition.

Proposition 2. *Suppose that $SD^{-2\sigma+1}$ has only real eigenvalues $\lambda_1, \lambda_2, \ldots, \lambda_n$. Then the inequality $\rho\left(\alpha\left(D^{\sigma-1}AD^{-\sigma} + \delta SD^{-2\sigma+1}\right)\right) < 1$ can be transformed into a simpler one:*

$$1 + \delta\gamma < 1/\alpha \qquad (6)$$

where γ is the maximum singular value of $SD^{-2\sigma+1}$ and δ is the penalty multiplier in Eq. (5).

Proof. See Appendix B.

Remark 3. In order to speed up the computation of singular values, we can use the randomized Singular Value Decomposition (SVD) [14]. Inequality (6) can

---
[3] $Z = \alpha\left(D^{\sigma-1}AD^{-\sigma} + \delta SD^{-2\sigma+1}\right)Z + (1-\alpha)Y.$

then be rewritten as $1+\delta(\gamma+\epsilon) < 1/\alpha$, where ϵ is the tolerance of the randomized SVD. The computational complexity of the randomized SVD is $C+O(n)$, where C is the cost of matrix-vector multiplications.

Algorithm 1 gives the outline of our novel Graph diffusion & PCA (GDPCA) framework derived from Propositions 1 and 2. GDPCA uses the following setup: $\mathcal{I}$ is the number of iterations, τ is the tolerance in GMRES, δ is a Lagrangian multiplier, σ is the parameter controlling the contribution of node degree and ϵ is the tolerance in randomized SVD.

Algorithm 1: GDPCA (Graph diffusion & PCA)

$\quad$ **INPUT:** $X, A, Y, \sigma, \alpha, \delta, \mathcal{I}, \tau, \epsilon$;

$\quad$ **INITIALIZE:**

$\quad$ $\bar{X}_i^T = X_i^T - \frac{1}{d}\sum_j^d X_j^T \forall i \in (1,\ldots,n);\ S = \frac{\bar{X}^T\bar{X}}{d-1}$

$\quad$ $\gamma = randomizedSVD(SD^{-2\sigma+1})$

$\quad$ **IF:** $1 + \delta(\gamma - \epsilon) < 1/\alpha$:

$\quad\quad$ $Z = PI(\alpha(D^{\sigma-1}AD^{-\sigma} + \delta SD^{-2\sigma+1}),(1-\alpha)Y,\mathcal{I})$

$\quad$ **ELSE:**

$\quad\quad$ $Z = GMRES((I - \alpha(D^{\sigma-1}AD^{-\sigma} + \delta SD^{-2\sigma+1})),(1-\alpha)Y,\tau,\mathcal{I})$

Note also that Proposition 1 simplifies to the known results of PRSSL [1] for the value $\delta = 0$. GDPCA can thus be seen as a generalization of PRSSL enriching the default random walk matrix $D^{\sigma-1}AD^{-\sigma}$ thanks to the sample covariance matrix S. Notice that S is retrieved from PCA loss in Problem (4) (see Appendix A). This enrichment of the binary weights ($A_{i,j} = 0$ or $A_{i,j} = 1$) by node covariance allows bypassing the *Sparse A* issue. Similarly, we assume that our framework solves the *Curse of dimensionality* issue thanks to the use of PCA loss.

4   Experiments

4.1   Datasets Description

In the experimental part of this work, we consider two types of datasets: datasets with an underlying graph structure, and datasets that are non-graph based. The latter allow us to test the flexibility of our framework.

Graph-Based Datasets. We consider the citation networks datasets of Cora, Citeseer, and Pubmed [29]. These datasets have bag-of-words [15] representation for each node (paper) features and a citation network between papers. The citation links are considered as edges in the adjacency matrix A. Each paper has a class label ($X_i \in \mathcal{C}_j$).

Non-graph Based Datasets. *Images.* We consider the standard MNIST image dataset [21] composed of square 28×28 pixel grayscale images of handwritten digits from 0 to 9. Besides, we flattened square pixels in 784 d-space features for this dataset. *Text data. Covid clinical trials (CCT) crawled dataset.* We consider a second non-graph based dataset which we prepared and processed from the ClinicalTrials resource[4] from summaries of evidence-based [22] clinical trials on COVID. This dataset is particularly important given the current need from medical experts on this topic. We analyzed 1001 xml files as follows:

1. the feature matrix X was generated from a bag-of-words model based on the descriptive fields "official_title", "brief_summary", "detailed_description", "eligibility";
2. the label matrix Y was generated from the field "masking", which takes values in (*Open, Blind*)[5], as it is one of the essential parameters of evidence-based medicine EBM [8]. The type of masking corresponds to the way of conducting clinical trials: the *Open* way is a less expensive and complicated procedure than the *Blind* one.

Note that the CCT dataset could be useful to other researchers who wish to improve even further the labeling of COVID clinical trials. The registration procedure of clinical trial is useful when authors forget to create masking tag for their work. Particularly after analyzing 1001 xml files, we found that from 3557 clinical trials 1518 of them do not have a masking tag.

As the non-graph based datasets do not have a predefined graph structure, we apply the K-nearest neighbours (KNN) [11] algorithm to generate the adjacency matrix. In Appendix C, we show on validation sets of MNIST and CCT datasets how the choice of distances and number of neighbours for the generation of the adjacency matrix by KNN influence GDPCA. We followed the strategy for train/validation/test splitting as in [32] for Pubmed, Citeseer, Cora and CCT, and as in [24] for MNIST.

The above datasets and code with GDPCA are available through a GitHub repository[6]. Table 1 provides a description of these datasets, where $LR = n_l/n$ is the learning rate with n_l the number of labeled nodes.

Table 1. Dataset statistic.

	CITESEER	CORA	PUBMED	CCT	MNIST
n	3327	2708	19717	2039	50000
e	4732	5492	44338	–	–
k	6	7	3	2	10
d	3703	1433	500	7408	784
LR	0.036	0.052	0.003	0.019	0.002
c_0	0.898	1.889	39.43	0.275	63.77

---

[4] https://clinicaltrials.gov/ct2/resources/download#DownloadMultipleRecords.
[5] In order to simplify the labeling process, we replaced the long description of masking by a shorter version (e.g. Single Blind (Participant, Investigator) by *Blind*).
[6] https://github.com/KamalovMikhail/GDPCA.

4.2 State-of-the-Art (SOTA) Algorithms

As some of the SOTA algorithms cannot be applied to all types of datasets, we consider specific SOTA algorithms depending on the datasets. For the graph-structured Citeseer, Cora and Pubmed datasets, we compare GDPCA to the LP [34] and ManiReg [4] linear graph diffusion algorithms and to the SemiEmb [31], Planetoid [32], GCN [19] and DeepWalk [23] graph convolution-based neural networks. For MNIST, we compared it to the transductive SVM (TSVM) [16] and KNN [11] linear algorithms, and to the GCN neural network. Finally, for CCT, we compared it to the linear LP [34], KNN [11], and PRSSL [1], and to GCN.

Accuracy for Non-reproduced Benchmarks. Since for training and estimation of the GDPCA framework, we use the train/validation/test split strategy for Citeseer, Pubmed, Cora and CCT datasets as in [32] we can use the accuracy of SOTA algorithms from work [32]. In particular, we can take the accuracy of LP [34], ManiReg [4], TSVM [16], SemiEmb [31], Planetoid [32] algorithms from work [32], and the GCN [19], DeepWalk [23] algorithm's accuracy from work [19]. Since for MNIST dataset we use the train/validation/test split strategy as in [24] we can use the value of accuracy of KNN [11] and TSVM[16] algorithms from work [24].

Algorithm Parameters for Reproduced Benchmarks. We trained LP, PRSSL, KNN and GCN on CCT and MNIST datasets with the best hyper-parameters defined in the articles describing these algorithms: LP [34] $RBF(\cdot)$ kernel function; GCN [19] 0.5 dropout rate, $5 \cdot 10^{-4}$ L2 regularization, 16 hidden units and 200 epochs; KNN parameters selected by Randomized Search [5] for Cora, Citeseer, Pubmed and CCT datasets.

For a fair model comparison between GDPCA, PRSSL and GCN, we replaced A by $A+I$ as was done in [19,32]. Also, for GDPCA and PRSSL we fixed $\alpha = 0.9$ and $\sigma = 1$ on all datasets as it was shown in [1] that these parameters provide the best accuracy result for PRSSL. We trained GDPCA on Cora, Citeseer and CCT with $\delta = 1$, $\mathcal{I} = 10$, $\tau = 10^{-3}$, $\epsilon = 10^{-3}$, and the same for MNIST and Pubmed but changing the value of δ to 10^{-3}. We selected these specific $\mathcal{I}$, ϵ, τ parameters by Random Search algorithm [5] as a trade-off between fast computation with GMRES and PowerIteration and accuracy on the validation set. Moreover, for MNIST and CCT we generated a synthetic adjacency matrix A by KNN with respect to the results from Appendix C. In particular, we generated synthetic adjacency matrices based on the following parameters of KNN for datasets: for CCT - *Dice* distance and 7 nearest neighbours; for MNIST - *Cosine* distance and 7 nearest neighbours. We used these synthetic adjacency matrices for the training of GDPCA, PRSSL and GCN algorithms.

4.3 Results

Accuracy Results. The aforementioned comparisons in terms of accuracy (%) are presented in Table 2 and Table 3. Table 2 shows that GDPCA outperforms

Table 2. Classification accuracy (%) comparison with linear algorithms.

Dataset	Cora	Citeseer	Pubmed	CCT	MNIST
TSVM [16]	57.5	64.0	62.2	–	83.2
KNN [11]	43.9	47.4	63.8	57.1	74.2
LP [34]	68.0	45.3	63.0	53.5	34.2
ManiReg [4]	59.5	60.1	70.7	–	–
PRSSL [1]	69.3	45.9	68.4	55.8	87.2
GDPCA	**77.7**	**73.1**	**76.1**	**61.1**	**88.4**

other SOTA linear algorithms, especially it is significantly better on the Cora, Citeseer and Pubmed, where it outperforms the others by 8%, 9% and 5% respectively. Moreover, Table 3 shows that our linear GDPCA framework provides performance that is close to the best neural network algorithms results. Note that GDPCA has a fixed explicit solution (5) as opposed to the neural network algorithms, which depend on the layer's weights initialization process. Furthermore, Table 2 and Table 3 show that GDPCA has a good performance on standard Cora, Citeseer, Pubmed and MNIST as well as on real dataset CCT.

Table 3. Classification accuracy (%) comparison with neural network algorithms.

Dataset	Cora	Citeseer	Pubmed	CCT	MNIST
SemiEmb [31]	59.0	59.6	71.1	–	–
DeepWalk [23]	67.2	43.2	65.3	–	–
Planetoid [32]	75.7	64.7	77.2	–	–
GCN [19]	**81.5**	70.3	**79.0**	55.2	81.4
GDPCA	77.7	**73.1**	76.1	**61.1**	**88.4**

Computational Complexity. We finish this experiment section by comparing the computational complexity of GDPCA with the SOTA algorithms that obtained the most similar performance, namely GCN[7] and Planetoid[8]. The algorithmic complexity of GDPCA is $\mathcal{O}(\mathcal{I}nk)$ in the case of PowerIteration, where e' is the number of non-zero elements in matrix $(D^{\sigma-1}AD^{-\sigma} + \delta SD^{-2\sigma+1})$, and $\mathcal{O}(\mathcal{I}nk)$ in the case of GMRES. Note that PowerIteration can be computed in the distributed over node regime [6, p. 135], and GMRES can be distributed over classes. The comparison of GDPCA framework with GCN and Planetoid algorithms in big-$\mathcal{O}$ notation is presented in Table 4. Figure 3 provides the time (in seconds) of 50 completed trainings on CPU(1.4 GHz quad-core Intel Core i5) for each algorithms. It shows a clear advantage of GDPCA over the GCN and Planetoid especially with GMRES, in terms of computational time.

[7] https://github.com/tkipf/gcn.
[8] https://github.com/kimiyoung/planetoid.

Table 4. Comparison of computational complexity, where l is the number of layers, n is the number of nodes, d is the number of features, r is the number sampled neighbors per node, k is the batch size; ϕ is the number of random walks; p is the walk length; w is the window size; m is a representation size; k is the number of classes.

Algorithm	GCN	GDPCA	PLANETOID
TIME	$O(led + lndm)$	$O(\mathcal{I}nk)$	$O(\phi npw(m + m\log n))$
MEMORY	$O(lnd + ld^2)$	$O(e')$	$O(nld^2)$

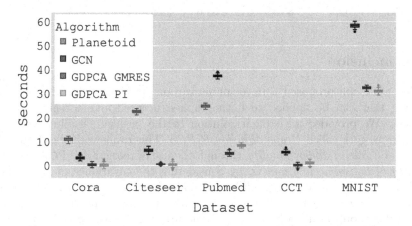

Fig. 3. Computational time of 50 completed trainings on CPU.

Significance of the Covariance Matrix. In this experiment, the aim is to verify that the use of the covariance matrix S actually leads to an improvement. In order to do so, we compare GDPCA with PRSSL ($\delta = 0$) and other values of δ, as well as with variants of GDPCA where S is replaced with the following efficient similarity matrices: $W_{COS} = \frac{[COS(X_i,X_j)]_{i,j=1}^{n,n}}{d-1}$ and $W_{RBF} = \frac{[RBF(X_i,X_j)]_{i,j=1}^{n,n}}{d-1}$. Table 5 displays the average accuracies of each variant along with their statistical significance evaluated with t-tests. It shows that using S in GDPCA is significantly better on the Cora, Citeseer and Pubmed datasets, where it outperforms the others at least by 7%, 8% and 3% respectively. Notice that Table 2 and Table 3 contain accuracy on a test set of fixed dataset splits: as in [32] for Citeseer, Cora, Pubmed and CCT datasets; as in [24] for MNIST dataset, and Table 5 has accuracy on test sets averaged over 50 random splits. All experiments mentioned above are available through a GitHub repository[9].

---

[9] https://github.com/KamalovMikhail/GDPCA.

Table 5. Average accuracy (%), ▲ denotes the statistical significance for $p < 0.05$.

Dataset	GDPCA $\delta = 1$ (S)	GDPCA $\delta = 10^{-3}$ (S)	PRSSL $\delta = 0$	GDPCA $\delta = 1$ (W_{COS})	GDPCA $\delta = 1$ (W_{RBF})
Cora	77.3 ▲	71.8	69.8	70.1	68.3
Citeseer	73.0 ▲	65.1	44.8	64.8	44.5
Pubmed	68.7	75.8 ▲	67.9	72.6	71.1
CCT	60.4 ▲	54.5	55.6	54.2	56.2
MNIST	62.5	85.3▲	82.6	60.6	59.2

5   Conclusion

In this work, we proposed a novel minimization problem for semi-supervised learning that can be applied to both graph-structured and non-graph based datasets. We provided an explicit solution to the problem, leading to a new linear framework called *Graph diffusion & PCA*. This framework allows to overcome the *Curse of dimensionality*, through the use of reorganized PCA, and the *sparsity of the adjacency matrix*, by considering the covariance matrix, which are both common issues in graph-based semi-supervised learning. We demonstrated the impact of these improvements in experiments on several datasets with and without an underlying graph structure. We also compared it to state-of-the art algorithms and showed that GDPCA clearly outperforms the other linear graph-based diffusion ones. As for the comparison with neural networks, the experiments showed that the performance are similar, while GDPCA has a significantly lower computational time in addition to providing an explicit solution. In future works, we plan to generalize GDPCA to a nonlinear case keeping the low computational complexity and improving classification performance. Also, we want to avoid the bottleneck that arises in the dense covariance matrix S, which can lead to high memory consumptions. Particularly, by distributed PI regimes [6, p. 135] and GMRES, we can directly compute covariance between nodes for a small distributed portion of nodes. This preserves the space consumption as opposed to the precomputed (S).

A   Proof of Proposition 1

Proof. This proof uses the same strategy as the proof of Proposition 2 in [1]. Rewriting Problem (4) in matrix form with the standard Laplacian $L = D - A$ and with $Z_{.i}, Y_{.i} \in \mathbb{R}^{n \times 1}$:

$$Q(Z) = 2 \sum_{i=1}^{k} Z_{.i}^T D^{\sigma-1} L D^{\sigma-1} Z_{.i}$$

$$+ \mu \sum_{i=1}^{k} (Z_{.i} - Y_{.i})^T D^{2\sigma-1} (Z_{.i} - Y_{.i}) - \delta \sum_{i=1}^{k} Z_{.i} S Z_{.i}^T$$

where $S = \bar{X}^T \bar{X}/(d-1) \in \mathbb{R}^{n \times n}$. Considering $\frac{Q(Z)}{\partial Z} = 0$:

$$2Z^T(D^{\sigma-1}LD^{\sigma-1} + D^{\sigma-1}L^T D^{\sigma-1}) + 2\mu(Z-Y)^T D^{2\sigma-1} - \delta Z^T(S+S^T) = 0$$

Multiplying by $D^{-2\sigma+1}$ and replacing $L = D - A$ results in:

$$Z^T(2I - 2D^{\sigma-1}AD^{-\sigma} + \mu I - 2\delta SD^{-2\sigma+1}) - \mu Y^T = 0$$

Taking out the μ over the parentheses and transposing the equation:

$$Z = \frac{\mu}{(2+\mu)}(I - \frac{2}{(2+\mu)}(D^{\sigma-1}AD^{-\sigma} + \delta SD^{-2\sigma+1}))^{-1}Y$$

Finally, the desired result is obtained with $\alpha = 2/(2+\mu)$.   □

B   Proof of Proposition 2

Proof. Apply Theorem 1 of sums of spectral radii [35] for the following inequality:

$$\rho(D^{\sigma-1}AD^{-\sigma} + \delta SD^{-2\sigma+1}) \leq \rho\left(D^{\sigma-1}AD^{-\sigma}\right) + \rho(\delta SD^{-2\sigma+1}) < 1/\alpha$$

based on the fact that spectral radius of a matrix similar to the stochastic matrix is equal to 1 (Gershgorin bounds):

$$1 + \delta\rho(SD^{-2\sigma+1}) < 1/\alpha$$

apply the Theorem 7 [26] for replacing $\rho(SD^{-2\sigma+1})$ by the γ maximum singular value of $SD^{-2\sigma+1}$ we obtain the desired result in (6).   □

C   Generation of Synthetic Adjacency Matrix

For selecting the best synthetic adjacency matrix for GDPCA, we have considered three standard distances, such as *Cosine, Minkowski, Dice* and the number of neighbours from 1 till 14 for KNN algorithm. The accuracy of GDPCA on above parameters on the validation set for MNIST and CCT datasets are shown in Fig. 4. Figure 4 shows that the best GDPCA accuracy on the validation set is obtained with the use of 7 neighbours and *Dice* distance for the CCT dataset is obtained with the use of 7 neighbours and *Cosine* distance for the MNIST dataset.

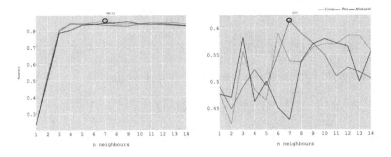

Fig. 4. Estimate different adjacency matrix for GDPCA.

References

1. Avrachenkov, K., Mishenin, A., Gonçalves, P., Sokol, M.: Generalized optimization framework for graph-based semi-supervised learning. In: Proceedings of the 2012 SIAM International Conference on Data Mining, pp. 966–974. SIAM (2012)
2. Baeza-Yates, R., Navarro, G.: Block addressing indices for approximate text retrieval. J. Am. Soc. Inf. Sci. **51**(1), 69–82 (2000)
3. Bair, E., Hastie, T., Paul, D., Tibshirani, R.: Prediction by supervised principal components. J. Am. Stat. Assoc. **101**(473), 119–137 (2006)
4. Belkin, M., Niyogi, P., Sindhwani, V.: Manifold regularization: a geometric framework for learning from labeled and unlabeled examples. J. Mach. Learn. Res. **7**, 2399–2434 (2006)
5. Bergstra, J., Bengio, Y.: Random search for hyper-parameter optimization. J. Mach. Learn. Res. **13**(2) (2012)
6. Bertsekas, D.P., Tsitsiklis, J.N.: Parallel and Distributed Computation: Numerical Methods, vol. 23. Prentice Hall, Englewood Cliffs (1989)
7. Chapelle, O., Scholkopf, B., Zien, A.: Semi-supervised learning. IEEE Trans. Neural Netw. **20**(3), 542 (2009). (chapelle, o. et al., eds.; 2006)[bibbook reviews]
8. Day, S.J., Altman, D.G.: Blinding in clinical trials and other studies. BMJ **321**(7259), 504 (2000)
9. Ding, C., He, X.: K-means clustering via principal component analysis. In: Proceedings of the Twenty-first International Conference on Machine Learning, p. 29 (2004)
10. Fisher, R.A.: The use of multiple measurements in taxonomic problems. Ann. Eugen. **7**(2), 179–188 (1936)
11. Fix, E.: Discriminatory analysis: nonparametric discrimination, consistency properties. USAF school of Aviation Medicine (1951)
12. Freund, R.M.: Quadratic functions, optimization, and quadratic forms (2004)
13. Grover, A., Zweig, A., Ermon, S.: Graphite: iterative generative modeling of graphs. In: International Conference on Machine Learning, pp. 2434–2444. PMLR (2019)
14. Halko, N., Martinsson, P.G., Tropp, J.A.: Finding structure with randomness: probabilistic algorithms for constructing approximate matrix decompositions. SIAM Rev. **53**(2), 217–288 (2011)
15. Harris, Z.S.: Distributional structure. Word **10**(2–3), 146–162 (1954)
16. Joachims, T.: Transductive inference for text classification using support vector machines. In: ICML, vol. 99, pp. 200–209 (1999)

17. Johnson, R., Zhang, T.: Graph-based semi-supervised learning and spectral kernel design. IEEE Trans. Inf. Theory **54**(1), 275–288 (2008)
18. Kamalov, M., Avrachenkov, K.: GenPR: generative PageRank framework for semi-supervised learning on citation graphs. In: Filchenkov, A., Kauttonen, J., Pivovarova, L. (eds.) AINL 2020. CCIS, vol. 1292, pp. 158–165. Springer, Cham (2020). https://doi.org/10.1007/978-3-030-59082-6_12
19. Kipf, T.N., Welling, M.: Semi-supervised classification with graph convolutional networks. In: 5th International Conference on Learning Representations. ICLR (2017)
20. Klicpera, J., Bojchevski, A., Günnemann, S.: Predict then propagate: graph neural networks meet personalized pagerank. arXiv preprint arXiv:1810.05997 (2018)
21. LeCun, Y., Bottou, L., Bengio, Y., Haffner, P.: Gradient-based learning applied to document recognition. Proc. IEEE **86**(11), 2278–2324 (1998)
22. Masic, I., Miokovic, M., Muhamedagic, B.: Evidence based medicine-new approaches and challenges. Acta Informatica Medica **16**(4), 219 (2008)
23. Perozzi, B., Al-Rfou, R., Skiena, S.: DeepWalk: online learning of social representations. In: Proceedings of the 20th ACM SIGKDD International Conference on Knowledge Discovery and Data Mining, pp. 701–710 (2014)
24. Rifai, S., Dauphin, Y.N., Vincent, P., Bengio, Y., Muller, X.: The manifold tangent classifier. Adv. Neural. Inf. Process. Syst. **24**, 2294–2302 (2011)
25. Ritchie, A., Scott, C., Balzano, L., Kessler, D., Sripada, C.S.: Supervised principal component analysis via manifold optimization. In: 2019 IEEE Data Science Workshop (DSW), pp. 6–10. IEEE (2019)
26. Rojo, O., Soto, R., Rojo, H.: Bounds for the spectral radius and the largest singular value. Comput. Math. Appl. **36**(1), 41–50 (1998)
27. Roli, F., Marcialis, G.L.: Semi-supervised PCA-based face recognition using self-training. In: Yeung, D.-Y., Kwok, J.T., Fred, A., Roli, F., de Ridder, D. (eds.) SSPR /SPR 2006. LNCS, vol. 4109, pp. 560–568. Springer, Heidelberg (2006). https://doi.org/10.1007/11815921_61
28. Saad, Y., Schultz, M.H.: GMRES: a generalized minimal residual algorithm for solving nonsymmetric linear systems. SIAM J. Sci. Stat. Comput. **7**(3), 856–869 (1986)
29. Sen, P., Namata, G., Bilgic, M., Getoor, L., Galligher, B., Eliassi-Rad, T.: Collective classification in network data. AI Mag. **29**(3), 93 (2008)
30. Walder, C., Henao, R., Mørup, M., Hansen, L.: Semi-Supervised Kernel PCA. IMM-Technical Report-2010-10, Technical University of Denmark, DTU Informatics, Building 321 (2010)
31. Weston, J., Ratle, F., Mobahi, H., Collobert, R.: Deep learning via semi-supervised embedding. In: Montavon, G., Orr, G.B., Müller, K.-R. (eds.) Neural Networks: Tricks of the Trade. LNCS, vol. 7700, pp. 639–655. Springer, Heidelberg (2012). https://doi.org/10.1007/978-3-642-35289-8_34
32. Yang, Z., Cohen, W., Salakhudinov, R.: Revisiting semi-supervised learning with graph embeddings. In: Proceedings of Machine Learning Research, vol. 48, pp. 40–48. PMLR, New York, 20–22 June 2016
33. Zhu, X., Ghahramani, Z.: Learning from labeled and unlabeled data with label propagation (2002)
34. Zhu, X., Ghahramani, Z., Lafferty, J.D.: Semi-supervised learning using gaussian fields and harmonic functions. In: Proceedings of the 20th International conference on Machine learning (ICML 2003), pp. 912–919 (2003)
35. Zima, M.: A theorem on the spectral radius of the sum of two operators and its application. Bull. Aust. Math. Soc. **48**(3), 427–434 (1993)

Exact Counting and Sampling of Optima for the Knapsack Problem

Jakob Bossek[1]([envelope]) [iD], Aneta Neumann[2] [iD], and Frank Neumann[2] [iD]

[1] Statistics and Optimization, University of Münster, Münster, Germany
bossek@wi.uni-muenster.de
[2] School of Computer Science, The University of Adelaide, Adelaide, Australia
{aneta.neumann,frank.neumann}@adelaide.edu.au

Abstract. Computing sets of high quality solutions has gained increasing interest in recent years. In this paper, we investigate how to obtain sets of optimal solutions for the classical knapsack problem. We present an algorithm to count exactly the number of optima to a zero-one knapsack problem instance. In addition, we show how to efficiently sample uniformly at random from the set of all global optima. In our experimental study, we investigate how the number of optima develops for classical random benchmark instances dependent on their generator parameters. We find that the number of global optima can increase exponentially for practically relevant classes of instances with correlated weights and profits which poses a justification for the considered exact counting problem.

Keywords: Zero-one knapsack problem · Exact counting · Sampling · Dynamic programming

1 Introduction

Classical optimisation problems ask for a single solution that maximises or minimises a given objective function under a given set of constraints. This scenario has been widely studied in the literature and a vast amount of algorithms are available. In the case of NP-hard optimisation problems one is often interested in a good approximation of an optimal solution. Again, the focus here is on a single solution.

Producing a large set of optimal (or high quality) solutions allows a decision maker to pick from structurally different solutions. Such structural differences are not known when computing a single solution. Computing a set of optimal solutions has the advantage that more knowledge on the structure of optimal solutions is obtained and that the best alternative can be picked for implementation. As the number of optimal solutions might be large for a given problem, sampling from the set of optimal solutions provides a way of presenting different alternatives.

Related to the task of computing the set of optimal solutions, is the task of computing diverse sets of solutions for optimisation problems. This area of

© Springer Nature Switzerland AG 2021
D. E. Simos et al. (Eds.): LION 2021, LNCS 12931, pp. 40–54, 2021.
https://doi.org/10.1007/978-3-030-92121-7_4

research has obtained increasing attention in the area of planning where the goal is to produce structurally different high quality plans [9–11,19]. Furthermore, different evolutionary diversity optimisation approaches which compute diverse sets of high quality solutions have been introduced [14,15,20]. For the classical traveling salesperson problem, such an approach evolves a diverse set of tours which are all a good approximation of an optimal solution [3].

Counting problems are frequently studied in the area of theoretical computer science and artificial intelligence [2,6–8]. Here the classical goal is to count the number of solutions that fulfill a given property. This might include counting the number of optimal solutions. Many counting problems are #P-complete [21] and often approximations of the number of such solutions, especially approximations on the number of optimal solutions are sought [5]. Further examples include counting the number of shortest paths between two nodes in graphs [13] or exact counting of minimum-spanning-trees [1]. For the knapsack problem (KP), the problem of counting the number of feasible solutions, i.e. the number of solutions that do not violate the capacity constraint, is #P-complete. As a consequence, different counting approaches have been introduced to approximately count the number of feasible solutions [4,18,22].

In this paper, we study the classical zero-one knapsack problem (KP). We develop an algorithm that is able to compute all optimal solutions for a given knapsack instance. The algorithm adapts the classical dynamic programming approach for KP in the way that all optimal solutions are produced implicitly. As the number of such solutions might grow exponentially with the problem size for some instances, we develop a sampling approach which samples solutions for the set of optimal solutions uniformly at random.

We carry out experimental investigations for different classes of knapsack instances given in the literature (instances with uniform random weights and instances with different correlation between weights and profits). Using our approach, we show that the number of optimal solutions significantly differs between different knapsack instance classes. In particular, for instances with correlated weights and profits – a group of great importance in practical applications – an exponential growth of optima is observed. In addition, we point out that changing the knapsack capacity slightly can reduce the number of optimal solutions from exponential to just a single solution.

The paper is structured as follows. In the next section, we introduce the task of computing the set of optimal solutions for the knapsack problem. Afterwards, we present the dynamic programming approach for computing the set of optimal solutions and show how to sample efficiently from the set of optimal solutions without having to construct the whole set of optimal solutions. In our experimental investigations, we show the applicability of our approach to a wide range of knapsack instances and point out insights regarding the number of optimal solutions for these instances. Finally, we finish with some concluding remarks.

2   Problem Formulation

We now introduce the problem of computing all optimal solutions for the knapsack problem. In the following, we use the standard notation $[n] = \{1, \ldots, n\}$ to express the set of the first n positive integers. The problem studied is the classical NP-hard *zero-one knapsack problem* (KP). We are given a knapsack with integer capacity $W > 0$ and a finite set of n items, each with positive integer weight w_i and associated integer profit or value v_i for $i \in [n]$. Each subset $s \subset [n]$ is called a *solution/packing*. We write

$$w(s) = \sum_{i \in s} w_i \text{ and } v(s) = \sum_{i \in s} v_i$$

for the total weight and value respectively. A solution is *feasible* if its total weight does not exceed the capacity. Let

$$\mathcal{S} = \{s \mid s \subset [n] \wedge w(s) \leq W\}$$

be the set of feasible solutions. The goal in the optimisation version of the problem is to find a solution $s^* \in \mathcal{S}$ such that

$$s^* = \arg\max_{s \in \mathcal{S}} v(s).$$

Informally, in the optimisation version of the KP, we strive for a subset of items that maximises the total profit under the constraint that the total weight remains under the given knapsack capacity.

Let $v_{\max}$ be the value of an optimal solution s^* and let

$$\mathcal{S}^* = \{s \mid s \in \mathcal{S} \wedge v(s) = v_{\max}\}$$

be the set of optimal solutions. In this work we study a specific counting problem which we refer to as #KNAPSACK* in the following. Here, the goal is to determine *exactly* the cardinality of the set $\mathcal{S}^*$. Note that this is a special case of the classic counting version #KNAPSACK where we aim to count the set of all feasible solutions $\mathcal{S}$, and $\mathcal{S}^* \subset \mathcal{S}$. In addition, we are interested in procedures to sample uniformly at random a subset of k out of $|\mathcal{S}^*|$ solutions with $k \leq |\mathcal{S}^*|$.

3   Exact Counting and Sampling of Optima

In this section we introduce the algorithms for the counting and sampling problems stated. We first recap the classic dynamic programming algorithm for the zero-one KP as it forms the foundation for our algorithm(s).

3.1   Recap: Dynamic Programming for the KP

Our algorithms are based on the dynamic programming approach for the optimisation version (see, e.g. the book by Kellerer et al. [12]). This well-known algorithm maintains a table V with components $V(i, w)$ for $0 \leq i \leq n, 0 \leq w \leq W$.

Here, component $V(i, w)$ holds the maximum profit that can be achieved with items up to item i, i.e., $\{1, \ldots, i\}$, and capacity w. The table is constructed bottom-up following the recurrence

$$V(i, w) = \max\{ \underbrace{V(i-1, w)}_{\text{(a) leave item } i}, \underbrace{V(i-1, w-w_i) + v_i}_{\text{(b) take item } i} \}$$

for $1 \leq i \leq n, 0 \leq w \leq W$. Essentially, the optimal value $V(i, w)$ is achieved by making a binary decision for every item $i \in [n]$ relying on pre-calculated optimal solutions to sub-problems with items from $\{1, \ldots, i-1\}$. The options are (a) either leaving item i where the optimal solution is realised by the maximum profit achieved with items $\{1, \ldots, i-1\}$ and capacity w. Option (b) deals with putting item i into the knapsack (only possible if $w_i < w$) gaining profit v_i at the cost of additional w_i units of weight. In consequence, the optimal profit $V(i, w)$ is the optimal profit with items from $\{1, \ldots, i-1\}$ and capacity $w - w_i$, i.e., $V(i, w) = V(i-1, w-w_i) + v_i$. Initialization follows

$$V(0, w) = 0 \qquad \forall 0 \leq w \leq W \tag{1}$$
$$V(i, w) = -\infty \qquad \forall w < 0 \tag{2}$$

which covers the base cases of an empty knapsack (Eq. (1)) and a negative capacity, i.e., invalid solution (Eq. (2)), respectively. Eventually, $V(n, W)$ holds the profit of an optimal solution.

3.2 Dynamic Programming for #KNAPSACK*

We observe that if $V(i-1, w) = V(i-1, w-w_i) + v_i$ we can achieve *the same maximum profit* $V(i, w)$ by both options (a) or (b). Analogously to $V(i, w)$ let $C(i, w)$ be the number of solutions with maximum profit given items from $\{1, \ldots, i\}$ and capacity w. Then there are three update options for $C(i, w)$ for $1 \leq i \leq n$ and $0 \leq w \leq W$ (we discuss the base case later): (a') Either, as stated above, we can obtain the same maximum profit $V(i, w)$ by packing or not packing item i. In this case, $C(i, w) = C(i-1, w) + C(i-1, w-w_i)$ since the item sets leading to $V(i-1, w)$ and $V(i-1, w-w_i)$ are necessarily disjoint by construction. Options (b') and (c') correspond to (a) and (b) leading to the recurrence

$$C(i, w) = \begin{cases} C(i-1, w) + C(i-1, w-w_i) & \text{if } V(i-1, w) = V(i-1, w-w_i) + v_i \\ C(i-1, w) & \text{if } V(i-1, w) > V(i-1, w-w_i) + v_i \\ C(i-1, w-w_i) & \text{otherwise.} \end{cases}$$

$$\tag{3}$$

Analogously to Eqs. (1) and (2) the bases cases

$$C(0, w) = 1 \quad \forall 0 \leq w \leq W \tag{4}$$
$$C(i, w) = 0 \quad \forall w < 0 \tag{5}$$

Algorithm 1: DP-algorithm for #KNAPSACK*

Input: Number of items n, capacity W

1 **for** $w \leftarrow 0$ **to** W **do**
2    $V(0,w) \leftarrow 0$;
3    $C(0,w) \leftarrow 1$;

4 **for** $i \leftarrow 0$ **to** n **do**
5    $V(i,0) \leftarrow 0$;
6    $C(i,0) \leftarrow 1$;

7 **for** $i \leftarrow 1$ **to** n **do**
8   **for** $w \leftarrow 1$ **to** W **do**
9   **if** $w_i > w$ **then**
10       $V(i,w) \leftarrow V(i-1,w)$;
11       $C(i,w) \leftarrow C(i-1,w)$;
12   **else**
13   **if** $V(i-1,w) = V(i-1,w-w_i) + v_i$ **then**
14        $V(i,w) \leftarrow V(i-1,w)$;
15        $C(i,w) \leftarrow C(i-1,w) + C(i-1,w-w_i)$;
16   **else if** $V(i-1,w) > V(i-1,w-w_i) + v_i$ **then**
17        $V(i,w) \leftarrow V(i-1,w)$;
18        $C(i,w) \leftarrow C(i-1,w)$;
19   **else**
20        $V(i,w) \leftarrow V(i-1,w-w_i) + v_i$;
21        $C(i,w) \leftarrow C(i-1,w-w_i)$;

22 **return** V, C

---

handle the empty knapsack (the empty-set is a valid solution) and the case of illegal items (no valid solution(s) at all). The base cases are trivially correct. The correctness of the recurrence follows inductively by the preceding argumentation. Hence, we wrap up the insights in the following lemma whose proof is embodied in the preceding paragraph.

Lemma 1. *Following the recurrence in Eq. (3) $C(i,w)$ stores the number of optimal solutions using items from $\{1,\ldots,i\}$ given the capacity w.*

Algorithm 1 shows pseudo-code for our algorithm. The algorithm directly translates the discussed recurrences for tables V and C. Note that the pseudo-code is not optimised for performance and elegance, but for readability.

Theorem 1. *Let S^* be the set of optimal solutions to a zero-one knapsack problem with n items and capacity W. There is a deterministic algorithm that calculates $|S^*|$ exactly with time- and space-complexity $O(n^2W)$.*

Proof. The correctness follows from Lemma 1. For the space- and time-complexity note that two tables with dimensions $(n+1) \times (W+1)$ are filled. For Table V the algorithm requires constant time per cell and hence time and space $O(nW)$. Table

Table 1. Exemplary knapsack instance (left) and the dynamic programming tables $V(i, w)$ (center) and $C(i, w)$ (right) respectively. Table cells highlighted in light-gray indicate components where two options are possible: either packing item i or not. Light-green cells indicate the total value of any optimal solution for $V(\cdot, \cdot)$ and $|S^*|$ for $C(\cdot, \cdot)$.

$W = 8$

i	w_i	v_i
1	3	3
2	8	10
3	2	3
4	2	4
5	2	3

$V(\cdot, \cdot)$

i \ w	0	1	2	3	4	5	6	7	8
0	0	0	0	0	0	0	0	0	0
1	0	0	0	3	3	3	3	3	3
2	0	0	0	3	3	3	3	3	10
3	0	0	3	3	3	6	6	6	10
4	0	0	4	4	7	7	7	10	10
5	0	0	4	4	7	7	10	10	10

$C(\cdot, \cdot)$

i \ w	0	1	2	3	4	5	6	7	8
0	1	1	1	1	1	1	1	1	1
1	1	1	1	1	1	1	1	1	1
2	1	1	1	1	1	1	1	1	1
3	1	1	1	2	2	1	1	1	1
4	1	1	1	1	1	2	2	1	2
5	1	1	1	1	1	2	3	1	3	4

C stores the number of solutions which can be exponential in the input size n as we shall see later. Therefore, $C(i, j) \leq 2^n$ and $O(\log(2^n)) = O(n)$ bits are necessary to encode these numbers. Thus, the addition in line 15 in Algorithm 1 requires time $O(n)$ which results in space and time requirement of $O(n^2 W)$. □

Note that the complexity reduces to $O(nW)$ if $|S^*| = poly(n)$. Furthermore, the calculation of row i only relies on values in row $i - 1$. Hence, we can reduce the complexity by a factor of n if only two rows of V and C are stored and the only value of interest is the number of optimal solutions.

For illustration we consider a simple knapsack instance with $n = 5$ items fully described by the left-most table in Table 1. Let $W = 8$. In this setting there exist four optima $s_1 = \{2\}$, $s_2 = \{1, 4, 5\}$, $s_3 = \{1, 3, 4\}$ and $s_4 = \{3, 4, 5\}$ with profit 10 each and thus $|S^*| = 4$. The dynamic programming tables are shown in Table 1 (center and right). For improved visual accessibility we highlight table cells where both options (packing item i or not) are applicable and hence an addition of the number of combinations of sub-problems is performed in C by the algorithm (cf. first case in the recurrence in Eq. 3).

3.3   Uniform Sampling of Optimal Solutions

The DP algorithm introduced before allows to count $|S^*|$ and hence to solve #KNAPSACK* exactly. The next logical step is to think about a sampler, i.e., an algorithm that samples uniformly at random from S^* even if S^* is exponential in size. In fact, we can utilize the tables C and V for this purpose as they implicitly encode the information on all optima. A similar approach was used by Dyer [4] to approximately sample from the (approximate) set of feasible solutions in his dynamic programming approach for #KNAPSACK. Our sampling algorithm however is slightly more involved due to a necessary case distinctions.

Algorithm 2: Uniform Sampling of Optima

Input: DP tables V and C (see Algorithm 1), number of items n, capacity W, desired number of solutions k.

1 $S \leftarrow \emptyset$;
2 **while** $k > 0$ **do**
3      $L \leftarrow \emptyset$;
4      $i \leftarrow n$;
5      $w \leftarrow W$;
6   **while** $i > 0$ **and** $w > 0$ **do**
7   **if** $w_i \le w \wedge V(i,w) = V(i-1,w) \wedge V(i,w) = V(i-1,w-w_i)+v_i$ **then**
8              $q \leftarrow \frac{C(i-1,w-w_i)}{C(i,w)}$;
9   Let r be a random number in $[0,1]$;
10   **if** $r < q$ **then**
11                 $L \leftarrow L \cup \{i\}$;
12                 $w \leftarrow w - w_i$;
13   **else if** $V(i,w) > V(i-1,w)$ **then**
14             $L \leftarrow L \cup \{i\}$;
15             $w \leftarrow w - w_i$;
16         $i \leftarrow i - 1$;
17     $S \leftarrow S \cup \{L\}$;
18     $k \leftarrow k - 1$;
19 **return** S

The algorithm starts at $V(n,W)$ respectively and reconstructs a solution $L \subset [n]$, initialized to $L = \emptyset$, bottom-up by making occasional random decisions. Assume the algorithm is at position $(i,w), 1 \le i \le n, 1 \le w \le W$. Recall (cf. Algorithm 1) that if $V(i-1,w) < V(i-1,w-w_i)+v_i$ we have no choice and we need to put item i into the knapsack. Likewise, if $V(i-1,w) > V(i-1,w-w_i)+v_i$, item i is guaranteed not to be part of the solution under reconstruction. Thus, in both cases, the decision is deterministic. If $V(i-1,w)$ equals $V(i-1,w-w_i)+v_i$, there are two options how to proceed: in this case with probability

$$\frac{C(i-1,w-w_i)}{C(i,w)}$$

item i is added to L and with the converse probability

$$\frac{C(i-1,w)}{C(i,w)} = 1 - \frac{C(i-1,w-w_i)}{C(i,w)}$$

item i is ignored. If i was packed, the algorithm proceeds (recursively) from $V(i-1,w-w_i)$ and from $V(i-1,w)$ otherwise. This process is iterated while $i > 0$ and $w > 0$. To sample k solutions we may repeat the procedure k times which results in a runtime of $O(kn)$. This is polynomial as long as k is polynomially bounded and of benefit for sampling from an exponential-sized set $\mathcal{S}^*$ if W is low and hence Algorithm 1 runs in polynomial time. Detailed pseudo-code of the procedure is given in Algorithm 2.

We now show that Algorithm 2 in fact samples each optimal solution uniformly at random from $\mathcal{S}^*$.

Theorem 2. *Let $\mathcal{S}^*$ be the set of optimal solutions for the knapsack problem and $s \in \mathcal{S}^*$ be an arbitrary optimal solution. Then the probability of sampling s using the sampling approach is $1/|\mathcal{S}^*|$.*

Proof. Let $s \in \mathcal{S}^*$ be an optimal solution. Note that after running Algorithm 1 we have $C(n, W) = |\mathcal{S}^*|$. For convenience we assume that $C(i, w) = 1$ for $w < 0$ to capture the case of invalid solutions. Consider the sequence of $1 \leq r \leq n$ decisions made while traversing back from position (n, W) until a termination criterion is met (either $i \leq 0$ or $w \leq 0$) in Algorithm 2. Let $q_i = \frac{a_i}{b_i}, i \in [r]$ be the decision probabilities in iterations $i \in [r]$. Here, q_i corresponds to q in line 8 of Algorithm 2 if there is a choice whether the corresponding item is taken or not. If there is no choice we can set $q_i = 1 = \frac{x}{x}$ with $x = C(i - 1, w)$ if the item is not packed and $x = C(i - 1, w - w_i)$ if the item is packed. A key observation is that (1) $b_1 = C(n, W)$, (2) $b_i = a_{i-1}$ holds for $i = 2, \ldots, r$ by construction of C and (3) $a_r = 1$ since the termination condition applies after r iterations (see base cases for C in Eq. 4 and Eq. 5). Hence, the probability to obtain s is

$$\prod_{i=1}^{r} q_i = \frac{a_1}{b_1} \cdot \frac{a_2}{b_2} \cdot \ldots \cdot \frac{a_{r-1}}{b_{r-1}} \cdot \frac{a_r}{b_r}$$

$$= \frac{a_1}{b_1} \cdot \frac{a_2}{a_1} \cdot \ldots \cdot \frac{a_{r-1}}{a_{r-2}} \cdot \frac{1}{a_{r-1}}$$

$$= \frac{1}{b_1}$$

$$= \frac{1}{C(n, W)}$$

$$= \frac{1}{|\mathcal{S}^*|}.$$

□

Theorem 3. *Let $\mathcal{S}^*$ be the set of optimal solutions for the knapsack problem with n items. Algorithm 2 samples k uniform samples of $\mathcal{S}^*$ in time $O(kn)$.*

Proof. The probabilistic statement follows Theorem 3. For the running time we note that each iteration takes at most n iterations each with a constant number of operations being performed. This is repeated k times which results in $O(kn)$ runtime which completes the proof. □

4 Experiments

In this section we complement the preceding sections with an experimental study and some derived theoretical insights. We first detail the experimental setup, continue with the analysis and close with some remarks.

4.1 Experimental Setup

The main research question is to get an impression and ideally to understand how the number of global optima develops for classical KP benchmark instances dependent on their generator-parameters. To this end we consider classical random KP benchmark generators as studied, e.g., by Pisinger in his seminal paper on hard knapsack instances [17]. All considered instance groups are generated randomly with item weights w_i sampled uniformly at random within the data range $\{L, \ldots, R\}$ with $L = 1$ and varying R for $i \in [n]$; in any case $L < R$. Item profits v_i are in most cases based on a mapping of item weights. The reader may want to take a look at Fig. 1 alongside the following description for visual aid.

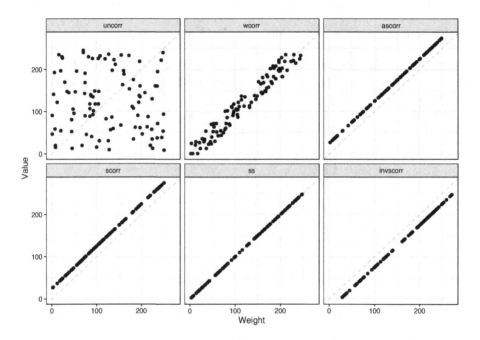

Fig. 1. Showcase of considered instance groups. We show each one instance with $n = 100$ items, $L = 1$ and $R = 250$. The gray dashed diagonal lines serves to aid recognising the differences between correlated instance groups.

Uncorrelated (uncorr). Here, both weights and profits are sampled uniformly at random from $\{L, \ldots, R\}$.

Weakly Correlated (wcorr). Weights w_i are distributed in $\{L, \ldots, R\}$ and profits v_i are sampled from $[w_i - R/10, w_i + R/10]$ ensuring $v_i \geq 1$. Here, the values are typically only a small percentage off the weights.

Almost Strongly Correlated (ascorr). Weights are distributed in $\{L, \ldots, R\}$ and v_i are sampled from $[w_j + R/10 - R/500, w_j + R/10 + R/500]$.

Strongly Correlated (scorr). Weights w_i are uniformly distributed from the set $\{L, \ldots, R\}$ while profits are corresponding to $w_i + R/10$. Here, for all items the profit equals the positive constant plus some fixed additive constant.

Subset Sum (susu). In this instance group we have $w_i = v_i \, \forall i \in [n]$, i.e., the profit equals the weight. This corresponds to strong correlation with additive constant of zero.

Inversely Strongly Correlated (invscorr). Here, first the profits v_i are sampled from $\{L, \ldots, R\}$ and subsequently weights we set $w_i = v_i + R/10$. This is the counterpart of strongly correlated instances.

Correlated instances may seem highly artificial on first sight, but they are of high practical relevance. In economics this situation arises typically if the profit of an investment is directly proportional to the investment plus some fixed charge (strongly correlated) with some problem-dependent noise (weakly/almost strongly correlated).

In our experiments we vary the instance group, the number of items $n \in \{50, 100, \ldots, 500\}$ and the upper bound $R \in \{25, 50, 100, 500\}$. In addition, we study different knapsack capacities by setting $D = 11$ and

$$W = \left\lfloor \frac{d}{D+1} \sum_{i=1}^{n} w_i \right\rfloor$$

for $d = 1, \ldots, D$ [16,17]. Intuitively – for most considered instance groups – the number of optima is expected to decrease on average for very low and very high capacities as the number of feasible/optimal combinations is likely to decrease. For each combination of these parameters, we construct 25 random instances and run the DP algorithm to count the number of optima.

Python 3 implementations of the algorithms and generators and the code for running the experiments and evaluations will be made available in a public GitHub repository upon acceptance. The experiments were conducted on a Mac-Book Pro 2018 with a 2,3 GHz Quad-Core Intel Core i5 processor and 16 GB RAM. The operating system was macOS Catalina 10.15.6 and python v3.7.5 was used. Random numbers were generated with the built-in python module `random` while `joblist` v0.16.0 served as a multi-core parallelisation backend.

4.2   Insights into the Number of Optima

Figure 2 depicts the distribution of the number of optima for each combination considered in the experiments via boxplots. The data is split row-wise by instance group and col-wise by R (recall that $L = 1$ in any case). Different box-colors indicate the knapsack capacity which – for sake of interpretability – is given in percentage of the total weight, i.e., $\lfloor 100 \cdot (\frac{d}{D+1}) \rfloor$. We observe different patterns dependent on the instance group. For uncorrelated instances (uncorr), we observe only few optima with single outliers reaching $2^8 = 256$ for $R = 50$ and large n.

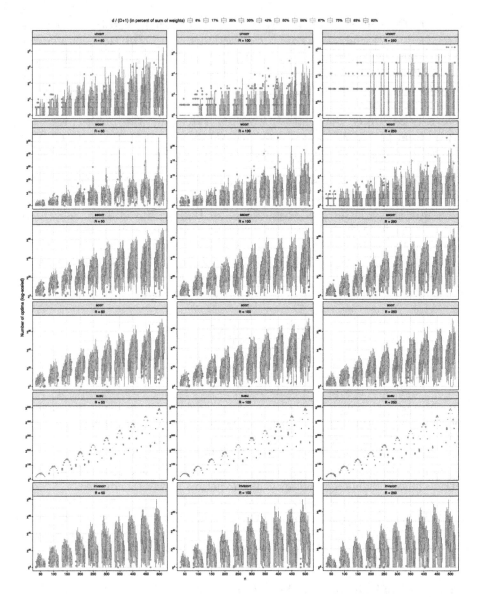

Fig. 2. Boxplots of the number of optima as a function of the number of items n. The data is split by instance group (rows) and the upper bound R (columns). Different colors indicate the knapsack capacity (shown as percentage of the sum of all items weights).

Median values are consistently below $2^4 = 16$. In line with expectation the numbers are highest for relatively small R and high n. In fact, the ratio

$$H = \frac{n}{R}$$

is a good indicator. It is the expected number of elements with weight $w = 1, \ldots, R$. In consequence, $H > 1$ and especially $H \gg 1$ indicates many elements with the same weight. In contrast, $H < 1$ indicates that on average there will be at most one element of weight $w = 1, \ldots, R$. For all correlated instances, i.e., scorr, ascorr, wcorr, invscorr and susu, we observe a very different pattern. Here, the number of optima grows exponentially with growing n given a fixed upper bound R. Even if H is low there is huge number of optima. By far the highest count of optima can be observed for subset sum (susu) instances where even peaks with up to ≈3% of all 2^n solutions are optimal. Here, the boxplots look degenerate, because the variance is very low. Recall that for this type of instance we have $w_i = v_i \, \forall i \in [n]$ and thus for each solution s the equality $w(s) = v(s)$ holds. In consequence we aim to maximally exploit the knapsack capacity.

To get a better understanding we consider a subset-sum type knapsack instance with $w_i \in \{1, \ldots, R\}, w_i = v_i, \forall i \in [n]$. Assume for ease of calculations that n is a multiple of R and there are exactly (n/R) items of each weight $w \in \{1, \ldots, R\}$, i.e., $|\{i \in [n] \mid w_i = w\}| = n/R$. Note that this corresponds to the expected number of w-weights if n such weights are sampled uniformly at random from $\{1, \ldots, R\}$. Consider $W = \frac{1}{2} \sum_{i=1}^{n} w_i$. Recall that given this instance, one way we can build an optimum $s \subset [n]$ with $w(s) = v(s) = W$ is by choosing each $\left(\frac{1}{2}\right) \cdot \left(\frac{n}{R}\right)$ items from each weight class, i.e., half of these items (note that there are many more combinations leading to profit W). With this we get

$$|\mathcal{S}^*| \geq \left(\frac{\frac{n}{R}}{\frac{n}{2R}}\right)^R \geq \left(\left(\frac{n}{R} \cdot \frac{2R}{n}\right)^{\frac{n}{2R}}\right)^R = 2^{\frac{n}{2}}.$$

Here we used to the well-known lower bound $\binom{n}{k} \geq \left(\frac{n}{k}\right)^k$ for the binomial coefficient. This simple bound establishes that we can expect at least $2^{n/2}$ optima for subset-sum instances if the capacity is set accordingly.

With respect to the knapsack capacity Fig. 2 also reveals different patterns. For inverse strongly correlated instances we observe a decreasing trend with increasing capacity. The vice versa holds for weakly, almost strongly and strongly correlated instances. This is in line with intuition as the size of the feasible search space also grows significantly.

However, note that in general the knapsack capacity can have a massive effect on the number of optima.

Theorem 4. *For every even n there exist a KP instance and a weight capacity W such that $|\mathcal{S}^*|$ is exponential, but $|\mathcal{S}^*| = 1$ for $W' = W + 1$.*

Proof. Consider an instance with n items (n even), where $w_i = v_i = 1$ for $i \in [n-1]$ and $w_n = \frac{n}{2} + 1$. Let $v_n > \frac{n}{2} + 1$. Now consider the knapsack capacity $W = \frac{n}{2}$. Then every subset of $\frac{n}{2}$ items from the first $n - 1$ items is optimal with total weight W and total value W while the n-th item does not fit into the knapsack. There are at least

$$\binom{n-1}{\frac{n}{2}} = \frac{(n-1)!}{(n/2)!(n-1-n/2)!}$$

$$\geq \frac{(n-1)!}{\left(\left(\frac{n}{2}\right)!\right)^2}$$

$$\geq \frac{\sqrt{2\pi}\sqrt{n-1}(n-1)^{n-1}2^n e^n}{2e^{n-1}4(2\pi)n^{n+1}}$$

$$= \frac{2^{n-3}e}{\sqrt{2\pi}} \cdot \underbrace{\left(\frac{n-1}{n}\right)^{n-1}}_{\geq e^{-1}} \cdot \frac{\sqrt{n}}{n^2} \cdot \underbrace{\sqrt{1-\frac{1}{n}}}_{\geq 1/\sqrt{2}}$$

$$\geq \frac{1}{\sqrt{\pi}} \cdot \frac{2^{n-4}}{n^{3/2}}$$

$$= \Omega(2^{n-4}/n^{3/2})$$

optima in this case. Here we basically used Stirling's formula to lower/upper bound the factorial expressions to obtain an exponential lower bound. Now instead consider the capacity $W' = W + 1$. The n-th item now fits into the knapsack which results in a unique optimum with weight $W' = \frac{n}{2} + 1$ and value $v_n > \frac{n}{2} + 1$ which cannot be achieved by any subset of light-weight items.   □

4.3 Closing Remarks

Knapsack instances with correlations between weights and profits are of high practical interest as they arise in many fixed charge problems, e.g., investment planning. In this type of instances item profits correspond to their weight plus/minus a fixed or random constant. Our experimental study suggests an exponential increase in the number of global optima for such instances which justifies the study and relevance of the considered counting and sampling problem.

5   Conclusion

We considered the problem of counting exactly the number of optimal solutions of the zero-one knapsack problem. We build upon the classic dynamic programming algorithm for the optimisation version. Our modifications allow to solve the counting problem in pseudo-polynomial runtime complexity. Furthermore, we show how to sample uniformly at random from the set of optimal solutions without explicit construction of the whole set. Computational experiments and derived theoretical insights reveal that for variants of problem instances with correlated weights and profits (a group which is highly relevant in real-world scenarios) and for a wide range of problem generator parameters, the number of optimal solutions can grow exponentially with the number of items. These observations support the relevance of the considered counting and sampling problems.

Future work will focus on (approximate and exact) counting/sampling of high-quality knapsack solutions which all fulfill a given non-optimal quality threshold. In addition, in particular if the set of optima has exponential size, it is desirable to provide the decision maker with a diverse set of high-quality solutions. Even though the introduced sampling is likely to produce duplicate-free samples if the number of solutions is exponential, it seems more promising to bias the sampling process towards a diverse subset of optima, e.g., with respect to item-overlap or entropy. This opens a whole new avenue for upcoming investigations.

References

1. Broder, A.Z., Mayr, E.W.: Counting minimum weight spanning trees. J. Algorithms **24**(1), 171–176 (1997)
2. Cai, J.Y., Chen, X.: Complexity Dichotomies for Counting Problems: Volume 1, Boolean Domain. Cambridge University Press, USA, 1st edn. (2017)
3. Do, A.V., Bossek, J., Neumann, A., Neumann, F.: Evolving diverse sets of tours for the travelling salesperson problem. In: Proceedings of the Genetic and Evolutionary Computation Conference, GECCO 2020, pp. 681–689. ACM (2020)
4. Dyer, M.: Approximate counting by dynamic programming. In: Proceedings of the Thirty-Fifth Annual ACM Symposium on Theory of Computing. STOC 2003, pp. 693–699. Association for Computing Machinery, New York, NY, USA (2003)
5. Dyer, M., Goldberg, L.A., Greenhill, C., Jerrum, M.: On the relative complexity of approximate counting problems. In: Jansen, K., Khuller, S. (eds.) APPROX 2000. LNCS, vol. 1913, pp. 108–119. Springer, Heidelberg (2000). https://doi.org/10.1007/3-540-44436-X_12
6. Fichte, J.K., Hecher, M., Meier, A.: Counting complexity for reasoning in abstract argumentation. In: Proceedings of the The Thirty-Third AAAI Conference on Artificial Intelligence, AAAI 2019, pp. 2827–2834. AAAI Press (2019)
7. Fichte, J.K., Hecher, M., Morak, M., Woltran, S.: Exploiting treewidth for projected model counting and its limits. In: Beyersdorff, O., Wintersteiger, C.M. (eds.) SAT 2018. LNCS, vol. 10929, pp. 165–184. Springer, Cham (2018). https://doi.org/10.1007/978-3-319-94144-8_11
8. Fournier, H., Malod, G., Mengel, S.: Monomials in arithmetic circuits: complete problems in the counting hierarchy. Comput. Complex. **24**(1), 1–30 (2015)
9. Katz, M., Sohrabi, S.: Reshaping diverse planning. In: Proceedings of the The Thirty-Fourth AAAI Conference on Artificial Intelligence, AAAI 2020, pp. 9892–9899. AAAI Press (2020)
10. Katz, M., Sohrabi, S., Udrea, O.: Top-quality planning: finding practically useful sets of best plans. In: Proceedings of the The Thirty-Fourth AAAI Conference on Artificial Intelligence, AAAI 2020, pp. 9900–9907. AAAI Press (2020)
11. Katz, M., Sohrabi, S., Udrea, O., Winterer, D.: A novel iterative approach to top-k planning. In: Proceedings of the Twenty-Eighth International Conference on Automated Planning and Scheduling, ICAPS 2018, pp. 132–140. AAAI Press (2018)
12. Kellerer, H., Pferschy, U., Pisinger, D.: Knapsack Problems. Springer, Berlin (2004). https://doi.org/10.1007/978-3-540-24777-7
13. Mihalák, M., Šrámek, R., Widmayer, P.: Approximately counting approximately-shortest paths in directed acyclic graphs. Theory Comput. Syst. **58**(1), 45–59 (2016)

14. Neumann, A., Gao, W., Doerr, C., Neumann, F., Wagner, M.: Discrepancy-based evolutionary diversity optimization. In: Proceedings of the Genetic and Evolutionary Computation Conference, GECCO 2018, pp. 991–998 (2018)
15. Neumann, A., Gao, W., Wagner, M., Neumann, F.: Evolutionary diversity optimization using multi-objective indicators. In: Proceedings of the Genetic and Evolutionary Computation Conference, GECCO 2019, pp. 837–845 (2019)
16. Pisinger, D.: Core problems in knapsack algorithms. Oper. Res. **47**(4), 570–575 (1999)
17. Pisinger, D.: Where are the hard knapsack problems? Comput. Oper. Res. **32**(9), 2271–2284 (2005)
18. Rizzi, R., Tomescu, A.I.: Faster FPTASes for counting and random generation of knapsack solutions. Inf. Comput. **267**, 135–144 (2019)
19. Sohrabi, S., Riabov, A.V., Udrea, O., Hassanzadeh, O.: Finding diverse high-quality plans for hypothesis generation. In: Proceedings of the 22nd European Conference on Artificial Intelligence, ECAI 2016. Frontiers in Artificial Intelligence and Applications, vol. 285, pp. 1581–1582. IOS Press (2016)
20. Ulrich, T., Thiele, L.: Maximizing population diversity in single-objective optimization. In: Proceedings of the Genetic and Evolutionary Computation Conference, GECCO 2011, pp. 641–648 (2011)
21. Valiant, L.G.: The complexity of computing the permanent. Theor. Comput. Sci. **8**, 189–201 (1979)
22. Štefankovič, D., Vempala, S., Vigoda, E.: A deterministic polynomial-time approximation scheme for counting knapsack solutions. SIAM J. Comput. **41**(2), 356–366 (2012)

Modeling of Crisis Periods in Stock Markets

Apostolos Chalkis[1,2], Emmanouil Christoforou[1,2(✉)], Theodore Dalamagas[1], and Ioannis Z. Emiris[1,2]

[1] ATHENA Research and Innovation Center, Marousi, Greece
{dalamag,emiris}@athenarc.gr
[2] Department of Informatics and Telecommunications,
National and Kapodistrian University of Athens, Athens, Greece
{achalkis,echristo}@di.uoa.gr

Abstract. We exploit a recent computational framework to model and detect financial crises in stock markets, as well as shock events in cryptocurrency markets, which are characterized by a sudden or severe drop in prices. Our method manages to detect all past crises in the French industrial stock market starting with the crash of 1929, including financial crises after 1990 (e.g. dot-com bubble burst of 2000, stock market downturn of 2002), and all past crashes in the cryptocurrency market, namely in 2018, and also in 2020 due to covid-19. We leverage copulae clustering, based on the distance between probability distributions, in order to validate the reliability of the framework; we show that clusters contain copulae from similar market states such as normal states, or crises. Moreover, we propose a novel regression model that can detect successfully all past events using less than 10% of the information that the previous framework requires. We train our model by historical data on the industry assets, and we are able to detect all past shock events in the cryptocurrency market. Our tools provide the essential components of our software framework that offers fast and reliable detection, or even prediction, of shock events in stock and cryptocurrency markets of hundreds of assets.

Keywords: Copula · Crisis detection · Stock market · Clustering · Financial portfolio · Bitcoin · Investment risk

1 Introduction

Modern finance has been pioneered by Markowitz who set a framework to study choice in portfolio allocation under uncertainty [6,8]. Within this framework, portfolios are characterized by their returns, and by their risk which is defined as the variance (or volatility) of the portfolios' returns. An investor would build a portfolio to maximize its expected return for a chosen level of risk. In normal times, stocks are characterized by somewhat positive returns and a moderate volatility, in up-market times (typically bubbles), by high returns and low volatility, and during financial crises, by strongly negative returns and high

© Springer Nature Switzerland AG 2021
D. E. Simos et al. (Eds.): LION 2021, LNCS 12931, pp. 55–65, 2021.
https://doi.org/10.1007/978-3-030-92121-7_5

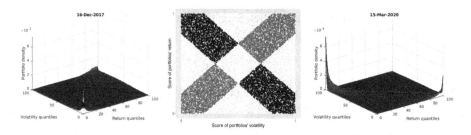

Fig. 1. Copulae that correspond to cryptocurrencies' states. Left, a normal period (16/12/2017) and right, a shock event due to Covid-19 (15/03/2020). The middle plot shows the mass of interest to characterize the market state.

volatility [1]. Thus, it is crucial to describe the time-varying dependency between portfolios' return and volatility. To capture this dependency, in [3] they rely on the copula representation of the portfolios distribution, which offers a very powerful tool. A copula is a bivariate probability distribution for which the marginal probability distribution of each variable is uniform. Following Markowitz' framework, the variables considered are the portfolios' return and volatility. Figure 1 illustrates a copula showing a positive (left) and negative (right) dependency between return and variance.

We illustrate the power of the framework in [3] to detect shocks in two different markets: the French industrial stock market and the market of digital assets. We use the daily returns of 30 French industrial assets to detect all reported financial crises after 1990. We also detect earlier crises, such as the crash of 1929 (Fig. 5). Interestingly, the indicator recognizes the period of military occupation of France (1940–45) as normal; this is related to the strict regulation during that period, which led to a paradoxical rise in nominal stock values [7,10]. Also, we use the daily returns of 12 cryptocurrencies with the longest history available to detect all shock events in the cryptocurrency market. The indicator detects successfully the 2018 cryptocurrency crash and the timeline of its most notable events, such as the crash of nearly all cryptocurrencies in the beginning of 2018 and the fall of Bitcoin's market capitalization and price in the end of 2018. Finally, it detects the shock event in early 2020 due to covid-19.

We validate the reliability of the framework by clustering based on probability distributions' distances. Our work is complementary to that in [4] The computed copulae form clusters that sort the value of the indicator, resulting to clusters of similar financial states (normal, crisis). We employ quadratic regression models to model the copula structure so as to capture several patterns of the mass of portfolios during different market states. We train our model using the French industry assets and we use it to detect shock events in the cryptocurrency market. Our trained model successfully detects all past events. It uses less than 10% of the information on the dependency between portfolios' return and volatility that is required in [3]. Lastly, the open-source implementation[1] of our methods provides

---

[1] https://github.com/TolisChal/crises_detection.

a software framework for shock event detection and modeling in stock markets of hundreds of assets.

The rest of the paper is organized as follows. Section 2 presents the computational framework [3] and uses real data to detect past crises and shock events. Section 3 exploits clustering and regression models to provide more sophisticated tools on crisis detection and modeling.

2   Detecting Shock Events with Copulae

In this section we present the computational framework in [3] and then we exploit it to detect past crises and crashes in two markets with different characteristics.

Let a portfolio x invest in n assets. The set of portfolios in which a long-only asset manager can invest can be represented by the canonical simplex $\Delta^{n-1} := \{(x_1, \ldots, x_n) \in \mathbb{R}^n \mid \sum_{i=1}^n x_i = 1, \text{ and } x_i \geq 0\} \subset \mathbb{R}^n$. Given a vector of asset returns $R \in \mathbb{R}^n$ and the variance-covariance matrix $\Sigma \in \mathbb{R}^{n \times n}$ of the distribution of asset returns, we say that any portfolio $x \in \Delta^{n-1}$ has return $f_{ret}(x, R) = R^T x$ and variance (volatility) $f_{vol}(x, \Sigma) = x^T \Sigma x$.

To capture the relationship between return and volatility in a given time period we approximate the *copula* between portfolios' return and volatility. Thus, we define two sequences of m bodies each, $\Delta^{n-1} \cap S_i := \{x \in \Delta^{n-1} \mid s_i \leq f_{ret}(x, R) \leq s_{i+1}\}$ and $\Delta^{n-1} \cap U_i := \{x \in \Delta^{n-1} \mid u_i \leq f_{vol}(x, \Sigma) \leq u_{i+1}\}$, $i \in [m]$. Moreover, we compute s_i, $u_i \in \mathbb{R}$ such that $\Delta^{n-1} \cap \text{vol}(S_i)$ and $\Delta^{n-1} \cap \text{vol}(U_i)$ are all equal to a small fixed portion of $\text{vol}(\Delta^{n-1})$ (e.g. 1%). Then, to obtain the copula one has to estimate all the ratios $\dfrac{\text{vol}(Q_{ij})}{\text{vol}(\Delta^{n-1})}$ where $Q_{ij} := \{x \in \Delta^{n-1} \mid s_i \leq f_{ret}(x, R) \leq s_{i+1} \text{ and } u_j \leq f_{vol}(x, \Sigma) \leq u_{j+1}\}$.

To compute these ratios, we leverage uniform sampling from Δ^{n-1} [12]. Let us consider up- and down- (main) diagonal bands: we define the indicator as the ratio of the down-diagonal over the up-diagonal band. The indicator is the ratio of the mass of portfolios in the blue area over the mass of portfolios in the red one in Fig. 1. When the value of the indicator is smaller than 1 then the copula corresponds to a normal period. Otherwise, it probably comes from a crisis period. Considering the scalability of this method, it can be applied for stock markets with a few thousands of assets, since the cost per uniformly distributed sample in Δ^{n-1} using the method in [12] is $O(n)$.

2.1   Shock Detection Using Real Data

We now use two data sets from two different asset sections. First, we use the daily returns of 30 French industrial asset returns[2]. Second, we use the daily returns of 12 out of the top 100 cryptocurrencies, ranked by CoinMarketCap's[3] market cap (cmc_rank) on 22/11/2020, having the longest available history (Table 1). We compute the daily return for each coin using the daily close price

---

[2] https://mba.tuck.dartmouth.edu/pages/faculty/ken.french/data_library.html.
[3] https://coinmarketcap.com/.

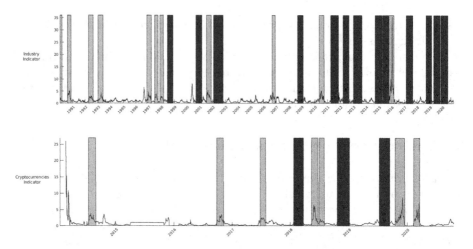

Fig. 2. Warning (yellow) and Crises (red) periods detected by the indicator. Top for industry assets (1990–2020), bottom for cryptocurrencies (2014–2020). (Color figure online)

obtained by CoinMarketCap, for several notable coins such as Bitcoin, Litecoin and Ethereum.

The indicator is estimated on copulae by drawing $500,000$ points. We compute the indicator per copula over a rolling window of $k = 60$ days and with a band of $\pm 10\%$ with respect to the diagonal. When the indicator exceeds 1 for more than 60 days but less than 100 days, we report the time interval as a "warning" (yellow color); see Fig. 5. When the indicator exceeds 1 for more than 100 days we report the interval as a "crisis" (red); see Fig. 2, 5, and 7. The periods are more than 60 days long to avoid detection of isolated events whose persistence is only due to the auto-correlation implied by the rolling window.

We compare results for industrial assets with the database for financial crises in Europe [5] from 1990 until 2020. The first warnings in 1990 correspond to the early 90's recession, the second crisis in 2000 to 2001 to the dot-com bubble burst, the warning and third crisis in 2001 and 2002 to the stock market downturn of 2002, and the fourth crisis in 2008 to 2009 corresponds to the sub-prime crisis.

Our cryptocurrencies indicator detects successfully the 2018 (great) cryptocurrency crash. The first shock event detected in 2018 (mid-January to late March) corresponds to the crash of nearly all cryptocurrencies, following Bitcoin's, whose price fell by about 65% from 6 January to 6 February 2018, after an unprecedented boom in 2017. Intermediate warnings (mid-May to early August) should correspond to cryptocurrencies collapses (80% from their peak in January) until September. The detected crash at the end of 2018 (November 2018 until early January 2019) corresponds to the fall of Bitcoin's market capitalization (below \$100 billion) and price by over 80% from its peak, almost one-third of its previous week value. Finally, the detected event in early 2020 corresponds to the shock event due to covid-19.

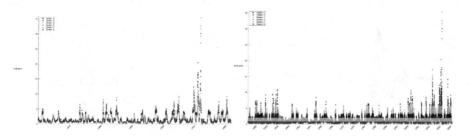

Fig. 3. Left, spectral clustering ($k = 6$) on EMD matrix. Right, k-medoids ($k = 6$) on copulae features. Clusters appear to contain similar indicator values.

3 Exploring the Dynamics of Copulae

Several clustering methods confirm the indicator's reliability. Then, we model the structure of copulae using a novel regression model to detect shock events.

3.1 Clustering of Copulae

In order to further evaluate our results we clustered the copulae of the industry returns. To confirm whether the copulae are able to distinguish different market states (normal, crisis and intermediate), as well as to validate the indicator, we experimented with clustering based on probability distributions distances. We select various number of clusters, such as 6 and 8, and show that the resulting clusters include copulae from similar market states, with similar indicator values.

To cluster the probability distributions distances of the copulae, we computed a distance matrix (D) between all copulae using the earth mover's distance (EMD) [13]. The EMD between two distributions is the minimum amount of work required to turn one distribution into the other. Here we use a fast and robust EMD algorithm, which appears to improve both accuracy and speed [11]. Then, we apply spectral clustering [9], a method to cluster points using the eigenvectors of the affinity matrix (A) which we derive from the distance matrix, computed by the radial basis function kernel, replacing the Euclidean distance with EMD, where $A_{ij} = exp(-D_{ij}^2/2\sigma^2)$, and for σ we chose the standard deviation of distances. Using the k largest eigenvectors of the laplacian matrix, we construct a new matrix and apply k-medoids clustering by treating each row as a point, so as to obtain k clusters. The results with $k = 6$ and $k = 8$ are shown on the indicators' values in Fig. 3, 8, and 9. Clusters appear to contain copulae with similar indicator values. Crisis and normal periods are assigned to clusters with high and low indicator values respectively. Therefore, the clustering of the copulae is proportional to discretising the values of the indicator.

Other experiments included clustering on features generated form the copulas, based on the indicator. We generate vector representations for each copula using the rates between all the possible combinations of the indicators' corners: for U_L, U_R being the upper left and right corner of a copula respectively,

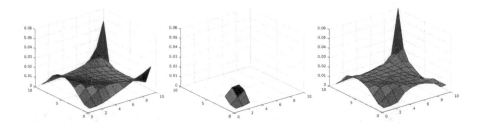

Fig. 4. Left: copula on 21/09/2014 using cryptocurrency returns. Middle: proportion of the mass of the left copula that our model uses as input. Right: copula that the model estimates. We trained our model with the industry asset returns.

and for L_L, L_R the lower left and right corners, the vector representation is $\left[\begin{smallmatrix} U_L & U_L & U_L & U_R & U_R & L_L \\ U_R & L_L & L_R & L_L & L_R & L_R \end{smallmatrix}\right]$. These representations allow us to use clustering, such as k-medoids. Results of the clustering also follow the values of the indicator as expected (Fig. 3, 10).

3.2 Modeling Copulae

We further explore the dynamics of copulae by modeling the mass distribution using a quadratic regression model.

We compute the 10×10 copulae of the industry data set. That is $N = 49\,689$ copulae in total while each consists of $10 \times 10 = 100$ cells. Each cell has a value while they all sum up to 1. For all copulae we pick a certain subset S of k cells, e.g. the 3×3 left up corner as Fig. 4 illustrates; that is $|S| = 9$. For each copula we represent the values of the k cells that belong to S as a vector $X \in \mathbb{R}^k$. Thus, in total we get the vectors $X_1, \ldots, X_N$. Then, for each cell that does not belong to S we fit a quadratic regression model. In particular, let $Y_{ij} \in \mathbb{R}$, $i = 1, \ldots, 100 - k$, $j = 1, \ldots, N$ the value of the i-th cell in the j-th copula. Then, we define the following models,

$$\text{model } \mathcal{M}_i : \quad \min_{\Sigma_i \succeq 0} \sum_{j=1}^{N} (Y_j - X_j^T \Sigma_i X_j)^2, \quad i = 1, \ldots, 100 - k, \tag{1}$$

where $\Sigma_i \succeq 0$ declares that the matrix Σ_i is a positive semidefinite matrix. To solve the optimization problems in Eq. (1) we use the `matlab` implementation of the *Trust Region Reflective Algorithm* [2], provided by function `lsqnonlin()`.

To illustrate the efficiency and transferability of our model, we train it on industry asset returns and use it to detect the shock events in the cryptocurrency market from Sect. 2.1; we use the copulae of the industry asset returns. For each copula the vector $X_i \in \mathbb{R}^9$ corresponds to the 3×3 left-down corner cell, as Fig. 4 (middle) shows. We exploit the model to estimate the 10×10 copula of each sliding window of the cryptocurrencies' returns. Finally, we compute the indicator of each estimated copula and plot the results in Fig. 6. Interestingly, the copulae that our model estimates suffice to detect all the past shock events that we also detect using the exact copulas, such as the 2018 cryptocurrency crash (see Sect. 2.1), except the first warning period (mid-May) of 2018.

Acknowledgements. This research is carried out in the context of the project "PeGASUS: Approximate geometric algorithms and clustering with applications in finance" (MIS 5047662) under call "Support for researchers with emphasis on young researchers: cycle B" (EDBM103). The project is co-financed by Greece and the European Union (European Social Fund-ESF) by the Operational Programme Human Resources Development, Education and Lifelong Learning 2014–2020. We thank Ludovic Calès for his precious guidance throughout this work.

A   Data

Table 1. Cryptocurrencies used to detect shock events in market.

Coin	Symbol	Dates
Bitcoin	BTC	28/04/2013–21/11/2020
Litecoin	LTC	28/04/2013–21/11/2020
Ethereum	ETH	07/08/2015–21/11/2020
XRP	XRP	04/08/2013–21/11/2020
Monero	XMR	21/05/2014–21/11/2020
Tether	USDT	25/02/2015–21/11/2020
Dash	DASH	14/02/2014–21/11/2020
Stellar	XLM	05/08/2014–21/11/2020
Dogecoin	DOGE	15/12/2013–21/11/2020
DigiByte	DGB	06/02/2014–21/11/2020
NEM	XEM	01/04/2015–21/11/2020
Siacoin	SC	26/08/2015–21/11/2020

62   A. Chalkis et al.

B   Crises Indicator

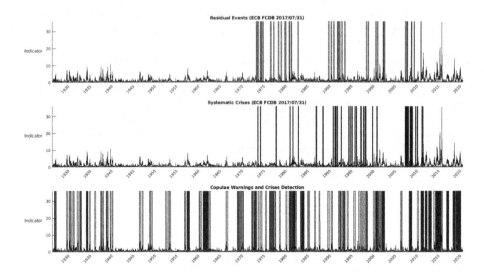

Fig. 5. Warnings (yellow) and Crises (red) detected by indicator (bottom) for industry assets, against real residual events (top) and systematic crises (middle). (Color figure online)

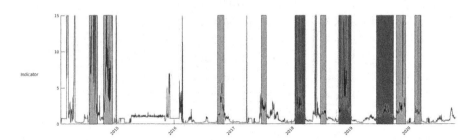

Fig. 6. The shock events we detect in the cryptocurrency market using the indicator from Eq. (1). Note that we trained the model using the daily returns of the French industry assets.

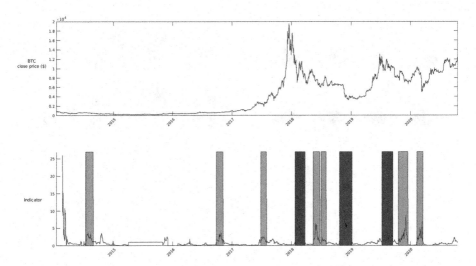

Fig. 7. Warning (yellow) and Crises (red) periods detected by indicator (bottom) for cryptocurrencies against BTC daily close price (top). (Color figure online)

C   Clustering of Copulae

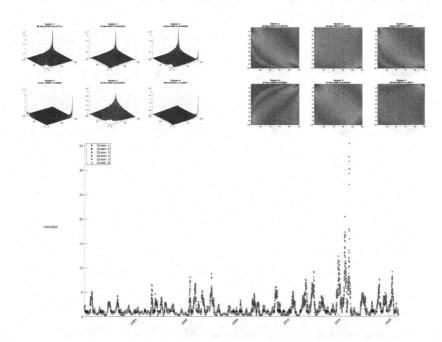

Fig. 8. Clustering of copulae using spectral clustering on EMD distances with $k = 6$.

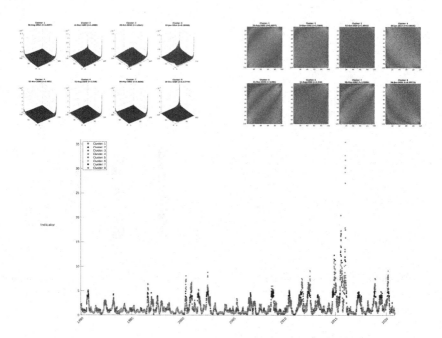

Fig. 9. Clustering of copulae using spectral clustering on EMD distances with $k = 8$.

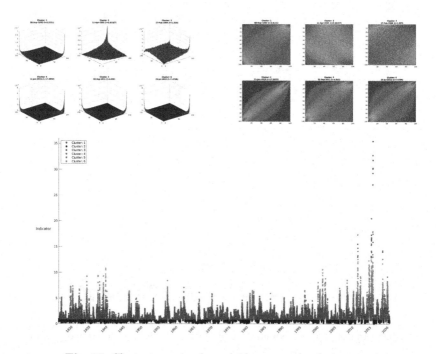

Fig. 10. Clustering using k-medoids on copulae features.

References

1. Billio, M., Getmansky, M., Pelizzon, L.: Dynamic risk exposures in hedge funds. Comput. Stat. Data Anal. **56**(11), 3517–3532 (2012). https://doi.org/10.1016/j.csda.2010.08.015, http://www.sciencedirect.com/science/article/pii/S0167947310003439
2. Branch, M.A., Coleman, T.F., Li, Y.: A subspace, interior, and conjugate gradient method for large-scale bound-constrained minimization problems. SIAM J. Sci. Comput. **21**(1), 1–23 (1999). https://doi.org/10.1137/S1064827595289108
3. Calès, L., Chalkis, A., Emiris, I.Z., Fisikopoulos, V.: Practical volume computation of structured convex bodies, and an application to modeling portfolio dependencies and financial crises. In: Speckmann, B., Tóth, C. (eds.) Proceedings International Symposium Computational Geometry (SoCG). Leibniz International Proceedings Informatics, vol. 99, pp. 19:1–15. Dagstuhl, Germany (2018). https://doi.org/10.4230/LIPIcs.SoCG.2018.19
4. Di Lascio, F.M.L., Durante, F., Pappadà, R.: Copula–based clustering methods. In: Úbeda Flores, M., de Amo Artero, E., Fernández Sánchez, J. (eds.) Copulas and Dependence Models with Applications, pp. 49–67. Springer, Cham (2017). https://doi.org/10.1007/978-3-319-64221-5_4
5. Duca, M.L., et al.: A new database for financial crises in European countries. Technical Report 13, European Central Bank & European Systemic Risk Board, Frankfurt, Germany (2017)
6. Kroll, Y., Levy, H., Markowitz, H.: Mean-variance versus direct utility maximization. Journal of Finance **39**(1), 47–61 (1984)
7. Le Bris, D.: Wars, inflation and stock market returns in France, 1870–1945. Financ. Hist. Rev. **19**(3), 337–361 (2012). https://doi.org/10.1017/S0968565012000170
8. Markowitz, H.: Portfolio selection. J. Financ. **7**(1), 77–91 (1952). https://doi.org/10.1111/j.1540-6261.1952.tb01525.x
9. Ng, A.Y., Jordan, M.I., Weiss, Y.: On spectral clustering: analysis and an algorithm. In: Proceedings 14th International Conference Neural Information Processing Systems: Natural and Synthetic, pp. 849–856. NIPS 2001, MIT Press, Cambridge, MA, USA (2001)
10. Oosterlinck, K.: French stock exchanges and regulation during world war II. Financ. Hist. Rev. **17**(2), 211–237 (2010). https://doi.org/10.1017/S0968565010000181
11. Pele, O., Werman, M.: Fast and robust earth mover's distances. In: IEEE 12th International Conference Computer Vision, pp. 460–467. IEEE, September 2009. https://doi.org/10.1109/ICCV.2009.5459199
12. Rubinstein, R., Melamed, B.: Modern Simulation and Modeling. Wiley, New York (1998)
13. Rubner, Y., Tomasi, C., Guibas, L.: The earth mover's distance as a metric for image retrieval. Int. J. Comput. Vis. **40**, 99–121 (2000)

Feature Selection in Single-Cell RNA-seq Data via a Genetic Algorithm

Konstantinos I. Chatzilygeroudis[1,2]($\boxtimes$), Aristidis G. Vrahatis[3],
Sotiris K. Tasoulis[3], and Michael N. Vrahatis[2]

[1] Computer Engineering and Informatics Department (CEID), University of Patras,
Patras, Greece
costashatz@upatras.gr
[2] Computational Intelligence Laboratory, Department of Mathematics,
University of Patras, Patras, Greece
vrahatis@math.upatras.gr
[3] Department of Computer Science and Biomedical Informatics,
University of Thessaly, Volos, Greece
arisvrahatis@uth.gr, stasoulis@uth.gr

Abstract. Big data methods prevail in the biomedical domain leading to effective and scalable data-driven approaches. Biomedical data are known for their ultra-high dimensionality, especially the ones coming from molecular biology experiments. This property is also included in the emerging technique of single-cell RNA-sequencing (scRNA-seq), where we obtain sequence information from individual cells. A reliable way to uncover their complexity is by using Machine Learning approaches, including dimensional reduction and feature selection methods. Although the first choice has had remarkable progress in scRNA-seq data, only the latter can offer deeper interpretability at the gene level since it highlights the dominant gene features in the given data. Towards tackling this challenge, we propose a feature selection framework that utilizes genetic optimization principles and identifies low-dimensional combinations of gene lists in order to enhance classification performance of any off-the-shelf classifier (e.g., LDA or SVM). Our intuition is that by identifying an optimal genes subset, we can enhance the prediction power of scRNA-seq data even if these genes are unrelated to each other. We showcase our proposed framework's effectiveness in two real scRNA-seq experiments with gene dimensions up to 36708. Our framework can identify very low-dimensional subsets of genes (less than 200) while boosting the classifiers' performance. Finally, we provide a biological interpretation of the selected genes, thus providing evidence of our method's utility towards explainable artificial intelligence.

Keywords: Feature selection · Optimization · Single-cell RNA-seq · High-dimensional data

1 Introduction

Almost two decades ago, the Human Genome Project [11] was completed, where the human genome was analyzed, offering the complete set of genetic informa-

© Springer Nature Switzerland AG 2021
D. E. Simos et al. (Eds.): LION 2021, LNCS 12931, pp. 66–79, 2021.
https://doi.org/10.1007/978-3-030-92121-7_6

tion. The evolution of DNA sequencing from this point has undoubtedly brought about a great revolution in the field of biomedicine [29]. Although new technologies and analysis tools are constantly emerging, their experimental data have ultra-high dimensionality hindering success of traditional methods [8]. Hence, data mining approaches in such data create several computational challenges that novel or updated existing computational methodologies can address. Indicatively, a gene expression experiment includes each sample measurements for the entire genome, which contains tens of thousands of genes.

Meanwhile, classification in gene expression profiles is a longstanding research field with remarkable progress in complex disease identification, and treatment [13]. Started with data from the microarrays high-throughput technology [6] and continued with sequencing data [35]. We are now in the single-cell sequencing era, which allows biological information to be extracted from individual cells offering a deeper analysis at the cellular level. An indicative undercase transcriptomics study has gene measurements simultaneously for the entire genome isolating hundreds or thousands or even millions of cells in recent years. Given that we obtain measurements of tens of thousands of genes for each cell, in the computational perspective, we have to manage single-cell RNA-sequencing (scRNA-seq) data with ultra-high complexity.

Several single-cell RNA-seq data challenges are addressed through classification methods under the Machine Learning family [26]. These methods shed light on various biological issues such as the new cell types of identification [27], the cellular heterogeneity dissection [16], the cell cycle prediction [28], the cell sub-populations [7], the cells classification [25] and much more [33]. Despite the remarkable progress and promising results in the above challenges, the increasing scRNA-seq data generation, and the related technologies improvements creates new challenges and the need for novel classification methods under the perspective of supervised learning.

The nature of scRNA-seq technology, that is to examine individual cells from specific tissues, creates a quite sparse counts matrix since for every cell usually exists a high fraction of genes which are not informative [31]. Two appropriate ways to deal with this inherent particularity are dimensionality reduction techniques and feature selection methods. Dimension reduction techniques in scRNA-seq data aim to transfer the original $\mathbb{R}^D$ cell's space, where D is the genes expression profiles, to a lower-dimensional $\mathbb{R}^K$ space, with $K \ll D$. Such methods have gained ground in recent years with promising results in visualization [34] as in classification performance tasks [26]. Indicatively, the t -distributed stochastic neighbor embedding [21] and uniform manifold approximation and projection [5] techniques are usually applied in scRNA-seq data to obtain low-dimensional embedding offering a better visualization to uncover the relationship among cells and their categories.

However, their major drawback is that the reduced-dimensional projected space does not contain information about each gene since the original space has been transformed. It does not allow us a further biological analysis and a deeper interpretation of a given case under study (disease, biological process).

In gene expressions data, feature selection or variable selection is selecting a subset of genes for model construction or results interpretation. It has been shown that feature selection in such data improves the classification performance and offers the potential for a better data interpretation in a given study since we know its dominant and redundant genes (features), which are related to the various class separation. There are numerous feature selection algorithms with promising results in gene expression data [1,12,20]. In scRNA-seq data, feature selection methods aim to identify uninformative genes (features) with no meaningful biological variation across cells (samples) [31]. Identifying the appropriate set of marker genes interprets the scRNA-seq data at the gene level with a deeper biological meaning [3,30].

Some studies considered the feature selection problem in such data as an optimization task from the mathematical perspective. In [19], the authors described a fitness function that incorporates both performance and feature size. Applying the Particle Swarm Optimization (PSO) method and the utilization of Convolutional Neural Networks, they offer promising results in classifying the different types of cancer based on tumor RNA-Seq gene expression data.

In [24], a feature selection approach is proposed for RNA-seq gene expression data. It reduces the irrelevant features by applying an ensemble L1-norm support vector machine methodology. Its classification performance in RNA-seq data shown promising results, especially in small n – large p problems, with n samples and p features (genes). scTIM [15] framework utilizes a multiobjective optimization technique aiming to maximize gene specificity by considering the gene-cell relationship while trying to minimize the number of selected genes. This model allows the new cell type discovery as well as the better cell categories separation. M3Drop [3], describes two feature selection methods for scRNA-seq data which isolate genes with the high proportion of zero values among their cells, also called the "dropouts" effect. It is a central feature of scRNA-seq due to the considerable technical and biological noise.

Despite the remarkable progress of feature selection methods in gene expressions, their adaptation in single-cell RNA-seq is at a very early stage. Given that these data have high complexity, dimensionality, and sparsity, lead us on the necessity of incorporating an optimization method for the appropriate feature (genes) selection. Our intuition here is that across the genome, several combinations of certain genes will be dominant in cell separation of a given experimental study.

In this paper, we propose a novel feature selection method and analysis, called Feature Selection via Genetic Algorithm (FSGA), that tackles the above challenges. FSGA utilizes genetic optimization principles and identifies low-dimensional sets of features. The aim here is to use a simple distance-based classifier (we use a KNN classifier) during the feature selection process in order to identify feature groups that are nicely separated in Euclidean space. This property is desirable in most classification methods, and thus we expect to boost the performance of any classifier. We showcase FSGA's effectiveness in two real scRNA-seq experiments with gene dimensions up to 36708. Our framework can identify very low-dimensional subsets of genes (less than 200) while boosting the classifiers' performance. The obtained results offer new insights in the single-cell RNA-seq data analysis offering the potential that variants of the proposed method can work also in other data types.

2   Approach

2.1   Problem Formulation

scRNA-seq datasets have very high-dimensional feature spaces, with features spaces going up to $D = 30K$ dimensions. We would like to find a group of K features where $K \ll D$ and we can achieve similar or even better classification performance. We represent the feature space as $\boldsymbol{x} = [x_1, x_2, \ldots, x_D]^T \in \mathbb{R}^D$ and define the problem of feature selection as (see also [1,12]):

$$\boldsymbol{f}^* = \operatorname*{argmax}_{\boldsymbol{f}} J(\boldsymbol{x}_f) \tag{1}$$

where $\boldsymbol{f} = [b_1, b_2, \ldots, b_D]^T \in \mathbb{B}^D$ with b_i being a Boolean[1] value whether we select the dimension i, $\boldsymbol{x}_f \in \mathbb{R}^K$ is the feature dimension vector where we keep only the dimensions as defined by $\boldsymbol{f}$, and J is training and evaluating the performance of a given feature vector.

2.2   Feature Selection via Genetic Algorithms

We choose to tackle this problem using a Genetic Algorithm (GA). GAs operate on a population of individuals and attempt to produce better individuals every generation. At each generation, a new population is created by selecting individuals according to their level of performance and recombining them together using operators borrowed from natural evolution. Offspring might also undergo a mutation operation. In more detail, any GA has the following generic steps:

1. INITIALIZATION OF THE POPULATION
2. EVALUATION OF THE POPULATION
3. SELECTION OF THE FITTEST INDIVIDUALS
4. CROSSOVER BETWEEN SOME OF THE SELECTED INDIVIDUALS
5. MUTATION OF SOME INDIVIDUALS
   The previous two steps produce a new population
6. GO BACK TO STEP 2

There a few critical parameters to choose so that a GA can be effective: (1) gene representation, (2) selection pressure, (3) crossover operation, (4) mutation operation, (5) initialization of the population and (6) performance measure. Below we detail our choices.

---

[1] We define $\mathbb{B}$ as the space of Boolean variables.

Gene Representation. In order to be able to use GAs for solving the problem as defined in Eq. (1), we use the vectors $f = [b_1, b_2, \ldots, b_D]^T \in \mathbb{B}^D$ for the gene representation. This is a natural choice for this task as changing the values in the gene will correspond in selecting different features for the classification [1,12,14,17].

Selection Operator. We adopt a selection operator that selects the top-50% individuals of the population (according to the performance measure). More sophisticated selection operators can be used here to improve performance. We also always insert the best individual back into the new population (thus making the algorithm elitist).

Crossover Operator. The crossover operator consists of combining two (2) individuals (called parents) to produce a new one (offspring). We randomly determine parts of the gene parent vectors to be swapped.

Mutation Operator. Each offspring individual can undergo a mutation operator. For each individual we randomly switch any dimension of its gene vector. So, with some probability we change which features the offspring keeps for the classification.

Initialization of the Population. One crucial aspect of the initial population is to push for as little as possible number of selected features, but not hurt performance. For this reason, we produce the initial population where for each individual each feature dimension has an 1‰ chance of being selected. This procedure produces populations with small number of selected features, but keeps diversity in which feature dimensions are being selected.

Objective Function (Performance Measure). When optimizing for the best features, in each run of the algorithm we split the datasets into three sets: (a) training set, (b) validation set, and (c) test set. The sets are roughly 60%, 20% and 20% of the size of the original dataset respectively (keeping the percentage of classes similar in each dataset). At each evaluation, we use the training set to train the KNN-classifier, and create an objective function of the form:

$$J(x_f) = 0.6 * \text{acc}_{\text{val}} + 0.4 * \text{acc}_{\text{train}} - P_{\text{sparseness}} \tag{2}$$

where acc_{val} is the accuracy of the classifier in the validation set, $\text{acc}_{\text{train}}$ is the accuracy of the classifier in the training set, and $P_{\text{sparseness}} = 10 * \sum_{i=1}^{D} b_i$ is a penalty score penalizing high dimensionality of the selected feature space. The proposed objective function is slightly different from the ones in the literature [1,12,14,17]; we are doing the weighted average of the validation and the training set accuracy. The reasoning behind this weighted average is to not let the algorithm overfit a specific part of the dataset. At the end of each generation, we report the accuracy on the test set (see Sect. 3), but the algorithm never uses this.

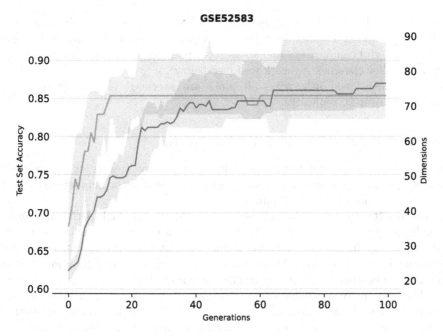

Fig. 1. Convergence of algorithm in GSE52583 dataset. Solid lines are the median over 20 replicates and the shaded regions are the regions between the 25-th and 75-th percentiles.

3   Experimental Analysis on scRNA-seq Datasets

We evaluated the classification performance of our FSGA method using two real transcriptomics datasets from single-cell RNA-seq studies. Datasets were obtained from Gene Expression Omnibus [10] and ArrayExpress [4]. More specific, the first dataset (accession number: GSE52583) [32] has transcriptomics experimental data profiles for 23, 228 genes. It is a transcriptome analysis of 201 distal mouse lung epithelial cells from four developmental stages. The second dataset (accession number: E-MTAB-2805) has studied expression patterns for 36078 genes at single cell level across the different cell cycle stages in 288 mouse embryonic stem cells [7].

The evaluation process was split into to three (3) parts: (a) evaluating the optimization process and whether our proposed scheme was converging to good individuals in all runs, (b) evaluating whether the produced features provide a good set of features for any classifier, and (c) try to determine whether the selected features have a biological meaning.

To tackle both challenges we chose to use a simple KNN classifier when optimizing for the best features. The KNN performs k-nearest-neighbor classification model [2] using the default parameters with Euclidean as distance measure as well KD-tree option as search method for $N = 5$ nearest neighbors. The rationale behind this choice is that a) KNN is fast, and b) a set of features that works well under KNN directly means that these features are nicely separated in Euclidean space. The first fact gave us the ability to run many replicates and have meaningful comparisons and statistics, while the second one makes it more likely for other classifiers to work well (see below).

If not mentioned otherwise, all plots are averaged (or taking median/percentiles) over 20 replicates.

3.1   Evaluation of Feature Selection Process

In order to evaluate the feature selection process, we keep track of the best individual of the optimization at each generation as well as the number of selected feature dimensions of the best individual.

The results show that the optimization is able to find high-performing individuals (see Fig. 1 and Fig. 2). In both datasets, we achieve a median accuracy score over 0.75 in the test set (this is the set that both the classifier and the optimizer have never seen). This showcases that our objective function is able to produce classifiers with nice generalization properties.

Moreover, the results demonstrate that the optimization process increases the dimensionality of the feature space as long as this helps the process get better performance. Once the performance stabilizes to a fixed value, the dimensionality of the feature space stops increasing. This is a desirable property of a feature selection process since we do not want it to keep adding dimensions if they do not help in the classification performance. The algorithm converges at around 77 dimensions for the GSE52583 dataset and around 165 dimensions for E-MTAB-2805 dataset (median values over 20 replicates). Our initialization process is crucial for achieving these results (see Sect. 2.2), as preliminary results with a population with individuals containing many dimensions did not manage to converge to low number of features.

3.2   Evaluation of Selected Features

In this section, we want to evaluate the quality of the selected features both quantitatively and qualitatively. For a principled analysis, we perform the following steps:

- For each run of our algorithm[2], we take the feature dimensions of the best individual at convergence;
- We take those feature dimensions and modify the datasets (i.e., we include only those input dimensions);
- Using the modified datasets we train three (3) different classification methods, namely KNN, LDA and SVM [9,23];
- We compare the performance of the algorithms using our selected features against training the same classifiers using all the feature dimensions (we use Accuracy and F1-score for comparisons);

---

[2] We have 20 runs/replicates.

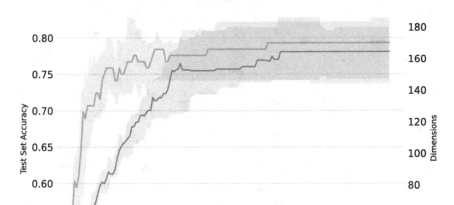

Fig. 2. Convergence of algorithm in E-MTAB-2805 dataset. Solid lines are the median over 20 replicates and the shaded regions are the regions between the 25-th and 75-th percentiles.

– All executions are done using the 10-fold cross validation process in 20 independent trials.

Parameter setting for all methods was chosen based on a fitting procedure in order to optimize their performance. Minor variations for the selected values do not affect the results significantly and thus an extensive analysis is excluded. All algorithms were run with the corresponding default parameters. We exclude an extensive parameter analysis of all classifiers since our aim was to highlight for each classifier its difference between the classification performance in the original and in the reduced feature space.

The results showcase that in almost all cases the feature space produced by our algorithm increases the performance of any classification method (see Fig. 3). In all cases, except when using LDA on the E-MTAB-2805 dataset, our feature selection approach boosts significantly the performance of all classifiers. Even in the worst case (LDA/E-MTAB-2805), the result of the classification is comparable with training in the full feature dimensions by using only 180 dimensions. The performance improvement to the SVM classifier is highlighting the effectiveness of our approach to generate separable feature spaces.

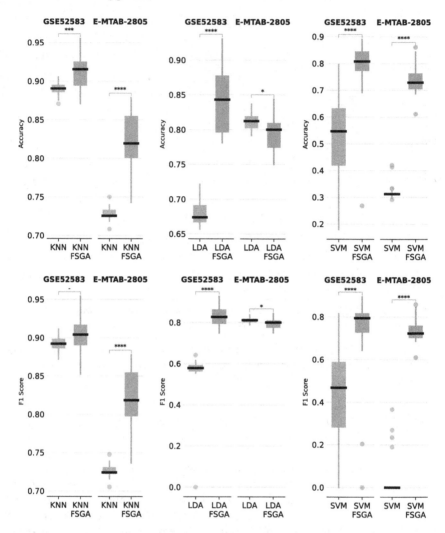

Fig. 3. Evaluation of selected features using different classifiers (20 replicates). We compare using two (2) metrics: accuracy (top row) and F1-Score (bottom row). For each algorithm, we show results before and after the usage of our feature selection algorithm for both datasets. The box plots show the median (black line) and the interquartile range ($25th$ and $75th$ percentiles); the whiskers extend to the most extreme data points not considered outliers, and outliers are plotted individually. The number of stars indicates that the p-value of the Mann-Whitney U test is less than 0.05, 0.01, 0.001 and 0.0001 respectively.

Figure 4 shows tSNE plots [22] of one typical feature selection run of the algorithm in the GSE52583 and the E-MTAB-2805 datasets respectively for qualitative inspection/verification. The plots showcase the effectiveness of the proposed method to find feature dimensions that can separate the classes in different regions of the space.

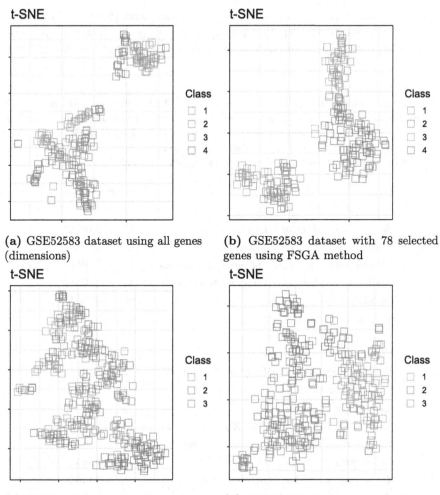

(a) GSE52583 dataset using all genes (dimensions)

(b) GSE52583 dataset with 78 selected genes using FSGA method

(c) E-MTAB-2805 dataset using all genes (dimensions)

(d) E-MTAB-2805 dataset with 180 selected genes using FSGA method

Fig. 4. 2D t-SNE visualizations are illustrated for comparisons between the original datasets and the datasets with reduced features using our FSGA method. Each point represents a cell sample, and each color represents a different cell type according to original data annotation. Our method shows its superiority by efficiently discriminating the cell classes in both datasets.

3.3 Biological Analysis

We further examine the selected genes for each dataset concerning their enrichment in Gene Ontology terms for various Biological Processes (see Table 1) using the Functional Annotation Tool David [18]. Through this analysis, we aim to examine how our list of selected genes relates to terms corresponding to the respective biological case under-study. Both datasets extract genes which are related to cellular functions. Both studies are relevant with these functions since their studies are related with the developmental stages of distal mouse lung epithelial cells and the different cell cycle stages in Mouse Embryonic Stem Cells.

Table 1. Enrichment analysis for GSE52583 and E-MTAB-2805 datasets using gene ontology terms of the selected features obtained from the proposed framework. The first column contains the gene ontology terms for various biological processes. The second column represents the number of genes that present enrich action in each term. The third column represents a modified Fisher's exact p-value.

Term	Count	P-Value
GSE52583		
Negative regulation of cellular process	55	9.0E-2
Cellular macromolecule metabolic process	127	1.2E-10
Cellular nitrogen compound metabolic process	96	1.9E-6
Positive regulation of cellular metabolic process	54	3.4E-6
Cellular catabolic process	31	8.7E-5
Cellular response to chemical stimulus	44	1.0E-3
Response to extracellular stimulus	6	3.8E-2
Regulation of secretion by cell	7	6.2E-2
Negative regulation of cell differentiation	7	9.4E-2
EM-TAB-2805		
Intracellular transport	11	2.5E-2
Establishment of localization in cell	13	2.8E-2
Positive regulation of cell communication	11	5.4E-2
Cell communication	31	7.0E-2
Regulation of cellular component size	6	9.0E-2
Circulatory system process	8	2.2E-3
Response to extracellular stimulus	6	3.8E-2
Regulation of secretion by cell	7	6.2E-2
Negative regulation of cell differentiation	7	9.4E-2

4   Discussion and Conclusion

Machine Learning tasks have become the first choice for gaining insight into large-scale and high-dimensional biomedical data. These approaches can tackle part of such data complexity offering a platform for effective and robust computational methods. Part of this complexity comes from a plethora of molecular biology experimental data having extremely high dimensionality. An indicative example is the single-cell RNA-seq (scRNA-seq), an emerging DNA sequencing technology with promising capabilities but significant computational challenges due to the large-scaled generated data.

Given that this technology offers the opportunity to understand various biological phenomena and diseases better, there is a need for novel computational methods to deal with this complexity and dimensionality. Dimensionality reduction methods are an appropriate choice, but they do not give us explanatory power at the gene level. A significant challenge here is identifying the feature list in terms of genes (dimensions), which will maintain or increase the performance of various machine learning tasks.

Highlighting the salient by eliminating the irrelevant features in a high dimensional dataset such as the high-throughput gene expression experiments, may lead to the strengthening of a traditional classifier's performance [24]. Also, given a features list which is dominant in terms of class separation in a classification process, we obtain a better understanding and interpretation of a given gene expressions dataset.

On the other hand, deep learning has gained ground in biomedical data mining methods. However, its inherent black-box feature offers a poor interpretability for a better understanding of such data. In the case of gene expressions where the data contain a record of tens of thousands of genes, it is crucial to find the genes and especially the combination of these genes, which will better capture the information contained in the data set. Also, through the developing of interpretable ML approaches offers the opportunity not only for a better data interpretation but also for finding the dominant genes which may need to be considered individually or in combination for their potential effect on the under-study case (e.g. a disease, a biological process).

Through our proposed feature selection method using a Genetic Algorithm, we provided evidence about its potential in single-cell RNA-sew analysis regarding the classification performance. Our intuition was that an optimal combination of genes could improve both the classification performance and the interpretability of a given data. The first is critical since even if this feature subset contains genes unrelated to each other, their combination might be highly correlated with the classification. The latter can contribute in the emerging explainable artificial intelligence field.

The obtained results offer new insights in the single-cell RNA-seq data analysis offering the potential that variants of the proposed method can work also in other data types. Our contribution, which is lies in the intuition that specific combinations of small gene groups have a key role in our scRNA-seq data, is partly confirmed by the above results.

References

1. Alba, E., Garcia-Nieto, J., Jourdan, L., Talbi, E.G.: Gene selection in cancer classification using PSO/SVM and GA/SVM hybrid algorithms. In: 2007 IEEE Congress on Evolutionary Computation, pp. 284–290. IEEE (2007)
2. Altman, N.S.: An introduction to kernel and nearest-neighbor nonparametric regression. Am. Stat. **46**(3), 175–185 (1992)

3. Andrews, T.S., Hemberg, M.: M3drop: dropout-based feature selection for scrnaseq. Bioinformatics **35**(16), 2865–2867 (2019)
4. Athar, A., et al.: Arrayexpress update-from bulk to single-cell expression data. Nucleic Acids Res. **47**(D1), D711–D715 (2019)
5. Becht, E., et al.: Dimensionality reduction for visualizing single-cell data using UMAP. Nat. Biotechnol. **37**(1), 38 (2019)
6. Brown, M.P., et al.: Knowledge-based analysis of microarray gene expression data by using support vector machines. Proc. Nat. Acad. Sci. **97**(1), 262–267 (2000)
7. Buettner, F., et al.: Computational analysis of cell-to-cell heterogeneity in single-cell RNA-sequencing data reveals hidden subpopulations of cells. Nat. Biotechnol. **33**(2), 155–160 (2015)
8. Chattopadhyay, A., Lu, T.P.: Gene-gene interaction: the curse of dimensionality. Ann. Transl. Med. **7**(24) (2019)
9. Chatzilygeroudis, K., Hatzilygeroudis, I., Perikos, I.: Machine learning basics. In: Intelligent Computing for Interactive System Design: Statistics, Digital Signal Processing, and Machine Learning in Practice, pp. 143–193 (2021)
10. Clough, E., Barrett, T.: The gene expression omnibus database. In: Mathé, E., Davis, S. (eds.) Statistical Genomics. MMB, vol. 1418, pp. 93–110. Springer, New York (2016). https://doi.org/10.1007/978-1-4939-3578-9_5
11. Collins, F.S., Morgan, M., Patrinos, A.: The human genome project: lessons from large-scale biology. Science **300**(5617), 286–290 (2003)
12. Dhaenens, C., Jourdan, L.: Metaheuristics for data mining. 4OR **17**(2), 115–139 (2019). https://doi.org/10.1007/s10288-019-00402-4
13. Dudoit, S., Fridlyand, J., Speed, T.P.: Comparison of discrimination methods for the classification of tumors using gene expression data. J. Am. Stat. Assoc. **97**(457), 77–87 (2002)
14. Estévez, P.A., Caballero, R.E.: A Niching genetic algorithm for selecting features for neural network classifiers. In: Niklasson, L., Bodén, M., Ziemke, T. (eds.) ICANN 1998. PNC, pp. 311–316. Springer, London (1998). https://doi.org/10.1007/978-1-4471-1599-1_45
15. Feng, Z., et al.: scTIM: seeking cell-type-indicative marker from single cell RNA-seq data by consensus optimization. Bioinformatics **36**(8), 2474–2485 (2020)
16. Hedlund, E., Deng, Q.: Single-cell RNA sequencing: technical advancements and biological applications. Mol. Aspects Med. **59**, 36–46 (2018)
17. Hong, J.H., Cho, S.B.: Efficient huge-scale feature selection with speciated genetic algorithm. Pattern Recogn. Lett. **27**(2), 143–150 (2006)
18. Huang, X., Liu, S., Wu, L., Jiang, M., Hou, Y.: High throughput single cell RNA sequencing, bioinformatics analysis and applications. In: Gu, J., Wang, X. (eds.) Single Cell Biomedicine. AEMB, vol. 1068, pp. 33–43. Springer, Singapore (2018). https://doi.org/10.1007/978-981-13-0502-3_4
19. Khalifa, N.E.M., Taha, M.H.N., Ali, D.E., Slowik, A., Hassanien, A.E.: Artificial intelligence technique for gene expression by tumor RNA-seq data: a novel optimized deep learning approach. IEEE Access **8**, 22874–22883 (2020)
20. Liang, S., Ma, A., Yang, S., Wang, Y., Ma, Q.: A review of matched-pairs feature selection methods for gene expression data analysis. Comput. Struct. Biotechnol. J. **16**, 88–97 (2018)
21. Linderman, G.C., Rachh, M., Hoskins, J.G., Steinerberger, S., Kluger, Y.: Fast interpolation-based t-SNE for improved visualization of single-cell RNA-seq data. Nat. Methods **16**(3), 243–245 (2019)
22. van der Maaten, L., Hinton, G.: Visualizing data using t-SNE. J. Mach. Learn. Res. **9**, 2579–2605 (2008)

23. McLachlan, G.J.: Discriminant Analysis and Statistical Pattern Recognition, vol. 544. John Wiley & Sons, New York (2004)

24. Moon, M., Nakai, K.: Stable feature selection based on the ensemble l 1-norm support vector machine for biomarker discovery. BMC Genom. 17(13), 65–74 (2016)

25. Poirion, O.B., Zhu, X., Ching, T., Garmire, L.: Single-cell transcriptomics bioinformatics and computational challenges. Front. Genet. 7, 163 (2016)

26. Qi, R., Ma, A., Ma, Q., Zou, Q.: Clustering and classification methods for single-cell RNA-sequencing data. Briefings Bioinform. 21(4), 1196–1208 (2020)

27. Regev, A., et al.: Science forum: the human cell atlas. Elife 6, e27041 (2017)

28. Scialdone, A., et al.: Computational assignment of cell-cycle stage from single-cell transcriptome data. Methods 85, 54–61 (2015)

29. Shendure, J., et al.: DNA sequencing at 40: past, present and future. Nature 550(7676), 345 (2017)

30. Taguchi, Y.: Principal component analysis-based unsupervised feature extraction applied to single-cell gene expression analysis. In: Huang, D.-S., Jo, K.-H., Zhang, X.-L. (eds.) ICIC 2018. LNCS, vol. 10955, pp. 816–826. Springer, Cham (2018). https://doi.org/10.1007/978-3-319-95933-7_90

31. Townes, F.W., Hicks, S.C., Aryee, M.J., Irizarry, R.A.: Feature selection and dimension reduction for single-cell RNA-seq based on a multinomial model. Genome Biol. 20(1), 1–16 (2019)

32. Treutlein, B., et al.: Reconstructing lineage hierarchies of the distal lung epithelium using single-cell RNA-seq. Nature 509(7500), 371 (2014)

33. Vrahatis, A.G., Tasoulis, S.K., Maglogiannis, I., Plagianakos, V.P.: Recent machine learning approaches for single-cell RNA-seq data analysis. In: Maglogiannis, I., Brahnam, S., Jain, L.C. (eds.) Advanced Computational Intelligence in Healthcare-7. SCI, vol. 891, pp. 65–79. Springer, Heidelberg (2020). https://doi.org/10.1007/978-3-662-61114-2_5

34. Wang, B., Zhu, J., Pierson, E., Ramazzotti, D., Batzoglou, S.: Visualization and analysis of single-cell RNA-seq data by kernel-based similarity learning. Nat. Methods 14(4), 414 (2017)

35. Witten, D.M., et al.: Classification and clustering of sequencing data using a Poisson model. Ann. Appl. Stat. 5(4), 2493–2518 (2011)

Towards Complex Scenario Instances for the Urban Transit Routing Problem

Roberto Díaz Urra[✉], Carlos Castro, and Nicolás Gálvez Ramírez

Casa Central, Universidad Técnica Federico Santa María, Valparaíso, Chile
{roberto.diazu,nicolas.galvez}@usm.cl, carlos.castro@inf.utfsm.cl

Abstract. Transportation systems are critical components of urban cities, by hugely impacting the quality of life of their citizens. The correct design of the bus routes network is a fundamental task for a successful system. Thus, Urban Transit Routing Problem (UTRP) aims to find a set of bus routes that minimises users travelling time and system operator costs. Most instances of the UTRP lack real-life demand data information, where well-known benchmark instances are very small w.r.t. current standards and/or they have been randomly generated. State-of-the-art relaxation techniques are based on inherent features of urban transport systems and they cannot significantly reduce the order of magnitude of complex instances. In this work, we propose a relaxation technique based on well-known clustering algorithms which preserves demand behaviour and road structure by keeping central and periphery locations connected. We also apply this proposal in a well-known complex study case, the RED system from Santiago of Chile, to encourage researchers to improve the approximation models for other big scenarios.

Keywords: UTRP · Public transport system · Route planning · Clustering · Relaxation

1 Introduction

Urban transportation systems are critical to the performance of most cities: they must be carefully planned to be capable to transport citizens to their destinations in a fast and comfortable manner. A poorly designed system will impact negatively the quality of life of its users [6].

Their design is traditionally split into various stages [1]: network design, frequency setting, timetable development, bus scheduling and driver scheduling. Each of them is considered a complex problem by itself and it is solved independently. The Urban Transit Routing Problem *UTRP*, an NP-Hard problem [7], is one of several formulations for the network design stage [4], where a set of bus routes must be selected for public transportation to minimise the user average travel time and the costs of the system operator.

D. E. Simos et al. (Eds.): LION 2021, LNCS 12931, pp. 80–97, 2021.
https://doi.org/10.1007/978-3-030-92121-7_7

A UTRP instance has three basics elements:

- **Road Network:** A weighted undirected graph that models the city road structure and the locations that must be covered by the bus routes.
- **Demand Matrix:** A structure whose function is to set the number of passengers travelling from one location to another within a time slot period.
- **Route Parameters:** A set of constraints bounding the number of solutions.

A UTRP solution is a route network that is represented by a vector set of connected locations, where each of them encodes a single bus route. These routes must allow users to reach their destinations as quickly as possible, i.e., by avoiding transfers between bus routes. Simultaneously, the routes set must have a low maintenance cost to ease system sustainability. Therefore, the UTRP is a multi-objective problem whose goals are to minimise both the average passenger travel time and the costs of the system operator [10].

A current issue in transportation research is the lack of public benchmark data. Two main reasons are identified [10]: (1) The data collection regarding public transportation is hard to obtain because passengers can not be easily traced and surveys could be very expensive and inaccurate, and (2) the evaluation and quality of solutions in real-life systems rely on diverse and highly ad-hoc metrics, thus they could be used in a reduced amount of scenarios.

In the UTRP scene, benchmark instances are very small, randomly generated [10] and/or based on assumptions on population density data [5]. Moreover, the biggest state-of-the-art instances are very small w.r.t. real-life scenarios. Thus, real-life adjusted approximations are desired, e.g., relaxing the number of bus stops and keeping demand distribution by zones. A basic reduction could be found in [14], where UTRP instances are relaxed by combining nodes placed on street junctions. This compression is similar to an inherent feature in most urban transport systems: several bus stops are spread in huge areas because of traffic direction or public convergence, but they could be seen as the same strategical point. However, this relaxation is not enough for complex scenarios.

A well-known complex scenario is the urban transport system from Santiago of Chile, called *Red Metropolitana de Movilidad (RED)*. It gained worldwide attention in 2007 after its premature implementation despite several design issues, e.g., most of the critical infrastructure and system conditions required were not in operation [11]. Today, *RED* bus system transports near 6.2 million users from 34 communes from Santiago Metropolitan Area. It covers around 680 km^2 in urban zones [13] and comprises 11.320 bus stops grouped in more than 800 demand zones.

In this work, we discuss a relaxation technique for UTRP instances, based on well-known clustering algorithms. Through this method, we simplify complex instances by reducing their road network structure and preserving their demand behaviour. We use as our study case the *RED* public transport system from Santiago, Chile. Finally, we analyse the performance difference between the different clustering algorithms tested and how they affect our proposal.

This paper is organised as follows: the next section discusses state-of-the-art UTRP instances and their roots. In Sect. 3, this work proposal is explained in detail. In Sect. 4, the designed proposal is applied to the selected study case, and a description of the available data for this procedure is done. In Sect. 5, the features of the proposal output are analysed. Finally, in Sect. 6, conclusions and future perspectives about the UTRP systems and their development are set.

2   UTRP Instance Benchmark Analysis

As stated in [5,10], most UTRP instances are fictitious, randomly generated and/or simplified. This leads to an inaccurate approximation of their real-life counterparts. Table 1 presents the most well-known UTRP instances and their features. Note that, all shown features can be seen as solving *complexity metrics*. Following we present an analysis focused on the applied methodology to generate the UTRP instances.

Table 1. UTRP instances summary.

Instance	Nodes	Links	Routes	Route length	LB_{ATT}	MST
Mandl [8]	15	21	4–8	2–8	10.0058	63
Mumford0 [10]	30	90	12	2–15	13.0121	94
Mumford1 [10]	70	210	15	10–30	19.2695	228
Mumford2 [10]	110	385	56	10–22	22.1689	354
Mumford3 [10]	127	425	60	12–25	24.7453	394
Nottingham100 [5]	100	187	40	10–25	19.2974	254
Edinburgh200 [5]	200	362	90	5–25	20.8211	476
NottinghamR [14]	376	656	69	3–52	7.1003	418

The most straightforward metrics are the number of nodes and links: while they tend to be higher, the instance search space and solution evaluation time will highly increase. Thus, the Mandl instance, the most used one for algorithm comparison, is the easiest benchmark case of the set; conversely, NottinghamR is the most complex one. However, all these instances are very small w.r.t. big cities that have thousands of bus stops or locations, e.g., RED system from Santiago of Chile which includes more than 10000 nodes.

The amount and length of the routes, which are commonly set by the network designer, critically constrain the feasible solution space. More and longer routes lead to a bigger search space; conversely, fewer and shorter routes hardly constrain the search space. Thus, diversification is directly affected, leading to either very slow or premature convergence.

LB_{ATT} and MST metrics are respectively the lower bounds of the average passenger travel time and the costs of the system operator. These metrics show the complexity of each instance in both solving criteria, i.e., small values in each of both metrics imply a major effort for solving.

How instances are generated is also relevant, because the metrics values could change significantly. For instance, LB_{ATT} and MST are differently approached in [5] and [14]. The former uses a small subset taken from the source data, which produces higher LB_{ATT} and smaller MST values due to bigger distances between locations and the small number of nodes, respectively. The latter includes information on each street junction, which increase the number of short-distance travels and, consequently, the number of nodes. Thus, lower LB_{ATT} and higher MST values are set.

Table 2 summarises the selected methods to generate each UTRP instance discussed in this work.

Table 2. Generation of basic elements of UTRP instances

Instance	Node splacement	Link information	Demand matrix	Comment
Mandl	Unknown method. Information based on Swiss Network [8]			Smallest instance
Mumford0 Mumford1 Mumford2 Mumford3	Random coordinates	Euclidean distance	Random uniform value	Do not fit real scenarios
Nottingham100 Edinburgh200	Randomly picked bus stops from UK National Public Transport Data Repository	Google distance API	Census Data	Bus stop random selection can leave low-density areas unconnected
NottinghamR	Snapped street junctions. Based on UK Ordnance Survey	ArcGIS function "Closest Facility"	Census Data	Real-life data, but relatively small w.r.t. big cities. Reduction is limited to mix nodes representing the same locations and streets

Mumford instances use random procedures for node placement and demand behaviour, thus their features hardly fit real scenarios. Nottingham100 and Edinburgh200 partially use real-life data to address the scenario fitting issue. However, the bus-stop random selection procedure tends to pick those located in high-density areas obviating periphery and low demanded zones. NottinghamR instance addresses both issues of their predecessors, but it is still small w.r.t. big cities that include thousands of bus stops and/or locations. The node grouping technique proposed in [14] is a weak effort for massive cases, i.e., it is unable to reduce the order of magnitude of the instances. Therefore, a relaxation technique for intractable and complex scenarios is required to address them with state-of-the-art solving algorithms.

3 Relaxing UTRP Instances

In this section, we present our relaxation proposal. We define basic concepts as the road network and demand matrix. Then, we show how to transform these elements.

3.1 Basic Definitions

Each UTRP instance includes a **road network** represented by an undirected graph, where each node maps to a location, i.e., bus stop, and each link maps to a road connecting two of them. Note that links use travel times, in minutes, as weight. Each location represents a bidirectional travelling spot.

Definition 1. *Given a node set* $N = \{n_1, \ldots, n_m\}$, *and a link set* $L = \{(n_i, n_j) \in N^2 \mid connected(n_i, n_j) \wedge n_i \neq n_j\}$. *Then, a **road network** R is an undirected graph* $R = (N, L)$ *such that each node* n_i *is a location and each link* $(n_i, n_j) \in L$ *is a direct connection between locations.*

The demand of passengers is represented by a square hollow matrix whose dimension is equal to the number of nodes. In this matrix, each entry is the number of passengers that travel from one location to another during a time frame. This matrix is also symmetric, i.e., the demand of any two locations is the same regardless of the travel direction.

Definition 2. *Given a road network* $R = (N, L)$ *with node set* $N = \{n_1, \ldots, n_m\}$. *We define* $D_{|N| \times |N|}$ *as the **Demand Matrix**. Each entry* $D_{xy} \in \mathbb{Z}$ *indicates the number of passengers that travel from* n_x *to* n_y. *Note that* $D_{xx} = 0$.

To formalise this work proposal; which relax the previously presented UTRP instance components: the **Road network** and the **Demand matrix**; some basic clustering definitions are introduced.

Definition 3. *Given a set of locations* S, *where each element* $s \in S$ *has a position* $pos(s)$. *The **centroid** of the set* S *is the average position between all their elements.*

$$C_S = \frac{1}{|S|} \sum_{s \in S} pos(s) \tag{1}$$

Definition 4. *Given a location set* S. *A clustering algorithm* α *generates a **set of clusters** A by including m location subsets.*

$$A = \left\{ a_i \;\middle|\; a_i \subset S \wedge a_i \cap a_j = \emptyset \;\forall i \neq j \wedge \bigcup_{1 \leq i \leq m} a_i = S \right\} \tag{2}$$

*Thus, each **cluster** $a_i \in A$, i.e., a location set, has its own centroid C_{a_i}. From here, we use the reduced notation $a \in A$ for each cluster; hence, we understand that $|A| = m$.*

Definition 5. *Similarly, given a location set* S, *a* ***set of zones*** Z *includes* q *location subsets.*

$$Z = \left\{ z_i \;\middle|\; z_i \subset S \;\wedge\; z_i \cap z_j = \emptyset \; \forall i \neq j \;\wedge\; \bigcup_{1 \leq i \leq q} z_i = S \right\} \qquad (3)$$

Thus, each ***zone*** $z_i \in Z$, *i.e., a location set, has its own centroid* C_{z_i}. *From here, we use the reduced notation* $z \in Z$ *for each cluster; hence, we understand that* $|Z| = q$.

Let us note that the main difference between a cluster (Eq. 2) and a zone (Eq. 3) is the former is generated by a clustering algorithm, meanwhile, the latter is part of the original data set by the network designer.

3.2   Relaxed Road Network

The **Relaxed Road Network (RRN)** is generated by applying the following criterion over the original road network: *Two centroids of two clusters are connected, if and only if there are a pair of locations, one from each cluster, connected on the source road network.* A basic RRN is shown in Example 1.

Definition 6. *Given a road network* $R = (N, L)$ *and a set of clusters* A *generated by a clustering algorithm* α. *Then,* $\hat{N}$ *is the set of centroids from each cluster generated by* α:

$$\hat{N} = \{C_a \mid a \in A\} \qquad (4)$$

and $\hat{L}$ *is a set of links between connected centroids from* $\hat{N}$.

$$\hat{L} = \{(C_a, C_{a^*}) \mid \exists (n_i, n_j) \in L \wedge n_i \in a \wedge n_j \in a^*\} \qquad (5)$$

Thus, let $\hat{R}$ *be a* ***RRN*** *represented by undirected graph such that* $\hat{R} = (\hat{N}, \hat{L})$.

Example 1. Let R be a road network with $|N| = 16$ locations and its set of connections L, as shown in Fig. 1. A clustering algorithm α groups the locations in a cluster set $A = \{a_1, a_2, a_3\}$. The centroid C_{a_2} is connected to the others centroids, as there are connections between at least one of their locations on the road information. The centroids C_{a_1} and C_{a_3} are disconnected because they have not locations connected between them.

3.3   Relaxed Demand Matrix

The demand information of each relaxed instance is represented by a double-weighted demand matrix called **Relaxed Demand Matrix (RDM)**. The RDM maps the original case demand data into the newly generated RRN scenario.

The first step for the RDM generation, is to compute a demand correction factor for each combination of centroids and zones. This *double-weight factor* is composed by a *distance-based weight* and a *location-based weight*.

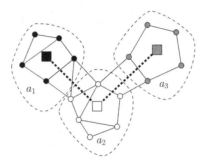

Fig. 1. RRN construction example. The circles are locations inside clusters and the squares are the centroids of each cluster. The resulting road network connects the white centroid (C_{a_2}) with the black (C_{a_1}) and the gray (C_{a_3}) centroids. These links are generated because there is at least one location directly connected between their clusters. (Color figure online)

The *location-based weight* allows to proportionally redistribute the demand of each zone because its locations might be split between many clusters. Thus, this factor is set as the proportion of the locations from a zone in a cluster w.r.t. all locations in that zone.

The *distance-based weight* corrects the demand contribution of each zone in a cluster. In each cluster, the centroid appropriately represents each nearby location. However, if the distance of a location w.r.t the centroid grows, it becomes more inaccurate to represent its demand behaviour.

Definition 7. *Given a clusters set A and a zones set Z. Let $dist(X,Y)$ and $dist_{max}$ be the distance between two locations X and Y, and the maximum distance parameter, respectively. For each cluster $a \in A$, its **double-weight factor** w.r.t. a zone $z \in Z$, is defined as the multiplication of both location-based and distance-based weights.*

$$dw(a,z) = \underbrace{\max\left\{1 - \frac{dist(C_a, C_z)}{dist_{max}}, 0\right\}}_{\textit{distance-based weight}} \cdot \underbrace{\left(\frac{|a \cap z|}{|z|}\right)}_{\textit{location-based weight}} \qquad (6)$$

Note that $dist_{max}$ is a parameter set by the network designer (see Sect. 4.3 for more details).

Therefore, the locations and zones whose demand influence is marginal; i.e., they are far enough from a cluster centroid; are consequently treated as outliers, as shown in Example 2.

Example 2. Let $\hat{R}$ be a RRN and $Z = \{z_1, z_2\}$ a zones set, as is shown in Fig. 2. Let us note that $\hat{R}$ is composed by a centroid set $\hat{N} = \{C_{a_1}, C_{a_2}\}$ linked to a cluster set $A = \{a_1, a_2\}$. For the purposes of this example, it is not necessary to show the connection links of the RRN.

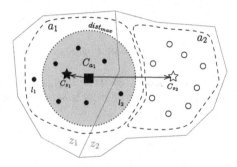

Fig. 2. Double-weight factor example. Circles represent locations: a_1 and a_2 elements are black-coloured and white-coloured, respectively. The grey circle centred on C_{a_1} (black square) with radius $dist_{max}$ bounds the demand contribution of those zones belonging to the cluster. Note that, each zone centroid is represented by a star figure. (Color figure online)

Note that, each zone contributes a different amount of elements, i.e., demand, to each cluster centroid. Thus:

1. The demand contribution of zone z_1 to the cluster a_1 is corrected by a distance-based weight $\in [0,1]$, because its distance w.r.t centroid C_{a_1} is lower than $dist_{max}$. Meanwhile, the location-based weight equals to $\frac{|a_1 \cap z_1|}{|z_1|} = \frac{6}{6} = 1$, because the a_1 cluster contains all the locations from the z_1 zone. Note that l_1 location is not an outlier, it belongs to a_1.
2. The z_2 zone does not contribute any demand to the a_1 cluster, i.e., the distance-based weight equals zero, because C_{z_2} lies outside the boundary fixed by $dist_{max}$ distance from C_{a_1}, despite the positive value of its location-based weight: $\frac{|a_1 \cap z_2|}{|z_2|} = \frac{1}{9}$. Note that, location l_2 is an outlier in the cluster assignment.
3. The z_2 zone contributes a proportioned demand to the cluster a_2, because $\frac{|a_2 \cap z_2|}{|z_2|} = \frac{8}{9} < 1$. For sake of simplicity, we obviate the effect from the distance-based weight.

The original demand data includes the travel information between all pairs of zones in a time window schedule divided into several time frames. For each time frame, there are different amounts of travels between zones introducing a new dimension to the demand matrix. However, UTRP relies on a unique time frame matrix. Thus, the temporal dimension needs to be reduced to a single time slot.

Definition 8. *Given a zone set Z, a demand time frames set T, and a non-symmetric demand matrix between zones by times frames $F_{|Z| \times |Z| \times |T|}$. Each matrix element F_{xyw} is the number of travels between zones z_x and z_y on time frame t_w. Note that F is a hollow matrix, i.e., $F_{xxw} = 0; \forall x$.*

For each pair of zones $(z_x, z_y) \in Z^2$, we select the time frame with the maximum amount of travels.

$$t' = \arg\max_{t \in T}(F_{xyt}) \tag{7}$$

*Thus, the **Simplified Demand Zone Matrix** $E_{|Z| \times |Z|}$ is symmetric and each element denotes the maximum amount of travels between two zones in all time frames.*

$$E_{xy} = \max\{F_{xyt'}, F_{yxt'}\} \tag{8}$$

Once all these elements are defined, we can combine them in order to generate the **RDM**.

Definition 9. *Given an RRN $\hat{R}$ composed by a centroid set $\hat{N} = \{C_{a_1}, \cdots, C_{a_m}\}$ linked to a cluster set $A = \{a_1, \cdots, a_m\}$, a zone set $Z = \{z_1, \cdots, z_k\}$ and a Simplified Demand Zone Matrix $E_{|Z| \times |Z|}$. The **Relaxed Demand Matrix** $D_{|A| \times |A|}$ contains the amount of travels between any node pair $(C_{a_x}, C_{a_y}) \in \hat{N}^2$*

$$D_{xy} = \sum_{z_i \in Z} \sum_{z_j \in Z} dw(a_x, z_i) \times dw(a_y, z_j) \times E_{ij}; \ \forall(a_x, a_y) \in A^2 \tag{9}$$

4 Experimental Configuration

In this section, we set up this work proposal w.r.t. the selected study case: RED network, the public transport system from Santiago of Chile. The information about the RED system is published annually by the DTPM (*Directorio de Transporte Público Metropolitano* and it includes bus-stop location and zone-grouped passenger demand data [3].

4.1 Clustering Algorithms

Clustering algorithms are strictly necessary to generate the RRN and RDM. Thus, the selection of the tested clustering algorithms and their features are summarised in Table 3. These algorithms are available in the *scikit-learn* toolkit [12].

The performance difference between the selected clustering algorithms is analysed through the similarity of their outputs, to observe strong variations triggered in the relaxation process. Note that, most selected algorithms are stochastic, then the comparison is made over ten executions of each algorithm. Two metrics are set for this task: (1) the Adjusted Rand index (ARI) [15] and (2) the Unique Match index (UMI). In both, if their values are closer to their maximum implies a bigger similarity.

The adjusted Rand index (ARI) (see Definition 10) measures the similarity between two clustering algorithms outputs by counting the number of agreements between them including an adjustment establishing random clustering as a baseline [15]. A numerical example for this metric is presented in Example 4.

Table 3. Summary of selected clustering algorithms

Algorithm	Parameters	Description
K-Means	Number of Clusters	It adjusts cluster centroids utilizing iteration, labelling near elements and then recalculating their positions
Agglomerative	Number of Clusters	It starts with single element clusters and successively mix those with the lowest average distance between their elements
Spectral	Number of Clusters	It performs a low-dimension embedding of the similarity matrix between elements, to then cluster the components of the eigenvectors
Gaussian Mixture	Number of Clusters	It creates several clusters based on non-spherical Gaussians distributions to fit the data sample. Then, each of them is filled with those elements with the highest probability to be generated between all clusters
Mean shift	Bandwidth	It updates a candidate list for centroids, which could fit as the mean of the elements within a given region. These candidates are filtered, in a post-processing stage, to eliminate near-duplicates
Affinity Propagation	Damping, sample preference	It defines node exemplars by checking the similarity between each pair of elements. Then, it iteratively updates the w.r.t. the values from other pairs. Nodes associated with the same exemplar are grouped in the same cluster
DBSCAN	Neighbourhood distance size	It finds core samples of high density, to then expands clusters from them. It better suites data with non-homogeneous density distribution
OPTICS	Minimum cluster membership	It resembles DBSCAN, but it automatically addresses neighbourhood distance size by means of a position ordering hierarchy

Definition 10. *Given a location set S and two sets of clusters A and B generated by clustering algorithms α and β, respectively. Let us define the following concepts:*

- *Let a be the number of pairs that are classified as belonging to the same cluster in A and B.*
- *Let b be the number of pairs that are in different clusters in both A and B.*
- *Let c be the number of pairs that are in the same cluster in A, but in different clusters in B.*
- *Let d be the number of pairs that are in different clusters in A, but in the same cluster in B.*

*Then, the value of the **Adjusted Rand Index** (ARI) for the cluster sets A and B is defined as follows.*

$$ARI(A, B) = \frac{2(ab - cd)}{(b + c)(c + d) + (b + d)(d + a)} \tag{10}$$

Note that, $a + b$ is the number of agreements, meanwhile $c + d$ is the number of disagreements. Thus, ARI outputs values in the $[-1, 1]$ range.

Example 3. Let $S = \{s_1, \ldots, s_6\}$ be a location set. Let us define a few possible clusters sets:

- $A_1 = \{\{s_1, s_2\}, \{s_3, s_4\}, \{s_5, s_6\}\}$
- $A_2 = \{\{s_5, s_6\}, \{s_1, s_2\}, \{s_3, s_4\}\}$
- $A_3 = \{\{s_4, s_5, s_6\}, \{s_1, s_2\}, \{s_3\}\}$
- $A_4 = \{\{s_5\}, \{s_1\}, \{s_2, s_3, s_4, s_6\}\}$
- $A_5 = \{\{s_1, s_3\}, \{s_5, s_6\}, \{s_2, s_4\}\}$

Note that, ARI will output bigger values for a higher number of agreements between cluster sets. Thus:

1. The value of $ARI(A_1, A_2) = +1.00$ because these cluster are a permutation of the same subsets. Therefore, they are a perfect match whose value is the maximum.
2. In the A_3 set, the s_4 location is switched to a different cluster. Thus, the value decreases (to $ARI(A_1, A_3) = +0.44$) as the number of agreements does.
3. Between the sets A_4 and A_5 there are not many resemblances, thus the ARI value is near zero ($ARI(A_4, A_5) = -0.06$). These cluster sets were randomly generated for this example.

The Unique Match Index (UMI) (Definition 11) is the proportion of unique matches of a centroid set w.r.t. another centroid set. A unique match occurs when only one centroid of the second set has a centroid of the first set as the nearest. An example of this metric can be seen in Example 4.

Definition 11. *Given two cluster sets $A = \{a_1, \ldots, a_n\}$ and $B = \{b_1, \ldots, b_n\}$ with centroid sets $C_A = \{c_{a_1}, \ldots, c_{a_n}\}$ and $C_B = \{c_{b_1}, \ldots, c_{b_n}\}$, respectively. Let $dist(x, y)$ be a distance function between locations/centroids x and y.*

Let us also define the $M(c, C_B) \subset C_B$ as a centroid subset whose elements have to $c \in C_A$ as the closest matching centroid. Thus, $|M(c, C_B)|$ is the number of matches of c in C_B.

$$M(c, C_B) = \left\{ z \mid z \in C_B, dist(z, c) < dist(z, c') \; \forall c' \in C_A, c' \neq c \right\} \tag{11}$$

*Then, the **Unique Match Index** between centroid sets C_A and C_B is defined as the proportion of centroids in C_A with a single match in C_B:*

$$UMI(C_A, C_B) = \frac{\left|\left\{c \mid c \in C_A, \ |M(c, C_B)| = 1\right\}\right|}{n} \tag{12}$$

Note that, UMI outputs values in the $[0, 1]$ range.

Example 4. Given two cluster sets $A = \{a_1, a_2, a_3, a_4\}$ and $B = \{b_1, b_2, b_3, b_4\}$ with 4 clusters each one, and their centroid sets $C_A = \{c_{a_1}, c_{a_2}, c_{a_3}, c_{a_4}\}$ and $C_B = \{c_{b_1}, c_{b_2}, c_{b_3}, c_{b_4}\}$, respectively. As shown in the Fig. 3, two different scenarios are analysed.

- The Fig. 3a states a full match between centroid sets. Here, each centroid has a unique match in both sets given by the small differences between both clustering outputs. Thus, the RRN produced by our technique will be very similar.
- The opposite scenario is shown in Fig. 3b. Thus, centroids C_{a_2} and C_{a_4} move away from their previous matches and they approach other centroids in C_B which already had a match. Here, the UMI value is not symmetric and can be equal to zero. This, our technique is likely to produce a different RRN for each centroid set.

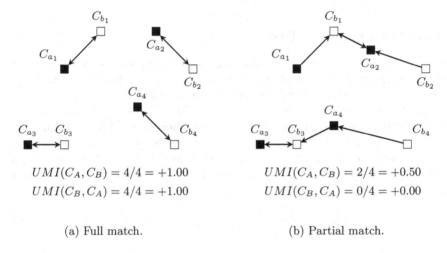

$$UMI(C_A, C_B) = 4/4 = +1.00$$
$$UMI(C_B, C_A) = 4/4 = +1.00$$

$$UMI(C_A, C_B) = 2/4 = +0.50$$
$$UMI(C_B, C_A) = 0/4 = +0.00$$

(a) Full match.   (b) Partial match.

Fig. 3. UMI example. A and B centroids are black-coloured and white-coloured squares, respectively. The arrows point to the closest centroid from the other set. Note that, the centroids pointed by a single arrow have a unique match and, consequently, increase the UMI value.

4.2   Relaxed Road Network

RED network information includes the position of 11.320 bus stops and 803 passenger zones. Each bus stop position is mapped to a geographical coordinate pair, i.e., latitude and longitude. All bus stops are spread in pairs to address bidirectional traffic. Regrettably, RED data does not include the required link set between locations. Before the RRN is generated, RED data is pre-processed as follows:

1. Each bus stop position is mapped to a 3D Euclidean coordinate by assuming the Earth as a sphere whose radius equals 6371 km. Thus, the Euclidean distance can be set between any two locations in the network.
2. Each bus stop pair must be unified to fulfil the bidirectional bus stop feature of the UTRP. The Agglomerative Clustering algorithm [9] is used to link clusters within a distance of 200 meters. Each generated centroid is, therefore, a unique bus stop that covers a wider area. This procedure reduces the number of locations by 62.3%, i.e., from 11.320 to 4.260 nodes.
3. The road network is set through the k-neighbours algorithm [9], by using $k = 7$ as the main parameter value. In this step, a fully connected network is generated, including no links through invalid extensions of terrain, e.g., hills, big parks, airports, etc. However, some links skip small urban features, e.g., bridges, that could not have a near road. Thus, there is no proper way to avoid these details without using geographical data.

In order to generate several relaxations, different versions of the RRN are created. In this way, for each selected clustering algorithm, three different RRN sizes were defined: 35, 90, and 135 clusters.

4.3   Relaxed Demand Matrix

The DTPM publishes the passenger demand information as matrices which include the average number of travels between each pair from the 803 predefined zones. The average calculation is done through five working days and it is presented as a schedule of 24 h divided into 48 time frames of half an hour each. These matrices were generated by tracing user behaviour through smart-card databases and GPS information from buses [3]. Thus, the unique parameter to be set, in order to perform the procedures described in Sect. 3.3, is the $dist_{max}$ value. This parameter is set as the average distance between all connected centroids on the RRN:

$$dist_{max} = \overline{dist} = \frac{1}{|\hat{L}|} \sum_{(C_i, C_j) \in \hat{L}} dist(C_i, C_j) \qquad (13)$$

All experiments of this work are implemented on Jupyter Notebook running Python 3.8.1 as the kernel. Tests are executed in a workstation with an AMD FX-8350 4.0 GHz CPU, 16 GB RAM and Windows 10 Pro 64-bit OS.

5   Experimental Highlights

In this section, we discuss the most relevant results obtained by applying the discussed proposal in the selected study case.

5.1   Clustering Algorithm Recomendation

From the clustering algorithm set proposed in Sect. 4.1, we recommend the use of the following: K-Means, Agglomerative, Spectral and Gaussian Mixture. They can easily fit the desired relaxed scenario and they generate fully connected networks given its main parameter: *Number of Clusters*. Dismissed algorithms are less versatile. Mean Shift and Affinity Propagation algorithms are discarded because their hyper-parameters must be carefully tuned to generate valid or different scenarios. In addition, DBSCAN and OPTICS are dismissed because they cannot handle the homogeneous distribution of bus stops after their noise-reduction procedure which isolates or overlook most peripheral areas.

In Fig. 4, an example of these behaviours is shown. While Agglomerative clustering generates a well-fitted RRN with 135 nodes, Affinity propagation needs a fine hyper-parameter tuning setting to generate a 61 node relaxation. Note that, the latter algorithm cannot generate instances with less or more density of nodes. Moreover, DBSCAN generates clusters with huge size differences, which may be not connected. This behaviour is more noticeable in peripheral zones due to this algorithm noise-reduction procedure. Through this recommendation, and the number of nodes for the relaxation given in Sect. 4.2, twelve new UTRP instances are generated.

5.2   UTRP Relaxing Scenarios

Each recommended algorithm could correctly perform and fit each selected relaxation scenario. As Fig. 5 shows, while bigger the number of clusters, the better the approximation to the original scenario. Let us note, the more complex relaxation (Fig. 5c) has one and two orders of magnitude less than the mixed location (Fig. 5d) and the original RED instance, respectively,

The fitting behaviour mentioned above is noticeable in peripheral zones, big urban structures and geographical features, e.g., airports, fields, hills, etc. Figure 6 shows how the urban elements are better detected with a bigger

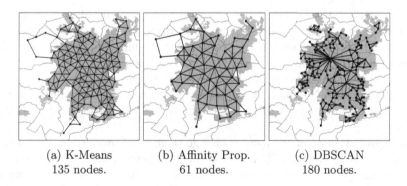

| (a) K-Means | (b) Affinity Prop. | (c) DBSCAN |
| 135 nodes. | 61 nodes. | 180 nodes. |

Fig. 4. Comparison of RRN generated by different clustering algorithms.

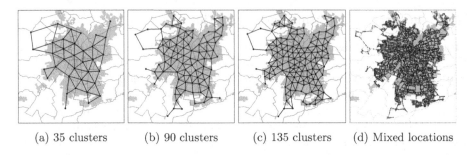

(a) 35 clusters   (b) 90 clusters   (c) 135 clusters   (d) Mixed locations

Fig. 5. RED instance relaxation by means of Gaussian Mixture. (Color figure oline)

cluster amount. Once again, note that the proposed relaxation is dramatically less complex than the original scenario.

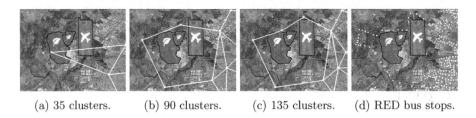

(a) 35 clusters.   (b) 90 clusters.   (c) 135 clusters.   (d) RED bus stops.

Fig. 6. Comparison between original RED scenario and its relaxation by means of Agglomerative clustering.

5.3   Similarity Between Clustering Algorithms

The features of instances previously discussed are consistent between the recommended algorithms. In Table 4, the values of the similarity metrics discussed in Sect. 4.1 are shown.

The average of both metrics is in the highest quartile of their respective ranges for each compared pair. The standard deviation also points to low data dispersal. Therefore, outputs are not identical, but quite similar, i.e., each RRN shares a similar configuration of clusters and centroids.

These similarities hold for all relaxation scenarios discussed and, even, in their complexity metrics (see Sect. 2). As shown in Table 5, the complexity metrics values of the generated UTRP instances are quite similar for each relaxation scenario.

As expected, most of the metrics change their value proportionally w.r.t. an increasing number of nodes. However, the LB_{ATT} metric value is almost constant between different levels of relaxation. This behaviour relies on the relaxation proposal which does a proper distribution of the demand for different RRN configurations.

Table 4. Average (top) and Standard deviation (bottom) of the similarity metrics applied between recommended algorithms for the 90 cluster relaxation.

ARI	Kmn	Agg	Spc	Gau	UMI	Kmn	Agg	Spc	Gau
Kmn	0.633	0.547	0.554	0.615	**Kmn**	0.823	0.811	0.766	0.817
	0.022	0.012	0.013	0.025		0.026	0.033	0.029	0.037
Agg	0.547	1.000	0.530	0.546	**Agg**	0.822	1.000	0.783	0.816
	0.012	0.000	0.006	0.019		0.026	0.000	0.028	0.032
Spc	0.554	0.530	0.854	0.559	**Spc**	0.764	0.746	0.924	0.767
	0.013	0.006	0.023	0.020		0.026	0.028	0.024	0.034
Gau	0.615	0.546	0.559	0.604	**Gau**	0.817	0.817	0.770	0.801
	0.025	0.019	0.020	0.032		0.032	0.023	0.037	0.039

Table 5. Summary of complexity metrics for the generated UTRP relaxed instances. The bus average speed used to set LB_{ATT} equals 19.31 km/h [2].

Crit.	$|N|$											
	35				90				135			
	$|L|$	$\overline{dist}$	LB_{ATT}	MST	$|L|$	$\overline{dist}$	LB_{ATT}	MST	$|L|$	$\overline{dist}$	LB_{ATT}	MST
Min	64	4745	35.70	426	187	2988	35.85	689	293	2430	35.87	814
Max	72	5389	36.11	475	197	3174	36.30	712	308	2518	37.16	844
Average	69	4946	35.89	444	193	3053	36.13	698	301	2461	36.24	828
Std Dev.	3.6	297.5	0.206	21.7	4.32	82.5	0.198	9.95	6.18	39.22	0.62	12.4

5.4 Demand Distribution

As previously stated, a critical feature of the relaxation procedure is to preserve the demand distribution, in order to generate a good approximation of the original scenario and to explain the similarities between recommended cluster algorithms. As Fig. 7 shows, the relaxed scenarios follow a very similar demand

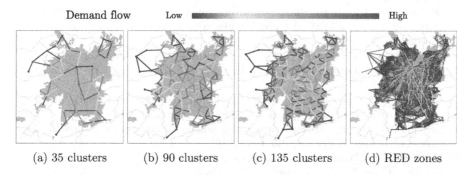

(a) 35 clusters   (b) 90 clusters   (c) 135 clusters   (d) RED zones

Fig. 7. Comparison of the demand distribution between generated instances using Spectral Clustering and the RED original data. Link colour is the flow of passengers travelling on the shortest route to their destination. (Color figure oline)

distribution w.r.t. the original scenario. Let us also note, the better resemblance of the demand distribution, while the number of clusters is bigger.

6   Conclusions and Perspectives

In this work, a clustering-based technique to relax high complexity UTRP instances is defined. This procedure reduces the size of the Santiago RED system instance, a well-known complex scenario, up to two orders of magnitude. The proposal also offers a high-quality approximation of the travel demand distribution and the road network features. Several clustering algorithms have also been tested, four of which are recommended because they allow inputting the number of clusters. Dismissed algorithms did not fit well to the study case: they discard peripheral locations as noise, they cannot classify homogeneous data and/or they need careful hyper-parameter tuning. The recommended algorithms also uphold the study case properties after the relaxation process by presenting similar output and performance, even when they are quite different from each other.

From this point onwards, several research paths are available such as mapping the solution of a relaxed instance back to the original scenario, revisiting RRN building by using a georeferencing system for better approximation, and testing algorithms relying on a hyper-parameter tuning procedure.

Acknowledgments. The work of Roberto Díaz-Urra was supported by the Chilean government: CONICYT-PCHA/MagisterNacional/2013-21130089.

References

1. Ceder, A., Wilson, N.H.: Bus network design. Transp. Res. Part B Methodol. **20**(4), 331–344 (1986)
2. Directorio de Transporte Público Metropolitano: Informe de gestión 2017 (2017). http://www.dtpm.gob.cl/index.php/documentos/informes-de-gestion. Accessed 22 June 2020
3. Directorio de Transporte Público Metropolitano: Matrices de viaje 2017 (2017). http://www.dtpm.gob.cl/index.php/documentos/matrices-de-viaje. Accessed 22 June 2020
4. Ibarra-Rojas, O.J., Delgado, F., Giesen, R., Muñoz, J.C.: Planning, operation, and control of bus transport systems: a literature review. Transp. Res. Part B Methodol. **77**, 38–75 (2015)
5. John, M.P.: Metaheuristics for designing efficient routes and schedules for urban transportation networks. Ph.D. thesis, Cardiff University (2016)
6. Lee, R.J., Sener, I.N.: Transportation planning and quality of life: where do they intersect? Transp. Policy **48**, 146–155 (2016)
7. Magnanti, T.L., Wong, R.T.: Network design and transportation planning: models and algorithms. Transp. Sci. **18**(1), 1–55 (1984)
8. Mandl, C.: Applied network optimization (1980)
9. Manning, C.D., Schütze, H., Raghavan, P.: Introduction to Information Retrieval. Cambridge University Press, Cambridge (2008)

10. Mumford, C.L.: New heuristic and evolutionary operators for the multi-objective urban transit routing problem. In: 2013 IEEE Congress on Evolutionary Computation, pp. 939–946 (2013)
11. Muñoz, J.C., Batarce, M., Hidalgo, D.: Transantiago, five years after its launch. Res. Transp. Econ. **48**, 184–193 (2014)
12. Pedregosa, F., et al.: Scikit-learn: machine learning in Python. J. Mach. Learn. Res. **12**, 2825–2830 (2011)
13. Red Metropolitana de Movilidad: Información del Sistema | Red Metropolitana de Movilidad (2020). http://www.red.cl/acerca-de-red/informacion-del-sistema. Accessed 22 June 2020
14. Heyken Soares, P., Mumford, C.L., Amponsah, K., Mao, Y.: An adaptive scaled network for public transport route optimisation. Public Transp. **11**(2), 379–412 (2019). https://doi.org/10.1007/s12469-019-00208-x
15. Vinh, N.X., Epps, J., Bailey, J.: Information theoretic measures for clusterings comparison: variants, properties, normalization and correction for chance. J. Mach. Learn. Res. **11**, 2837–2854 (2010)

Spirometry-Based Airways Disease Simulation and Recognition Using Machine Learning Approaches

Riccardo Di Dio[1,2(✉)], André Galligo[1,2], Angelos Mantzaflaris[1,2],
and Benjamin Mauroy[2]

[1] Université Côte d'Azur, Inria, Nice, France
`riccardo.di_dio@univ-cotedazur.fr`
[2] Université Côte d'Azur, CNRS, LJAD, VADER Center, Nice, France

Abstract. The purpose of this study is to provide means to physicians for automated and fast recognition of airways diseases. In this work, we mainly focus on measures that can be easily recorded using a spirometer. The signals used in this framework are simulated using the linear bicompartment model of the lungs. This allows us to simulate ventilation under the hypothesis of ventilation at rest (tidal breathing). By changing the resistive and elastic parameters, data samples are realized simulating healthy, fibrosis and asthma breathing. On this synthetic data, different machine learning models are tested and their performance is assessed. All but the Naive bias classifier show accuracy of at least 99%. This represents a proof of concept that Machine Learning can accurately differentiate diseases based on manufactured spirometry data. This paves the way for further developments on the topic, notably testing the model on real data.

Keywords: Lung disease · Machine Learning · Mathematical modeling

1 Introduction

Having a fast and reliable diagnosis is a key step for starting the right treatment on time; towards this goal, Machine Learning (ML) techniques constitute potential tools for providing more information to physicians in multiple areas of medicine. More specifically, as far as respiratory medicine is concerned, there is a recent blooming of publications regarding the investigations of Artificial Intelligence (AI), yet the majority of them refers to computer vision on thoracic X-Rays or MRI [9]. However, for lung diseases using Pulmonary Function Tests (PFTs) recent studies have only scratched the surface of their full potential, by coupling spirometry data with CT scans for investigating Chronic Obstructive Polmunary Diseases COPD on large datasets like COPDGene [3]. In our study, only normal ventilation is used allowing diagnosis also for children. Our aim is to provide a first proof-of-concept and provide the first positive results that could lead to fast, accurate and automated diagnosis of these diseases, similarly e.g. to Cystic Fibrosis (CF) where Sweat chloride test is a central asset [8].

© Springer Nature Switzerland AG 2021
D. E. Simos et al. (Eds.): LION 2021, LNCS 12931, pp. 98–112, 2021.
https://doi.org/10.1007/978-3-030-92121-7_8

Normally, ML models are trained and tested on data and labels are provided by medical doctors. However, the originality of this study consists in using mathematical equations to simulate the ventilation following the directives of IEEE [2], then this data will be used to train the ML models. The obtained volume flows respect the expectations for both healthy and not healthy subjects. Using a synthetic model with a low number of parameters allows us to have everything under control.

During this study, the lungs are modeled as elastic balloons sealed in the chest wall and the airways are modeled as rigid pipes, this allows to play with few parameters to simulate healthy and not healthy subjects and create synthetic volumetric data of tidal breathing. This data is then split and used to train and test different ML models for diagnosis. The accuracies reached during the study are very high, however, the choice of the parameters and the restriction of synthetic data allowed for promising results. Further tests are needed with real data to validate the accuracy reached. Nevertheless, this study points out that not every classifier is suited for this task.

A brief introduction to human lungs and its physiology is given in 1.1, then Sect. 1.2 shows how ventilation has been modeled. In Sect. 2.1 is shown how the dataset has been realized, Sect. 2.3 reflects the training of the models on the dataset and finally in Sect. 3 and 4 the results are exposed and discussed.

1.1 Lung Ventilation

The respiratory system can be split in two different areas, the bronchial tree (also referred as *conducting zone*) and the acini, the *respiratory zone*. The very beginning of the bronchial tree is composed of the trachea which is directly connected to the larynx, the mouth and the nose. The trachea can be seen as the *root* of our tree, which then ramifies into two bronchi that will split again and again until about 23 divisions [21]. During each division, the dimensions of the children are smaller compared to the parent, according to Weibel's model, the reduction factor between each split is around 0.79 [7,12–14]. After the very first ramification, the two bronchi leads to the left and the right lung. Inside the lungs, the ventilation takes place. The lungs are inflated and deflated thanks to the respiratory muscles. Their role is to transport the air deep enough in the lung so that the gas exchanges between air and blood could occur efficiently.

Some diseases affect the physiological behavior of the bronchial tree and of the lungs. In asthma, a general shrinking of the bronchi happens and the patient feels a lack of breath due to the increased total resistance of the bronchial tree. In mechanical terms, the patient will need a greater muscular effort in order to provide adequate pressure for restoring a normal flow within the lungs.

In cystic fibrosis, there is an accumulation of mucus within the bronchial tree that will impact the capacity of the lungs to inflate and deflate, hence its rigidity. Normally, it is harder to breath for patients with cystic fibrosis because of the increased rigidity of their lungs.

The lungs mechanical properties can be used to build a mathematical model that can mimic the respiratory system.

1.2   Mathematical Modeling

The more tractable model to mimic the lung mechanics and ventilation is to mimic separately its resistive and elastic parts. We represent the resistive tree using a rigid tube with given length l and radius r. The resistance R of such a tube can be calculated by using Poiseuille's law that depends on the air viscosity μ_{air} [11].

$$R = \frac{8\mu_{air}l}{\pi r^4} \qquad (1)$$

The elastic part of the lung can be mimicked with an elastic balloon with elastance property E. Figure 1a is a representation of such a model. However, for this study, the model used is slightly more sophisticated to get a better representation of the distribution of the ventilation, see Fig. 1b. The profile of the pressure used to mimic the muscular action is taken from L. Hao et al. [2] and represents a standard for tidal breathing. It is necessary to highlight that the hypotheses of linearity used in this model are respected in the regime of tidal breathing [18].

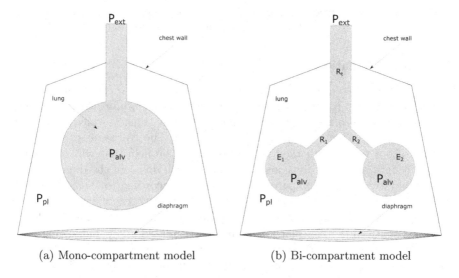

(a) Mono-compartment model   (b) Bi-compartment model

Fig. 1. (a) Mono-compartment model of the lung. The Bronchial tree is collapsed in the tube having the total resistance R and the alveoli are mimicked by balloons characterized by their elastance E. (b) Parallel bi-compartment model. This model better respects the anatomy of the respiratory system.

The fundamental equation that links the resistance R and the elastance E to the Pressure P, Volume V and its derivative in time $\dot{V}$ can be easily derived from the Mono-compartment model:

$$P_{ext} - P_{alv} = \Delta P(t) = R\dot{V}(t) \qquad (2)$$

$$P_{alv} - P_{pl} = P_{el}(t) = EV(t) \tag{3}$$

P_{el} represents the pressure drop between the acini and the pleural space, this depends on the elastance of the compartment. ΔP is the air pressure drop between the airways opening and the acini and it takes into account the resistance of the airways and the parenchyma. The total pressure drop of the model is the sum of the two contributions:

$$P(t) = P_{el} + \Delta P(t) \tag{4}$$

This equation holds true regardless of whether $P(t)$ is applied at airways' opening or at the outside of the elastic compartment [18].

In the parallel model, Fig. 1b, there are two governing equations, one for each compartment, respectively of volume V_1 and V_2:

$$\begin{cases} P(t) = E_1 V_1(t) + (R_1 + R_t)\dot{V}_1(t) + R_t \dot{V}_2(t) \\ P(t) = E_2 V_2(t) + (R_2 + R_t)\dot{V}_2(t) + R_t \dot{V}_1(t) \end{cases} \tag{5}$$

where R_1 and R_2 refers to the resistances of each bronchi, R_t is the resistance of the trachea and E_1 and E_2 are the elastances of the left and right lung, respectively, see Fig. 1b. These are the parameters of the model. $V_1(t)$ and $V_2(t)$ are the volumes associated to each lung and $P(t)$ is the muscular pressure that drives the lung ventilation.

Let us take the derivative of each equation:

$$\begin{cases} \dot{P}(t) = E_1 \dot{V}_1(t) + (R_1 + R_t)\ddot{V}_1(t) + R_t \ddot{V}_2(t) \\ \dot{P}(t) = E_2 \dot{V}_2(t) + (R_2 + R_t)\ddot{V}_2(t) + R_t \ddot{V}_1(t) \end{cases} \tag{6}$$

We substitute $\ddot{V}_2$ from Eq. (6)a into Eq. (6)b and replace $\dot{V}_2$ with its expression derived in Eq. (5)a. The equation for compartment 1 alone is:

$$\begin{aligned} R_2 \dot{P}(t) + E_2 P(t) = {} & \left[R_1 R_2 + R_t(R_1 + R_2) \right] \ddot{V}_1(t) \\ & + \left[(R_2 + R_t)E_1 + (R_1 + R_t)E_2 \right] \dot{V}_1(t) + E_1 E_2 V_1(t) \end{aligned} \tag{7}$$

Because the model is symmetric, the equation for compartment 2 is the same as (7) with inverted indexes 1 and 2. Then remembering that $V(t) = V_1(t) + V_2(t)$ the referral equation for the bi-compartment parallel model is:

$$\begin{aligned} (R_1 + R_2)\dot{P}(t) + (E_1 + E_2)P(t) = {} & \left[R_1 R_2 + R_t(R_1 + R_2) \right] \ddot{V}(t) \\ & + \left[(R_2 + R_t)E_1 + (R_1 + R_t)E_2 \right] \dot{V}(t) \\ & + E_1 E_2 V(t) \end{aligned} \tag{8}$$

2   Methods

2.1   Creation of the Dataset

It is possible to mimic the behavior of healthy subjects by setting physiological values of $R_{eq} = R_t + \frac{R_1 R_2}{R_1 + R_2}$ and $E_{eq} = \frac{E_1 E_2}{E_1 + E_2}$. In the literature, they are set to: $R_{eq} = 3\ cmH_2O/L/s$ and $E_{eq} = 10\ cmH_2O/L$ [2]. In this work, we mimic cystic fibrosis by doubling the healthy elastance (doubling the rigidity of the balloons): $E_{eq} = 20\ cmH_2O/L$, and asthmatic subjects by setting $R_{eq} = 5\ cmH_2O/L/s$. It is possible to follow the characteristic approach of electrical analysis [1] in which complex differential equations are studied in the frequency domain through the Laplace transform. The Laplace transform of Eq. (8) is:

$$H(s) = \frac{s(R_1 + R_2) + (E_1 + E_2)}{s^2 \left[R_1 R_2 + R_t(R_1 + R_2) \right] + s \left[(R_2 + R_t)E_1 + (R_1 + R_t)E_2 \right] + E_1 E_2} \tag{9}$$

Figure 2a shows the module and phase of the transfer function of the system in the cases of healthy, fibrosis and asthma, while Fig. 2b shows the responses of each system to physiological $P(t)$.

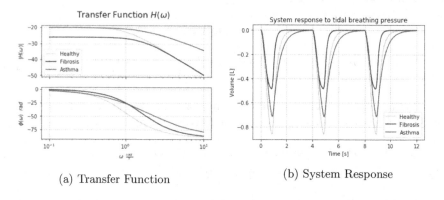

(a) Transfer Function   (b) System Response

Fig. 2. (a) Three different transfer functions, in the subplot above there is the module of the transfer function: $|H(\omega)|$ and below the phase: $\phi(\omega)$. Increasing the rigidity affects the response of the system at lower frequencies whereas increasing the total resistance affects higher frequencies. Tidal breathing happens at around 0.25 Hz being in the middle of the cutting frequence of $H(\omega)$. Consequently the output of the Volume changes. Figure (b) represents the output of the system (Volumetric signal) for one sample for each class.

Gaussian noise with mean $\mu = 0$ and standard deviation $\sigma = 0.5$ for the R_{eq} parameter and $\sigma = 5$ for the E_{eq} parameter, is added to R_{eq} and E_{eq} to mimic physiological diversity among different subjects as showed in Fig. 3a, there are 1000 samples for each class.

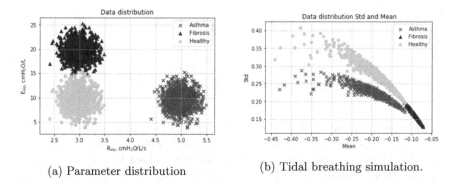

(a) Parameter distribution

(b) Tidal breathing simulation.

Fig. 3. (a) Synthetic data distribution, the three different clusters are well visible in this space. (b) Mean and Std features taken from volumetric signals of tidal breathing. These signals are the output of the model as explained in Eq. (8)

2.2 Training Machine Learning Algorithms

Before talking about the training, a short summary for each classifier used is reported. The implementation has been done using Python and the open-source library scikit-learn.

Naive Bayes. Naive Bayes classifier is a classifier that naively apply the Bayes theorem. A classifier is a function f that take an example $\boldsymbol{x} = (x_1, x_2, ..., x_n)$ where x_i is the ith feature and transform it in a class y. According to Bayes theorem the probability P of an example $\boldsymbol{x}$ being class y is:

$$P(y \mid \boldsymbol{x}) = \frac{P(\boldsymbol{x} \mid y)P(y)}{P(\boldsymbol{x})} \tag{10}$$

Assuming that all the attributes are independent, the likelihood is:

$$P(\boldsymbol{x} \mid y) = P(x_1, x_2, ..., x_n \mid y) = \prod_{i=1}^{n} P(x_i \mid y), \tag{11}$$

hence, rewriting the posterior probability:

$$P(y \mid \boldsymbol{x}) = \frac{P(y) \prod_{i=1}^{n} P(x_i \mid y)}{P(\boldsymbol{x})} \tag{12}$$

because $P(\boldsymbol{x})$ is constant with regards of y, Eq. 12 can be rewritten and used to define the Naive Bayes (NB) classifier:

$$P(y \mid \boldsymbol{x}) \propto P(y) \prod_{i=1}^{n} P(x_i \mid y)$$

$$\Downarrow \qquad\qquad (13)$$

$$\hat{y} = \arg\max_{y} P(y) \prod_{i=1}^{n} P(x_i \mid y),$$

Albeit the hypothesis of independence among attributes is never respected in real world, this classifier still has very good performance. Indeed, it has been observed that its classification accuracy is not determined by the dependencies but rather by the distribution of dependencies among all attributes over classes [6,22].

Logistic Regression. Despite its name, this is actually a classification algorithm. This is a linear classifier used normally for binary classification even though it can be extended to multiclass through different techniques like OvR (One versus Rest) or multinomial [19]. In our work, the *newton-cg* solver has been used together with *multinomial* multiclass. In this configuration, the ℓ_2 regularization is used and the solver learns a true multinomial logistic regression model using the cross-entropy loss function [15]. Using these settings allows the estimated probabilities to be better calibrated than the default "one-vs-rest" setting, as suggested in the official documentation of scikit-learn [16].

When multinomial multiclass is used, the posterior probabilities are given by a softmax transformation of linear functions of the feature variables [15]:

$$P(y_k|\boldsymbol{x}) = \frac{e^{\boldsymbol{w}_k^T \boldsymbol{x}}}{\sum_j e^{\boldsymbol{w}_j^T \boldsymbol{x}}} \qquad (14)$$

where $\boldsymbol{w} \in \mathbb{R}^n$ is the vector of trainable weights, $\boldsymbol{x}$ is the feature vector and y is the class label. Using 1-of-K encoding scheme, it is possible to define a matrix $\boldsymbol{T}$ composed by n rows (being N the total number of features for each class) and k columns (being K the total number of classes) [15]. In our case $N = 2$ and $K = 3$. Each vector $\boldsymbol{t_n}$ will have one in the position of its class and zeros all over the rest. In this scenario, the *cross-entropy* loss function to minimize for the multinomial classification regularized with ℓ_2 is:

$$\min_{\boldsymbol{w}} \left(\frac{1}{2}\boldsymbol{w}^T\boldsymbol{w} - \sum_{n=1}^{N}\sum_{k=1}^{K} t_{nk} \ln(\hat{y}_{nk}) \right) \qquad (15)$$

being $\hat{y}_{nk} = P(y_k|\boldsymbol{x_n})$.

Perceptron. For linear separable datasets, Perceptron can achieve perfect performances because it guarantees to find a solution, hence the learning rate η is not essential and by default is set to 1.0 in scikit-learn. In our implementation, the loss function is the number of mislabelled samples and it is not regularized. The weights of the model are updated on mistakes as follows:

$$w_{j+1} = w_j + \eta\big(y^{(i)} - \hat{y}^{(i)}\big)x_j^{(i)} \tag{16}$$

where i is the sample, j is the feature, y is the target and $\hat{y}$ its respective prediction [19]. The weights are updated with Stochastic Gradient Descent (SGD) optimizer, meaning that the gradient of the loss is estimated for each sample at a time and the model is updated along the way [17].

Support-Vector Machines. This is one of the most robust supervised ML algorithms, it is used for both regression and classification problems and it can be used in a non-linear fashion thanks to kernel tricks [10]. In SVMs, we used the Radial Basis Function (rbf) kernel:

$$K(X, X') = e^{\gamma\|X - X'\|^2} \tag{17}$$

When implementing this function, there are 2 parameters required:

- γ, which is the term in the expression of the rbf and it is the coefficient of multiplication for the euclidean distance.
- C, which is the error Cost, this is not directly related to the kernel function, instead it is the penalty associated with misclassified instances.

Setting these parameters together is important for achieving good results. In our case, a vast selection of pairs reports similar results, as shown in Fig. 4.

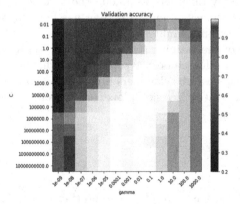

Fig. 4. Heat map for setting the best pair of γ and C in rbf kernel function for SVMs, brightest colors correspond to highest validation accuracy.

Random Forest. Random Forest (RF) is an ensemble of decision tree estimators in which each estimator classifier has been used with 100 trees and Gini impurity as criterion. The class prediction is performed by averaging the probabilistic prediction of each estimator instead of using the voting system as implemented in its original publication [4]. Random Forests follows the exhaustive search approach for the construction of each tree, where the main steps are listed in Algorithm 1 [5].

Algorithm 1. Pseudocode for tree construction - **Exhaustive search**

1: Start at the root node
2: **for** each X **do**
3:     Find the set S that minimizes the sum of the node impurities
       in the two child nodes and choose the split $S^* \in X^*$ that gives the
       minimum overall X and S
4: **if** Stopping criterion is reached **then**
5:     Exit
6: **else**
7:     Apply step 2 to each child node in turn

2.3 Training

Once the dataset is ready, simulations are performed and significant statistical features are extracted from the signals (examples of output signals of the system are shown in Fig. 2b). Here, for facilitating the graphical representation, two features are extracted: mean and standard deviation. These features have enough information to correctly differentiate among the three classes. Before training each of the previous models, standard scaling has been fit on the training set and applied on both training and test sets. The dataset has been split randomly by keeping the size of the training set at 80% of the total dataset. Hence performances have been evaluated on 800 samples after having checked the correct balance among the classes. The classifiers previously reported have been trained and tested and their relative decision plots can be seen in Fig. 7.

3 Results

3.1 Lung Model

The parallel model used is a good representation of the lung when detailed geometrical characteristics are not important to model. Working with this model allows to control the resistance and elastance of the respiratory system, allowing the simulation of certain diseases. However, it is important to ensure that the results given by our model are coherent with reality. Because of this, the output signals of volumes and flows have been observed and visually compared with real signals, see Fig. 2b. The flow Φ has been calculated as $\Phi(t) = \partial V(t)/\partial t$. Pressure-Volume plots and Flow-Volume plots have been evaluated for each class, see Fig. 5. In typical Pressure-Volume plots a decreasing of compliance is manifested as a shift on the right of the loop. However the model that we are using is pressure driven. In a pressure driven model, when the compliance decreases, the pressure control yields less and less volume for the same pressure level, causing a lowering of the loop curve. Asthmatic subjects on the other hand are simulated as having same compliance as healthy subjects but greater resistance, and because of this, their flow is lower than the others in a pressure driven model as visible in Fig. 5b.

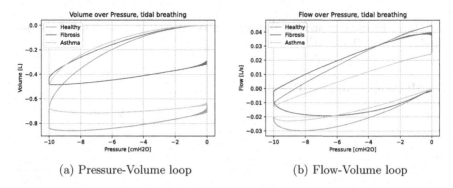

(a) Pressure-Volume loop   (b) Flow-Volume loop

Fig. 5. (a) Pressure-Volume loop and (b) Flow-volume loop of synthetic data for healthy and diseases.

The model performs a transformation $(R_{eq}, E_{eq}) \in \mathbb{R}^2 \to \mathbb{S}$, with $\mathbb{S} \subseteq \mathbb{R}$, which is the signal space, in our case the signal is $V(t) \in \mathbb{S}$. By extracting features from the signal space, we pass to $\mathbb{R}^N$, where N is the number of features extracted. Here, we have extracted two features, therefore we have a mapping $\mathbb{R}^2 : (R_{eq}, E_{eq}) \to \mathbb{R}^2 : (\mu, \sigma)$ where μ and σ are the extracted features, respectively the mean and the standard deviation of the signal. The image of this mapping is particularly useful to set limits in the prediction of the AI. Indeed, for measurements that are not contained in the image, and thus inconsistent, it makes sense to have a separate treatment that will alert when a measurement must be discarded and replaced by a new one because it is not physiological. To achieve this, a physiological region has been defined (gray area in Fig. 6a), the boundary of the rectangle region (the physiological set) is passed to the system and the output path is then patched again to form a polygon of acceptable measurements (gray area in Fig. 6b). Out of this patch of acceptable measurements, the data will not be passed to the AI for prediction and a message of "wrong acquisition" will be displayed. In Fig. 6, the data distribution is obtained with standard deviation $\sigma(R_{eq}) = 1 \; cmH_2O/L/s$ and $\sigma(E_{eq}) = 3.5 \; cmH_2O/L$ to allow for a better spreading and overlapping. Moreover, because we want the elliptic patch of each class to be comprehensive of almost all its possible samples, we fix the width and the height of the ellipse to be three times the respective standard deviation.

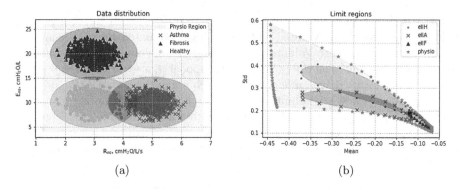

Fig. 6. Data distribution and visual representation of each set in the (R_{eq}, E_{eq}) space (a) and in the (μ, σ) space (b)

3.2   Machine Learning Results

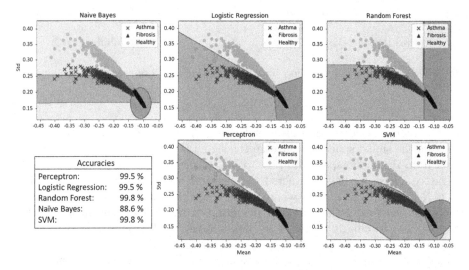

Fig. 7. Decision regions and accuracy of the implemented classifiers. The regions should be shortened according to the limitations indicated in Fig. 6b.

Multiple Machine Learning algorithms have been tested against synthetic data retrievable using spirometers. Their performances are shown in Fig. 7 in terms of accuracy and decision boundaries. In this section, there are some considerations regarding the different classifiers used. By looking at the data distribution in Fig. 3b, we observe that a linear separation will not be perfect but still good. Multinomial logistic regression and Perceptron have been used and compared for such a linear separation. Their performances are then compared with other commonly used non-linear classifiers; the first to be tried has been Naive Bayes because of its interpretability and ease of use, however its poor performance led us to try more sophisticated models like SVM with rbf kernel and RF. These latter classifiers lead to great performance, even if they have local errors close to the decision region boundaries. Nevertheless, these regions are characterized by spurious areas located on both bottom corners in the case of SVM (Fig. 7 SVM subplot) and the right-up region in the case of RF.

- **Naive Bayes:** It is possible to see how this algorithm is not suited for this kind of multi-class classification. Indeed, this classifier is usually used for binary classification (it is used a lot for spam detection). In this particular dataset, Naive Bayes fails in detecting a border between healthy and asthmatic subjects and the found boarder is not significant compared with other classifiers.
- **Logistic Regression:** This classifier is one of the most used in biomedical applications, both for its easy comprehension and its great performances when the classes are linearly separable. In this case, classes are not linearly separable. Nevertheless, this keeps a good level of generalization without renouncing to ease of usage, comprehension and good performances.
- **Random Forest:** RF is probably one of the most powerful classifiers. It is used also in biomedical applications for its good performances and its resistance to overfitting. It is an ensemble method where multiple decision trees (DTs) are singularly trained. Finally the average of the predictions of all the estimators will be used to make the decision of the RF classifier. DTs are used in medicine because of their clarity in the decision, however with RF there is a loss of this explainability in the decision, caused by an enhancing of complexity due to the ensemble.
- **Perceptron:** This is a linear model as long as only one layer is provided. It is powerful and performs very well in this particular situation. It can be useful to increase the deepness of its structure once there are lots of features and their relationships are not of easy interpretation. Also, in contrast with logistic regression it can be easily used for non-linearity.
- **Support Vector Machine:** This model is widely used in a lot of applications. In this work, the Radial Basic Function kernel has been used. Its non-linear nature allows to follow better the separation between healthy and asthma. In this dataset, this is the most performant model to distinguish these two classes. However, like the Naive Bayes it creates a green area below the red and blue zone that are incorrect.

In contrast with Deep Learning, ML models are normally faster to train. Figure 8a shows the differences of training times among the used classifiers. Even if the timings are very small, it is interesting to see, for instance, how fast the Perceptron is compared to Random Forest (RF). This is an interesting property for large datasets. Figure 8b shows the Receiver Operating Characteristic (ROC) curves and the respective calculated Area Under the Curve (AUC) for each classifier. ROC curves have been adapted for multiclass using the macro averaging technique. As expected, SVM and RF outperforms the other classifiers, however, using this metric is possible to observe the difference between Logistic Regression and Perceptron. The origin of this difference is probably on the separation between asthma and fibrosis as observable in Fig. 7.

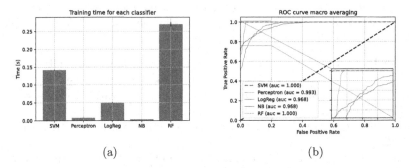

(a)   (b)

Fig. 8. (a) Different timings of training for each classifiers. Trainings are performed using the CPU runtime on Colab. (b) Macro averaging of Receiver Operating Characteristic curve for each curve, zoom on the upper left part. RF and SVM are overlapped and their value is fixed on 1.

It is worth to mention that we carried out supplementary simulations using data more spread than in Fig. 3a, which resulted in even more overlaps between the different classes (see Fig. 6). In these cases, the decision regions of each classifier stay very similar to those shown in Fig. 7 with a resulting lower accuracy due to the overlapping.

4   Conclusions and Outlook

As a brief recap, in this work the following has been done:

A second order ODE mathematical model of the lung has been used to generate synthetic data of asthma, cystic fibrosis and healthy subjects. This data has been used to train Machine Learning models. The models have been evaluated on different synthetic data sampled from the same distribution of the training set. A solution has been proposed for non physiological measurements, see Subsect. 3.1. Finally, differences among the classifiers have been studied in terms of accuracy, ROC curves and training timings.

We elaborated on the potential use of modern ML techniques to diagnose diseases of the human respiratory system. The direct conclusion of this work is the ability of ML algorithms to distinguish among linear separable clusters in the $\mathbb{R}^2$ (R_{eq}, E_{eq}) space, also in the non-linear feature space of $f(g(R_{eq}, E_{eq}))$ where $g : \mathbb{R}^2 \rightarrow \mathbb{S}$ is the parallel lung model and $f : \mathbb{S} \rightarrow \mathbb{R}^N$ is the feature extraction with N being the number of features.

One limitation of our work is the training and testing based entirely on simulated data and the utilization of the same pressure profile for all the classes. After the present proof-of-concept, future work will include training on simulated data and testing on acquired real data. Moreover, the usage of depth camera will be investigated to extract tidal breathing patterns [20]. Furthermore, the pressure was assumed to be uniform throughout the lungs and there was no difference in the application of the pressure among the three considered cases. This condition is in principle not respected because some patients can increase their muscle effort in order to keep a satisfactory ventilation. However, it is possible to assist the patients and train them to follow a specific pattern while breathing in the spirometer.

To conclude, the different ML models presented are proven in principle reliable, therefore they could provide the physicians with real-time help for the diagnosis decision.

Acknowledgements. This project has received funding from the European Union's Horizon 2020 research and innovation programme under the Marie Curie grant agreement No 847581 and is co-funded by the Région SUD Provence-Alpes-Côte d'Azur and IDEX UCA JEDI.

References

1. Otis, A.B., et al.: Mechanical factors in distribution of pulmonary ventilation. J. Appl. Physiol. **8**(4), 427–443 (1956)
2. Hao, L., et al.: Dynamic characteristics of a mechanical ventilation system with spontaneous breathing. IEEE Access **7**, 172847–172859 (2019). https://doi.org/10.1109/ACCESS.2019.2955075
3. Bodduluri, S., et al.: Deep neural network analyses of spirometry for structural phenotyping of chronic obstructive pulmonary disease. JCI Insight **5**(13) (2020). https://doi.org/10.1172/jci.insight.132781
4. Breiman, L.: Random forests. Technical Report (2001)
5. Di Dio, R.: Analyzing movement patterns to facilitate the titration of medications in late stage parkinson's disease. Master's thesis, Politecnico di Torino (2019)
6. Domingos, P., Pazzani, M.: On the optimality of the simple Bayesian classifier under zero-one loss. Mach. Learn. **29**(2), 103–130 (1997). https://doi.org/10.1023/A:1007413511361
7. Weibel, E.R.: Geometry and dimensions of airways of conductive and transitory zones. In: Morphometry of the Human Lung, Springer, Berlin (1963)
8. Philip, M., et al.: Diagnosis of cystic fibrosis: consensus guidelines from the cystic fibrosis foundation. J. pediatr. **181**, S4–S15 (2017). https://doi.org/10.1016/j.jpeds.2016.09.064
9. Gonem, S.: Applications of artificial intelligence and machine learning in respiratory medicine. Thorax **75**(8), 695–701 (2020). https://doi.org/10.1136/thoraxjnl-2020-214556
10. Hofmann, T., Schölkopf, B., Smola, A.J.: Kernel methods in machine learning. Ann. Statist. **36**(3), 1171–1220 (2008). https://doi.org/10.1214/009053607000000677

11. Pfitzner, J.: Poiseuille and his law. Anaesthesia **31**(2), 273–275 (1976). https://doi.org/10.1111/j.1365-2044.1976.tb11804.x
12. Horsfield, K., Cumming, G.: Morphology of the bronchial tree in man. J. Appl. Physiol. **24**(3), 373–383 (1968)
13. Mauroy, B., Filoche, M., Weibel, E.R., Sapoval, B.: An Optimal Bronchial tree may be dangerous. Nature **427**(6975), 633–636. https://doi.org/10.1038/nature02287
14. Tawhai, M.H., Hunter, P., Tschirren, J., Reinhardt, J., McLennan, G., Hoffman, E.A.: Ct-based geometry analysis and finite element models of the human and ovine bronchial tree. J. Appl. Physiol. **97**(6), 2310–2321 (2004)
15. Schulz, H.: Pattern Recognition and Machine Learning (2011)
16. Scikit learn documentation - logistic regression. https://scikit-learn.org/stable/modules/linear_model.html#logistic-regression
17. Scikit learn documentation - perceptron. https://scikit-learn.org/stable/modules/linear_model.html#perceptron
18. Bates, T., Jaspm, H.: Lung Mechanics, an inverse modeling approach. Cambridge University Press, Cambridge (2009)
19. Raschka, S.: Python Machine Learning. Packt, Maharashtra (2018)
20. Wang, Y., Hu, M., Li, Q., Zhang, X.P., Zhai, G., Yao, N.: Abnormal respiratory patterns classifier may contribute to large-scale screening of people infected with COVID-19 in an accurate and unobtrusive manner (2020)
21. Weibel, E.R.: The Pathway for Oxygen. Harvard University Press, Cambridge (1984)
22. Zhang, H.: The optimality of naive Bayes. In: Technical report (2004). www.aaai.org

Long-Term Hypertension Risk Prediction with ML Techniques in ELSA Database

Elias Dritsas[✉], Nikos Fazakis, Otilia Kocsis, Nikos Fakotakis,
and Konstantinos Moustakas

Department of Electrical and Computer Engineering, University of Patras,
26504 Rion, Greece
dritsase@ceid.upatras.gr, {fazakis,okocsis,moustakas}@ece.upatras.gr,
fakotaki@upatras.gr
http://www.vvr.ece.upatras.gr/en/

Abstract. Hypertension is a leading risk factor for cardiovascular diseases (CVDs) which in their turn are among the main causes of death worldwide and public health concern, with heart diseases being the most prevalent ones. The early prediction is considered one of the most effective ways for hypertension control. Based on the English Longitudinal Study of Ageing (ELSA) [2], a large-scale database of ageing participants, a dataset is engineered to evaluate the long-term hypertension risk of men and women aged older than 50 years with Machine Learning (ML). We evaluated a series of ML prediction models concerning AUC, Sensitivity, Specificity and selected the stacking ensemble as the best performer. This work aims to identify individuals at risk and facilitate earlier intervention to prevent the future development of hypertension.

Keywords: Hypertension · Risk prediction · Supervised learning

1 Introduction

Hypertension condition, usually referred to as High Blood Pressure (HBP), is related to increased blood pressure. It may occur as a primary condition to ageing people or as a secondary condition triggered by some other disease usually occurring in the kidneys, arteries, heart or endocrine system. Isolated systolic hypertension, an elevation in systolic but not diastolic pressure, is the most prevalent type of hypertension in those aged 50 or more, occurring either from the beginning or as a development after a long period of systolic-diastolic hypertension, with or without treatment. The effect of age on blood pressure increase is mainly related to structural changes in the arteries (e.g., high arteries stiffness). In many cases, the condition is undiagnosed [25] and, the early identification of

---

Partially supported by the SmartWork project (GA 826343), EU H2020, SC1-DTH-03-2018 - Adaptive smart working and living environments supporting active and healthy ageing.

D. E. Simos et al. (Eds.): LION 2021, LNCS 12931, pp. 113–120, 2021.
https://doi.org/10.1007/978-3-030-92121-7_9

hypertension risk is necessary, as it is a primary risk factor for cardiovascular diseases.

HBP is a multifactorial disease with various risk factors [25]. Individuals with family history are at increased risk to develop hypertension, too. Moreover, the blood vessels gradually lose some of their elastic quality with ageing, which can considerably increase blood pressure. Concerning gender, men are more likely to suffer from HBP until the age of 64, while women are more likely to get HBP after the age of 65 and older. The HBP also relates to race, i.e., dark-skinned people are at higher risk of developing HBP. Besides, sedentary lifestyle increases the risk of getting high blood pressure [22]. Another crucial factor is nutrition. An unhealthy diet, especially with high sodium, calories, trans-fat and sugar, carries an additional risk for HBP [23]. In [27], authors examine the importance of early life displacement, nutrition and adult-health status in hypertension.

Overweight or obese individuals [21] are also at increased risk of developing HBP. Moreover, the excessive use of alcohol can dramatically increase blood pressure, raising the risk of hypertension [31]. Smoking, either active or passive, can temporarily raise blood pressure and lead to damaged arteries. Another significant factor concerns stress. In particular, extensive stress may increase blood pressure [5] due to harmful lifestyle habits such as poor diet, physical inactivity, and tobacco or alcohol consumption [32]. Septoe et al. [32] assessed the health behaviours in hypertensive older people in the ELSA database. They found that hypertensives are less likely to smoke than non-hypertensives, but are more likely to drink heavily and be sedentary. Finally, other chronic conditions, such as high cholesterol, diabetes [8,35] and sleep apnea [14] may increase the risk of developing HBP which is particularly common in the case of people with resistant hypertension [4].

Machine learning (ML) is a branch of artificial intelligence (AI) that is often utilized in the literature for screening or risk assessment in the cases of various chronic health conditions, such as diabetes, hypertension, cardiovascular diseases etc. In this study, we focus on hypertension disease for which numerous risk prediction tools have been designed. Moreover, in the relevant literature, ML techniques are suggested to tackle the limitations of risk score systems which are validated in different populations or cohorts in works such as [15,29].

The main contributions of this work are: i) the construction of a balanced dataset that is derived from the ELSA database, ii) a comparative evaluation of different ML models, and iii) the proposal of a stacking ensemble for the long-term hypertension risk prediction of older people aged at least 50 years. Besides, the generated dataset may contribute to the prognosis of hypertension as we choose to monitor the attributes' values of individuals who, in reference waves, have not been diagnosed with hypertension. Finally, these models are part of the predictive AI tools integrated in the SmartWork system, which aims to sustain workability of older office workers [17]. Specifically, the workability modeling will be based on personalized models obtained from integration of the health condition of specific patient models. Hypertension is one of the chronic conditions that will be considered in SmartWork. For the workability modeling, several other elements should be taken into account, as analyzed in [16], however, they are out of the scope of the current study.

The rest of this paper is organized as follows. Section 2 presents the main parts of the methods for the long-term risk prediction of hypertension. In particular, the design of a training and testing dataset, feature selection and the experiments set up are analyzed. Section 3 concludes the paper.

2   Methods for the Long-Term Risk Prediction

The importance of HBP as a risk factor for cardiovascular disease [4] has driven the implementation of risk prediction tools, as summarized in Table 1, and several ML-based approaches for hypertension prediction [9,26].

Table 1. HBP risk scoring systems.

Model	Tool Risk factors included
Framingham hypertension risk score [15]	Age, Gender, BMI, Systolic Blood Pressure (SBP), Diastolic Blood Pressure (DBP), parental hypertension, Smoking
A Hypertension Risk Score for Middle-Aged and Older Adults [18]	Age, SBP or DBP, Smoking, family history of hypertension, T2DM, BMI, the age–DBP interaction, Gender, Physical activity
A point-score system for predicting incident hypertension [3]	Gender, Age, Family history of premature CVDs, SBP, DBP
Chien et al. [7]: Chinese and Taiwan HBP clinical risk model	Age, Gender, BMI, SBP and DBP
Lim et al. 2013, [24]: Korean risk model for incident hypertension	Age, Gender, Smoking, SBP, DBP, parental hypertension, BMI
Fava et al. 2015, [10]: Swedish risk model	Age, Gender, heart rate, obesity (BMI >30Kg/m^2), diabetes, hypertriglyceridemia, prehypertension, family history of hypertension, sedentary behaviour, alcohol, marital status, work type, (white collar), smoking

In this work, we aim to correctly classify each instance in ELSA Database as either *hypertensive* or *non-hypertensive* and achieve high sensitivity and Area Under Curve (AUC) through supervised ML, meaning that the hypertension class can be predicted correctly. Note that, the hypertension or non-hypertension class labels are indicated by the follow-up assessment after 2-years. In the following sections, we are demonstrating the main components of the process.

2.1   Training and Test Dataset

In the ELSA database, a hybrid data collection is conducted, using both questionnaires (non-invasive way) and clinical tests (invasive way) [20]. Except from questionnaires, tests were also performed by clinical nurses, including blood tests, blood pressure measurements, and other frailty-related tests (e.g., mobility). The training and test dataset for the hypertension prediction model was derived from the ELSA database, based on waves 2, 4 and 6 as reference ones, and the corresponding waves 3, 5 and 7 for the follow-up assessment. Note that, participants that already had HBP at the reference waves were excluded from the engineered dataset. However, the population distributions were not well aligned with the reported prevalence of hypertension for the respective age groups. Thus, the engineered dataset was balanced using random undersampling, in order to be inline with the European level of the HBP prevalence namely, 27% for people younger than 44 years, 40% for those in the age group of 45–54, 60% for those in the age group of 55–64, 78% for those in the age group of 65–74, and over 80% for those older than 75 years [34]. The distributions of participants per age group in the balanced dataset satisfying similar to the aforementioned criteria, are shown in Table 2.

Table 2. Distribution per age group of newly diagnosed HBP at 2-years follow-up in the balanced dataset.

	HBP	50–54	55–59	60–64	65–69	70–74	75+	Total
Ref Wave 2	No	58	137	62	56	46	63	422
F-up wave 3	Yes	23	82	37	44	36	49	271
Ref wave 4	No	83	92	123	76	50	55	479
F-up wave 5	Yes	33	55	74	59	39	43	303
Ref wave 6	No	43	53	70	47	55	59	327
F-up wave 7	Yes	17	32	42	37	43	46	217
All waves	No	184	282	255	179	151	177	1228
	Yes	73	169	153	140	118	138	791

2.2 Feature Selection

The initial features set considered for the training of the ML-based models included 106 variables, with 61 being categorical and 45 numeric attributes, among those collected at the reference waves of the ELSA dataset. To avoid overfitting, we reduce the dataset dimensionality, and thus complexity, considering only the relevant features to the hypertension class. From the wrapper feature selection methods, we used the stepwise backward elimination using Logistic model with Ridge regularization (L2-penalty) [30], namely, $Loss = \sum_{i=1}^{M}(h_i \boldsymbol{f}_i \boldsymbol{b} - log(1 + e^{\boldsymbol{f}_i \boldsymbol{b}})) + \lambda \sum_{i=1}^{p} b_i^2$, where h_i is the target binary variable for the (non-)hypertension class. Also, $\boldsymbol{b} = \begin{bmatrix} b_1, b_2, \ldots, b_p \end{bmatrix}^T$ are the regression weights attached to features row vector $\boldsymbol{f}_i = \begin{bmatrix} f_{i1}, f_{i2}, f_{i3}, \ldots, f_{ip} \end{bmatrix}^T, i = 1, 2, \ldots, M$ (M denotes the dataset size with $M \gg p$), and $\lambda \geq 0$ is the regularization parameter that controls the significance of the regularization term.

The final list of features was reduced to 18 attributes, among which the most important are: age, gender, weight(Kg) and BMI, ethnicity, cholesterol levels (total, HDL and LDL-mg/dL), work behaviour (sedentary versus non-sedentary work), drinking and smoking habits, systolic and diastolic blood pressure (mmHg), heart rate (bpm), physical activity, diagnosis of other chronic conditions (e.g. diabetes, cholesterol), perceived job pressure, and self-assessed health status.

2.3   Performance Evaluation of ML Models

In this section, the performance of several ML models is evaluated using 3-cross validation experimentation setup on the engineered balanced dataset. Logistic Regression (LR) [28], Naïve Bayes (NB) [19], k-Nearest Neighbors with $k = 5$ (5NN), Decision Trees (DT), Random Forests (RFs) [6] and a Stacking ensemble were applied. As base classifiers for the stacking ensemble the Logistic Regression with Ridge estimator (LRR) and the RFs models were used, while as a meta-classifier LRR was also applied [12]. Moreover, using the Wald test statistic [13], we determined the discrimination ability of models based on the ELSA dataset. All tests indicated the significance of the calculated Areas Under Roc Curve (AUC), as all p-values were 0 (<0.05).

Table 3. Performance of ML models for hypertension risk prediction.

	LR [18]	LRR	NB	5NN	DT	RFs	Stacking
AUC	0.807	0.816	0.758	0.653	0.692	0.798	0.823
	(0.76 0.85)	(0.77 0.86)	(0.71 0.81)	(0.63 0.75)	(0.63 0.75)	(0.70 0.85)	(0.78 0.87)
Sens	0.819	0.782	0.702	0.513	0.815	0.689	0.756
	(0.76 0.87)	(0.72 0.83)	(0.64 0.76)	(0.45 0.56)	(0.76 0.86)	(0.63 0.75)	(0.70 0.81)
Spec	0.702	0.748	0.710	0.702	0.611	0.794	0.786
	(0.62 0.78)	(0.67 0.82)	(0.62 0.79)	(0.62 0.78)	(0.52 0.70)	(0.71 0.86)	(0.71 0.85)

In Table 3, the first LR method applied on the engineered dataset is based on [18] and is used as a benchmark comparison model. It utilizes the same features of this work except for the hypertension family history variable, as it is not available in ELSA. The second LRR model has considered the features selected with stepwise backward elimination, as discussed in Sect. 2.2; the same holds true for the rest of the ML models. Several performance metrics are utilized to assess the risk prediction performance of previous ML models. AUC, Sensitivity (Sens) and Specificity (Spec) along with their deviations [33] are considered to assess the risk prediction performance of previous ML models in hypertension screening. The results, summarized in Table 3, reveal the heterogeneity in performance among different ML algorithms in the ELSA dataset. Finally, in the current dataset, the quantitative analysis of the risk models showed that the predictive ability of ensemble machine learning with stacking is promising and superior in terms of AUC against the single ML models.

3   Conclusions

In conclusion, we employed ML to develop a high precision prediction model for hypertension prognosis in the elderly using the stacking ensemble methodology. In comparison with other works, the main limitation of this work is the size of the dataset. As ELSA database is on-going, more data on hypertension condition will be available for experimentation in the future studies. Also, the direct comparison of the suggested ML-based risk models was also limited by the lack of some relevant information (such as microalbumin, urine albumin creatinine ratio) or different populations between our dataset and other published studies [1,20]. However, the current analysis, in combination with diagnostic tools, may assist clinicians to make personalized lifestyle modifications or treatment decisions to prevent or delay the development of hypertension. As future work, handling of missing values with techniques such as [11] can be explored and thus further improve the prediction abilities of the proposed methodology. Moreover, the semi-supervised learning schemes can also be exploited in order to take advantage of the vast unlabeled instances available in ELSA and thus further improving knowledge extraction from the engineered dataset.

References

1. AlKaabi, L.A., Ahmed, L.S., Al Attiyah, M.F., Abdel-Rahman, M.E.: Predicting hypertension using machine learning: Findings from qatar biobank study. PLoS One **15**(10), e0240370 (2020)
2. Banks, J., et al.: ELSA English Longitudinal Study of Ageing English Longitudinal Study of Ageing: Waves 0-9, 1998-2019 (2021). https://doi.org/10.5255/UKDA-SN-5050-22, https://beta.ukdataservice.ac.uk/datacatalogue/doi/?id=5050
3. Bozorgmanesh, M., Hadaegh, F., Mehrabi, Y., Azizi, F.: A point-score system superior to blood pressure measures alone for predicting incident hypertension: tehran lipid and glucose study. J. Hypertens. **29**(8), 1486–1493 (2011)
4. Carey, R.M., et al.: Resistant hypertension: detection, evaluation, and management: a scientific statement from the American heart association. Hypertension **72**(5), e53–e90 (2018)
5. Chamik, T., Viswanathan, B., Gedeon, J., Bovet, P.: Associations between psychological stress and smoking, drinking, obesity, and high blood pressure in an upper middle-income country in the African region. Stress Health **34**(1), 93–101 (2018)
6. Chang, W., et al.: A machine-learning-based prediction method for hypertension outcomes based on medical data. Diagnostics **9**(4), 178 (2019)
7. Chien, K.L., et al.: Prediction models for the risk of new-onset hypertension in ethnic Chinese in Taiwan. J. Hum. Hypertens. **25**(5), 294–303 (2011)
8. De Boer, I.H., et al.: Diabetes and hypertension: a position statement by the American diabetes association. Diabetes Care **40**(9), 1273–1284 (2017)
9. Echouffo-Tcheugui, J.B., Batty, G.D., Kivimäki, M., Kengne, A.P.: Risk models to predict hypertension: a systematic review. PloS one **8**(7), e67370 (2013)
10. Fava, C., et al.: A genetic risk score for hypertension associates with the risk of ischemic stroke in a Swedish case-control study. Eur. J. Hum. Genet. **23**(7), 969–974 (2015)

11. Fazakis, N., Kostopoulos, G., Kotsiantis, S., Mporas, I.: Iterative robust semi-supervised missing data imputation. IEEE Access **8**, 90555–90569 (2020). https://doi.org/10.1109/ACCESS.2020.2994033
12. Fitriyani, N.L., Syafrudin, M., Alfian, G., Rhee, J.: Development of disease prediction model based on ensemble learning approach for diabetes and hypertension. IEEE Access **7**, 144777–144789 (2019)
13. Goksuluk, D., Korkmaz, S., Zararsiz, G., Karaagaoglu, E.: easyROC: an interactive web-tool for ROC curve analysis using R language environment. R J. **8**, 213–230 (2016). https://doi.org/10.32614/RJ-2016-042
14. Hou, H., et al.: Association of obstructive sleep apnea with hypertension: a systematic review and meta-analysis. J. Glob. Health **8**(1), 010405 (2018)
15. Kivimaki, M., et al.: Validating the Framingham hypertension risk score: results from the Whitehall ii study. Hypertension **54**(3), 496–501 (2009)
16. Kocsis, O., et al.: Conceptual architecture of a multi-dimensional modeling framework for older office workers. In: Proceedings of the 12th ACM International Conference on PErvasive Technologies Related to Assistive Environments, pp. 448–452 (2019)
17. Kocsis, O., et al.: Smartwork: designing a smart age-friendly living and working environment for office workers. In: Proceedings of the 12th ACM International Conference on PErvasive Technologies Related to Assistive Environments, pp. 435–441 (2019)
18. Kshirsagar, A.V., et al.: A hypertension risk score for middle-aged and older adults. J. Clin. Hypertens. **12**(10), 800–808 (2010)
19. Kublanov, V.S., Dolganov, A.Y., Belo, D., Gamboa, H.: Comparison of machine learning methods for the arterial hypertension diagnostics. Appl. Bionics Biomech. **2017** (2017)
20. LaFreniere, D., Zulkernine, F., Barber, D., Martin, K.: Using machine learning to predict hypertension from a clinical dataset. In: 2016 IEEE symposium series on computational intelligence (SSCI), pp. 1–7. IEEE (2016)
21. Landi, F., et al.: Body mass index is strongly associated with hypertension: Results from the longevity check-up 7+ study. Nutrients **10**(12), 1976 (2018)
22. Lee, P.H., Wong, F.K.: The association between time spent in sedentary behaviors and blood pressure: a systematic review and meta-analysis. Sports Medicine **45**(6), 867–880 (2015)
23. Lelong, H., Galan, P., Kesse-Guyot, E., Fezeu, L., Hercberg, S., Blacher, J.: Relationship between nutrition and blood pressure: a cross-sectional analysis from the nutrinet-santé study, a french web-based cohort study. Am. J. Hypertens. **28**(3), 362–371 (2015)
24. Lim, N.K., Son, K.H., Lee, K.S., Park, H.Y., Cho, M.C.: Predicting the risk of incident hypertension in a Korean middle-aged population: Korean genome and epidemiology study. J. Clinic. Hypertens. **15**(5), 344–349 (2013)
25. Lionakis, N., Mendrinos, D., Sanidas, E., Favatas, G., Georgopoulou, M.: Hypertension in the elderly. World J. Cardiol. **4**(5), 135 (2012)
26. López-Martínez, F., Schwarcz, A., Núñez-Valdez, E.R., Garcia-Diaz, V.: Machine learning classification analysis for a hypertensive population as a function of several risk factors. Exp. Syst. Appl. **110**, 206–215 (2018)
27. McEniry, M., Samper-Ternent, R., Flórez, C.E., Cano-Gutierrez, C.: Early life displacement due to armed conflict and violence, early nutrition, and older adult hypertension, diabetes, and obesity in the middle-income country of Colombia. J. Aging Health **31**(8), 1479–1502 (2019)

28. Nusinovici, S., et al.: Logistic regression was as good as machine learning for predicting major chronic diseases. J. Clinic. Epidemiol. **122**, 56–69 (2020)
29. Otsuka, T., et al.: Development of a risk prediction model for incident hypertension in a working-age Japanese male population. Hypertens. Res. **38**(6), 419–425 (2015)
30. Pereira, J.M., Basto, M., da Silva, A.F.: The logistic lasso and ridge regression in predicting corporate failure. Proc. Econ. Finance **39**, 634–641 (2016)
31. Santana, N.M.T., et al.: Consumption of alcohol and blood pressure: results of the elsa-brasil study. PLoS One **13**(1), e0190239 (2018)
32. Steptoe, A., McMunn, A.: Health behaviour patterns in relation to hypertension: the english longitudinal study of ageing. J. Hypertens. **27**(2), 224–230 (2009)
33. Trevethan, R.: Sensitivity, specificity, and predictive values: foundations, pliabilities, and pitfalls in research and practice. Front. Public Health **5**, 307 (2017)
34. Wolf-Maier, K., et al.: Hypertension prevalence and blood pressure levels in 6 European countries, Canada, and the united states. Jama **289**(18), 2363–2369 (2003)
35. Zaki, N., Alashwal, H., Ibrahim, S.: Association of hypertension, diabetes, stroke, cancer, kidney disease, and high-cholesterol with COVID-19 disease severity and fatality: a systematic review. Diab. Metab. Syndr. Clinic. Res. Rev. **14**(5), 1133–1142 (2020)

An Efficient Heuristic for Passenger Bus VRP with Preferences and Tradeoffs

Suhendry Effendy, Bao Chau Ngo, and Roland H. C. Yap[(✉)]

National University of Singapore, 13 Computing Drive, Singapore, Singapore
{effendy,ngobc,ryap}@comp.nus.edu.sg

Abstract. One category of vehicle routing problems (VRP) involving groups of people are School Bus Routing Problems (SBRP). In this paper, we investigate a form of SBRP where passengers can be dropped off at a number of possible locations. If the location is not the home address, they continue on foot. Different drop-off choices allow for different set of passengers to be dropped off together which affects the total vehicle driving distance. There is an inevitable trade-off between two goals, reducing driving versus walking distance. Unlike typical SBRP, the set of passengers are not known well in advance and there is little time to compute a solution, hence, efficiency is more important than optimality. In this short paper, we propose a model with an efficient incremental algorithm for such problems. We demonstrate the efficacy with experiments on real-world datasets with quick solving while balancing trade-offs.

1 Introduction

Vehicle Routing Problems (VRP) [3] are generally NP-hard problems common in logistics and transportation. There are a wealth of VRP variants; Kilby and Shaw [3] highlights many real-world problems that have not been studied in academia. In this short paper, we are concerned with novel VRPs exemplified by the following example. Consider a transportation problem—a number of employees are transported home from a depot by buses. The set of passengers vary daily due to differing shifts or schedules among the employees. Passengers can either be dropped off at their home address or at some preferred (close-by) stops. If the passenger is not dropped off at their home, they can walk home from the stop. Flexibility in choosing stops allows multiple passengers to be dropped off together at a stop even though they have different home addresses—this has the advantage of reducing the bus driving distance. Two trade-offs arise: (i) the fewer the drop-off locations, the shorter the bus travel distance and time; but (ii) it means some passengers have to walk. The primary goal of the logistics provider is to reduce costs, i.e. driving distance. Conversely, the goal of individual passengers would be to reduce any walking distance. These two objectives naturally conflict. A practical transportation service needs to balance this trade-off (CHALLENGE #1). Additionally, in our problem, the set of passengers is not

D. E. Simos et al. (Eds.): LION 2021, LNCS 12931, pp. 121–127, 2021.
https://doi.org/10.1007/978-3-030-92121-7_10

known until a registration cut-off time. To allow for more passengers, the cut-off and departure times are close in time. This requires the routing algorithm to be efficient (CHALLENGE#2) with a strict computation time limit, trading optimality for faster running time. Our problem is a variation of the School Bus Routing Problem (SBRP) [1,6] but not in a school bus setting. It adds passenger preferences with a more complex goal requiring balancing opposing objectives so that both logistics operator and passengers are "happy".

In this paper, we propose a general form of set cover which maps to our SBRP problem. This leads to an efficient incremental algorithm to find solutions with different number of stops which in turn allows for a trade-off analysis between driving distance and individual walking distance. We show our incremental algorithm can be on par with VRP solvers in solution quality while being considerably faster on real-world datasets. We also show that trade-offs giving a value proposition for both the service provider and passengers can be efficiently obtained.

2   Related Work

The School Bus Routing Problem (SBRP) is originally about school bus transport for students where multiple students can be assigned to bus stops. Several surveys on SBRP are available [1,6]. SBRP differs from other VRPs as passengers/students are clustered to stops. Many formulations divide the problem into a stop selection phase and a VRP generation phase. We differ from most studied SBRPs as it is not a school bus setting with additional requirements.

The General Vehicle Routing Problem (GVRP) [2] is an extension of the classical VRP where the destination nodes are partitioned into multiple clusters. The goal of GVRP is to find a set of routes which visit each cluster exactly once [5,7–9]. Both SBRP and GVRP have clusters, but in SBRP (and our problem) the clusters are not necessarily mutually exclusive.

Recently, Lewis, et al. [4] proposed a heuristic-based iterated local search algorithm for SBRP. It differs from our problem as they do not consider passenger preferences. In this paper, we also need to deal with the trade-off between different objective functions, thus, a fine-grain algorithm is more preferable for our case. Unlike many SBRP, we need to deal with passenger preferences and have a more complex optimization goal trading off median/average passenger walking time with driving distance, whereas in typical SBRP, minimizing the number of buses or driving distance is common.

3   Preliminaries

Let $C = \{c_1, c_2, \ldots\}$ be a set of passengers. Each passenger c has a set of preferred stops represented by a well-ordered set $T^{(c)} = \{t_1^{(c)}, t_2^{(c)}, \ldots\}$ with ordering $t_i^{(c)} \prec_{(c)} t_j^{(c)}$ iff $i < j$. Let $S \subset \bigcup_c T^{(c)}$ be the set of chosen stops.

Definition 1. *Set S is **satisfying** if and only if $T^{(c)} \cap S \neq \varnothing$ for all $c \in C$.*

Let $v(c, S)$ denotes the *preferred stop* of c with respect to S, namely, $v(c, S) = \min(T^{(c)} \cap S)$ with respect to the ordering $\prec_{(c)}$ on S. Let $\Upsilon(C, S)$ denotes the set of all chosen stops of C with respect to S, i.e. $\Upsilon(C, S) = \{ v(c, S) : c \in C \}$.

Definition 2. *Set S is **fully-utilized** iff S is satisfying and $|S| = |\Upsilon(C, S)|$.*

Each stop in a fully-utilized S is a preferred stop of some passenger(s) in C.

The vehicles routes to visit all locations in a fully-utilized set S from a depot d is represented by $\mathcal{R}(S) = \{R_1, R_2, \dots\}$. The total distance of all the routes in $\mathcal{R}(S)$ is $\text{cost}(S)$, and the distance between location a and b is denoted by $\delta(a, b)$.

4 An Incremental Algorithm

Our problem is an SBRP with the addition of preferences and two challenges: (i) (CHALLENGE#1) a trade-off analysis to find a solution which can suit both the logistics provider and the passengers; (ii) (CHALLENGE#2) a strict computational time limit. We present a hybrid local search algorithm which uses a number of heuristics. It gives a set of k chosen stops and routes *for each* integer k, allowing a trade-off analysis between vehicle driving distance and passenger walking distance to be performed. It is efficient as we have a low polynomial time and incremental algorithm on k.

The algorithm starts with an *initial chosen set of stops* which is obtained by aggregating all the most preferred stop from each passenger, i.e. $S_0 = \bigcup \{t_1^{(c)}\}$. Observe that S_0 is fully-utilized so all the preferences are met. No stop selection or trade-off analysis is needed with S_0 since all passengers get their best choice. Next, we present a hybrid algorithm which computes a new set of stops S_i where $|S_i| < |S_{i-1}|$. Any off-the-shelf VRP algorithm is used to obtain the initial $\mathcal{R}(S_0)$. An incremental local search is used to obtain S_i and $\mathcal{R}(S_i)$ for subsequent $i > 0$.

We *repeatedly decrease* the size of S with local moves REMOVEONE or IMPLODE (greedily picking the one with lower cost) and produce the routes at each iteration. Each local move decreases the size of the current set of stops by one so that a route is generated for each S_i. The algorithm terminates when it cannot further reduce the size of S or when some other stopping criterion is met. We propose the following two local moves.

A. REMOVEONE*(S_i) Operation.* Find a fully-utilized set S_{i+1} such that $S_{i+1} = S_i \setminus \{x^*\}$ where $x^* \in S_i$ and $\text{cost}(S_{i+1})$ is minimum among all possible S_{i+1}, so $S_{i+1} \subset S_i$.

This operation considers all S_{i+1} that can be obtained by removing exactly one stop from S_i. For a tuple $\langle x \rangle$ where $x \in S_i$, the corresponding route and cost of removing x are constructed incrementally from $\mathcal{R}(S_i)$ and $\text{cost}(S_i)$.

Let p_i, x, p_j be three locations in $S_i \cup \{d\}$ that are visited consecutively by the same vehicle in $\mathcal{R}(S_i)$. The new route is obtained by decoupling x from $\mathcal{R}(S_i)$, i.e. alter the route $p_i \to x \to p_j$ in $\mathcal{R}(S_i)$ into $p_i \to p_j$ in $\mathcal{R}(S_{i+1})$. The cost of $S_i \setminus \{x\}$ can be computed as follows.

$$\text{cost}(S_i \setminus \{x\}) = \text{cost}(S_i) - \delta(p_i, x) - \delta(x, p_j) + \delta(p_i, p_j) \tag{1}$$

The tuple $\langle x^* \rangle$ is $\langle x \rangle$ with the least $\text{cost}(S_i \setminus \{x\})$.

The time complexity for this operation is $O(|S|^2 |C|)$ as there are $O(|S_i|)$ possible tuples and $O(|S||C|)$ for each tuple to check the fully-utilized constraint.

B. IMPLODE(S_i) *Operation.* Find a fully-utilized set S_{i+1} such that $S_{i+1} = (S_i \setminus \{x^*, y^*\}) \cup \{z^*\}$ where $x^*, y^* \in S_i$, $x^* \neq y^*$, $z^* \notin S_i$ and $\text{cost}(S_{i+1})$ is minimum among all possible S_{i+1}. Note that with this move, the resulting S_{i+1} is not a subset of S_i.

This operation consists of two remove operations and one insert operation. For each tuple $\langle x, y, z \rangle$ where $x, y \in S_i$ and $z \notin S_i$, the corresponding route and cost for $(S_i \setminus \{x, y\}) \cup \{z\}$ are constructed incrementally from $\mathcal{R}(S_i)$ and $\text{cost}(S_i)$ in three steps, i.e. remove x, remove y, insert z, one step at a time.

The cost of removing x and y (and their routes) can be computed similarly to the REMOVEONE operation. Let $S_i'' = S_i \setminus \{x, y\}$ and $\mathcal{R}(S_i'')$ be its routes. The location z is inserted into $\mathcal{R}(S_i'')$ such that the resulting cost is minimum while the route changes are minimal, i.e. z is only inserted between any two consecutive locations in $\mathcal{R}(S_i'')$. Let p_i and p_j be two locations that are visited consecutively by the same vehicle in $\mathcal{R}(S_i'')$. The cost of inserting z can be computed as follows.

$$\text{cost}(S_i'' \cup \{z\}) = \min_{\langle p_i, p_j \rangle} \{ \text{cost}(S_i'') + \delta(p_i, z) + \delta(z, p_j) - \delta(p_i, p_j) \} \quad (2)$$

To insert z between p_i and p_j, simply alter the route $p_i \to p_j$ into $p_i \to z \to p_j$. The tuple $\langle x^*, y^*, z^* \rangle$ is the tuple $\langle x, y, z \rangle$ with the least $\text{cost}((S_i \setminus \{x, y\}) \cup \{z\})$.

The time complexity to perform this operation naively is $O(|S|^3 |T||C|)$ as there are $O(|S_i|^2 |T|)$ possible tuples of $\langle x, y, z \rangle$ and $O(|S||C|)$ for each tuple to check the fully-utilized constraint. However, observe that any passenger who becomes unsatisfied due to the removal of location x and y (i.e. c such that $T^{(c)} \cap (S_i \setminus \{x, y\}) = \varnothing$) needs to have z as their preferred stop to make the set satisfying. This allows the following optimization which only considers x, y, and z which are neighbours to each other, e.g., within a certain distance. Let the number of such tuples be $\eta(S, T)$, then the time complexity to perform the IMPLODE operation can be reduced to $O(\eta(S, T)|S||C|)$. In our evaluation with real-world datasets, the $\eta(S, T)$ is much smaller compared to the number of all possible tuples.[1]

As S_0 and any other S_i for $i > 0$ in the subsequent iterations are fully-utilized sets, then any valid routes to visit all the stops in each of these sets is a valid solution for the corresponding SBRP. Additional constraints on the routes such as capacity and time-window can also be integrated into the local moves as well.

---

[1] To illustrate, compare the number of bus stops in a neighbourhood within a city. An example from the P1 dataset (Sect. 5) experiments is that on average, there are 8.5M possible tuples but only 5K tuples in $\eta(S, T)$, showing that the REMOVEONE optimization is effective.

Table 1. Summary of the evaluation on real-world datasets.

Dataset	k_0	D_0	k_{min}	D_{min}	Median W_{min}	Average W_{min}
P1	318	814.9 km	139	675.2 km (82.9%)	202.0 m	205.5 m
P2	127	407.7 km	71	359.4 km (88.2%)	199.4 m	184.6 m
P3	107	367.2 km	69	328.2 km (89.4%)	27.6 m	135.5 m
P4	214	575.2 km	104	496.9 km (86.4%)	138.0 m	171.0 m

k_0 and D_0 are the number of stops and the total driving distance for the initial solution, respectively. k_{min}, D_{min}, and W_{min} are the number of stops, the total driving distance, and the walking distance when the number of drop-off locations is minimum, respectively.

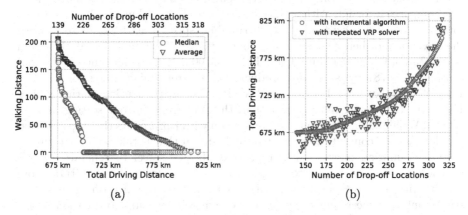

(a)   (b)

Fig. 1. Evaluation on the P1 dataset. (a) Trade-off between the total driving distance and walking distance. (b) Comparing solution quality (driving distance) of the incremental algorithm versus solving each sub-problem as a VRP instance.

5   Evaluation

We evaluate with real-world datasets. As a general VRP solver, we used jsprit[2] for $\mathcal{R}(S_0)$ (we found it gave shorter routing than OR-Tools on our instances). The real-world dataset is an employee home bus transportation problem where buses transport all the employees from a depot to their home addresses (details are confidential to the company). There are 4 different problem instances (P1 to P4). The number of passengers are: P1: 318, P2: 127, P3: 107, and P4: 214. The preferred drop-off locations for each passenger is their home address and all bus stops within 500 meters from their home, prioritized by distance, which also gives the walking preferences. There are 2631 relevant bus stops in total in each of P1-P4. We note that this is not a small VRP instance given the number of stops. All buses have 1.5 hours to complete all routes.

Table 1 summarizes the results. Overall, our algorithm reduces the number of stops by at least 35% and the total driving distance by at least 10% in these real-world instances. Figure 1 shows the evaluation result on the P1 dataset (other

---

[2] https://github.com/graphhopper/jsprit

datasets have a similar trend). Figure 1a shows the trade-off between the total driving distance and walking distance (summarized by median/average); each dot is a set of chosen stops of a certain size. To illustrate the trade-off analysis in Fig. 1a, consider the following analysis—find the set of stops such that the cost function $\alpha D_i + \beta W_i$ is minimized where D_i, W_i are the total driving distance and walking distance from the stops. Suppose $\alpha = 0.001$, $\beta = 1$, and using median walking distance for W_i, the minimum can be found when $|S| = 227$. We note that 60 REMOVEONE and 119 IMPLODE operations are used for P1.

To evaluate the quality of the produced routes, we run each chosen set of stops S_i from our algorithm on jsprit, comparing each route cost. Figure 1b shows this comparison for P1. Although our incremental algorithm is only greedy, the total driving distance is on par with running jsprit on each set of stops with the advantage of being substantially more efficient. Our incremental algorithm takes 175 s of CPU-time, on a Core(TM) i7-6700 3.40 GHz machine—the initial routing takes 78 s by jsprit and producing all the remaining routes for P1 ($k = 139\ldots317$) takes 97 s. The repeated VRP solver method using jsprit requires 16494 s of CPU-time, much larger than the time limit (in our setting, it is < 1800 s). We highlight also that the comparison with jsprit omits the stop selection problem, so the computation cost will be higher if that is taken into account.

The evaluation shows that using a general VRP solver to generate a route for each set of stops can be too expensive, e.g., two orders of magnitude more running time in our experiments. This is not practical for our problem given the strict running time requirement. The problem solving and coordination of the entire transportation schedule with all parties needs to fit within the cutoff to the departure time. More importantly, a general-purpose VRP solver does not deal with the stops selection; thus, it is insufficient to solve the problem. The comparison is simply to show that our heuristics give good enough route quality at a low running time.

6   Conclusion

In this short paper, we solve a real-world SBRP with complex features having preferences and the balancing of opposing costs, reducing vehicle costs versus overall passenger satisfaction. There is also a strict computational limit which requires the solution to be efficient. We present a formulation using a set cover variation which this problem can be mapped into. We give an efficient hybrid greedy local search which incrementally solves our SBRP varying the number of stops. Our algorithm can also handle typical VRP side constraints, e.g., capacity and time windows. The evaluation on the real-world dataset shows that a balanced trade-off can reduce the total driving distance at different median/average walking distances. Our algorithm is also shown to be efficient. We believe this is a challenging problem combining novel features different from traditional approaches. We successfully combine a general-purpose VRP solver with (additional) local search, complex objectives and constraints. Our setting and solution are also practical and applicable in many group transportation settings such as worker transportation.

Acknowledgements. This research is supported by the National Research Foundation Singapore under its AI Singapore Programme (Award Number: [AISG-100E-2018-009])

References

1. Ellegood, W.A., Solomon, S., North, J., Campbell, J.F.: School bus routing problem: contemporary trends and research directions. Omega **95** (2020). https://doi.org/10.1016/j.omega.2019.03.014
2. Ghiani, G., Improta, G.: An efficient transformation of the generalized vehicle routing problem. Eur. J. Oper. Res. **122**(1), 11–17 (2000). https://doi.org/10.1016/S0377-2217(99)00073-9
3. Kilby, P., Shaw, P.: Vehicle routing. In: Rossi, F., Van Beek, P., Walsh, T. (eds.) Handbook of Constraint Programming. Elsevier, New York (2006)
4. Lewis, R., Smith-Miles, K.: A heuristic algorithm for finding cost-effective solutions to real-world school bus routing problems. J. Discrete Algorithms **52**, 2–17 (2018). https://doi.org/10.1016/j.jda.2018.11.001
5. Moccia, L., Cordeau, J.F., Laporte, G.: An incremental tabu search heuristic for the generalized vehicle routing problem with time windows. J. Oper. Res. Soc. **63**(2), 232–244 (2012). https://doi.org/10.1057/jors.2011.25
6. Park, J., Kim, B.I.: The school bus routing problem: a review. Eur. J. Oper. Res. **202**(2), 311–319 (2010). https://doi.org/10.1016/j.ejor.2009.05.017
7. Pop, P.C., Kara, I., Marc, A.H.: New mathematical models of the generalized vehicle routing problem and extensions. Appl. Math. Model. **36**(1), 97–107 (2012). https://doi.org/10.1016/j.apm.2011.05.037
8. Pop, P.C., Matei, O., Sitar, C.P.: An improved hybrid algorithm for solving the generalized vehicle routing problem. Neurocomputing **109**, 76–83 (2013). https://doi.org/10.1016/j.neucom.2012.03.032
9. Pop, P.C., Zelina, I., Lupşe, V., Sitar, C.P., Chira, C.: Heuristic algorithms for solving the generalized vehicle routing problem. Int. J. Comput. Commun. Control **6**(1), 158–165 (2011). https://doi.org/10.15837/ijccc.2011.1.2210

Algorithm for Predicting the Quality of the Product Based on Technological Pyramids in Graphs

Damir N. Gainanov[1,2], Dmitriy A. Berenov[1], and Varvara A. Rasskazova[2(✉)]

[1] Ural Federal University, Ekaterinburg, Russia
`berenov@dc.ru`
[2] Moscow Aviation Institute (National Research University), Moscow, Russia

Abstract. In this paper, the problem of the quality of the product is investigated in the conditions when the reassignment can be organized in the process of realization of a technological route. The information on the completed technological routes forms a training sample for the pattern recognition problem and the choice of the technological route for the continuation of the production process is carried out taking into account the expected quality indicators of the final product. To reduce the dimensionality of the problem, a given set of executed technological routes is divided into discrete classes, in each of which an algorithm for constructing a decision tree can be implemented. The paper gives a formal description of the developed algorithm for the node of the decision tree and a polynomial heuristic dichotomy algorithm in a multi-class pattern recognition problem is proposed for it. Computational experiments are carried out to confirm the effectiveness of the proposed algorithm by comparing the obtained solution with the exact solution.

Keywords: Predictive analysis · Graph theory · Heuristic algorithm · Metallurgical production

1 Introduction

The paper proposes a mathematical model for the study of the applied problem of reassigning the technological route in the process of its execution for discrete production. Consideration of this class of problems is motivated by its importance for the optimization of production costs. For example, in a meaningful sense, the manufacture of defective products can be stopped earlier if the receipt of the defect was predicted or, more generally, any running process can be reassigned if its continuation in the planned version can lead to the receipt of the product of inadequate quality. This approach significantly reduces the expenditure of both time and material resources. Thus, the task of reassigning the technological route can be interpreted as the problem of product quality forecasting, which will be made as a result of the execution of this route. In this formulation, the problem can be investigated using methods of analysis

© Springer Nature Switzerland AG 2021
D. E. Simos et al. (Eds.): LION 2021, LNCS 12931, pp. 128–141, 2021.
https://doi.org/10.1007/978-3-030-92121-7_11

of infeasible systems of conditions, well-established in the application to pattern recognition, see, for example, [1–3].

The mathematical model of the applied problem under study can be adequately described in the language of graph theory. The technological route can be represented as a directed path in the graph, the vertices of which are the aggregates of technological lines. Then the manufacture of each individual unit of production is described by a certain technological route, while the production process itself on each unit (vertex of the graph) is characterized by a certain set of values of the parameters of this process. As a result of processing the obtained data, the problem is reduced to the solving of the classical problem of pattern recognition in a geometric setting, when it is necessary to build a decision rule for the assignment of the input vector to one of the classes. To solve the problem of pattern recognition in its geometric setting, various algorithms can be used; they are described, for example, in [5,6,8,9]. An alternative covering algorithm from [11] can also be applied to solve the problem. Full-scale research on general optimization methods for solving manufacturing problems could be also find in [12].

During the long-term operation of the production under consideration, a large amount of historical data is accumulated, which characterizes the realized technological processes for each manufactured product. These data in the model under consideration are formed as directed graph routes along with sets of values of technological parameters of routes. In general, a whole network of multi-class pattern recognition problems is created on the set of such data on the executed technological processes, the solutions of which will be used to optimize the assignment of technological routes. In practical use of the constructed model, in the process of executing of each technological route when reaching certain nodal points (nodal vertices of the graph), the previously constructed decision rules of the corresponding recognition problems are applied, which predict the quality of a unit of production that will be obtained with further continuation of a given technological process. Depending on the result of this forecast, one of the possible decisions will be made: continuation of the planned technological route, reassigning the route to another, more promising route, or stopping the route. The implementation of this strategy can significantly improve the quality of products, as well as reduce production costs due to the reduction of production defects.

This paper presents an algorithm for solving a multi-class problem of pattern recognition in its geometric setting, the implementation of which determines the classification of the expected products. The developed algorithm is based on the principles of constructing logical decision trees and a polynomial heuristic dichotomy algorithm used for each nonterminal vertex of the decision tree.

2   Basic Definitions

Let us consider a discrete production, consisting of a number of technological processes which are executing by technological aggregates and lines.

Let $\mathcal{A} = \{A_1, \ldots, A_n\}$ be the set of technological aggregates involved in production.

The product unit (PU) is understood as an indivisible part of the output or input products obtained by the aggregate or production line. For instance, in the metallurgical production typical examples of PU are: the steel produced in steelmaking aggregate and released in the steel ladle; the slabs obtained by the machine for continuous casting of steel; hot-rolled coil which are produced as a final result of the hot rolling mill; cold-rolled steel coil which are produced as a final result of the work of the cold rolling mill.

Definition 1. *The directed graph $\overrightarrow{G} = (\mathcal{A}, E)$ with the set of vertices $\mathcal{A}$ and the set of arcs $E \subseteq \mathcal{A}^2$ is called the infrastructural graph if $(A_1, A_2) \in E$ if and only if the output PU of the aggregate A_1 can serve as the input PU for the aggregate A_2.*

Definition 2. *The technological pyramid $\mathrm{Pir}\left(\overrightarrow{G}, v\right)$ with the root vertex $v, v \in \mathcal{A}$ of the infrastructural graph $\overrightarrow{G} = (\mathcal{A}, E)$ is the subgraph generated by the set of vertices $\left\{v \cup \overrightarrow{G}(v) \cup \overrightarrow{G}^2(v) \cup \cdots \cup \overrightarrow{G}^k(v))\right\}$ and such that every directed path of the graph $\overrightarrow{G}$ starting at vertex v is lying entirely in this subgraph.*

Here $\overrightarrow{G}^k(v)$ denotes the set of all vertices $v', v' \in \mathcal{A}$ of the graph $\overrightarrow{G}$ such that there exists a simple directed path from vertex v to vertex v' of length k.

Definition 3. *The vertex $v \in \mathcal{A}$ in the infrastructural graph (or in its technological pyramid) is called a fork-vertex if $\left|\overrightarrow{G}(v')\right| > 1$.*

Definition 4. *The vertex $v \in \mathcal{A}$ in the infrastructural graph (or in its technological pyramid) is called the terminal vertex if there are no arcs in the graph (or in the technological pyramid respectively) which leaves from this vertex.*

Definition 5. *Any simple directed path in the infrastructural graph $\overrightarrow{G} = (\mathcal{A}, E)$ of the form*

$$\mathbf{P}_i = \left(A_{i_1}, A_{i_2}, \ldots, A_{i_{k(i)}}\right),$$

where $A_{i_j} \in \mathcal{A}$ and $\left(A_{i_j}, A_{i_{j+1}}\right) \in E$ for all $A_{i_j}, A_{i_{j+1}} \in \mathbf{P}_i$, is called the technological route (TR) of the production under consideration.

In Definition 5 the number of arcs of the infrastructural graph which are included in the sequence of the directed path—the number $(k(i) - 1)$—describes the length of the corresponding TR $\mathbf{P}_i$. Wherein the set of all TRs $\mathcal{P} = \{\mathbf{P}_1, \mathbf{P}_2, \ldots\}$ forms the technological base of production under consideration.

We denote by $\mathrm{PU} = \{pu_i : i = [1, n]\}$ the set of all possible product units of production under consideration.

Let each product unit $pu_i : i = [1, n]$ be characterized by a set of parameters $P_i = \{p_{i_1}, p_{i_2}, \ldots, p_{i_{n(i)}}\}$, where $i = [1, n]$ and $n(i) \in [1, K]$—is the number of different parameters for the given product unit.

Definition 6. *The sequence*

$$AI_i = \left(A_{i1}, P_{i1}\left(AI_i\right), \ldots, A_{is}, P_{is}\left(AI_i\right)\right),$$

where $P_{ij}\left(AI_i\right)$ is the set of parameter values for pu_{ij} in a particular implementation of the technological route AI_i, is called the executed technological route (ETR).

Let $U, U \subset \mathcal{A}$ be the set of terminal vertices of some technological pyramid $\mathrm{Pir}\left(\overrightarrow{G}, v\right)$ with root node $v, v \in \mathcal{A}$, of the infrastructural graph $\overrightarrow{G} = (\mathcal{A}, E)$. For each terminal vertex $u \in U, U \subset \mathcal{A}$, there is a certain product unit $pu_i \in PU$ which is the output product unit for this vertex, and there may be several such product units depending on the types of ETRs as a result of which these product units were received.

Definition 7. *An ETR is called a productive ETR if the output product unit of the terminal vertex A_{i_s} of this ETR AI_i—let us denote this vertex as $\mathrm{term}(AI_i)$—is one of the types of the final product deliverable to the market.*

In the framework of this paper, the generalized problem of assigning a TR is considered, in which it is supposed that it is possible to control the choice of the further passage to processing a product unit in the fork-vertices of the infrastructural graph $\overrightarrow{G}$. This means that the TR in the process of its execution can be reassigned in order to increase production efficiency and reduce the level of rejection.

3 Formulation of the Problem

As a result of the production activity of the production under consideration, the set of executed technological routes will be generated at the current time moment t:

$$\mathcal{P}_{\mathrm{ETR}}(t) = \{AI_i \colon i \in [1, q(t)]\}. \tag{1}$$

Consider the set (1) of all productive ETRs.

For each productive ETR AI_i one can define two parameters for PU of its terminal vertex $pu_i = pu\left(\mathrm{term}\left(AI_i\right)\right)$: $\mathrm{Price}(pu_i)$ is the market price of the pu_i, and $\mathrm{C}(pu_i)$ is the production cost of the product unit pu_i.

Let there be a set of productive ETRs such that the initial sections to the fork-vertex are coincident in the part of the passage of aggregates. Then each productive ETR can be represented as a sequence:

$$AI_i = (BI_i, CI_i), i \in [1, q(t)],$$

where BI_i is the ETR from the initial vertex v_1 to the considered fork-vertex v' and CI_i—ETR from the vertex v' to the terminal vertex $\mathrm{term}(AI_i)$.

Since each ETR AI_i passages a certain technological route—we denote such route as $\mathbf{P}(AI_i)$—then the set of all ETRs can be divided into several classes

$$\mathcal{P}_{\text{ETR}}(t) = \mathcal{P}_{\text{ETR}}^{(1)}(t) \cup \mathcal{P}_{\text{ETR}}^{(2)}(t) \cup \ldots \cup \mathcal{P}_{\text{ETR}}^{(l)}(t) \tag{2}$$

such as AI_i and AI_j belong to the same class if and only if $\mathbf{P}(AI_i) = \mathbf{P}(AI_j)$.

We denote by $\mathbf{P}_i = \mathbf{P}\left(\mathcal{P}_{\text{ETR}}^{(i)}(t)\right)$ a technological route which is common for all ETRs from $\mathcal{P}_{\text{ETR}}^{(i)}(t)$. Then the problem is to determine which of TRs $\mathbf{P}_i$ should be chosen for further passage when reaching the fork-vertex v'.

For each class $\mathcal{P}_{\text{ETR}}^{(i)}(t)$ from (2) and some fork-vertex v, which belongs to TR $\mathbf{P}_i = \mathbf{P}\left(\mathcal{P}_{\text{ETR}}^{(i)}(t)\right)$, one can construct the sample $Z(\mathbf{P}_i, v)$. This sample contains start-parts of all ETRs AI_j such that $AI_j \in \mathcal{P}_{\text{ETR}}^{(i)}(t)$ from the beginning and until fork-vertex v under consideration:

$$Z(\mathbf{P}_i, v) = \left\{ (\text{Ef}(AI_j), i, BI_j) : AI_j \in \mathcal{P}_{\text{ETR}}^{(i)}(t) \right\}, \tag{3}$$

where

$$\text{Ef}(AI_j) = \frac{\text{Price}(\text{term}(AI_j)) - \text{C}(\text{term}(AI_j))}{\text{C}(\text{term}(AI_j))}.$$

Let's break the set of $\text{Ef}(AI_j)$ into several intervals $E_1, E_2, \ldots, E_m$ each of which indicates the class of product quality obtained by implementation of ETR AI_j.

Next, we represent the sample (3) in the form of the corresponding multidimensional vectors

$$\bar{a}_j = (a_{j0}, a_{j1}, a_{jm_1}, \ldots, a_{jn}),$$

where $a_{j0} \in \{E_1, \ldots, E_m\}$ is the value of the ETR's efficiency, a_{j1} is the identifier of the technological route of this ETR. We assign the vector $a_j = (a_{j1}, \ldots, a_{jn})$ to the class K_i, if $a_{j0} \in E_i$. Then each a_j vector will be assigned to one of the classes $K_1, \ldots, K_m$ and the well-known problem of pattern recognition in geometric formulation arises.

3.1   The Concept of Decision Tree Construction

A set of n-dimensional vectors is given

$$A = \left\{ (a_{i_1}, \ldots, a_{i_n}) : i \in [1, N] \right\},$$

and its partition into m classes

$$A = A_1 \cup A_2 \cup \ldots \cup A_m.$$

It is required to construct a decision rule for assigning the vector a_i to one of the classes. The solution will be sought in the class of logical decision trees given by a directed binary tree $\overrightarrow{G} = (V, E)$ with root vertex $v_0 \in V$.

The binary tree $\overrightarrow{G} = (V, E)$ defines the process of sequentially separating of the sample A into two subsamples at the vertices of degree 2 so that each terminal vertex v_i corresponds to a subset $A_{v_i} \subseteq A$, which can be assigned to one of the classes $class_{v_i} \in [1, m]$. In the case under consideration, linear functions will be used to separate the subsample at each vertex of the decision tree.

If v is a vertex of degree 2 in the graph $\overrightarrow{G}$, then a vector n_v and a scalar variable $\mathcal{E}_v$ are given for it, such that A_v is separated into two subsamples of A'_v and A''_v according to the following rule:

$$A'_v = \left\{ a_i \in A_v : \langle n_v , a_i \rangle \le \mathcal{E}_v \right\},$$

$$A''_v = \left\{ a_i \in A_v : \langle n_v , a_i \rangle > \mathcal{E}_v \right\},$$

and for the root-vertex v_0 we should have:

$$A_{v_0} = A.$$

It is required to construct the decision tree $\overrightarrow{G} = (V, E)$ with minimal number of vertices, and at each terminal vertex $v \in V$ we have:

$$p(v) = \frac{\left| \{a_i \in A_v : a_i \in class_v\} \right|}{|A_v|} \ge p_{\min}, \tag{4}$$

that is, the fraction of vectors belonging to the some class $class_v$ is not less than a given value $p_{\min}$. If $p_{\min} = 1$ then each terminal vertex corresponds to the vectors of one particular class.

The rule (4) acts if $|A_v| \ge K_{\min}$. If $|A_v| < K_{\min}$ then the process of further separating of the sample A_v is not performed and the vertex v is declared terminal, and the rule (4) may not be executed. In other words, for $|A_v| < K_{\min}$ the sample A_v is not representative enough for constructing a further decision rule.

3.2   Algorithm for Constructing the Decision Function for the Node of Decision Tree

Suppose that we have a vertex $v \in V$ for which A_v is given. Suppose we have a partition

$$A_v = (A_v \cap A_1) \cup \ldots \ldots \cup (A_v \cap A_m), \tag{5}$$

in which there are m' non-empty sets. If $m' = 1$ then the vertex v is terminal and $p(v) = 1$, if $2 \le m' \le m$ then we will sequentially calculate the values:

$$p_i(v) = \frac{\left| \{A_v \cap A_i\} \right|}{|A_v|}, \quad i \in [1, m].$$

If there exists $i_0 \in [1, m]$ such that $p_{i_0}(v) \ge p_{\min}$ then the vertex v is terminal and the class $class_v = i_0$, if $|A_v| < K_{\min}$ then the vertex v is terminal and

$$class_v = \arg\max_i \left\{ |p_i(v)| : i \in [1, m'] \right\}.$$

Consider the case

$$\begin{cases} 2 \le m' \le m \,, \\ |A_v| \ge K_{\min} \,, \\ p_i(v) < p_{\min} \ \forall \ i \in [1, m] \,, \end{cases}$$

and denote by

$$I = \{i\colon A_v \cap A_i \ne \emptyset \,, \ i \in [1, m]\} \,.$$

Let some vector n_v and a scalar value $\mathcal{E}_v$ be assigned. Then the vertex v is associated with two vertices v_1 and v_2 that are children of the vertex v in the constructed decision tree such that:

$$A_{v_1} = \{a_j \in A_v\colon \langle n_v\,, a_j\rangle \le \mathcal{E}_v\}\,,$$

$$A_{v_2} = \{a_j \in A_v\colon \langle n_v\,, a_j\rangle > \mathcal{E}_v\}\,.$$

Let

$$p(A_{v_1}) = \left(\frac{|A_{v_1} \cap A_1|}{|A_{v_1}|}\,, \ \dots\,, \ \frac{|A_{v_1} \cap A_n|}{|A_{v_1}|} \right)\,,$$

$$p(A_{v_2}) = \left(\frac{|A_{v_2} \cap A_1|}{|A_{v_2}|}\,, \ \dots\,, \ \frac{|A_{v_2} \cap A_n|}{|A_{v_2}|} \right)\,.$$

Consider the following value

$$\operatorname{discrim}(A_v\,, n_v\,, \mathcal{E}_v) = \sum_{i \in I} \left| \frac{|A_{v_1} \cap A_i|}{|A_{v_1}|} - \frac{|A_{v_2} \cap A_i|}{|A_{v_2}|} \right|\,.$$

The value $\operatorname{discrim}(A_v\,, n_v\,, \mathcal{E}_v)$ will be called the separating force of the function

$$f(a) = a \cdot n_v - \mathcal{E}_v$$

concerning the subsample A_v. The meaning of this notion is that the stronger the vectors from the classes A_i of the training sample are separated in the half-space obtained by dividing the space by a hyperplane

$$f(a) = a \cdot n_v - \mathcal{E}_v = 0\,,$$

the more the function $f(a)$ separates vectors from the training sample into classes.

Thus, the formulation is natural, where it is required to find $n_v \in \mathbb{R}^n$ and $\mathcal{E}_v \in \mathbb{R}$ for the sample (5) such that the value of quantity $\operatorname{discrim}(A_v, n_v, \mathcal{E}_v)$ reaches its maximum. The naturalness of such formulation is also confirmed by the fact that for $m = 2$ the best solution is achieved for $\operatorname{discrim}(A_v, n_v, \mathcal{E}_v) = 2$, which corresponds to a linear separation into classes $A_v \cap A_1$ and $A_v \cap A_2$ by the hyperplane $f(a) = n_v \cdot a - \mathcal{E} = 0$.

For an arbitrary subset $A' \subseteq A$ we introduce the notation for the center of the subsample

$$C(A') = \frac{1}{|A'|} \sum_{i \in A'} \{a_i\colon a_i \in A'\}\,,$$

and
$$A(I) = \{a_i \colon a_i \in A, i \in I\} \ .$$

Let be given the partition $I = I_1 \cup I_2$, where $I_1 \neq \emptyset, I_2 \neq \emptyset$. Consider the interval $[C(I_1), C(I_2)] \subset \mathbb{R}^n$. Let $n(I_1, I_2)$ be the normal vector

$$\frac{C(I_2) - C(I_1)}{\|C(I_2) - C(I_1)\|} \ ,$$

then we divide the interval $[C(I_1), C(I_2)]$ into M parts, where the length of each part is

$$\frac{\|C(I_2) - C(I_1)\|}{M} \ .$$

We consider the $(M-1)$ separating functions $f_j(a) = a \cdot n_v - \mathcal{E}_j$, which are passing sequential through all $(M-1)$ dividing points of the interval $[C(I_1), C(I_2)]$. We will search the best option for the separating force among these functions:

$$j_0 = \arg\max \{\operatorname{discrim}(A_v, n_v, \mathcal{E}_j) \colon j \in [1, M-1]\} \ .$$

It is easy to see that for j_0 we have $A_{v_1} \neq \emptyset, A_{v_2} \neq \emptyset$.

We denote by

$$\operatorname{discrim}(I_1, I_2) = \operatorname{discrim}(A_v, n_v, \mathcal{E}_{j_0}) \ .$$

In the case of $C(A(I_1)) = C(A(I_2))$ any two most distant points from the sample A_v are chosen and for the interval which connects these points it is used the same procedure for constructing $(M-1)$ separating planes and choosing the best of them. In the general case it is assumed that all partitions of the form $I = I_1 \cup I_2$ are searched, and the chosen partition is such that $\operatorname{discrim}(I_1, I_2)$ reaches its maximum.

Consider the partition $I = I_1 \cup I_2$ and denote by

$$a(I_1, I_2) = \sum\sum \{(a_s - a_t) \colon a_s \in A_v \cap I_i \ \forall \, i \in I_1, a_t \in A_v \cap I_j \ \forall \, j \in I_2\} \ .$$

The most practical efficiently seems the algorithm for separating the sample A_v in the vertex v which chooses the partition $I = I_1 \cup I_2$, for which $|a(I_1, I_2)|$ is maximal among all partitions $I = I_1 \cup I_2$. Then the choice of the vector $n(I_1, I_2)$ and the scalar value $\mathcal{E}(I_1, I_2)$ can be made according to the procedure described above for the obtained fixed partition $I = I_1 \cup I_2$.

We introduce the notation:

$$a_{ij} = \sum\sum \{(a_s - a_t) \colon a_s \in A_i, a_t \in A_j\}, i, j \in [1, m] \ ,$$

then for $I = I_1 \cup I_2$ we have

$$a(I_1, I_2) = \sum\sum \{a_{ij} \colon i \in I_1, j \in I_2\} \ . \tag{6}$$

Proceeding from the relation (6) it is possible to significantly reduce the amount of computation when choosing the optimal partition $I = I_1 \cup I_2$ for the sample A_v using the previously calculated values $a_{ij} \ \forall \, i, j \in I$.

3.3   Algorithm for Constructing an Optimal Partition of a Set of Classes

We consider the problem of finding the partition $I = I_1 \cup I_2$ that delivers the maximum of the function $|a\,(I_1, I_2)|$. We denote by K_i the number of points in the set $\{a_s : a_s \in A_i\}$, that is,

$$K_i = |\{a_s : a_s \in A_i\}| \ .$$

It is easy to see that

$$a_{ij} = \sum_{a_s \in A_i} \sum_{a_t \in A_j} (a_s - a_t) = \sum_{a_s \in A_i} K_j \cdot a_s - \sum_{a_t \in A_j} K_i \cdot a_t$$

$$= K_i \cdot K_j \cdot \Big(C\,(A_i) - C\,(A_j) \Big) \ .$$

Thus, the problem reduces to the following.

For a given set of points $A \subset \mathbb{R}^n$ and its partition $A = A_1 \cup A_2 \cup \ldots A_k$ it is required to find the partition $I = I_1 \cup I_2$, where $I = [1, k]$, such that the function

$$|a\,(I_1, I_2)| = \left| \sum_{i \in I_1} \sum_{j \in I_2} K_i \cdot K_j \cdot (C\,(A_i) - C\,(A_j)) \right|$$

takes the maximal value among all possible partitions $I = I_1 \cup I_2$. Obviously, the problem obtained has a much smaller dimension than in the original formulation. Since in the formulation mentioned above only centers of the subsets $A_i, i \in [1, k]$ participate, then we will further simplify the formulation of the problem.

Suppose that are given a finite set of points $C = \{c_i, i = [1, k]\}$ (which serve as an analogue of the centers of the subsets in previous formulation) and a set of natural numbers $n_i, i \in [1, k]$, (which serve as an analogue of the number of points in the subsets of $A_i, i \in [1, k]$ in the previous formulation). Then the problem is formulated as follows:

$$\begin{cases} \left| \sum_{i \in I_1} \sum_{j \in I_j} n_i \cdot n_j \cdot (c_i - c_j) \right| \longrightarrow \max\,, \\ I = I_1 \cup I_2 \ . \end{cases} \qquad (7)$$

The following particular cases of the problem (7) are also of interest. Let $n_i = n$ for all $i \in [1, k]$. Then the formulation of the problem takes the form:

$$\begin{cases} \left| \sum_{i \in I_1} \sum_{j \in I_2} n_i \cdot n_j \cdot (c_i - c_j) \right| = n^2 \left| \sum_{i \in I_1} \sum_{j \in I_j} (c_i - c_j) \right| \longrightarrow \max\,, \\ I = I_1 \cup I_2 \ . \end{cases} \qquad (8)$$

The next simplification is to consider the case $n_i = 1$ for all $i \in [1, k]$. Then the formulation of the problem takes the form:

$$\begin{cases} |I_2| \sum_{i \in I_1} c_i - |I_1| \cdot \sum_{i \in I_2} c_i \longrightarrow \max\,, \\ I = I_1 \cup I_2 \ . \end{cases} \qquad (9)$$

We formulate the problem (9) in the following form:

$$\begin{cases} |I_2| \cdot |I_1| \dfrac{\sum\limits_{i \in I_1} c_i}{|I_1|} - |I_1| \cdot |I_2| \cdot \dfrac{\sum\limits_{i \in I_2} c_i}{|I_2|} = |I_1| \cdot |I_2| \cdot \left(\dfrac{1}{|I_1|} \cdot \sum\limits_{i \in I_1} c_i - \dfrac{1}{|I_2|} \cdot \sum\limits_{i \in I_2} c_i \right) \longrightarrow \max, \\ I = I_1 \cup I_2 . \end{cases}$$

Finally, the problem of the species under discussion has a good geometric interpretation in the following formulation.

Let

$$C(I') = \frac{1}{|I'|} \cdot \sum_{i \in I} c_i \text{ for all } I' \subset I, I = [1, k]$$

$$\begin{cases} \|C(I_1) - C(I_2)\| \longrightarrow \max, \\ I = I_1 \cup I_2 . \end{cases} \tag{10}$$

Let us consider in detail the problem of (10). We use the standard notation $\operatorname{conv} A$ for the convex hull of the set A. For the set of points A it is obvious that

$$C(I_1) \in \operatorname{conv} A,$$
$$C(I_2) \in \operatorname{conv} A,$$

and, consequently,

$$[C(I_1), C(I_2)] \subset \operatorname{conv} A.$$

Proposition 1. *Let a finite set* $C = \{c_i : i \in [1, k]\}, C \subset \mathbb{R}^n$ *be given. Let* $c_{i_1}, c_{i_2} \in C$ *be two vectors such that*

$$\|c_{i_1} - c_{i_2}\| = \max_{i,j \in I, i \neq j} \|c_i - c_j\| .$$

Then, if $I_1 \cup I_2 = I$ *is the optimal solution of the problem (10), then*

$$\|C(I_1) - C(I_2)\| \leq \|c_{i_1} - c_{i_2}\| . \tag{11}$$

Proof. Let us prove that under the conditions of this statement that $\|c_{i_1} - c_{i_2}\|$ is the diameter of the set A, which means

$$\|c_{i_1} - c_{i_2}\| = \max_{a,b \in \operatorname{conv} A} \rho(a, b) ,$$

where $\rho(a, b)$ is the Euclidean distance between points a, b in the space $\mathbb{R}^n$.

Assume in contrary, that for some pair of points $a, b \in \operatorname{conv} C$, such that $[a, b]$ is the diameter of the $\operatorname{conv} C$, at least one of these points is not extreme. For definiteness, suppose $b \notin \operatorname{vert}(\operatorname{conv} C)$, where by the $\operatorname{vert}(\cdot)$ are denoted the extreme points of the $\operatorname{conv} C$.

Consider the interval $[a, b]$. Since b is not an extreme point, there is a segment $[c, d]$ for which the point b is interior and $[c, d] \subset \operatorname{conv} C$. Then we obtain a triangle with vertices a, c, d in which b is an interior point of the interval $[c, d]$. All sides of this triangle lie in the set $\operatorname{conv} A$.

It is easy to see that in this case one of the inequalities holds: $\rho(a,c) > \rho(a,b)$ or $\rho(a,d) > \rho(a,b)$ in contradiction with the maximality of $\rho(a,b)$.

Thus, the points c_{i_1}, c_{i_2} lie on the ends of some diameter of the set $\mathrm{conv}C$ and, since $C(I_1) \in \mathrm{conv}A$ and $C(I_2) \in \mathrm{conv}A$, then $[C(I_1), C(I_2)] \subset \mathrm{conv}A$ and, therefore, the inequality (11) holds.

Further we will construct a heuristic algorithm for solving the problem (10), that is for finding the partition $I = I_1 \cup I_2$ such that the value $\rho(C(I_1), C(I_2))$ reaches its maximum. Let's denote this algorithm as $\mathcal{L}$. Let's also denote as $\mathcal{A}$ a full search algorithm which finds an exact solution of the problem (10). We will compare the quality of the approximate solution of the problem (10) found using heuristic algorithm $\mathcal{L}$ and the exact one found by $\mathcal{A}$.

First let's describe the full search algorithm $\mathcal{A}$ for finding exact solution for the partition $I = I_1 \cup I_2$.

Algorithm $\mathcal{A}$.

1. We consider all partitions $I = I_1 \cup I_2$.
2. For each partition suppose:

$$C_{I_1} = \frac{1}{|I_1|} \sum_{i \in I_1} C_i \text{ and } C_{I_2} = \frac{1}{|I_2|} \sum_{i \in I_2} C_i.$$

3. If the length of the interval $[C_{I_1}, C_{I_2}] \neq 0$ then the normal vector n_{I_1,I_2} is constructed and the value $\mathcal{E}_{I_1,I_2}$ is fixed such that discrim $(A_v, n_{I_1,I_2}, \mathcal{E}_{I_1,I_2})$ is maximal (search through all hyperplanes perpendicular to n_{I_1,I_2} and bypassing $[C_{I_1}, C_{I_2}]$ for M steps from the point C_{I_1} to the point C_{I_2}).
4. Among all partitions of the form $I = I_1 \cup I_2$ one need to choose a partition, for which the value discrim $(A_v, n_{I_1,I_2}, \mathcal{E}_{I_1,I_2})$ (from the item 3) reaches its maximum.

Algorithm $\mathcal{L}$ splits the projection of the set C on the direction $[a,b]$ with maximum distanced points $a, b \in C$. Moreover this algorithm checks only partitions where maximum intervals covering points of each subset are not intersected. It is clear that there are not more than $(k-1)$ such partitions.

Let $\overline{C} = \{\overline{c}_1, \overline{c}_2, \ldots, \overline{c}_k\}$ be the projection of the set C on the direction $[a,b]$ with maximum distanced points. Then $\overline{C}$ could be linear ordered by any coordinate of its points, which is not equal to zero for all points from $\overline{C}$. So, the set $\overline{C}$ contains not more than k different points. Let $\{\overline{c}_1, \overline{c}_2, \ldots, \overline{c}_m\}$ be the mentioned ordered set of different points.

Let's consider $(m-1)$ points of the form $b_i = 0.5 \cdot \overline{c}_i + 0.5 \cdot \overline{c}_{i-1}, i = [1, m-1]$. Then we can state for algorithm $\mathcal{L}$ to check only partitions of the form:

$$I_1(b_i) = \left\{ i \in [1,m]: \langle \overline{c}_i, \overrightarrow{ab} \rangle < \langle \overline{c}_i, b_i \rangle \right\},$$

$$I_2(b_i) = \left\{ i \in [1,m]: \langle \overline{c}_i, \overrightarrow{ab} \rangle > \langle \overline{c}_i, b_i \rangle \right\}.$$

Thus, algorithm $\mathcal{L}$ will check not more than $(m-1)$ different partitions for $\overline{C}$ and, sequentially, not more than $(k-1)$ different partitions for initial set C.

For algorithm $\mathcal{A}$ one need to check 2^k pairs of subsets of C, which generates 2^{k-1} different partitions of the form $I = I_1 \cup I_2$. Then for each mentioned pair one need to execute $\mathcal{O}(n)$ iterations to calculate closeness between subsets of the pair $\rho(C(I_1), C(I_2))$. Thus, the complexity of $\mathcal{A}$ is $\mathcal{O}(n \cdot 2^m)$.

At the same time algorithm $\mathcal{L}$ needs $\mathcal{O}(n \cdot m^2)$ iterations to extract the pair of maximum distanced points of C, and $\mathcal{O}(n \cdot m)$ iterations to calculate projections of points of the initial set on the direction with maximum distanced points. Finally, this algorithm needs $\mathcal{O}(m^2)$ iterations to define the best partition of the initial set into two subsets. Note that variable n not active on this step of the complexity's estimate, because we mean one-dimensional projection $\overline{C}$ of the initial set C of n-dimensional points. Thus, the complexity of $\mathcal{L}$ is $\mathcal{O}(n \cdot m^2)$.

To compare the quality of different solutions, each partition could be performed by binary couple $\alpha = (\alpha_1, \alpha_2, \ldots, \alpha_m), \alpha_i \in \{0, 1\}$ for all $i \in [1, m]$. In this case $\alpha_i = 0$ if and only if $i \in I_1$ and $\alpha_i = 1$ if and only if $i \in I_1$.

Let two partitions $I = I_1 \cup I_2$ and $I' = I_1' \cup I_2'$ be given. Let one need to calculate the closeness between these partitions performed by corresponding binary couples α and α'.

For any binary couple $\alpha = (\alpha_1, \alpha_2, \ldots, \alpha_m)$ it is defined a complementary couple $\overline{\alpha} = (\overline{\alpha}_1, \overline{\alpha}_2, \ldots, \overline{\alpha}_m)$ such that $\overline{\alpha}_i = 1 - \alpha_i$ for all $i \in [1, m]$. Let $hamming(\alpha, \alpha')$ denotes Hamming's distance between corresponding binary couples α and α', that is

$$hamming(\alpha, \alpha') = |\{i \colon \alpha_i \neq \alpha_i', i \in [1, m]\}| \ .$$

When comparing results of implementation of two algorithms $\mathcal{A}$ and $\mathcal{L}$ we will use the same set of points C and calculate the closeness between exact (global optimal) partition $I = I_1 \cup I_2$ obtained by $\mathcal{A}$ and partition $I' = I_1' \cup I_2'$, which is the best one found by algorithm $\mathcal{L}$. To calculate the closeness mentioned above we will use the naturalness formula:

$$\rho(\alpha, \alpha') = \min\{hamming(\alpha, \alpha'), hamming(\overline{\alpha}, \alpha')\} \ , \tag{12}$$

where α, α' are binary couples corresponding to partitions $I = I_1 \cup I_2$ and $I' = I_1' \cup I_2'$.

Computational experiments were carried out using 100 random sets of 32 points in two-dimensional space $\mathbb{R}^2$. These experiments show rather interesting results despite the difference between computational complexity of algorithms $\mathcal{A}$ and $\mathcal{L}$. The best partitions obtained by these algorithms are very close one each other in means of criterion (12)—the value of closeness is equal to 94,9%. Thus, algorithm $\mathcal{L}$ is polynomial with complexity $\mathcal{O}(n \cdot m^2)$, but solutions obtained by this algorithm could be very close to exact ones obtained by exponential algorithm.

What about applied interest, it should be note that obtained results show that proposed polynomial algorithm $\mathcal{L}$ could be used for solving problem (10) with huge size of input data.

Let's have a look to the differences between the proposed algorithm for constructing a decision tree and others known algorithms.

In the well-known algorithm Principal Direction Divisive Partitioning (PDDP) ([14]) the problem of constructing a binary tree is characterized so that it solves the problem of unsupervised clustering (of documents) and therefore does not have any starting partition of the original sample into a set of classes. So, our quality criteria for solving the dichotomy problem for the binary tree node is not applicable for PDDP algorithm. But in our case this quality criteria plays very important role.

Well known machine learning algorithm C5.0 ([15–17]) allows one to build a decision tree in forecasting problems has become in a sense a standard. The main difference of algorithm proposed in the present work in comparison to the approach in the algorithm C5.0, is that the algorithm of the C5.0 at each node proposes division of the sample using only one parameter, which is chosen as the most effective for the sample corresponding to the current node of the decision tree. This approach has its advantages, especially in applications where the interpretation of the obtained decision rule is important. For example, it is of great importance in the problems of pattern recognition in the field of medical diagnostics. In the technological problems of recognition with a large number of parameters and very large sample volumes considered in this paper, the use of axis-parallel splits can lead to inaccurate classification and artificial and excessive increase in the size of the resulting decision trees, and therefore the construction of an effective heuristic dichotomy algorithm with separating planes of the general (non parallel) position is of considerable interest.

Conclusion

To solve the problem of the quality of discrete production an approach has been developed, in which the research is reduced to solving the problem of pattern recognition in a geometric formulation. A heuristic algorithm for constructing a decision rule is developed which is aiming to find the best possible partition of training sample corresponding to each vertex of decision tree. Substantial reducing of the computational complexity of considered algorithm is reached. Thus, for the production under consideration, a large number of problems of pattern recognition are constructed for each of which a decision tree is constructed.

An important feature of the approach is that the set $\mathcal{P}(t)$ of ETRs is continuously expanding, thus providing all the new data to improve the decision rule. To achieve the effectiveness of the proposed approach in practice it is necessary to carry out additional training as soon as a new portion of the ETRs arrives. This will ensure the continuous improvement of the decision rules and consequently the improvement of production efficiency. Algorithms proposed in the present paper were implemented for solving applied problem on the optimization of the assignment of technological routes for metallurgical production.

References

1. Eremin, I.I.: Improper Models of Optimal Planning. Library of Mathematical Economics, Nauka, Moscow (1988)
2. Eremin, I.I., Mazurov, V.D., Astafév, N.N.: Improper Problems of Linear and Convex Programming. Library of Mathematical Economics, Nauka, Moscow (1983)
3. Khachai, M.Yu., Mazurov, Vl.D., Rybin, A.I.: Committee constructions for solving problems of selection, diagnostics, and prediction. Proc. Steklov Inst. Math. **1**, 67–101 (2002)
4. Gainanov, D.N., Berenov, D.A.: Big data technologies in metallurgical production quality control systems. In: Proceedings of the Conference on Big Data and Advanced Analitycs, pp. 65–70. Minsk State University Press, Minsk, Belarus' (2017)
5. Gainanov, D.N.: Combinatorial Geometry and Graphs in the Analysis of Infeasible Systems and Pattern Recognition. Nauka, Moscow (2014)
6. Gainanov, D.N.: Graphs for Pattern Recognition. Infeasible Systems of Linear Inequalities. DeGruyter (2016)
7. Gainanov, D.N.: Combinatorial properties of infeasible systems of linear inequalities and convex polyhedra. Math. Not. **3**(38), 463–474 (1985)
8. Mazurov, Vl.D.: Committees Method in Problems of Optimization and Classification. Nauka, Moscow (1990)
9. Mazurov, Vl.D., Khachai, M.Yu.: Committees of systems of linear inequalities. Autom. Rem. Contr. **2**, 43–54 (2004)
10. Khachai, M.Yu.: On the estimate of the number of members of the minimal committee of a system of linear inequalities. J. Comput. Math. Math. Phys. **11**(37), 1399–1404 (1997)
11. Gainanov, D.N.: Alternative covers and independence systems in pattern recognition. Pattern Recognit. Image Anal. **2**(2), 147–160 (1992)
12. Liu, X., Pei, J., Liu, L., Cheng, H., Zhou, M., Pardalos, P.M.: Optimization and Management in Manufacturing Engineering. SOIA, vol. 126. Springer, Cham (2017). https://doi.org/10.1007/978-3-319-64568-1
13. Gainanov, D.N., Matveev, A.O.: Lattice diagonals and geometric pattern recognition problems. Pattern Recognit. Image Anal. **3**(1), 277–282 (1991)
14. Boley, D.: Principal direction divisive partitioning. Data Min. Knowl. Disc. **2**(4), 325–344 (1998)
15. Quinlan, J.R.: Learning with continuous classes. In: Proceedings of the 5th Australian Joint Conference on Artificial Intelligence, pp. 343–348 (1992)
16. Quinlan, J.R.: C4.5 – Programs for Machine Learning. Morgan Kaufman Publisher Inc., San Mateo (1993)
17. Quinlan, J.R.: Improved use of continuous attributes in C4.5. J. Artif. Intell. Res. **4**, 77–90 (1996)

Set Team Orienteering Problem
with Time Windows

Aldy Gunawan[1]([✉]), Vincent F. Yu[2,3], Andro Nicus Sutanto[2],
and Panca Jodiawan[2]

[1] School of Computing and Information Systems, Singapore Management University,
80 Stamford Road, Singapore 178902, Singapore
`aldygunawan@smu.edu.sg`
[2] Department of Industrial Management, National Taiwan University of Science
and Technology, 43, Sec. 4, Keelung Road, Taipei 106, Taiwan
`vincent@mail.ntust.edu.tw`
[3] Center for Cyber-Physical System Innovation, National Taiwan University
of Science and Technology, 43, Sec. 4, Keelung Road, Taipei 106, Taiwan

Abstract. This research introduces an extension of the Orienteering
Problem (OP), known as Set Team Orienteering Problem with Time
Windows (STOPTW), in which customers are first grouped into clusters. Each cluster is associated with a profit that will be collected if at
least one customer within the cluster is visited. The objective is to find
the best route that maximizes the total collected profit without violating time windows and time budget constraints. We propose an adaptive
large neighborhood search algorithm to solve newly introduced benchmark instances. The preliminary results show the capability of the proposed algorithm to obtain good solutions within reasonable computational times compared to commercial solver CPLEX.

Keywords: Orienteering problem · Time windows · Adaptive large
neighborhood search

1 Introduction

The Orienteering Problem (OP) was first introduced by [9] for which a set of
nodes is given, each with a score. The objective is to determine a path, limited
in length or travel time, that visits a subset of nodes and maximizes the sum
of the collected scores. The Orienteering Problem (OP) has received a lot of
attentions since many researchers have worked on it as well as its applications
and extensions [2], such as the inventory problem [10], Capacitated Team OP [8],
and Set OP [1].

The Set OP (SOP) was presented by [1]. The main difference lies on grouping
nodes into clusters and each cluster is associated with a profit. This profit is
collected by visiting at least one node in the respective cluster. Due to the time
budget constraint, only a subset of clusters can be visited on a path. Various

© Springer Nature Switzerland AG 2021
D. E. Simos et al. (Eds.): LION 2021, LNCS 12931, pp. 142–149, 2021.
https://doi.org/10.1007/978-3-030-92121-7_12

applications of the SOP can be found in mass distributions, yet it may benefit less to visit all customers within a particular district. Therefore, only delivering products to one customer and letting other customers within the same district to collect from the visited customer will actually help the distributor in terms of travelling time or travelled distance. [1] proposed a matheuristic algorithm and applied it in the context of the mass distribution problem. [7] introduced the applications of the SOP in the travel guide problems. Variable Neighborhood Search (VNS) is proposed to solve SOP. The Team Orienteering Problem with Time Windows (TOPTW) is an extension of the Team OP [5] where the visit on each node is constrained by a given time window. This TOPTW has been studied in the past few years. For more details, please refer to [2].

Our work introduces another extension of the SOP and TOPTW - namely, the Set Team OP with Time Windows (STOPTW). STOPTW considers both multiple paths and time windows. The visits to nodes are also bounded by the given time windows. We thus propose an adaptive large neighborhood search algorithm (ALNS) to solve newly introduced benchmark instances. The preliminary results of our experiments show the capability of the proposed algorithm to generate good solutions within reasonable computational times.

2 Problem Description

Given a complete directed graph $G = (N, A)$ where N represents a set of nodes, $N = \{n_0, n_1, \ldots, n_{|N|}\}$, A represents a set of arcs $A = \{a_{ij}\}$, and n_0 and $n_{|N|}$ are the start and end nodes, respectively. Given a pair of nodes n_i and n_j, there exists an arc a_{ij} with cost c_{ij}. In SOP, all nodes are grouped as clusters as disjoint sets $s_0, s_1, \ldots, s_m$, with $S = \{s_0, s_1, \ldots, s_m\}$, $s_i \cap s_j = \emptyset$ for $i \neq j, 0 \leq i, j \leq |N|$, and each node n_i is associated with exactly one particular set in S. All disjoint sets $s_0, s_1, \ldots, s_m$ have associated profits $p_0, p_1, \ldots, p_m$ for visiting at least one node within the set. We note that s_0 and s_m represent the starting and ending sets with $p_0 = p_m = 0$, respectively. The basic mathematical model of the SOP is presented in [7] while the TOPTW mathematical model can be referred to [2].

In the context of the STOPTW, each node has a non-negative service time s_i and time window $[E_i, F_i]$, where E_i and F_i are the earliest and the latest start times of service at node i, respectively. A visit beyond the time window F_i is not allowed while an early visit is possible with additional waiting times before entering at the earliest start time E_i. Each node i can only be visited once and must be visited within its respective time window. Each set s_j can only be visited at most once as well. Given h paths, each path must start its visit from the start node and also return to the end node. The objective of the STOPTW is to determine h paths, limited by a given time budget T_{max} and time windows, such that each path visits a subset of S and maximizes the total collected profit.

3 Proposed Algorithm

The initial solution is generated based on the nearest distance criterion. Each path h starts from node n_0 and follows up by visiting the nearest unvisited nodes. This is done until it reaches T_{max} or no more nodes can be visited. The time window constraint has to be considered, and the path ends at node n_0.

The initial solution is further improved by the proposed Adaptive Large Neighborhood Search (ALNS) (Algorithm 1). This proposed algorithm is adopted from a similar algorithm for solving another combinatorial optimization problem, namely the vehicle routing problem [3].

The main idea is to use a set of DESTROY operators for removing nodes from the current solution and to use a set of REPAIR operators for reinserting them into more profitable positions. A particular score is assigned to each selected operator in order to assess its performance upon generating a new neighborhood solution. The better the new generated solution is, the higher is the score given to the corresponding operator.

Let Sol_0, Sol^*, and Sol' be the current solution, the best found solution so far, and the starting solution at each iteration, respectively, we first set Sol_0, Sol^*, and Sol' to be the same as the generated initial solution. The current temperature ($Temp$) is set to the initial temperature (T_0) and will decrease by α after η_{SA} iterations. The number of iterations $Iter$ is set to zero. Let $R = \{R_r | r = 1, 2, \ldots, |R|\}$ be a set of DESTROY operators and $I = \{I_i | i = 1, 2, \ldots, |I|\}$ be a set of REPAIR operators. All operators $j(j \in R \cup I)$ initially have the same weight w_j and probability p_j to be selected, based on:

$$p_j = \begin{cases} \frac{w_j}{\sum_{k \in R} w_k} & \forall j \in R \\ \frac{w_j}{\sum_{k \in I} w_k} & \forall j \in I \end{cases} \tag{1}$$

ALNS adopts the Simulated Annealing (SA) acceptance criteria, under which a worse solution may be accepted with a certain probability [6]. Therefore, each of the operator's score s_j is adjusted by:

$$s_j = \begin{cases} s_j + \delta_1, & \text{if the new solution is the} \\ & \text{best found solution so far} \\ s_j + \delta_2, & \text{if the new solution improves} \\ & \text{the current solution} \\ s_j + \delta_3, & \text{if the new solution does not} \\ & \text{improve the current solution,} \\ & \text{but it is accepted} \end{cases} \quad \forall j \in R \cup I \tag{2}$$

with $\delta_1 > \delta_2 > \delta_3$. The operator's weight w_j is then adjusted by following:

$$w_j = \begin{cases} (1-\gamma)w_j + \gamma \frac{s_j}{\chi_j}, & \text{if } \chi_j > 0 \\ (1-\gamma)w_j, & \text{if } \chi_j = 0 \end{cases} \quad \forall j \in R \cup I \tag{3}$$

where γ refers to the reaction factor ($0 < \gamma < 1$) to control the influence of the recent success of an operator on its weight, and χ_j is the frequency of using operator j.

At each iteration, a certain number of nodes are removed from Sol_0 by using a selected DESTROY operator. The removed nodes are then reinserted into Sol_0 by applying another selected REPAIR operator. Sol_0 is directly accepted if its objective function value is better than Sol^* or Sol'; otherwise, it will only be accepted with probability $e^{\frac{-(Sol_0-Sol')}{Temp}}$. Each operator's score s_j is then updated according to (2). After η_{ALNS} iterations, each operator's weight w_j is updated by (3), and its probability p_j is updated according to (1). ALNS is terminated when there is no solution improvement after θ successive temperature reductions.

Algorithm 1: ALNS pseudocode

---

1 $Sol_0, Sol^*, Sol' \leftarrow$ Initial Solution
2 $Temp \leftarrow T_0$
3 ITER $\leftarrow 0$
4 FOUNDBESTSOL $\leftarrow$ False
5 Set s_j and w_j such that p_j is equally likely
6 **while** NOIMPR $< \theta$ **do**
7   REMOVEDNODES $\leftarrow 0$
8   **while** REMOVEDNODES $< \pi$ **do**
9          $Sol_0 \leftarrow$ Destroy(R_r)
10   UpdateRemovedNodes(REMOVEDNODES, R_r)
11   **end**
12   **while** REMOVEDNODES > 0 **do**
13         $Sol_0 \leftarrow$ Repair(I_i)
14   UpdateRemovedNodes(REMOVEDNODES, I_i)
15   **end**
16   AcceptanceCriteria($Sol_0, Sol^*, Sol', Temp$)
17   Update s_j
18   **if** ITER mod $\eta_{ALNS} = 0$ **then**
19   Update w_j and p_j
20   **end**
21   **if** ITER mod $\eta_{SA} = 0$ **then**
22   **if** FOUNDBESTSOL $=$ *False* **then**
23   NOIMPR $\leftarrow$ NOIMPR $+ 1$
24   **end**
25   **else**
26   NOIMPR $\leftarrow 0$
27   **end**
28   FOUNDBESTSOL $\leftarrow$ False
29         $Temp \leftarrow Temp \times \alpha$
30   **end**
31   ITER $\leftarrow$ ITER $+ 1$
32 **end**
33 **Return** Sol^*

Four DESTROY and six REPAIR operators used in the proposed ALNS are:

Random removal (R_1): select q nodes randomly and remove them from the current solution. REMOVEDNODES is increased by q.

Worst removal (R_2): remove the node with the smallest removal profit. The removal profit is defined as the difference in objective function values between including and excluding a particular node.

Shaw removal (R_3): remove a node that is highly related with other removed nodes in a predefined way. In other words, it tries to remove some similar nodes, such that it is easier to replace the positions of one another during the repair process. The last removed node is denoted as node i, while the next candidate of the removed node is denoted as node j. The relatedness value (φ_j) of node j to node i is calculated by:

$$\varphi_j = \begin{cases} \phi_1 c'_{ij} + \phi_2 t'_{ij} + \phi_3 l_{ij} + \phi_4 |P_i - P_j|, & \text{if } i \in S \\ \phi_1 c''_{ij} + \phi_2 t''_{ij} + \phi_3 l_{ij} + \phi_4 |D_i - D_j|, & \text{if } i \in C \end{cases} \quad (4)$$

Unvisited removal (R_4): this operator removes selected nodes that are not visited due to the time windows violation. When selecting nodes, random numbers are generated to determine whether they will be removed or not.

Greedy insertion (I_1): insert a removed node to a position resulting in the highest insertion profit (i.e., the difference in objective function value after and before inserting a node to a particular position).

Regret insertion (I_2): the regret value is calculated by the difference in Total Profit when node j is inserted in the best position and in the second best position. The idea is to select a node that leads to the largest regret value if it is not inserted into its best position. In other words, this operator tries to insert the node that one will regret the most if it is not inserted now.

Greedy visit insertion (I_3): the insertion is decided by the changes in the number of visited nodes for every inserted node. Since we consider time windows, after inserting a particular node, there will be some nodes that cannot be visited again. Here, we try to find an insertion with the highest number of visited nodes.

Random insertion (I_4): the insertion is decided by choosing a random position in the current solution and trying to insert any removed nodes into that position.

First feasible position insertion (I_5): this operator is adopted from [4]. Every removed node is inserted into the first position that makes the solution feasible, one at a time.

Last feasible position insertion (I_6): it works similarly to the previous operator. The main difference lies on the position of inserting it. It should start from last node of the feasible solution.

4   Computational Results

We first modified a set of TOPTW instances that are taken from Solomon's dataset - namely, Set A. There are 29 instances ($c100, r100$, and $rc100$) where each instance contains 100 nodes. We group nodes into clusters using a method proposed by [1]. The number of clusters is set to 20% of the total number of nodes. After randomly inserting nodes into each cluster, the cluster profit is calculated by adding all profits from all respective nodes in a particular cluster. Another set of larger instances, Set B, is introduced by modifying the above-mentioned instances. A hundred more nodes are added with respective parameters, such as service times, locations, profits, etc. This experiment is performed on a Windows 7 professional computer with Intel core i7-4790 CPU @3.60 GHz processor with 16.00 GB RAM. AMPL is utilized to run the mathematical programming using CPLEX, while Microsoft Visual C++ 2019 is used to code our ALNS algorithm. The obtained results are compared to those of the commercial software CPLEX (Table 1). The profit and computational (CPU) times are based on 10 runs of ALNS. We also calculate the gap (%) between CPLEX and ALNS results.

Table 1. Total profit comparison between ALNS and CPLEX when solving Set A

Instance	CPLEX		ALNS		
	Profit	CPU time	Profit	CPU time	Gap (%)
$c100$	1808.89	753.22	1666.44	109.40	2.79
$r100$	1387.92	3300.81	1223.80	109.87	2.58
$rc100$	1601.50	3600	1389.23	108.31	6.11

For solving Set A instances, our proposed ALNS is comparable to CPLEX. Here, our main purpose is to test the current performance of ALNS and emphasize the CPU time, which is much lower than the one of CPLEX. We note that ALNS outperforms CPLEX for 5 instances - namely, r102, r104, r107, r108, and rc104. The average gap in terms of the solution quality is 3.62%. We report the performance of ALNS in solving Set B. The 29 instances are served by four different numbers of vehicles, from one to four vehicles. CPLEX is also used to solve those instances with the maximum CPU times of 2 h (7200 s). The results are

Table 2. Total profit comparison between ALNS and CPLEX when solving Set B

Number of vehicles	Instance	CPLEX		ALNS		
		Profit	CPU time	Profit	CPU time	Gap (%)
1	$c100$	243.33	6405.1	282.56	1233.35	−26.85
	$r100$	133	6602.95	201.25	1657.66	−89.15
	$rc100$	224	7200	219	1708.86	−19.6
2	$c100$	561.11	7200	458	972	15.6
	$r100$	313.5	7200	371.75	1198.46	−29.19
	$rc100$	394.13	7200	383.75	1128.79	−4.58
3	$c100$	687.78	7200	646.67	1046.7	3.36
	$r100$	406.92	6682.44	533.08	980.55	−37.98
	$rc100$	466.38	7200	552.75	1032.34	−24.85
4	$c100$	802.22	7200	833.22	893.25	−8.43
	$r100$	451.33	7200	700.75	829.2	−69.26
	$rc100$	483.5	7200	720.75	860.85	−61.2

summarized in Table 2. We observe that ALNS outperforms CPLEX for solving larger instances. This can be seen from the calculated average gaps, which are -50.63%, -8.5%, -21.53%, and -48.16% for one, two, three, and four vehicles, respectively. For most instances, CPLEX is unable to obtain the optimal solutions, and therefore we only report the best found solutions within 2 h of CPU time.

5   Conclusion

This research introduces the Set Team Orienteering Problem with Time Windows (STOPTW) as a new extension of TOPTW, where customers are grouped into clusters and a profit is associated with each cluster. The profit collection will only happen if at least one customer is visited in a particular cluster. The objective of STOPTW is to maximize the total collected cluster profit without violating any time windows. We propose an Adaptive Large Neighborhood Search (ALNS) algorithm to solve STOPTW, while CPLEX is used to obtain optimal solutions for comparison purposes. The computational study shows that our algorithm outperforms CPLEX in solving newly larger introduced instances. More development on the algorithm can be considered as future research. Other destroy and removal operators can also be developed to provide better solutions. Since STOPTW is a new problem, other (meta)heuristics can also be considered.

Acknowledgment. This research was partially supported by the Ministry of Science and Technology of Taiwan under grant MOST 108-2221-E-011-051-MY3 and the Center for Cyber-Physical System Innovation from The Featured Areas Research Center Program within the framework of the Higher Education Sprout Project by the Ministry of Education (MOE) in Taiwan.

References

1. Archetti, C., Carrabs, F., Cerulli, R.: The set orienteering problem. Eur. J. Oper. Res. **267**, 264–272 (2018)
2. Gunawan, A., Lau, H.C., Vansteenwegen, P.: Orienteering problem: a survey of recent variants, solution approaches and applications. Eur. J. Oper. Res. **255**(2), 315–332 (2016)
3. Gunawan, A., Widjaja, A.T., Vansteenwegen, P., Yu, V.F.: Vehicle routing problem with reverse cross-docking: an adaptive large neighborhood search algorithm. In: Lalla-Ruiz, E., Mes, M., Voß, S. (eds.) ICCL 2020. LNCS, vol. 12433, pp. 167–182. Springer, Cham (2020). https://doi.org/10.1007/978-3-030-59747-4_11
4. Hammami, F., Rekik, M., Coelho, L.C.: A hybrid adaptive large neighborhood search heuristic for the team orienteering problem. Comput. Oper. Res. **123**, 105034 (2020)
5. Labadie, N., Mansini, R., Melechovský, J., Calvo, R.W.: The team orienteering problem with time windows: an LP-based granular variable neighborhood search. Eur. J. Oper. Res. **220**(1), 15–27 (2012)
6. Lutz, R.: Adaptive large neighborhood search (2015)

7. Pěnička, R., Faigl, J., Saska, M.: Variable neighborhood search for the set orienteering problem and its application to other orienteering problem variants. Eur. J. Oper. Res. **276**(3), 816–825 (2019)
8. Tarantilis, C.D., Stavropoulou, F., Repoussis, P.P.: The capacitated team orienteering problem: a bi-level filter-and-fan method. Eur. J. Oper. Res. **224**(1), 65–78 (2013)
9. Tsiligirides, T.: Heuristic methods applied to orienteering. J. Oper. Res. Soc. **35**(9), 797–809 (1984). https://doi.org/10.1057/jors.1984.162
10. Vansteenwegen, P., Mateo, M.: An iterated local search algorithm for single-vehicle cyclic inventory. Eur. J. Oper. Res. **237**(3), 802–813 (2014)

Reparameterization of Computational Chemistry Force Fields Using GloMPO (Globally Managed Parallel Optimization)

Michael Freitas Gustavo[1,2] and Toon Verstraelen[1(✉)]

[1] Center of Molecular Modeling, Ghent University, Ghent, Belgium
toon.verstraelen@ugent.be
[2] Software for Chemistry and Materials B.V., Amsterdam, Netherlands

Abstract. This paper introduces GloMPO (Globally Managed Parallel Optimization), an optimization framework which manages traditional optimization algorithms in real-time. For particularly difficult optimization tasks (like the reparameterization of computational chemistry models), classical approaches converge to poor minima. By managing an optimization task, GloMPO: 1) improves the efficiency with which an iteration budget is used; 2) provides better answers 60% to 80% of the time as compared to traditional approaches; and 3) is often able to identify several independent and degenerate minima.

Keywords: Computational chemistry · Black-box optimization · Global optimization · ReaxFF · Reparameterization

1 Introduction

The computational modeling of chemical systems has become a fundamentally important technology in the development of new chemical compounds and materials. The most important quantity required to study a given system is the potential energy surface (PES). A PES represents the energy of a system, typically as a function of atomic positions. These positions can be absolute, like in a Cartesian coordinate system, or relative, like expressing the position of two atoms as the distance between them.

The PES can be derived exactly from quantum mechanics and physical constants, but direct solution of these equations is not possible. Over the years, very many approaches have been developed to estimate the PES. All methods attempt to balance the computational cost of the calculation with the answer's accuracy.

These methods can be grouped into three broad categories: 1) *ab-initio* methods, which introduce some simplifying assumptions to the direct equations; 2) semi-empirical methods, which introduce simplifications to some aspects of the quantum mechanical equations through small parameterized models; or 3) fully empirical methods, which entirely substitute quantum mechanical equations

© Springer Nature Switzerland AG 2021
D. E. Simos et al. (Eds.): LION 2021, LNCS 12931, pp. 150–156, 2021.
https://doi.org/10.1007/978-3-030-92121-7_13

with parameterized models. In order to tackle large systems, fully empirical approaches are required as semi-empirical or *ab-initio* ones are too expensive.

ReaxFF [1] is an example of a fully empirical model which includes hundreds of global parameters, parameters for chemical elements and pairs, triplets and quadruplets of elements; often there is no clear physical interpretability for what these parameters represent. However, these sorts of models are the current state-of-the-art in simulating chemical reactions computationally on a large scale.

Determining the appropriate values for each parameter involves adjusting them until the differences between model predictions and some set of training data are minimized. In practice, the error functions, which combine all the deviations between training data and model into a sigle value, have proven to be extremely difficult to optimize. Not only are the problems very high dimensional, but they are also expensive to evaluate because they involve computations of properties of many chemical arrangements. Many such calculations involve a geometry optimization of the arrangements of atoms before chemical properties can be extracted. These optimizations are not deterministic and can lead to different arrangements and, hence, properties; in this way the error function can show stochastic behavior. The complexity of evaluation also hinders the calculation of analytical derivatives of the cost function. Finally, the PES can explode rapidly to very large or non-numerical values if atoms move very close together. This means that the error function is very oscillatory, and fluctuates over several orders of magnitude.

Optimization techniques typically used within the community have, to date, not been tremendously sophisticated because the expense, high-dimensionality, and black-box nature of the problems put severe constraints on the types of algorithms which can be applied. For example, SOPPE [2] (Sequential One-Parameter Parabolic Extrapolation) was the default approach for several years and typically required an expert to make manual fine-tuning adjustments.

Recently, more sophisticated algorithms such as CMA-ES, MCFF and OGOLEM have been applied with some success [3], however, one continues to face issues of stability, convergence to (bad) local minima, and computational efficiency. One particular hurdle has been the construction of the error function itself. While conceptually simple, it requires interfacing several pieces of software and strong programming skills. This represents a barrier for most researchers in the field.

2   GloMPO Package

To address these challenges, our team aims to create a standardized reparameterization interface which is simple enough to encourage broad adoption. ParAMS [4] is a new tool which is able to construct an error function for several computational chemistry models, including ReaxFF, GFN-xTB, DFTB and others. This deals with the 'upstream' work of the reparameterization problem. In this work we introduce GloMPO (Globally Managed Parallel Optimization) to handle the optimization and 'downstream' post-processing aspects.

GloMPO is a novel optimization framework which manages several child
optimizers in real-time. It aims to ensure that function evaluations are used
efficiently (i.e. not wasted exploring bad minima), and that all available computational resources are applied to the optimization task (since some algorithms
or codes are not adequately parallelized).

Figure 1 illustrates GloMPO's function. Given any global minimization task,
GloMPO starts several child optimizers which represent instances of any traditional optimization algorithm. GloMPO monitors the progress of each child and
compares them. It is empowered to terminate ones which appear to be converging to poor local minima (usually defined as a larger value than already seen by
another child). Terminated optimizers are replaced with new instances. When
good iterations are found by one child, GloMPO shares them with the others.
If the children can use this information in some way (for example, by sampling
the suggested point in subsequent evaluations), this can accelerate convergence.

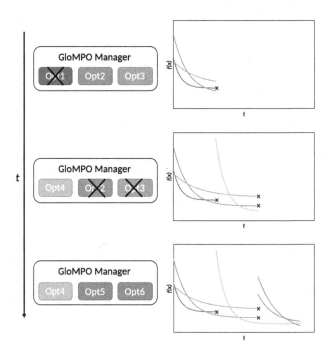

Fig. 1. Schematic of GloMPO's supervision of an optimization task over time. As child
optimizers appear to have converged to values higher than the best seen thus far, they
are shutdown and replaced with new instances.

By managing several independent optimizer instances, the distribution of
sampled points is spread over a wider area. In the context of ReaxFF, this often
produces answers which are virtually degenerate in cost function value, but far
apart in parameter space. Having access to several such solutions is invaluable

to computational chemists as they try and identify deficiencies in their training sets, and select one which works best for their application.

GloMPO is open-source and publicly available (see end). It supports running children as threads and processes, and further supports a second layer (of threads or processes) for parallelized function evaluations for optimizers which use multiple evaluations per iteration. GloMPO is compatible with any optimization task and optimization algorithm, and all major decision criteria are customizable. Finally, by providing a single interface to optimization regardless of the task or type of optimizer used, and given its modular design and customizability, GloMPO can be configured to manage complex optimization workflows through a single standardized interface.

3   Results

To quantify the effect of GloMPO's management, a benchmark test was run on the 66D Rastrigin, 20D Schwefel, 20D Deceptive, and 4D Shubert functions. 'Normal optimization' is defined here as repeated application of a traditional optimization algorithm (in this case CMA-ES [5]) to a minimization task. We compare GloMPO to repeated optimization because this is the traditional workflow for hard-to-optimize problems like the reparameterization of ReaxFF. Researchers, anticipating the difficulties they will face, often repeat optimizations and accept the lowest value of the set. We similarly define the best minimum of normal optimization as the lowest value found by any optimizer. In competition are GloMPO optimizations which operate on the same task, using the same type of child optimizers, and limited to the same iteration budget. In other words, the only point of difference between the two approaches is GloMPO's ability to replace optimizers exploring poor minima with new optimizer instances.

Overall, 4800 tests were conducted in this way. The distribution of answers found by normal repeated optimization and GloMPO are shown in Fig. 2. GloMPO demonstrates a clear improvement in both the mean final value and overall distribution of explored points; this performance proved to be robust across test configurations which changed the number of optimizers, available iteration budget, and other GloMPO settings. Overall, GloMPO identified a better answer than normal optimization $(70 \pm 13)\%$ of the time. GloMPO configurations which actively shared good evaluations with one another performed particularly well, winning $(86 \pm 6)\%$ of their comparisons.

GloMPO was also used to reparameterize two ReaxFF force fields. The first was a 12 parameter model describing cobalt [6] in various configurations, and the second was an 87 parameter model for disulfide compounds [7]. For statistical significance the optimization of each set was repeated 10 times.

Figure 3a shows the averages and standard deviations of the ten minima found by normal and GloMPO optimizations for the cobalt trainings. It again shows an improvement in quality of answers obtained by GloMPO. Not only is the average answer improved, the standard deviation is much tighter, in other words GloMPO has a higher chance of producing a good minimum as compared

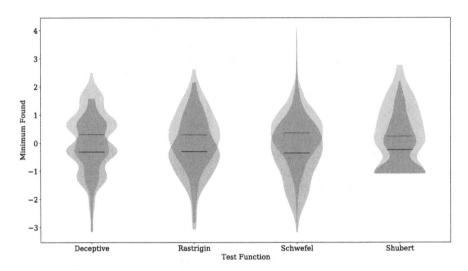

Fig. 2. Violin plot of final solutions from normal sequential optimization (solid red) and GloMPO managed (hatched blue) on four common global optimization test functions. Mean values are shown by the corresponding thick lines. Minima have been normalized to make the four functions visually comparable. Configurations were run 100 times each across 48 different configurations. GloMPO managed the same optimizer types as the normal optimization runs, CMA-ES. (Color figure online)

to normal optimization. Of particular note for the cobalt optimizations is the identification of degenerate sets; parameter sets whose error values are relatively close, but are located far from one another in parameter space. GloMPO on average found 3.4 degenerate sets during each optimization, sometimes as many as 6 in a single run. This is compared to normal optimizations which only averaged 2.4.

In the case of the disulfide model, the dimensionality was much higher and the error surface was poorly conditioned. This proved to be a tougher challenge than the cobalt trainings, and normal optimization approaches failed to consistently identify good solutions. Figure 3b shows the difference between the GloMPO minimum and normal optimization minimum across the 10 conducted runs. In this case, GloMPO performed better in eight out of the ten trials.

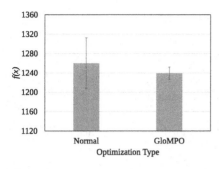
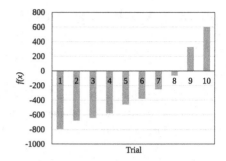

(a) Averaged minima found by GloMPO and normal optimization across ten runs each for a twelve parameter ReaxFF force field modeling cobalt. Error bars show the standard deviation of the minima sets.

(b) Differences between normal optimization and GloMPO minima found during reparameterizations of the ReaxFF disulfide training set. Both normal optimization and GloMPO used a maximum of 1 700 000 function evaluations distributed over their individual optimizers. Final values were sorted from lowest to highest before taking the difference. Negative values indicate a GloMPO victory.

Fig. 3. Comparison between normal and GloMPO optimization of two ReaxFF models. Repeated ten times each.

4 Conclusion

GloMPO is able to find better solutions to global optimization problems between 57% to 83% of the time when compared to traditional approaches. These figures are robust to changes in configuration and optimization task. GloMPO-generated solutions are on average lower, and have smaller variance, than their unmanaged counterparts. It also provides several qualitative advantages:

1. GloMPO prevents the inefficient use of function evaluations by preventing optimizers from exploring minima which are worse than that already seen by another optimizer.
2. GloMPO ensures that all computational resources are applied to the optimization problem.
3. GloMPO is fully customizable in terms of its decision making, and can be applied to any optimization problem using any combination of child algorithms. This flexibility means it can be leveraged as workflow manager.
4. GloMPO provides a standardized interface to optimization.
5. GloMPO provides standardized outputs regardless of the types of child optimizers used. The resulting log of trajectories can be analyzed and help in better conditioning of the optimization task.

Software Availability and Requirements

Project name: GloMPO
Project home page: www.github.com/mfgustavo/glompo
Operating system: Platform independent
Programming language: Python >3.6
Other requirements: >AMS2020.1 for ReaxFF interface
License: GPL-3.0

Acknowledgments. The authors thank the Flemish Supercomputing Centre (VSC), funded by Ghent University, FWO and the Flemish Government for use of their computational resources. Funding for the project was provided by the European Union's Horizon 2020 research and innovation program under grant agreement No 814143. T.V. is furthermore supported by the Research Board of Ghent University (BOF).

References

1. van Duin, A.C.T., Dasgupta, S., Lorant, F., Goddard, W.A.: ReaxFF: a reactive force field for hydrocarbons. J. Phys. Chem. A. **105**(41), 9396–9409 (2001). https://doi.org/10.1021/jp004368u
2. Larsson, H.R., van Duin, A.C.T., Hartke, B.: Global optimization of parameters in the reactive force field ReaxFF for SiOH. J. Comput. Chem. **34**, 2178–2189 (2013). https://doi.org/10.1002/jcc.23382
3. Shchygol, G., Yakovlev, A., Trnka, T., van Duin, A.C.T., Verstraelen, T.: ReaxFF parameter optimization with Monte-Carlo and evolutionary algorithms: guidelines and insights. J. Chem. Theory Comput. **15**(12), 6799–6812 (2019). https://doi.org/10.1021/acs.jctc.9b00769
4. Komissarov, L., Rüger, R., Hellström, M., Verstraelen, T.: ParAMS: parameter optimization for atomistic and molecular simulations. J. Chem. Inf. Model. (2021). https://doi.org/10.1021/acs.jcim.1c00333
5. Hansen, N., Ostermeier, A.: Completely derandomized self-adaptation in evolution strategies. Evol. Comput. **9**(2), 159–195 (2001). https://doi.org/10.1162/106365601750190398
6. LaBrosse, M.R., Johnson, J.K., van Duin, A.C.T.: Development of a transferable reactive force field for cobalt. J. Phys. Chem. A. **114**(18), 5855–5861 (2010). https://doi.org/10.1021/jp911867r
7. Müller, J., Hartke, B.: ReaxFF reactive force field for disulfide mechanochemistry, fitted to multireference ab initio data. J. Chem. Theory Comput. **12**(8), 3913–3925 (2016). https://doi.org/10.1021/acs.jctc.6b00461

Towards Structural Hyperparameter Search in Kernel Minimum Enclosing Balls

Hanna Kondratiuk[1,2]([✉]) and Rafet Sifa[1]([✉])

[1] Fraunhofer IAIS, Sankt-Augustin, Germany
{hanna.kondratiuk,rafet.sifa}@iais.fraunhofer.de
[2] University of Bonn, Bonn, Germany

Abstract. In this paper we attempt to provide a structural methodology for an informed hyper-parameter search when fitting Kernel Minimum Enclosing Balls (KMEBs) to datasets. Our approach allows us to control the number of resulting support vectors, which can be an important aspect for practical applications. To address this problem, we particularly focus on searching the width of Gaussian kernel and introduce two methods that are based on Greedy Exponential Search (GES) and Divide and Conquer (DaC) approaches. Both algorithms in case of non-convergence return the approximate result for the width value corresponding to the closest bound for the number of support vectors. We evaluate our method on standard benchmark datasets for prototype extraction using a Frank-Wolfe algorithm to fit the balls and conclude distance choices that yield descriptive results. Moreover, we compare the number of execution of the fitting algorithm and the number of iterations it took for our methods to result in convergence.

Keywords: Minimum enclosing balls · Kernel methods · Prototype extraction

1 Introduction

In the scope of data prototyping, the choice of the hyperparameters in KMEBs is crucial. Selected prototypes should not only be representative, but also quantifiable and interpretable at the same time in order to assist human analyst in decision making [4]. In order to achieve that, for the method proposed in [1], one needs to specify the usually low number of prototypes to compute as per interpritability for human analyst.

In [3] it was observed, that while finding KMEBs using Frank-Wolfe algorithm and Gaussian Kernel, the number of support vectors grows with the decreasing

---

Supported by the Competence Center for Machine Learning Rhine Ruhr (ML2R) which is funded by the Federal Ministry of Education and Research of Germany (grant no. 01—S18038B).

D. E. Simos et al. (Eds.): LION 2021, LNCS 12931, pp. 157–166, 2021.
https://doi.org/10.1007/978-3-030-92121-7_14

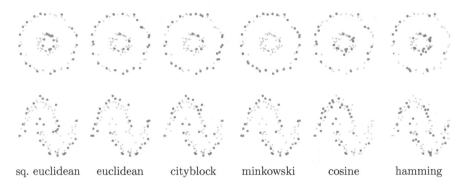

sq. euclidean   euclidean   cityblock   minkowski   cosine   hamming

Fig. 1. Toy example illustrating the idea of utilizing Kernel Minimum Enclosing Balls for prototyping using Gaussian kernel with a number of distances. Chosen prototypes are highlighted here in green and in blue is highlighted the data that is used to compute the prototypes. Prototypes are the support vectors, that are the points from dataset that correspond to the non-zero components of the Lagrange multipliers μ. Here we can observe, that change of the distance of the kernel retrieves different prototypes. (Color figure online)

values of gaussian width parameter σ, however not monotonically and with noticable oscillations. This fact can still be used to control the number of support vectors in the result of the Frank-Wolfe algorithm. As would be empirically shown in the scope of the paper, it retrieves meaningful prototypes for a number of toy and real-world datasets.

One illustration of the KMEBs algorithm providing the meaningful prototypes is shown at Fig. 1, retrieving the critical points or capturing the structure of that data. The data consists of one hundred points with noise ratio set to respectively 0.05 and 0.1 for two circles and moons examples. There were exactly thirty prototypes produced with KMEBs algorithm depending on the chosen distance. Here we may notice, that having chosen the ratio of number of prototypes to number of points set to 0.3, the algorithm is able to represent the structure of the data. The topology of circles and moons is clearly captured for squared euclidean distance, euclidean and hamming distance. Miskowski distance of the power 4 and cityblock distance is capturing the moon and outer circle topology as well, however number of the prototypes in the inner circle stay underrepresented for both cases.

On this example we can clearly see the practical application to specify and fix the number of the prototypes that we aspire to compute via KMEBs. The scope of use of the method is not only restricted in prototyping, but also can be used in a number of fields. For instance, it can be used to find the σ parameter in a novelty detection tasks, when the only parameter is a pre-defined ratio of number of support vectors to the number of data points.

---

1 Input: K, k, t_{max}
2 Initialize: $\mu_0 = \frac{1}{n}\mathbf{1}$
3 **for** $t = 0, \ldots, t_{max}$ **do**
4     $\eta = \frac{2}{t+2}$
5   Choose i $v_i \in \mathbb{R}^n$ s.t. $i = \text{argmin}_{j \in \{1,\ldots,n\}} [2K\mu_t - k]_j$
6     $\mu_{t+1} = \mu_t + \eta(v_i - \mu_t)$
7 Output: $\mu_{t_{max}}$

Algorithm 1: Frank-Wolfe Algorithm for finding KMEBs. The Frank-Wolfe algorithm is used in order to solve the problem in the following setting. Given a initial guess μ_0 for the solution, each iteration of the algorithm determines which direction $\nu_t \in \Delta^{n-1}$ minimizes the inner product $\nu^\top \nabla \mathcal{D}_{kernel}(\mu)$ and applies a conditional gradient update $\mu_{t+1} = \mu_t + \eta_t(\nu_t - \mu_t)$ with decreasing step size $\eta_t = \frac{2}{t+2} \in [0, 1]$.

2   Overview of the Problem

Minimum Enclosing Balls (MEB) problem is posed as

$$c_*, r_* = \underset{c, r}{\text{argmin}}\ r^2 \qquad \left\| x_i - c \right\|^2 - r^2 \leq 0 \qquad i \in [1, \ldots, n].$$

where x_i is the point from the dataset, c is the center of the ball and r is ball's radius.

The kernelized MEB problem is derived as shown in [1] by using Karush-Kuhn-Tucker conditions and kernel trick, and it takes form of:

$$\mu_* = \underset{\mu}{\text{argmin}}\ \mu^\top K \mu - \mu^\top k$$

where the goal is to find a Lagrange multiplier μ that resides in the standard simplex $\Delta^{n-1} = \{\mu \in \mathbb{R}^n | \mu \succeq 0, \mu^\top \mathbf{1} = 1\}$, $K \in \mathbb{R}^{n \times n}$ is a kernel matrix, k contains kernel matrix's diagonal (i.e. $k = \text{diag}[K]$) and $\mu \in \mathbb{R}^n$ contains Lagrange multipliers.

Support vectors in the context of KMEBs problem are the data points that correspond to the non-zero components of Lagrange multipliers vector $\mu_{t_{max}}$, where $\mu_{t_{max}}$ is the result of the iterative Frank-Wolfe algorithm summarized in Algorithm 1. The function of support vectors number for Gaussian kernel is defined as

$$p(t_{max}, \sigma, d) = \sum_{i=0}^{n} \text{sgn}(\mu_{t_{max}_i})$$

with sign function taking only non-negative values as the output of the Frank-Wolfe algorithm is essentially traversing through the simplex vertices.

Given the fixed number p^* of support vectors, we attempt to find such σ that for fixed number of iterations inside of Frank-Wolfe algorithm t^*_{max} and provided distance function d^* holds:

$$p(t^*_{max}, \sigma, d^*) = p^* \tag{1}$$

The dependence of number of support vectors on σ is not monotone, hence it introduces additional challenges in hyper-parameter search. Firstly, as the function of support vectors numbers returns different discrete values, that eliminates the usage of traditional optimization methods based on the assumption of existing derivative. Secondly, there is no theoretical guarantee, that for the desired number of support vectors p^*: $\exists \sigma : p(t^*_{max}, \sigma, d^*) = p^*$.

Choosing p^* or the ratio of number of prototypes to number of points is dataset and application dependent. The ratio of number of prototypes to number of points is defined as

$$r = \frac{p^*}{n} \in \left(\frac{\min_{\forall \sigma > 0} p(\sigma)}{n}, \frac{\max_{\forall \sigma > 0} p(\sigma)}{n} \right)$$

with number of support vectors $p^* \in [1, \min(n, t^*_{max})]$ and n being the data size. For instance, in the behaviour prototyping scope, we would specify the number of support vectors p^* according to the number of prototypes we want to extract, that is usually a small natural number greater than 1. In novelty detection application, the ratio r is defined in between $\frac{1}{n}$ and $\frac{\min(n, t^*_{max})}{n}$, with second value corresponding to using all the points possible as support vectors (that is an extreme case) and $\frac{1}{n}$ corresponding to the other extreme of computing only one support vector. In the scope of classification, the ratio of support vectors r can be chosen using cross-validation while for novelty detection using the false positive rate.

However, we note, that it is not guaranteed that there exists parameter σ such that $p = 1$ or $p = \min(n, t^*_{max})$. On the contrary, there are examples of the datasets when the minimum number of support vectors over high-power exponential interval is greater than 1 and the maximum number of support vectors is smaller than $\min(n, t^*_{max})$ for infinitely small σ. Despite that, there is a strong empirical evidence, that the intermediate values of natural number sequence exist:

$$\exists \sigma > 0 : \quad \forall p^* \in \{ j \in \mathbb{N}^* : \min_{\forall \sigma > 0} p(\sigma) \le j \le \max_{\forall \sigma > 0} p(\sigma) \} : p(t^*_{max}, \sigma, d^*) = p^*$$

as we may observe in the experimental section.

Given those facts, we propose two depth-restricted algorithms: Greedy Exponential Search (GES) and Divide and Conquer (DaC), attempting to solve the root finding problem for (1) and returning the approximate solution for the case when the algorithm does not converge in l_{max} iterations.

```
1  Input: p*, distance d*, t_max, depth l_max, number of intervals to spawn c
2  Initialize: exponential interval, svs_diff = max(n, t*_max)
3  for t = 0, ..., l_max do
4  │    svs_prev = 0, σ_prev = 0
5  │    for σ in exponential interval do
6  │    │    svs = p(t*_max, σ, d*)
7  │    │    if svs = p* then
8  │    │    └    return σ
9  │    │    if svs ≠ 1 AND svs_prev = 1 then
10 │    │    └    σ_higher = σ, svs_higher = svs
11 │    │    if svs_prev ≠ min(n, t*_max) AND svs = min(n, t*_max) then
12 │    │    └    σ_lower = σ_prev, svs_lower = svs_prev
13 │    │    if |svs - p*| < svs_diff then
14 │    │    │    σ_exp = σ
15 │    │    └    svs_diff = |svs - p*|
16 │    └    svs_prev = svs, σ_prev = σ
17 │    exponential interval = tsi(σ_exp, σ_higher, c) ⋃ tsi(σ_exp, σ_lower, c)
18 return σ_exp
```

Algorithm 2: Greedy Exponential Search. After initializing the desired number of support vectors p^*, distance d^*, number of Frank-Wolfe iterations t^*_{max}, the algorithms depth l_{max} and number of intervals c at line 1 to spawn around the found sigma. At line 2 we initialize *exponential interval* of i.e. $[10^{40}, 10^{39}, \ldots, 10^{-40}]$ and initialize the difference between desired support vectors number p^* and $p(t^*_{max}, \sigma, d^*)$ for current sigma. We denote this difference by svs_{diff} and initialize it as the maximum possible difference of support vectors, that is $\max(n, t^*_{max})$. For depth l_{max} we would repeat the same operation enclosed in for loop with line 3. At line 5 we traverse through the initial interval where for each sigma we compute the number of support vectors svs (line 6) and define the 'higher' and 'lower' sigma values in lines 10 and 12 that correspond to the extreme cases of sigma parameter. In line 14 we equalize σ_{exp} value to the σ that had the nearest support vector number to the desired p^*. Thus, we can assure that we keep the closest value of the support vectors. At line 17 we span the union of two sided intervals for the closest value for σ_{exp} corresponding to the closest value to p^*. Then this interval is used at depth 1 in line 5 and the search continues.

3   Proposed Approach

We concentrate our attempts to approach the problem by narrowing it down to Gaussian kernel with chosen distance function d^* and the assumption, that with the growth of σ the number of support vectors decreases with noticeable oscillations. The proposed algorithms are summarized at Algorithm 2 and Algorithm 3, and both of them provide the approximate solution in case of exceeding the pre-defined depth parameter l_{max}. GES and DaC also rely on the assumption that $t^*_{max} \gg 1$, $n \gg 1$, $\max_{\forall \sigma > 0} p(t^*_{max}, \sigma, d^*) = \min(n, t^*_{max})$

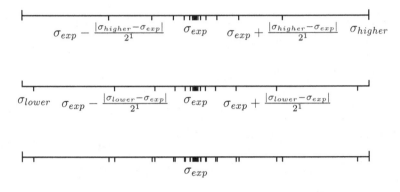

Fig. 2. Two-sided interval (tsi) search idea utilized in GES

and $\min_{\forall \sigma > 0} p(t^*_{max}, \sigma, d^*) = 1$, as there is no prior knowledge what are the values $\min_{\forall \sigma > 0} p(\sigma)$. If $\max_{\forall \sigma > 0} p(t^*_{max}, \sigma, d^*) = \min(n, t^*_{max})$ and $\min_{\forall \sigma > 0} p(t^*_{max}, \sigma, d^*) = 1$ does not hold, then we can replace 1 by $\min_{\forall \sigma > 0} p(\sigma)$ in line 9 and $\min(n, t^*_{max})$ by $\max_{\forall \sigma > 0} p(t^*_{max}, \sigma, d^*)$ in line 11 in both Algorithm 2 and Algorithm 3. Finding the maximum and minimum values is possible during the first step of going over initial *exponential interval*.

As the initialization step, both of the algorithms use the powers of 10 as the way to quickly identify the region when kernel provides the meaningful values for the prototyping with the intervals. It can be seen at lines 5 and called *exponential interval* in GES and *interval* in DaC respectively. Gaussian kernel can take both extremes, from being **1** matrix corresponding to the high values of σ and only one prototype, to identity matrix, corresponding to the low sigma value and maximum number of support vectors, that has empirically shown to be smaller or equal to the $\min(t_{max}, n)$.

The Greedy Exponential Search (GES) at each iteration saves the new appearances of the number of support vectors and two critical values of sigma $\sigma_{higher}, \sigma_{lower}$, corresponding to the minimum number of support vectors (that is 1) and maximum number of support vectors. Then such value σ_{exp} is found that corresponds to the support vectors number closest to p^*. The visual representation of the *tsi* can be found at Fig. 2. For the σ' two-sided exponential intervals are constructed for $\sigma_{higher}, \sigma_{lower}$, and the resulting sigma interval for the next step is taken as the union of the two. The two-sided interval *tsi* spawns the bi-sectional division over the interval $\sigma_{exp} \pm 2^{-i}|\sigma_{higher} - \sigma_{exp}|$ for $tsi(\sigma_{exp}, \sigma_{higher}, c)$ and $\sigma_{exp} \pm 2^{-i}|\sigma_{lower} - \sigma_{exp}|$ for $tsi(\sigma_{exp}, \sigma_{lower}, c)$ $\forall i \in \{1 \ldots c\}$. With the unity of those two intervals the next iteration step of GES is performed until the resulting σ^* is found or l_{max} iterations have passed. The following approach is inspired by the observation in [2] that for support vector machines searching through exponentially growing sequences of the hyperparameters is a practical

```
1  Input: p*, distance, t_max, depth l_max, number of equal intervals to spawn g
2  Initialize: interval, svs_diff = max(n, t*_max), count = 0
3  for t = 0, ..., l_max do
4  |   svs_prev = 0, σ_prev = 0
5  |   for σ in interval do
6  |   |   svs = p(t*_max, σ, d*), count = count + 1
7  |   |   if svs = p* then
8  |   |   |_  return σ
9  |   |   if svs ≠ 1 AND svs_prev = 1 then
10 |   |   |_  σ_higher = σ, svs_higher = svs
11 |   |   if svs_prev ≠ min(n, t*_max) AND svs = min(n, t*_max) then
12 |   |   |_  σ_lower = σ_prev, svs_lower = svs_prev
13 |   |   if |svs - p*| < svs_diff then
14 |   |   |   σ_closest = σ
15 |   |   |_  svs_diff = |svs - p*|
16 |   |_  svs_prev = svs, σ_prev = σ
17 |   if |p(t*_max, σ_higher, d*) - p*| > |p(t*_max, σ_lower, d*) - p*| then
18 |   |_  σ'_higher = (σ_higher + σ_lower)/2, direction = True
19 |   else
20 |   |_  σ'_lower = (σ_higher + σ_lower)/2, direction = False
21 |   if p* ∉ [p(σ'_higher), p(σ'_lower)] then
22 |   |   if direction then
23 |   |   |_  σ'_lower = (σ_higher + σ_lower)/2
24 |   |   else
25 |   |   |_  σ'_higher = (σ_higher + σ_lower)/2
26 |   |_  interval = linspace(σ_higher, σ_lower, g + 5 · count)
27 |   else
28 |   |_  interval = linspace(σ_higher, σ_lower, g)
29 return σ_closest;
```

Algorithm 3: Divide and Conquer. In the similar fashion, we define the interval borders as meaningful values (that are not corresponding to the extreme 1 support vector and $\min(n, t^*_{max})$ support vectors) at lines 10 and 12, and save in $\sigma_{closest}$ the value of σ that yields the smaller difference svs_{diff}. Then in line 17 we define to which direction it makes sense to adhere and cut the respective interval in a half in 18 and 20. If the desired svs number is not encapsulated in the interval $[p(\sigma'_{higher}), p(\sigma'_{lower})]$, then we assume that we come one step back and jump to another side of the interval with breaking it in more pieces as shown in 26 as it hints at the non-monotonic behavior.

method to identify the fitting ones. Opposed to uniform grid search, it spawns the bi-sectional search of depth c only around the σ values that are corresponding to support vector values that are greater than 1, lower than $\min(n, t^*_{max})$ and that full-fill the condition on line 13 of GES. By full-filling that condition we ensure that it would span the tsi around the sigma closest to the desired number p^*, hence we call this algorithm Greedy Exponential Search.

The summary of the second proposed algorithm is shown at Algorithm 3. The Divide and Conquer approach is concentrated on the assumption, that we acquire quicker convergence if we cut the interval in two, with traversing through the half-interval where respective number of support vectors is closer to the p^*. We find the values $\sigma_{higher}, \sigma_{lower}$ such that $1 < p(t^*_{max}, \sigma, d^*) < \min(t^*_{max}, n)$, that correspond to the highest and lowest support vector numbers in the interval. Over this interval we perform the bisection to two intervals. We choose the left half-interval if $|p(t^*_{max}, \sigma'_{higher}, d^*) - p^*| > |p(t^*_{max}, \sigma'_{lower}, d^*) - p^*|$ and right interval otherwise, where σ'_{lower} and σ'_{higher} are two values defining the half-interval. Then we check, whether p^* is in the interval of $[p(\sigma'_{higher}), p(\sigma'_{lower})]$. If the interval chosen does not include p^*, then the algorithm returns for one step and chooses the other side of the interval. After that, the half-interval is divided into g equal parts and this division is passed as an interval to the next iteration. There is a possibility that the interval limits show p^* is not included, but the inner values of the interval may contain p^* (due to the non-monotone and oscillating behavior). That is why in the algorithm depth value *count* is introduced in order to account for the possibility. It is used in case if the support vector number p^* does not lie in the subdivision to enlarge the number of linspace divisions.

GES can we also viewed as complementary approach to the DaC, as they both concentrate on different strategies. Namely, GES is going more "wide" than "deep" using the *tsi*, as it makes use of spanning the intervals around the obtained sigma values corresponding to support vectors that become iteratively closer to the desired value p^*. For DaC the algorithm rather searches the σ value "deep" reducing the interval in a half until while the number of support vectors of interval sides includes the desired number of support vectors p^*. However, both algorithms might not reach the resulting σ as the both intervals at some step l may not even contain p^*.

4   Results

In our experimental set-up the initialization of the interval was $[10^{-20}, 10^{-19}, \ldots, 10^{19}, 10^{20}]$, and the target value of support vectors is shown at the table as p^*, the maximum depth of the algorithm was fixed to $l_{max} = 10$, as l we assign the number of the runs before the algorithm converged and number of intervals spawned is $g = 10$, $c = 20$. The number of iterations in Frank-Wolfe was set to $t^*_{max} = 2000$ and seed was fixed to 120 for random noise in Moon and Circles conceptual examples. At the table we can observe that Divide and Conquer is inclined to have a greater number of runs of Frank-Wolfe algorithm comparing to Greedy Exponential Search. Coherently, it finds σ corresponding to the targeted value of support vectors $p(\sigma) = p^*$ on almost all the example datasets, including the toy datasets shown in the introduction section (Moons, Circles), as well as benchmark datasets as MNIST, Faces and Faces Olivetti. On the example of the experiments, we may conclude that the algorithm converges in all the cases but with squared euclidean distance on MNIST and Olivetti Faces. Greedy Exponential Search in the case of reaching depth $l = 10$ still

Table 1. Comparison of the DaC and GES algorithms for each of the used distances. Parameter l is the depth, when the solution was found, with 11 meaning that the algorithm did not converge with $l_{max} = 10$ iterations and returns approximate result. k is the number of calls to Frank-Wolfe algorithm and $p(\sigma)$ is the number of support vectors for the returned sigma. In the case of convergence $p(\sigma) = p^*$

Data	p^*	sqeucl.			eucl.			manh.			mink.			cos			hamming		
		l	k	$p(\sigma)$	l	k	$p(\sigma)$	l	k	$p(\sigma)$	l	k	$p(\sigma)$	l	k	$p(\sigma)$	l	k	$p(\sigma)$
Moons	30	7	1954	30	3	78	30	10	119264	30	5	857	30	5	1380	30	5	886	30
Circles	30	3	88	30	6	334	30	3	83	30	6	335	30	11	390	31	5	886	30
MNIST	10	11	2240	11	3	193	10	3	56	10	3	189	10	3	247	10	3	200	10
Faces	10	3	76	10	3	269	10	3	59	10	3	155	10	3	89	10	3	83	10
Olivetti	10	11	2190	11	3	179	10	6	335	10	6	827	10	1	15	10	3	152	10

(a) DaC

Data	p^*	sqeucl.			eucl.			manh.			mink.			cos			hamming		
		l	k	$p(\sigma)$	l	k	$p(\sigma)$	l	k	$p(\sigma)$	l	k	$p(\sigma)$	l	k	$p(\sigma)$	l	k	$p(\sigma)$
Moons	30	4	256	30	4	271	30	3	147	30	4	273	30	4	258	30	5	345	30
Circles	30	4	271	30	4	222	30	6	314	30	4	268	30	3	151	30	5	345	30
MNIST	10	11	763	11	11	801	13	7	394	10	11	763	11	3	155	10	11	763	11
Faces	10	5	230	10	4	197	10	5	316	10	7	358	10	3	188	10	4	276	10
Olivetti	10	5	234	10	4	187	10	6	357	10	2	103	10	1	15	10	4	196	10

(b) GES

has comparatively low number of Frank-Wolfe iterations, that however results in more frequent approximate solutions, such as, for instance, euclidean distance case for MNIST (Table 1). The resulting prototypes computed on MNIST and Faces datasets are shown at Fig. 3. For each distance choice algorithms extract exactly 10 prototypes in accordance with $p^* = 10$ in Table 1.

Fig. 3. Comparison of prototypes using the squared euclidean, euclidean, cityblock, minkowski, cosine and hamming distances obtained from GES and DaC algorithms. Hamming distance provides the most meaningful prototypes in MNIST, that is explained by the nature of the data as data images are really sparse and the fact that hamming distance calculates the number of the different entries in the data. Minkowski distance identifies few types of number 5, euclidean distance identifies all the classes apart from class 8. In Faces dataset, on contrary, the most meaningful prototypes are provided by euclidean, squared euclidean, cityblock and minkowski distances, allowing to have a variety of prototypes to preview a dataset.

5 Future Work

As a further research objective, we plan to optimize the presented algorithms, expand the number of the benchmark datasets, investigate different use cases (such as novelty detection [5]) and address the problem not only with the Gaussian kernel, but also with Distance-Based and Categorical (i.e. k^0, k^1) kernels.

References

1. Bauckhage, C., Sifa, R.: Joint selection of central and extremal prototypes based on kernel minimum enclosing balls. In: Proceedings DSAA (2019)
2. Hsu, C.W., Chang, C.C., Lin, C.J., et al.: A practical guide to support vector classification (2003)
3. Kondratiuk, H., Sifa, R.: Towards an empirical and theoretical evaluation of gradient based approaches for finding kernel minimum enclosing balls. In: Proceedings DSAA. IEEE (2020)
4. Sifa, R.: Matrix and Tensor Factorization for Profiling Player Behavior. LeanPub, Victoria (2019)
5. Sifa, R., Bauckhage, C.: Novelty discovery with kernel minimum enclosing balls. In: Kotsireas, I.S., Pardalos, P.M. (eds.) LION 2020. LNCS, vol. 12096, pp. 414–420. Springer, Cham (2020). https://doi.org/10.1007/978-3-030-53552-0_37

Using Past Experience for Configuration of Gaussian Processes in Black-Box Optimization

Jan Koza[1]([✉]) [iD], Jiří Tumpach[2], Zbyněk Pitra[3] [iD], and Martin Holeňa[4] [iD]

[1] Faculty of Information Technology, Czech Technical University, Prague, Czechia
kozajan@fit.cvut.cz
[2] Faculty of Mathematics and Physics, Charles University, Prague, Czechia
tumpach@cs.cas.cz
[3] Faculty of Nuclear Sciences and Physical Engineering, Czech Technical University, Prague, Czechia
[4] Academy of Sciences, Institute of Computer Science, Prague, Czechia
martin@cs.cas.cz

Abstract. This paper deals with the configuration of Gaussian processes serving as surrogate models in black-box optimization. It examines several different covariance functions of Gaussian processes (GPs) and a combination of GPs and artificial neural networks (ANNs). Different configurations are compared in the context of a surrogate-assisted version of the Covariance Matrix Adaptation Evolution Strategy (CMA-ES), a state-of-the-art evolutionary black-box optimizer. The configuration employs a new methodology, which consists of using data from past runs of the optimizer. In that way, it is possible to avoid demanding computations of the optimizer only to configure the surrogate model as well as to achieve a much more robust configuration relying on 4600 optimization runs in 5 different dimensions.

The experimental part reveals that the lowest rank difference error, an error measure corresponding to the CMA-ES invariance with respect to monotonous transformations, is most often achieved using rational quadratic, squared exponential and Matérn $\frac{5}{2}$ kernels. It also reveals that these three covariance functions are always equivalent, in the sense that the differences between their errors are never statistically significant. In some cases, they are also equivalent to other configurations, including the combination ANN-GP.

Keywords: Black-box optimization · Gaussian processes · Artificial neural networks · Covariance functions · Surrogate modeling

1 Introduction

Gaussian processes (GPs) [35] are stochastic models increasingly important in machine learning. They are frequently used for regression, providing an estimate

© Springer Nature Switzerland AG 2021
D. E. Simos et al. (Eds.): LION 2021, LNCS 12931, pp. 167–182, 2021.
https://doi.org/10.1007/978-3-030-92121-7_15

not only of the expected value of the dependent variable, but also of the distribution of its values. They are also encountered in classification and play an important role in optimization, especially black-box optimization [20,22], where the distribution of values provides search criteria that are alternatives to the objective function value, such as the probability of improvement or expected improvement. Moreover, the importance of GPs in machine learning incited attempts to combine them with the kind of machine learning models most frequently employed in the last decades – artificial neural networks (ANNs), including deep networks [5,6,8,20,21,40]. The importance of this research direction is supported by recent theoretical results concerning the relationships of asymptotic properties of important kinds of ANNs to properties of GPs [27,31,32].

In optimization, the alternative search criteria provided by a GP are useful primarily if GPs serve as surrogate models for *black-box objective functions*, i.e. pointwise observable functions for which no analytic expression is available and the function values have to be obtained either empirically, e.g. through measurements or experiments, or by means of simulations. *Surrogate modeling*, aka *metamodeling* [4,10,11,26,33,34,36] is an approach used in black-box optimization since the late 1990s to deal with situations when the evaluation of a black-box objective is expensive, e.g., in time-consuming simulations [13], or due to costly experiments [2]. The surrogate model is a regression model trained on the data evaluated so far using the black-box objective function. Predictions of such model are used instead of the original function to save the expensive evaluations.

Using GPs as surrogate models in black-box optimization brings a specific problem not known from using them directly as optimizers (e.g. in [22]), nor from GP models for regression or classification. It is connected with assessing the quality of the surrogate model approximation for different choices of GP covariance functions. As will be recalled below in Subsect. 2.1, the covariance function of a GP is crucial for modeling relationships between observations corresponding to different points in the input space. This problem consists in the fact that although the quality of the GP depends solely on its predictions and the true values of the black-box objective function, to obtain the prediction needs in addition a complete run of an optimizer, which is typically much more demanding. The problem is particularly serious when combing GPs and ANNs because then separate runs are needed for all considered combinations of GP covariance functions with the considered ANN topologies. That motivated the research reported in this paper, in which we have investigated 10 GP configurations, differing primarily through the choice of the covariance function: Instead of running the optimizer specifically for the configuration task, we used comprehensive data from its 4600 previous runs.

The next section recalls the fundamentals of Gaussian processes and their integration with neural networks. In Sect. 3, the employed past optimization data and the algorithm with which they were obtained are introduced. The core part of the paper is Sect. 4, which brings experimental results of configuring GPs based on those data.

2   Gaussian Processes and Their Neural Extension

2.1   Gaussian Processes

A *Gaussian process* on a set $\mathcal{X} \subset \mathbb{R}^d, d \in \mathbb{N}$, is a collection of random variables $(f(\mathbf{x}))_{\mathbf{x} \in \mathcal{X}}$, any finite number of which has a joint Gaussian distribution [35]. It is completely specified by a *mean function* $\mu : \mathcal{X} \to \mathbb{R}$, typically assumed constant, and by a *covariance function* $\kappa : \mathcal{X} \times \mathcal{X} \to \mathbb{R}$ such that for $\mathbf{x}, \mathbf{x}' \in \mathcal{X}$,

$$\mathbb{E}f(\mathbf{x}) = \mu, \quad \mathrm{cov}(f(\mathbf{x}), f(\mathbf{x}')) = \kappa(\mathbf{x}, \mathbf{x}'). \tag{1}$$

Therefore, a Gaussian process is usually denoted $\mathcal{GP}(\mu, \kappa)$ or $\mathcal{GP}(\mu, \kappa(\mathbf{x}, \mathbf{x}'))$.

The value of $f(\mathbf{x})$ is typically accessible only as a *noisy observation* $y = f(\mathbf{x}) + \varepsilon$, where ε is a zero-mean Gausssian noise with a variance $\sigma_n > 0$. Then

$$\mathrm{cov}(\mathbf{y}, \mathbf{y}') = \kappa(\mathbf{x}, \mathbf{x}') + \sigma_n^2 \mathbb{I}(\mathbf{x} = \mathbf{x}'), \tag{2}$$

where $\mathbb{I}$(proposition) equals for a true proposition 1, for a false proposition 0.

Consider now the prediction of the random variable $f(\mathbf{x}_\star)$ in a point $\mathbf{x}_\star \in \mathcal{X}$ if we already know the observations $y_1, \ldots, y_n$ in points $\mathbf{x}_1, \ldots, \mathbf{x}_n$. Introduce the vectors $\mathbf{x} = (\mathbf{x}_1, \ldots, \mathbf{x}_n)^\top$, $\mathbf{y} = (y_1, \ldots, y_n)^\top = (f(\mathbf{x}_1) + \varepsilon, \ldots f(\mathbf{x}_n) + \varepsilon)^\top$, $\boldsymbol{\mu}(\mathbf{x}) = (\mu(\mathbf{x}_1), \ldots, \mu(\mathbf{x}_\lambda))$, $\mathbf{k}_\star = (\kappa(\mathbf{x}_1, \mathbf{x}_\star), \ldots, \kappa(\mathbf{x}_n, \mathbf{x}_\star))^\top$ and the matrix $K \in \mathbb{R}^{n \times n}$ such that $(K)_{i,j} = \kappa(\mathbf{x}_i, \mathbf{x}_j) + \sigma_n^2 \mathbb{I}(i = j)$. Then the probability density of the vector $\mathbf{y}$ of observations is

$$p(\mathbf{y}; \boldsymbol{\mu}, \kappa, \sigma_n^2) = \frac{\exp\left(-\frac{1}{2}(\mathbf{y} - \boldsymbol{\mu}(\mathbf{x}))^\top K^{-1}(\mathbf{y} - \boldsymbol{\mu}(\mathbf{x}))\right)}{\sqrt{(2\pi)^n \det(K)}}, \tag{3}$$

where $\det(A)$ denotes the determinant of a matrix A. Furthermore, as a consequence of the assumption of Gaussian joint distribution, also the conditional distribution of $f(\mathbf{x}_\star)$ conditioned on $\mathbf{y}$ is Gaussian, namely

$$\mathcal{N}(\mu(\mathbf{x}_\star) + \mathbf{k}_\star^\top K^{-1}(\mathbf{y} - \boldsymbol{\mu}(\mathbf{x})), \kappa(\mathbf{x}_\star, \mathbf{x}_\star) - \mathbf{k}_\star^\top K^{-1} \mathbf{k}_\star). \tag{4}$$

According to (2), the relationship between the observations $\mathbf{y}$ and $\mathbf{y}'$ is determined by the covariance function κ. In the reported research, we have considered 9 kinds of covariance functions, listed below. In their definitions, the notation $r = \|\mathbf{x}' - \mathbf{x}\|$ is used, and among the parameters of κ, aka hyperparameters of the GP, frequently encountered are $\sigma_f^2, \ell > 0$, called *signal variance* and *characteristic length-scale*, respectively. Other parameters will be introduced for each covariance function separately.

(i)   *Linear:* $\kappa_{\mathrm{LIN}}(\mathbf{x}, \mathbf{x}') = \sigma_0^2 + \sigma_f^2 \mathbf{x}^\top \mathbf{x}'$, with a bias σ_0^2.

(ii) *Quadratic* is the square of the linear covariance: $\kappa_{\mathrm{Q}}(\mathbf{x}, \mathbf{x}') = (\sigma_0^2 + \sigma_0^2 \mathbf{x}^\top \mathbf{x}')^2$.

(iii) *Rational quadratic:* $\kappa_{\mathrm{RQ}}(\mathbf{x}, \mathbf{x}') = \sigma_f^2 \left(1 + \frac{r^2}{2\alpha\ell^2}\right)^{-\alpha}$, with $\alpha > 0$.

(iv) *Squared exponential:* $\kappa_{\mathrm{SE}}(\mathbf{x}, \mathbf{x}') = \sigma_f^2 \exp\left(-\frac{r^2}{2\ell^2}\right)$.

(v) *Matérn* $\frac{5}{2}$: $\kappa_{\mathrm{Mat}}^{\frac{5}{2}}(\mathbf{x}, \mathbf{x}') = \sigma_f^2 \left(1 + \frac{\sqrt{5}r}{\ell} + \frac{5r^2}{3\ell^2}\right) \exp\left(-\frac{\sqrt{5}r}{\ell}\right)$.

(vi) *Neural network:*

$$\kappa_{\mathrm{NN}}(\mathbf{x}, \mathbf{x}') = \sigma_f^2 \arcsin\left(\frac{2\tilde{\mathbf{x}}^\top \Sigma \tilde{\mathbf{x}}'}{\sqrt{(1 + 2\tilde{\mathbf{x}}^\top \Sigma \tilde{\mathbf{x}})(1 + 2\tilde{\mathbf{x}}'^\top \Sigma \tilde{\mathbf{x}}')}}\right) \qquad (5)$$

where $\tilde{\mathbf{x}} = (1, \mathbf{x}), \tilde{\mathbf{x}}' = (1, \mathbf{x}')$ and $\Sigma \in \mathbb{R}^{d+1 \times d+1}$ is positive definite. The name of this covariance function is due to the fact that for a simple ANN with d inputs connected to one output, computing the function $f(\mathbf{x}) = \mathrm{erf}(w_0 + \sum_{j=1}^d w_j x_j)$ with $\mathrm{erf}(z) = \frac{2}{\sqrt{\pi}} \int_0^z e^{-t^2} \mathrm{d}t$, if the weights are random variables and the vector $(w_0, w_1, \ldots, w_d)$ has the distribution $\mathcal{N}(0, \Sigma)$, then

$$\mathbb{E}f(\mathbf{x})f(\mathbf{x}') = \kappa_{\mathrm{NN}}(\mathbf{x}, \mathbf{x}'). \qquad (6)$$

(vii) The isotropic version of a covariance function with *spatially varying length-scales* $\ell_j : \mathcal{X} \to (0, +\infty), j = 1, \ldots, d$, originally proposed by Gibbs [15]:

$$\kappa_{\mathrm{Gibbs}}(\mathbf{x}, \mathbf{x}') = \sigma_f^2 \left(\frac{2\ell(\mathbf{x})\ell(\mathbf{x}')}{\ell^2(\mathbf{x}) + \ell^2(\mathbf{x}')}\right)^{\frac{d}{2}} \exp\left(-\frac{r^2}{\ell^2(\mathbf{x}) + \ell^2(\mathbf{x}')}\right). \qquad (7)$$

(viii) One *composite covariance function*, namely the sum of κ_{SE} and κ_{Q}:
$\kappa_{\mathrm{SE+Q}}(\mathbf{x}, \mathbf{x}') = \kappa_{\mathrm{SE}}(\mathbf{x}, \mathbf{x}') + \kappa_{\mathrm{Q}}(\mathbf{x}, \mathbf{x}')$.

(ix) *Spectral mixture* of Q components, $Q \in \mathbb{N}$ [39, 40]:

$$\kappa_{SM}(\mathbf{x}, \mathbf{x}') = \sum_{q=1}^Q c_q \sqrt{(2\pi\ell_q^2)^d} \prod_{j=1}^d \exp(-2\pi^2\ell_q^2(x_j - x_j')^2) \cos(2\pi(\boldsymbol{\mu}_q)_j(x_j - x_j')),$$

$$(8)$$

where $c_q \in \mathbb{R}, \ell_q > 0, \boldsymbol{\mu}_q \in \mathbb{R}^d, q = 1, \ldots, Q$. Its name is due to the fact that the function S defined $S(s) = \sqrt{(2\pi\ell_q^2)^d} \exp(-2\pi^2\ell^2\|\mathbf{s}\|^2)$ is the spectral density of κ_{SE}.

2.2   GP as the Output Layer of a Neural Network

The approach integrating GPs into an ANN as its output layer has been independently proposed in [6] and [40]. It relies on the following two assumptions:

1. If n_I denotes the number of the ANN input neurons, then the ANN computes a *mapping* net *of* n_I-*dimensional input values into the set* $\mathcal{X}$ on which is the GP. Consequently, the number of neurons in the last hidden layer equals the dimension d, and the ANN maps an input v into a point $x = \mathrm{net}(v) \in \mathcal{X}$, corresponding to an observation $f(x + \varepsilon)$ governed by the GP (Fig. 1). From the point of view of the ANN inputs, we get $\mathcal{GP}(\mu(\mathrm{net}(v)), \kappa(\mathrm{net}(v), \mathrm{net}(v')))$.
2. The GP mean μ *is assumed to be a known constant*, thus not contributing to the GP hyperparameters and independent of net.

Due to the assumption 2., the GP depends only on the parameters θ^κ of the covariance function. As to the ANN, it depends on the one hand on the vector θ^W of its weights and biases, on the other hand on the network architecture, which we will treat as fixed before network training starts.

Consider now n inputs to the neural network, $v_1, \ldots, v_n$, mapped to the inputs $x_1 = \mathrm{net}(v_1), \ldots, x_n = \mathrm{net}(v_n)$ of the GP, corresponding to the observations $\mathbf{y} = (y_1, \ldots, y_n)^\top$. Then the log-likelihood of θ is

$$\mathcal{L}(\theta) = \ln p(\mathbf{y}; \mu, \kappa, \sigma_n^2) \tag{9}$$
$$= -\tfrac{1}{2}(\mathbf{y} - \mu)^\top K^{-1}(\mathbf{y} - \mu) - \ln(2\pi) - \tfrac{1}{2}\ln \det(K), \tag{10}$$

where μ is the constant assumed in 2., and

$$(K)_{i,j} = \kappa(\mathrm{net}(v_i), \mathrm{net}(v_j)). \tag{11}$$

Let model training, searching for the vector $(\theta^\kappa, \theta^W)$, be performed in the simple but, in the context of neural networks, also the most frequent way – as gradient descent. The partial derivatives forming $\nabla_{(\theta^\kappa, \theta^W)}\mathcal{L}$ can be computed as:

$$\frac{\partial \mathcal{L}}{\partial \theta_\ell^\kappa} = \sum_{i,j=1}^{n} \frac{\partial \mathcal{L}}{\partial K_{i,j}} \frac{\partial K_{i,j}}{\partial \theta_\ell^\kappa}, \tag{12}$$

$$\frac{\partial \mathcal{L}}{\partial \theta_\ell^W} = \sum_{i,j,k=1}^{n} \frac{\partial \mathcal{L}}{\partial K_{i,j}} \frac{\partial K_{i,j}}{\partial x_k} \frac{\partial\, \mathrm{net}(v_k)}{\partial \theta_\ell^W}. \tag{13}$$

In (12), the partial derivatives $\frac{\partial \mathcal{L}}{\partial K_{i,j}}$, $i, j = 1, \ldots, n$, are components of the matrix derivative $\frac{\partial \mathcal{L}}{\partial K}$, for which the calculations of matrix differential calculus [30] together with (3) and (9) yield

$$\frac{\partial \mathcal{L}}{\partial K} = \frac{1}{2}\left(K^{-1}\mathbf{y}\mathbf{y}^\top K^{-1} - K^{-1}\right). \tag{14}$$

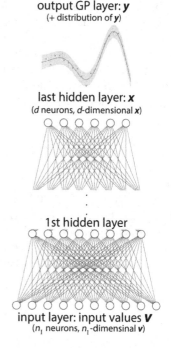

output GP layer: y
(+ distribution of *y*)

last hidden layer: x
(*d* neurons, *d*-dimensional *x*)

1st hidden layer

input layer: input values V
(n_1 neurons, n_1-dimensinal *v*)

Fig. 1. Schema of the integration of a GP into an ANN as its output layer.

3   Gaussian Processes as Black-Box Surrogate Models

3.1   Combining GPs with the Black-Box Optimizer CMA-ES

In the reported research, we used the state-of-the-art black-box evolutionary optimizer *Covariance Matrix Adaptation Evolution Strategy (CMA-ES)* [16,19]. The CMA-ES performs unconstrained optimization on $\mathbb{R}^d$, iteratively sampling populations of points from the d-dimensional distribution $\mathcal{N}(m, \sigma^2 C)$, and uses a part of the sampled points, corresponding to the lowest objective function values, to update the distribution. Hence, it updates the expected value m, serving as the current estimate of the function optimum, the matrix C and the step-size σ. The CMA-ES is invariant with respect to monotonous transformations of the objective function. Hence, to make use of objective function evaluations in a set of points, it needs to know only their ordering. Details of the algorithm can be found in [16,19]. Using GPs as surrogate models for CMA-ES has a long tradition [3,25,29,37,38]. In particular the *doubly trained surrogate (DTS) CMA-ES* [3] belongs to the most successful surrogate-assisted CMA-ES versions, together with [1,17], which use polynomials, and [28], which uses a ranking SVM.

In the DTS-CMA-ES, the criterion according to which the points for evaluation by the true black-box objective function BB are selected, is the *probability of improvement (POI)*. Using the notation introduced in Sect. 2.1 and considering (4) implies that the probability of improvement of the Gaussian process $(f(\mathbf{x}))_{\mathbf{x} \in \mathcal{X}}$ below some threshold $T > 0$ in a point $\mathbf{x}_\star \in \mathcal{X}$ is

$$\mathrm{PoI}_f(\mathbf{x}_\star | T) = P(f(\mathbf{x}_\star) \leq T) = \phi \left(\frac{T - \mu(\mathbf{x}_\star) - \mathbf{k}_\star K^{-1}(\mathbf{y} - \mu)}{\sqrt{\kappa(\mathbf{x}_\star, \mathbf{x}_\star) - \mathbf{k}_\star^\top K^{-1} \mathbf{k}_\star}} \right), \qquad (15)$$

where ϕ denotes the cumulative distribution function of $\mathcal{N}(0, 1)$.

The algorithm DTS-CMA-ES consecutively trains two GPs $(\mathbf{f}_1(\mathbf{x}))_{\mathbf{x} \in \mathcal{X}}$ and $(\mathbf{f}_2(\mathbf{x}))_{\mathbf{x} \in \mathcal{X}}$ to find an evaluation of the population $\mathcal{P} = \mathbf{x}_1, \ldots, \mathbf{x}_\lambda$. It performs *active learning* of points to be evaluated by the true black-box objective function BB, and the ordering ρ_f according to which the points in the population are chosen for the evaluation by the BB is given by the probability of improvement:

$$\rho_f : \mathcal{X}^\lambda \to \Pi(\lambda), (\forall (\mathbf{x}_1, \ldots, \mathbf{x}_\lambda) \in \mathcal{X}^\lambda)$$
$$(\rho_f(\mathbf{x}_1, \ldots, \mathbf{x}_\lambda))_i < (\rho_f(\mathbf{x}_1, \ldots, \mathbf{x}_\lambda))_j \Rightarrow \mathrm{PoI}_f(\mathbf{x}_i | y_{\min}) \geq \mathrm{PoI}_f(\mathbf{x}_j | y_{\min}), \quad (16)$$

where $\Pi(\lambda)$ denotes the set of permutations of $1, \ldots, \lambda$ and $y_{\min}$ is the minimal BB value found so far during the current optimization run. As to the number of actively learned points, it is based on the *ranking difference error (RDE)* between both trained GPs. The RDE of a GP f_2 with respect to a GP f_1 for $\mathbf{x} = (\mathbf{x}_1, \ldots, \mathbf{x}_\lambda) \in \mathcal{X}^\lambda$ considering $k \in \mathbb{N}$ best points:

$$\mathrm{RDE}_{\leq k}(\mathbf{f}_1(\mathbf{x}), \mathbf{f}_2(\mathbf{x})) = \frac{\sum_{i,(\rho_{\mathbf{f}_1}(\mathbf{x}))_i \leq k} |(\rho_{\mathbf{f}_2}(\mathbf{x}))_i - (\rho_{\mathbf{f}_1}(\mathbf{x}))_i|}{\max_{\pi \in \Pi(\lambda)} \sum_{i=1}^{k} |i - \pi^{-1}(i)|}. \qquad (17)$$

The way in which the DTS-CMA-ES performs active learning of the points evaluated by the BB between two CMA-ES iterations (in the CMA-ES evolutionary terminology called generations) is described in Algorithm 1, which uses an *archive* $\mathcal{A}$ of points previously evaluated in this way and the notation $\mathbf{f}_i(\mathbf{x}) = (f_i(\mathbf{x}_1), \ldots, f_i(\mathbf{x}_\lambda)), i = 1, 2$ and $k(\mathcal{A}) = \max\{h : |\mathcal{T}_h| \leq N_{\max}\}$, where

$$\mathcal{T}_h = \bigcup_{j=1}^{\lambda} \{\mathbf{x} \in N_h(\mathbf{x}_j; \mathcal{A}) : m_{\sigma^2 C}(\mathbf{x}, \mathbf{x}_j) < r_{\max}\} \text{ for } h = 1, \ldots, |\mathcal{A}|, \quad (18)$$

$N_h(\mathbf{x}; \mathcal{A})$ denotes the set of h nearest neighbors of x in the archive $\mathcal{A}$, $r_{\max} > 0, N_{\max} \in \mathbb{N}, N_{\max} \geq \lambda$ and $m_{\sigma^2 C}$ is the Mahalanobis distance with respect to $\sigma^2 C$:

$$m_{\sigma^2 C}(\mathbf{x}, \mathbf{x}') = \sqrt{(\mathbf{x} - \mathbf{x}')^\top \sigma^{-2} C^{-1}(\mathbf{x} - \mathbf{x}')}. \quad (19)$$

Algorithm 1. Using the surrogate model between two CMA-ES generations in DTS-CMA-ES

Require: $\mathbf{x}_1, \ldots, \mathbf{x}_\lambda \in \mathcal{X}$, archive $\mathcal{A}$, μ, σ and C – parameters of the CMA-ES, $N_{\max} \in$
    $\mathbb{N}$ with $N_{\max} \geq \lambda, r_{\max} > 0, \beta, \epsilon_{\min}, \epsilon_{\max}, \alpha_{\min}, \alpha_{\max} \in (0, 1)$
1: **if** this is the 1st call of the algorithm in the current CMA-ES run, **then**
2:     set $\alpha = \epsilon = 0.05$
3: **else**
4:     take over the returned values of α, ϵ from its previous call in the run
5: **end if**
6: Train a Gaussian process $\mathbf{f}_1$ on $\mathcal{T}_{k(\mathcal{A})}$, estimating the parameters of κ through
    maximization of the likelihood (3)
7: Evaluate $BB(\mathbf{x}_j)$ for $\mathbf{f}_j$ not yet BB-evaluated and such that $(\rho_{\mathbf{f}_1}(\mathbf{x}_1, \ldots, \mathbf{x}_\lambda))_j \leq$
    $\lceil \alpha\lambda \rceil$
8: Update $\mathcal{A}$ to $\mathcal{A} \cup \{(\mathbf{x}_j : (\rho_{\mathbf{f}_1}(\mathbf{x}_1, \ldots, \mathbf{x}_\lambda))_j \leq \lceil \alpha\lambda \rceil\}$
9: Train a Gaussian process $\mathbf{f}_2$ on $\mathcal{T}_{k(\mathcal{A})}$, estimating the parameters of κ through
    maximization of the likelihood (3)
10: For x_j that have already been BB-evaluated, update $\mathbf{f}_2(\mathbf{x}_j) = BB(\mathbf{x}_j)$
11: Update ϵ to $(1 - \beta)\epsilon + \beta \, \mathrm{RDE}_{\leq \lceil \frac{\lambda}{2} \rceil}(\mathbf{f}_1(\mathbf{x}_1, \ldots, \mathbf{x}_\lambda), \mathbf{f}_2(\mathbf{x}_1, \ldots, \mathbf{x}_\lambda))$ and α to $\alpha_{\min} +$
     $\max(0, \min(1, \frac{\epsilon - \epsilon_{\min}}{\epsilon_{\max} - \epsilon_{\min}}))(\alpha_{\max} - \alpha_{\min})$
12: For every j such that $(\rho_{\mathbf{f}_1}(\mathbf{x}_1, \ldots, \mathbf{x}_\lambda))_j > \lceil \alpha\lambda \rceil$, update $\mathbf{f}_2(\mathbf{x}_j)$ to $\mathbf{f}_2(\mathbf{x}_j) -$
     $\min\{\mathbf{f}_2(\mathbf{x}_{j'}) : (\rho_{\mathbf{f}_1}(\mathbf{x}_1, \ldots, \mathbf{x}_\lambda))_{j'} > \lceil \alpha\lambda \rceil\} + \min\{\mathbf{f}_2(\mathbf{x}_{j'}) : (\rho_{\mathbf{f}_1}(\mathbf{x}_1, \ldots, \mathbf{x}_\lambda))_{j'} \leq$
     $\lceil \alpha\lambda \rceil\}$
13: Return $\mathbf{f}_2(\mathbf{x}_1), \ldots, \mathbf{f}_2(\mathbf{x}_\lambda), \epsilon, \alpha$

3.2   Using Data from Past CMA-ES Runs

The black-box optimization sequences for the configurations of GPs have been collected from 4600 runs of the open-source implementation of the DTS-CMA-ES algorithm for the 23 nonlinear noiseless benchmark functions available on the platform for comparing continuous optimizers COCO [7,18]. The COCO noiseless function collection also contains a linear benchmark, the linear slope. However, the GP surrogate model achieved for it on test data a mean squared error 0 or nearly 0 (of the order 10^{-4} or lower) with 5 from the 9 covariance functions considered in Sect. 2, thus configuring it for this benchmark makes no sense. Each of those benchmarks has been considered in the dimensions $d = 2, 3, 5, 10, 20$, and for each combination of benchmark function and dimension, 40 runs are available. More important than the number of runs, however, is the number of their generations because data from each generation apart the first can be used for testing all those surrogate models that could be trained with data from the previous generations. The benchmarks and their numbers of available generations for individual dimensions are listed in Table 1.

Table 1. Noiseless benchmark functions of the platform comparing continuous optimizers (COCO) [7,12] and the number of available generations of DTS-CMA-ES runs for each of them in each considered dimension

Benchmark function name		Available generations in dimension				
		2	3	5	10	20
Separable						
1	Sphere	4689	6910	11463	17385	25296
2	Separable Ellipsoid	6609	9613	15171	25994	55714
3	Separable Rastrigin	7688	11308	17382	27482	42660
4	Büche-Rastrigin	8855	13447	22203	31483	49673
Moderately ill-conditioned						
5.	Attractive Sector	16577	25200	38150	45119	72795
6.	Step Ellipsoid	7103	9816	24112	34090	56340
7.	Rosenbrock	7306	11916	21191	32730	71754
8.	Rotated Rosenbrock	7687	12716	24084	35299	71017
Highly ill-conditioned						
9.	Ellipsoid with High Conditioning	6691	9548	15867	25327	59469
10.	Discus	6999	9657	15877	25478	45181
11.	Bent Cigar	10369	18059	28651	34605	56528
12.	Sharp Ridge	7760	11129	20346	30581	48154
13.	Different Powers	6653	10273	17693	31590	61960
Multi-modal with global structure						
14.	Non-separable Rastrigin	7855	11476	19374	28986	44446
15.	Weierstrass	9294	13617	24158	27628	40969
16.	Schaffers F7	9031	13960	24244	34514	56247
17.	Ill-Conditioned Schaffers F7	9598	13404	25802	31609	53836
18.	Composite Griewank-Rosenbrock	9147	16268	24456	34171	53536
Multi-modal weakly structured						
19.	Schwefel	9081	13676	24219	33753	53104
20.	Gallagher's Gaussian 101-me Points	7645	12199	18208	25366	43186
21.	Gallagher's Gaussian 21-hi Points	7629	11086	17881	26626	44971
22.	Katsuura	8751	11233	17435	25030	37366
23.	Lunacek bi-Rastrigin	8983	13966	19405	29762	44420

Due to the way how the DTS-CMA-ES works, the collected sequences have several specific properties:

(i) The first GP $(\mathbf{f}_1(\mathbf{x}))_{\mathbf{x} \in \mathcal{X}}$ trained after the g-th generation of the CMA-ES has as training data only pairs $(\mathbf{x}, BB(\mathbf{x}))$ in which the true value $BB(\mathbf{x})$ was obtained before the g-th generation. It does not depend on the results of the CMA-ES in the g-th and later generations.

(ii) The second GP $(\mathbf{f}_2(\mathbf{x}))_{\mathbf{x} \in \mathcal{X}}$ trained after the g-th generation of the CMA-ES has as training data pairs $(\mathbf{x}, BB(\mathbf{x}))$ in which the true value $BB(\mathbf{x})$ was obtained before the g-th generation, as well as the pairs $(\mathbf{a}, BB(\mathbf{a}))$ in which $\mathbf{a}$ is one of the points $\mathbf{a}_1, \ldots, \mathbf{a}_{\lceil \alpha_g \lambda \rceil}$ selected with active learning among the points $\mathbf{x}_1, \ldots, \mathbf{x}_\lambda$ generated in the g-th generation. It does not depend on the results of the CMA-ES in the $g+1$-st and later generations.

(iii) The CMA-ES in the $g+1$-st generation depends on the values $BB(\mathbf{a})$ in the points $\mathbf{a} \in \{\mathbf{a}_1, \ldots, \mathbf{a}_{\lceil \alpha_g \lambda \rceil}\}$ selected with active learning and on the values $\mathbf{f}_2(\mathbf{x})$ for $\mathbf{x} \in \{\mathbf{x}_1, \ldots, \mathbf{x}_\lambda\} \backslash \{\mathbf{a}_1, \ldots, \mathbf{a}_{\lceil \alpha_g \lambda \rceil}\}$. Due to (ii), it indirectly depends also on $(\mathbf{x}, BB(\mathbf{x}))$ with x generated in an earlier than g-th generation.

4   Empirical Investigation of GP Configurations

In the experiments described below, we used two different libraries implementing Gaussian processes. For those without an ANN, we used Matlab's toolbox GPML that accompanies the book of Rasmussen and Williams [35]. Except the spectral mixture kernel, which was implemented in Python using GPyTorch library as well as the combination of GP and ANN [14].

4.1   Experimental Setup

We compare Gaussian processes with nine different covariance functions and one surrogate model combining Gaussian processes and neural networks. For pure GPs, we covered all kernels described in Sect. 2.1 and we used a simple linear kernel for the GP with ANN.

As to the combination ANN-GP, we decided to train it on the same set of training data $\mathcal{T}_{k(\mathcal{A})}$ as was used in steps 6 and 9 of Algorithm 1. Due to the condition (18), this set is rather restricted and allows training only a rather restricted ANN. Therefore, we decided to use a multilayer perceptron with a single hidden layer, thus a topology (n_I, n_H, n_O), where n_I is the dimension of the training data, i.e. $n_I \in \{2, 3, 5, 10, 20\}$, and

$$n_H = n_O = \begin{cases} 2 & \text{if } n_I = 2, \\ 3 & \text{if } n_I = 3, 5, \\ 5 & \text{if } n_I = 10, 20. \end{cases} \tag{20}$$

As the activation function for both the hidden and output layer, we chose a logistic sigmoid.

We trained the weights and biases of the neural network together with the parameters of the Gaussian process as proposed in [40] and outlined in Sect. 2.2. As a loss function, we used the Gaussian log-likelihood and optimized the parameters with Adam [23] for a maximum of 1000 iterations. We also kept a 10% validation set out of the training data to monitor overfitting, and we selected the model with the lowest L_2 validation error during the training.

4.2   Results

The results are presented first in a function-method view in Table 2, then in a dimension-method view in Table 3. The interpretation of the results in both tables is the same. A Non-parametric Friedman test was conducted on RDEs across all results for particular functions and function types in Table 2, and for a particular combination of dimensions and function types in Table 3. If the null hypothesis of the equality of all ten considered methods was declined, a post hoc test was performed, and its results were corrected for multiple hypotheses testing using the Holm method. In the tables, we highlight bold those methods that are not significantly different from the method achieving the lowest RDE. Additionally, the number $n \in \mathbb{N}$ of other methods that achieved a significantly worse RDE is reported using the $*^n$ notation.

If we look closely at the results in Table 2, we can see that the lowest values of the average RDE are achieved using the rational quadratic covariance function in most cases. Specifically, the rational quadratic kernel performs best on 16 functions, Matérn $\frac{5}{2}$ on 3, squared exponential on 2, and Quadratic on 2 functions out of the total of 23. Interestingly, the two functions on which the Quadratic kernel is the best performing are functions 5 and 6, which are both based on quadratic functions [12]. This suggests that the quadratic kernel can indeed adequately capture the functions with the global structure of quadratic polynomial. On the other hand, the Matérn $\frac{5}{2}$ kernel is the best performing on functions 19, 22, and 23, which are highly multimodal and therefore hard to optimize.

Similar results can also be seen in Table 3, where the RDEs are grouped by the input space dimension. The rational quadratic covariance function is the best in the 18 cases, Matérn $\frac{5}{2}$ in 4, Quadratic in 2, and squared exponential once. Again Matérn $\frac{5}{2}$ covariance function was the best kernel when solving heavily multimodal weakly structured problems. We also analyzed the correlation between the dimensionality of the problem and the resulting average RDE for every considered method. For this purpose, we computed Pearson's and Spearman's correlation coefficients. Both coefficients detected a statistically significant positive correlation in the linear, quadratic, Gibbs, and spectral mixture kernels. The correlation of squared exponential kernel was found significant only by the Pearson's coefficient. Interestingly, both methods discovered a significant negative correlation for the neural network covariance function.

However, it is important to note here that the best three covariance matrices: rational quadratic, squared exponential, and Matérn, $\frac{5}{2}$ are statistically equivalent in all cases according to the Friedman-posthoc test with Holm correction.

Table 2. Comparison of average RDE values for different surrogate models depending on a particular benchmark function. There are nine different covariance functions for Gaussian processes and the ANN-GP combination with a linear kernel. Those methods that were not significantly different from the best performing are marked in bold. The number in the upper index indicates the number of methods that performed significantly worse.

Function		κ_{LIN}	κ_{Q}	κ_{RQ}	κ_{SE}	$\kappa^{\frac{5}{2}}_{Mat}$	κ_{NN}	κ_{Gibbs}	κ_{SE+Q}	κ_{SM}	ANN κ_{LIN}
Separable Functions	1	.238	.193	**.056*6**	**.063*6**	**.070*4**	.168	**.130*1**	.161	.225	.242
	2	.231	.181	**.096*5**	**.098*5**	**.099*5**	.187	**.153*2**	.222	.280	**.128*2**
	3	.236	**.185*1**	**.142*5**	**.147*4**	**.160*4**	**.185*1**	.202*1	.224	.503	.300
	4	.251	**.170*2**	**.134*6**	**.160*3**	**.153*4**	**.189*1**	**.213*1**	.224	.471	.352
	All	.239	**.182*2**	**.107*7**	**.117*7**	**.121*7**	**.182*1**	**.175*4**	.208	.370	.256
Moderate conditioning	5	**.209*2**	**.138*5**	**.203*3**	.215*2	**.208*2**	**.199*3**	.246	.254	.440	.351
	6	**.231*1**	**.185*3**	**.194*2**	**.224*1**	**.206*1**	**.192*3**	.249	.243	.438	**.200*2**
	7	.231	**.183**	**.129*4**	**.138*3**	**.143*2**	.182	**.184*1**	.198	**.187*1**	.307
	8	.206	**.179**	**.128*4**	**.140*3**	**.140*3**	.175	**.170*3**	**.167*1**	.263	.307
	All	**.219*1**	**.171*4**	**.163*5**	**.179*4**	**.174*4**	**.187*3**	**.212*2**	.216	.342	.291
High conditioning and unimodal	9	.235	.179	**.095*5**	**.101*5**	**.099*5**	.182	**.141*1**	.209	.295	**.142*1**
	10	.243	**.152*1**	**.105*4**	**.114*4**	**.115*4**	.196	**.144*2**	.241	.357	**.147*2**
	11	**.183**	**.166*2**	**.145*4**	**.153*3**	**.151*3**	.196	**.187*2**	.236	.290	.257
	12	.231	.202	**.116*4**	**.208*1**	**.220*1**	**.169*1**	**.177*1**	.210	.375	**.198*1**
	13	.234	.211	**.140*5**	**.165*4**	**.162*4**	**.187*2**	**.204*2**	.227	.310	.289
	All	**.225*1**	**.182*2**	**.120*7**	**.148*6**	**.149*6**	.186*2	**.171*5**	.224	.326	**.205*2**
Multi-modal adequate global structure	14	.242	**.172*3**	**.144*4**	**.141*4**	**.173*2**	**.182*2**	**.200*1**	.215	.488	.316
	15	**.221*1**	**.189*2**	**.172*3**	**.169*3**	**.173*3**	**.185*2**	**.207*2**	.234	.492	.551
	16	.242	**.199*2**	**.160*5**	**.211*2**	**.185*2**	**.189*2**	**.227*1**	.245	.463	.391
	17	.259	**.203*2**	**.157*4**	**.207*2**	**.182*4**	**.191*2**	**.210*2**	.245	.480	.380
	18	**.232*1**	**.206*2**	**.143*5**	**.154*4**	**.148*4**	**.200*2**	.249	**.217*1**	.537	.397
	All	**.239*2**	.194*4	**.155*7**	**.177*5**	**.172*5**	.189*4	**.218*2**	**.231*2**	.493	.408
Multi-modal weak global structure	19	.237	**.164*3**	**.161*2**	**.174*1**	**.155*4**	.191	**.218**	.216	.383	**.210*1**
	20	.214	**.197*1**	**.145*4**	**.170*2**	**.152*4**	**.184*2**	**.169*2**	.206	.372	.389
	21	.203	.215	**.128*6**	**.155*3**	**.145*5**	.192	**.160*2**	.224	.391	.347
	22	.237	**.208*1**	**.168*3**	**.156*3**	**.147*3**	**.213*1**	**.210*1**	**.206*1**	.543	.589
	23	.229	**.209*1**	**.144*3**	**.127*4**	**.118*4**	**.197*1**	.217	**.220**	.545	.446
	All	**.224*2**	**.199*2**	**.149*7**	**.156*7**	**.144*7**	**.195*2**	**.195*2**	**.214*2**	.442	.397

Table 3. Comparison of average RDE values for different surrogate models depending on the type of benchmark function and the input space dimension. There are nine different covariance functions for Gaussian processes and the ANN-GP combination with a linear kernel. Those methods that were not significantly different from the best performing are marked in bold. The number in the upper index indicates the number of methods that performed significantly worse.

	Method / Function	GP									ANN
		κ_{LIN}	κ_{Q}	κ_{RQ}	κ_{SE}	$\kappa_{Mat}^{\frac{5}{2}}$	κ_{NN}	κ_{Gibbs}	κ_{SE+Q}	κ_{SM}	κ_{LIN}
Dimensions 2	SEP	.172	**.162**	**.097**^{*5}	**.106**^{*4}	**.110**^{*4}	.179	**.135**^{*1}	.198	.280	.252
	MOD	**.160**^{*2}	**.154**^{*2}	**.174**^{*2}	**.167**^{*2}	.201	.189	**.182**^{*1}	.194	.286	.341
	HC	**.157**^{*2}	.177	**.129**^{*4}	**.135**^{*4}	**.147**^{*3}	.212	**.142**^{*4}	.222	.227	.204
	MMA	**.187**^{*2}	**.179**^{*2}	**.155**^{*3}	**.170**^{*2}	**.179**^{*2}	**.200**^{*2}	**.195**^{*2}	.231	.437	.416
	MMW	**.184**^{*2}	**.182**^{*2}	**.137**^{*3}	**.140**^{*4}	**.148**^{*3}	**.205**	**.169**^{*2}	.221	.414	.443
	All	.172^{*3}	.172^{*3}	**.139**^{*6}	**.144**^{*6}	**.157**^{*4}	.198^{*2}	**.165**^{*4}	.215^{*1}	.333	.334
Dimensions 3	SEP	.204	**.175**^{*1}	**.114**^{*5}	**.117**^{*5}	**.123**^{*5}	.196	**.163**^{*1}	.218	.378	.258
	MOD	**.201**	**.160**^{*3}	**.172**^{*3}	**.177**^{*3}	**.173**^{*3}	.198	**.187**^{*1}	.226	.342	.313
	HC	.210	**.173**^{*1}	**.132**^{*4}	**.141**^{*3}	**.143**^{*3}	.199	**.149**^{*3}	.234	.264	**.174**^{*1}
	MMA	.227	**.198**^{*2}	**.159**^{*4}	**.164**^{*4}	**.180**^{*3}	**.193**^{*2}	**.201**^{*2}	.232	.458	.353
	MMW	.215	**.197**^{*1}	**.148**^{*4}	**.155**^{*2}	**.145**^{*5}	.208	**.204**^{*1}	.213^{*1}	.427	.387
	All	.214^{*1}	.182^{*4}	**.145**^{*7}	**.153**^{*6}	**.154**^{*6}	.199^{*1}	.180^{*4}	.223^{*1}	.360	.278^{*1}
Dimensions 5	SEP	.241	.189^{*1}	**.111**^{*7}	**.118**^{*5}	**.121**^{*5}	.193	.171^{*1}	.226	.421	.270
	MOD	.249	**.174**^{*3}	**.157**^{*5}	**.168**^{*3}	**.159**^{*4}	**.195**^{*1}	.205	.224	.387	.297
	HC	.241	**.180**^{*1}	**.119**^{*6}	**.136**^{*3}	**.136**^{*3}	**.185**^{*1}	**.148**^{*3}	.240	.337	**.202**^{*1}
	MMA	.251	**.189**^{*2}	**.149**^{*5}	**.167**^{*4}	**.169**^{*4}	**.183**^{*2}	**.206**^{*2}	.240	.478	.373
	MMW	.241	**.205**^{*1}	**.140**^{*4}	**.143**^{*4}	**.136**^{*6}	**.202**^{*1}	**.185**^{*2}	.209	.459	.363
	All	.247^{*1}	.185^{*4}	**.136**^{*7}	**.148**^{*7}	**.146**^{*7}	.190^{*4}	.177^{*4}	.229^{*1}	.403	.289^{*1}
Dimensions 10	SEP	.271	.203^{*1}	**.123**^{*4}	**.117**^{*5}	**.127**^{*4}	**.188**^{*1}	**.183**^{*1}	.216	.456	.293
	MOD	.256	**.195**^{*2}	**.151**^{*5}	**.167**^{*4}	**.155**^{*5}	**.192**^{*2}	.243	.228	.407	.298
	HC	.264	**.198**^{*1}	**.111**^{*5}	**.159**^{*1}	**.156**^{*2}	**.179**^{*1}	**.166**^{*1}	.230	.412	**.206**^{*1}
	MMA	.274	**.204**^{*2}	**.159**^{*5}	**.175**^{*4}	**.162**^{*4}	**.187**^{*3}	**.224**^{*1}	.233	.538	.422
	MMW	.246	.206	**.139**^{*5}	**.146**^{*4}	**.131**^{*6}	**.183**^{*2}	.211	.215	.454	.363
	All	.264^{*1}	.200^{*3}	**.138**^{*7}	**.155**^{*6}	**.149**^{*6}	.185^{*4}	.205^{*3}	.226^{*1}	.443	.313^{*1}
Dimensions 20	SEP	.288	**.196**	**.109**^{*4}	**.132**^{*4}	**.120**^{*4}	**.168**^{*3}	.250	**.188**	.445	.348
	MOD	**.228**	**.197**^{*1}	**.143**^{*2}	.173	**.145**^{*2}	**.174**^{*2}	**.245**	**.206**	.523	.353
	HC	**.253**	**.184**	**.111**^{*1}	.171	**.164**^{*1}	.155	**.249**	**.197**	.471	**.255**
	MMA	.258	**.201**^{*2}	**.155**^{*4}	**.206**^{*2}	**.170**^{*2}	**.184**^{*2}	.266	**.220**	.596	.480
	MMW	**.234**	**.204**	**.181**	**.198**	**.160**	**.179**	**.205**	**.213**	**.538**	**.436**
	All	.249	.194^{*3}	**.140**^{*6}	**.180**^{*4}	**.156**^{*5}	**.169**^{*4}	.244	.203^{*2}	.513	.375

Finally, the combination ANN-GP with the linear kernel was statistically equivalent to the best-performing kernel on six different functions. In those cases, it performed better than just GP with the linear kernel. However, in the rest of the benchmark functions, the linear covariance function produced better results without the ANN.

5   Conclusion and Future Work

The paper addressed the configuration of Gaussian processes serving as surrogate models for black-box optimization, primarily the choice of the GP covariance function. To this end, it employs a novel methodology, which consists in using data from past runs of a surrogate-assisted version of the state-of-the-art evolutionary black-box optimization method CMA-ES. This allowed to avoid demanding runs of the optimizer only with the purpose to configure the surrogate model, as well as to achieve a much more robust configuration relying on 4600 optimization runs in 5 different dimensions.

The results in Tables 2 and 3 reveal that the lowest RDE values are most often achieved using the covariance functions rational quadratic, squared exponential, and Matérn $\frac{5}{2}$. Moreover, these three covariance functions are always equivalent, in the sense that among the RDE values achieved with them, no pair has been found significantly different by the performed Friedman-posthoc test with Holm correction for simultaneous hypotheses testing. Occasionally, they are also equivalent to other configurations, e.g. to the covariance functions linear or quadratic, or even to the combination ANN-GP. To our knowledge, the comparison of GP covariance functions presented in this paper is the first in the context of optimization data. Therefore, it is not surprising that the conclusions drawn from it are not identical to conclusions drawn from comparisons performed with other kinds of data [6,9,24,39,40].

The paper also reports our first attempt to use a combined ANN-GP model for surrogate modeling in black-box optimization. Such combined models will be the main direction of our future research in this area. We want to systematically investigate using different GP covariance functions as well as different ANN topologies, including the direction of deep Gaussian processes [5,8,20,21], in which only the topology is used from an ANN, but all neurons are replaced by GPs. Moreover, in addition to the selection of the training set used in DTS-CMA-ES and described in Algorithm 1, we want to consider also alternative ways of training set selection, allowing to train larger networks. Finally, we intend to perform research into transfer learning of surrogate models: An ANN-GP model with a deep neural network will be trained on data from many optimization runs, such as those employed in this paper, and then the model used in a new run of the same optimizer will be obtained through additional learning restricted only to the GP and last 1–2 layers of the ANN.

Acknowledgment. The research reported in this paper has been supported by the Czech Science Foundation (GAČR) grant 18-18080S and was also partially supported by SVV project number 260 575. Computational resources were supplied by the project "e-Infrastruktura CZ" (e-INFRA LM2018140) provided within the program Projects of Large Research, Development and Innovations Infrastructures.

References

1. Auger, A., Brockhoff, D., Hansen, N.: Benchmarking the local metamodel CMA-ES on the noiseless BBOB'2013 test bed. In: GECCO 2013, pp. 1225–1232 (2013)
2. Baerns, M., Holeňa, M.: Combinatorial Development of Solid Catalytic Materials. Design of High-Throughput Experiments, Data Analysis, Data Mining. Imperial College Press/World Scientific, London (2009)
3. Bajer, L., Pitra, Z., Repický, J., Holeňa, M.: Gaussian process surrogate models for the CMA evolution strategy. Evol. Comput. **27**, 665–697 (2019)
4. Booker, A., Dennis, J., Frank, P., Serafini, D., Torczon, V., Trosset, M.: A rigorous framework for optimization by surrogates. Struct. Multidiscip. Optim. **17**, 1–13 (1999)
5. Bui, T., Hernandez-Lobato, D., Hernandez-Lobato, J., Li, Y., Turner, R.: Deep Gaussian processes for regression using approximate expectation propagation. In: ICML, pp. 1472–1481 (2016)
6. Calandra, R., Peters, J., Rasmussen, C., Deisenroth, M.: Manifold Gaussian processes for regression. In: IJCNN, pp. 3338–3345 (2016)
7. The COCO platform (2016). http://coco.gforge.inria.fr
8. Cutajar, K., Bonilla, E., Michiardi, P., Filippone, M.: Random feature expansions for deep Gaussian processes. In: ICML, pp. 884–893 (2017)
9. Duvenaud, D.: Automatic model construction with Gaussian processes. Ph.D. thesis, University of Cambridge (2014)
10. El-Beltagy, M., Nair, P., Keane, A.: Metamodeling techniques for evolutionary optimization of computaitonally expensive problems: promises and limitations. In: Proceedings of the Genetic and Evolutionary Computation Conference, pp. 196–203. Morgan Kaufmann Publishers (1999)
11. Emmerich, M., Giotis, A., Özdemir, M., Bäck, T., Giannakoglou, K.: Metamodel-assisted evolution strategies. In: PPSN VII, pp. 361–370. ACM (2002)
12. Finck, S., Hansen, N., Ros, R., Auger, A.: Real-parameter black-box optimization benchmarking 2010: presentation of the noisy functions. Technical report, INRIA, Paris Saclay (2010)
13. Forrester, A., Sobester, A., Keane, A.: Engineering Design via Surrogate Modelling: A Practical Guide. Wiley, New York (2008)
14. Gardner, J.R., Pleiss, G., Bindel, D., Weinberger, K.Q., Wilson, A.G.: GPyTorch: blackbox matrix-matrix Gaussian process inference with GPU acceleration (2019)
15. Gibbs, M.: Bayesian Gaussian processes for regression and classification. Ph.D. thesis, University of Cambridge (1997)
16. Hansen, N.: The CMA evolution strategy: a comparing review. In: Lozano, J.A., Larrañaga, P., Inza, I., Bengoetxea, E. (eds.) Towards a New Evolutionary Computation, vol. 192, pp. 75–102. Springer, Heidelberg (2006). https://doi.org/10.1007/3-540-32494-1_4
17. Hansen, N.: A global surrogate assisted CMA-ES. In: GECCO 2019, pp. 664–672 (2019)

18. Hansen, N., Auger, A., Ros, R., Merseman, O., Tušar, T., Brockhoff, D.: COCO: a platform for comparing continuous optimizers in a black box setting. Optim. Meth. Softw. **35** (2020). https://doi.org/10.1080/10556788.2020.1808977
19. Hansen, N., Ostermaier, A.: Completely derandomized self-adaptation in evolution strategies. Evol. Comput. **9**, 159–195 (2001)
20. Hebbal, A., Brevault, L., Balesdent, M., Talbi, E., Melab, N.: Efficient global optimization using deep gaussian processes. In: IEEE CEC, pp. 1–12 (2018). https://doi.org/10.1109/CEC40672.2018
21. Hernández-Muñoz, G., Villacampa-Calvo, C., Hernández-Lobato, D.: Deep Gaussian processes using expectation propagation and Monte Carlo methods. In: ECML PKDD, pp. 1–17, paper no. 128 (2020)
22. Jones, D., Schonlau, M., Welch, W.: Efficient global optimization of expensive black-box functions. J. Global Optim. **13**, 455–492 (1998)
23. Kingma, D., Ba, J.: Adam: a method for stochastic optimization (2014). Preprint arXiv:1412.6980
24. Kronberger, G., Kommenda, M.: Evolution of covariance functions for Gaussian process regression using genetic programming. In: Moreno-Díaz, R., Pichler, F., Quesada-Arencibia, A. (eds.) EUROCAST 2013. LNCS, vol. 8111, pp. 308–315. Springer, Heidelberg (2013). https://doi.org/10.1007/978-3-642-53856-8_39
25. Kruisselbrink, J., Emmerich, M., Deutz, A., Bäck, T.: A robust optimization approach using kriging metamodels for robustness approximation in the CMA-ES. In: IEEE CEC, pp. 1–8 (2010)
26. Leary, S., Bhaskar, A., Keane, A.: A derivative based surrogate model for approximating and optimizing the output of an expensive computer simulation. J. Global Optim. **30**, 39–58 (2004)
27. Lee, J., Bahri, Y., Novak, R., Schoenholz, S., Pennington, J., et al.: Deep neural networks as Gaussian processes. In: ICLR, pp. 1–17 (2018)
28. Loshchilov, I., Schoenauer, M., Sebag, M.: Intensive surrogate model exploitation in self-adaptive surrogate-assisted CMA-ES (saACM-ES). In: GECCO 2013, pp. 439–446 (2013)
29. Lu, J., Li, B., Jin, Y.: An evolution strategy assisted by an ensemble of local Gaussian process models. In: GECCO 2013, pp. 447–454 (2013)
30. Magnus, J., Neudecker, H.: Matrix Differential Calculus with Applications in Statistics and Econometrics. Wiley, Chichester (2007)
31. Matthews, A., Hron, J., Rowland, M., Turner, R.: Gaussian process behaviour in wide deep neural networks. In: ICLR, pp. 1–15 (2019)
32. Novak, R., Xiao, L., Lee, J., Bahri, Y., Yang, G., et al.: Bayesian deep convolutional networks with many channels are Gaussian processes. In: ICLR, pp. 1–35 (2019)
33. Ong, Y., Nair, P., Keane, A., Wong, K.: Surrogate-assisted evolutionary optimization frameworks for high-fidelity engineering design problems. In: Jin, Y. (ed.) Knowledge Incorporation in Evolutionary Computation, vol. 167, pp. 307–331. Springer, Heidelberg (2005). https://doi.org/10.1007/978-3-540-44511-1_15
34. Rasheed, K., Ni, X., Vattam, S.: Methods for using surrogate modesl to speed up genetic algorithm optimization: informed operators and genetic engineering. In: Jin, Y. (ed.) Knowledge Incorporation in Evolutionary Computation, pp. 103–123. Springer, Heidelberg (2005). https://doi.org/10.1007/978-3-540-44511-1
35. Rasmussen, E., Williams, C.: Gaussian Processes for Machine Learning. MIT Press, Cambridge (2006)
36. Ratle, A.: Kriging as a surrogate fitness landscape in evolutionary optimization. Artif. Intell. Eng. Des. Anal. Manuf. **15**, 37–49 (2001)

37. Ulmer, H., Streichert, F., Zell, A.: Evolution strategies assisted by Gaussian processes with improved pre-selection criterion. In: IEEE CEC, pp. 692–699 (2003)
38. Volz, V., Rudolph, G., Naujoks, B.: Investigating uncertainty propagation in surrogate-assisted evolutionary algorithms. In: GECCO 2017, pp. 881–888 (2017)
39. Wilson, A., Adams, R.: Gaussian process kernels for pattern discovery and extrapolation. In: ICML, pp. 2104–2112 (2013)
40. Wilson, A., Hu, Z., Salakhutdinov, R., Xing, E.: Deep kernel learning. In: ICAIS, pp. 370–378 (2016)

Travel Demand Estimation in a Multi-subnet Urban Road Network

Alexander Krylatov[1,2]($\boxtimes$) (iD) and Anastasiya Raevskaya[1] (iD)

[1] Saint Petersburg State University, Saint Petersburg, Russia
a.krylatov@spbu.ru
[2] Institute of Transport Problems RAS, Saint Petersburg, Russia

Abstract. Today urban road network of a modern city can include several subnets. Indeed, bus lanes form a transit subnet available only for public vehicles. Toll roads form a subnet, available only for drivers who ready to pay fees for passage. The common aim of developing such subnets is to provide better urban travel conditions for public vehicles and toll-paying drivers. The present paper is devoted to the travel demand estimation problem in a multi-subnet urban road network. We formulate this problem as a bi-level optimization program and prove that it has a unique solution under quite a natural assumption. Moreover, for the simple case of a road network topology with disjoint routes, we obtain important analytical results that allow us to analyze different scenarios appearing within the travel demand estimation process in a multi-subnet urban road network. The findings of the paper contribute to the traffic theory and give fresh managerial insights for traffic engineers.

Keywords: Bi-level optimization · Travel demand estimation · Multi-subnet urban road network

1 Introduction

An urban road area of a modern city is a multi-subnet complex composited network, which has been permanently growing over the past 40 years due to the worldwide urbanization process [15]. The increasing dynamics of motorization leads to various negative outcomes such as congestions, accidents, decreasing of average speed, traffic jams and noise, lack of parking space, inconveniences for pedestrians, pollution and environmental damage [29]. The continuing growth of large cities challenges authorities, civil engineers, and researchers to face a lot of complicated problems at all levels of management [24]. The efficient traffic management is appeared to be the only way for coping with these problems since capacities of the actual road networks today are often close to their limits and

The work was jointly supported by a grant from the Russian Science Foundation (No. 20-71-00062 Development of artificial intelligence tools for estimation travel demand values between intersections in urban road networks in order to support operation of intelligent transportation systems).

D. E. Simos et al. (Eds.): LION 2021, LNCS 12931, pp. 183–197, 2021.
https://doi.org/10.1007/978-3-030-92121-7_16

already cannot be increased. Therefore, the development of intelligent systems for decision-making support in the field of large urban road network design seems to be of crucial interest [1,28]. However, the reliability of decisions made by such systems highly depends on the precision of travel demand information [9,10], while the travel demand estimation is a highly sophisticated problem itself [7,8].

In the present paper, we concentrate on a bi-level formulation of the travel demand estimation problem, which employs the user equilibrium assignment principle [6,17]. Clear consideration of various alternative bi-level travel demand estimation problems with user equilibrium assignment in the lower level is given in [26]. Since those bi-level programs appeared to be NP-hard, researchers proposed a single-level optimization model based on the stochastic user equilibrium concept to estimate path flows and hence, the travel demand [4,23]. However, there also exists an approach that estimates the set of routes and their flows under the static user-equilibrium assignment principle [2]. Other methods face NP-hard bi-level travel demand estimation problems via heuristics or approximation [16,20]. Inspired by the recent findings on a composited complex network of multiple subsets [21,25], and traffic assignment in a network with a transit subnetwork [12,13,27], this paper investigates the travel demand estimation problem in a multi-subnet composited urban road network.

In the paper we consider a multi-subnet urban road network under arc-additive travel time functions. Section 2 presents a multi-subnet urban road network as a directed graph, while Sect. 3 is devoted to travel demand estimation in such kind of network. We formulate the demand estimation problem in a multi-subnet urban road network as a bi-level optimization program and prove that it has a unique solution under quite a natural assumption. Section 4 gives important analytical results for a road network with disjoint routes, which are directly applied in Sect. 5 for demand estimation by toll road counters. Actually, the simple case of a road network topology in Sect. 5 allows us to analyze different scenarios appearing within the estimation process. Section 6 contains the conclusions.

2   Multi-subnet Urban Road Network

Let us consider a multi-subnet urban road network presented by a directed graph $G = (V, E)$, where V represents a set of intersections, while $E \subseteq V \times V$ represents a set of available roads between the adjacent intersections. If we define S as the ordered set of selected vehicle categories, then $G = G_0 \cup \bigcup_{s \in S} G_s$, where $G_0 = (V_0, E_0)$ is the subgraph of public roads, which are open to public traffic, and $G_s = (V_s, E_s)$ is the subgraph of roads, which are open only for the s-th category of vehicles, $s \in S$. Denote $W \subseteq V \times V$ as the ordered set of pairs of nodes with non-zero travel demand $F_0^w > 0$ and/or $F_s^w > 0$, $s \in S$, for any $w \in W$. W is usually called as the set of origin-destination pairs (OD-pairs). Any set of sequentially linked arcs initiating in the origin node of OD-pair w and terminating in the destination node of the OD-pair w we call *route* between the OD-pair w, $w \in W$. The ordered sets of all possible routes between nodes of the OD-pair w, $w \in W$, we denote as R_0^w for the subgraph G_0 and R_s^w for the subgraph G_s, $s \in S$. Demand $F_s^w > 0$ for any $s \in S$ and $w \in W$ is assigned between the available public routes R_0^w and routes for vehicles of s-th category R_s^w. Thus, on the one hand, $\sum_{r \in R_s^w} p_r^w = P_s^w$, where p_r^w is the traffic flow of the s-th category vehicles through the route $r \in R_s^w$, while P_s^w is the overall traffic flow of the s-th category vehicles through the routes R_s^w. On the other hand, the difference $(F_s^w - P_s^w)$ is the traffic flow of the s-th category vehicles, which can be assigned between the available public routes R_0^w for any $s \in S$ and $w \in W$, since P_s^w satisfies the following condition: $0 \leq P_s^w \leq F_s^w$ for any $s \in S$, $w \in W$. Therefore, demand $F_0^w > 0$ is assigned between the available public routes R_0^w together with the traffic flow $\sum_{s \in S}(F_s^w - P_s^w)$: $\sum_{r \in R_0^w} f_r^w = F_0^w + \sum_{s \in S}(F_s^w - P_s^w)$, where f_r^w is the traffic flow through the public route $r \in R_0^w$ between nodes of OD-pair $w \in W$.

Let us introduce differentiable strictly increasing functions on the set of real numbers $t_e(\cdot)$, $e \in E$. We suppose that $t_e(\cdot)$, $e \in E$, are non-negative and their first derivatives are strictly positive on the set of real numbers. By x_e we denote traffic flow on the edge e, while x is an appropriate vector of arc-flows, $x = (\ldots, x_e, \ldots)^{\mathrm{T}}$, $e \in E$. Defined functions $t_e(x_e)$ are used to describe travel time on arcs $e \in E$, and they are commonly called arc *delay*, *cost* or *performance* functions. In this paper we assume that the travel time function of the route $r \in R_0^w \cup \bigcup_{s \in S} R_s^w$ between OD-pair $w \in W$ is the sum of travel delays on all edges belonging to this route. Thus, the travel time through the route $r \in R_0^w \cup \bigcup_{s \in S} R_s^w$ between OD-pair $w \in W$ can be defined as the following sum:

$$\sum_{e \in E} t_e(x_e)\delta_{e,r}^w \quad \forall r \in R_0^w \cup \bigcup_{s \in S} R_s^w, w \in W,$$

where, by definition,

$$\delta_{e,r}^w = \begin{cases} 1, \text{ if edge } e \text{ belongs to the route } r \in R_0^w \cup \bigcup_{s \in S} R_s^w, \\ 0, \text{ otherwise.} \end{cases} \quad \forall e \in E, w \in W,$$

while, naturally,

$$x_e = \sum_w \sum_{s \in S} \sum_{r \in R_s^w} p_r^w \delta_{e,r}^w + \sum_w \sum_{r \in R_0^w} f_r^w \delta_{e,r}^w, \quad \forall e \in E,$$

i.e., traffic flow on the arc is the sum of traffic flows through those routes, which include this arc.

3   Demand Estimation in a Multi-subnet Road Network

The first consideration of the travel demand estimation problem as an inverse traffic assignment problem was made in very general terms under the assumption that the locations of OD-pairs, as well as a matrix of route choice, are given [5].

However, the traffic assignment search is the optimization problem, where locations of OD-pairs and travel demand values are believed to be given, while an equilibrium assignment pattern is to be found [19]. Hence, the inverse traffic assignment search should be the optimization problem, where the equilibrium assignment pattern $\bar{x}$ is given, while locations of OD-pairs and travel demand values are to be found [11]. Therefore, let us introduce the following mapping:

$$\chi(F) = \arg\min_{x} \sum_{e \in E} \int_0^{x_e} t_e(u)du, \tag{1}$$

subject to

$$\sum_{r \in R_0^w} f_r^w = F_0^w + \sum_{s \in S} F_s^w, \quad \forall w \in V \times V, \tag{2}$$

$$f_r^w \geq 0 \quad \forall r \in R_0^w, w \in V \times V, \tag{3}$$

where, by definition,

$$x_e = \sum_w \sum_{r \in R_0^w} f_r^w \delta_{e,r}^w \quad \forall e \in E. \tag{4}$$

Mapping $\chi(F)$ applies the optimization program (1)–(3) with definitional constraint (4) to the feasible travel demand patterns F. In other words, if one applies mapping $\chi(F)$ to a feasible travel demand pattern F then the corresponding equilibrium traffic assignment pattern x will be obtained [14,18]. Indeed, the solution of (1)–(3) satisfies user-equilibrium behavioral principle, formulated by J. G. Wardrop: *"The journey times in all routes actually used are equal and less than those that would be experienced by a single vehicle on any unused route"* [22]. Thus, the travel demand estimation search as an inverse traffic assignment problem can be formulated as follows [11]:

$$\min_{F} ||\chi(F) - \bar{x}||$$

subject to

$$F_0^w \geq 0 \quad \forall w \in V \times V,$$
$$F_s^w \geq 0 \quad \forall w \in V \times V, s \in S,$$

where mapping $\chi(F)$ is given by the optimization program (1)–(3) with definitional constraint (4).

Let us develop the *travel demand estimation problem* for a multi-subnet urban road network. For this purpose, we refer to the principle, like the user-equilibrium one, which should be satisfied by the equilibrium traffic assignment pattern in a multi-subnet urban road network: *"The journey times in all routes actually used are equal and less than those that would be experienced by a single vehicle on any unused route, as well as the journey times in all routes actually used in any subnet less or equal than the journey times in all routes actually used in a public road network"*. Thus, we introduce the following mapping:

$$\hat{\chi}(F) = \arg\min_{x} \sum_{e \in E} \int_0^{x_e} t_e(u)du, \tag{5}$$

with constraints $\forall\, w \in W$

$$\sum_{r \in R_s^w} p_r^w = P_s^w, \quad \forall s \in S, \tag{6}$$

$$\sum_{r \in R_0^w} f_r^w = F^w + \sum_{s \in S}\left(F_s^w - P_s^w\right), \tag{7}$$

$$p_r^w \geq 0 \quad \forall r \in R_s^w, \forall s \in S, \tag{8}$$

$$f_r^w \geq 0 \quad \forall r \in R_0^w, \tag{9}$$

$$0 \leq P_s^w \leq F_s^w \quad \forall s \in S, \tag{10}$$

with definitional constraint

$$x_e = \sum_w \sum_{s \in S} \sum_{r \in R_s^w} p_r^w \delta_{e,r}^w + \sum_w \sum_{r \in R_0^w} f_r^w \delta_{e,r}^w, \quad \forall e \in E. \tag{11}$$

Mapping $\hat{\chi}(F)$ applies the optimization program (5)–(10) with definitional constraint (11) to the feasible travel demand patterns F. In other words, if one applies mapping $\hat{\chi}(F)$ to a feasible travel demand pattern F then the corresponding equilibrium traffic assignment pattern x for a multi-subnet urban road network will be obtained. Thus, the travel demand estimation search in a multi-subnet urban road network as an inverse traffic assignment problem can be formulated as follows:

$$\min_F \|\hat{\chi}(F) - \bar{x}\| \tag{12}$$

subject to

$$F_0^w \geq 0 \quad \forall w \in V \times V, \tag{13}$$

$$F_s^w \geq 0 \quad \forall w \in V \times V, s \in S, \tag{14}$$

where mapping $\hat{\chi}(F)$ is given by the optimization program (5)–(10) with definitional constraint (11).

Theorem 1. *Travel demand estimation problem* (12)–(14) *with mapping* $\hat{\chi}(F)$ *given by the optimization program* (5)–(10) *and definitional constraint* (11) *has the unique solution, which satisfies the following conditions:*

$$if \sum_{e \in E} t_e(\bar{x}_e)\delta_{e,r_s^w}^w \begin{cases} \leq \sum_{e \in E} t_e(\bar{x}_e)\delta_{e,r_0^w}^w, & then\; P_s^w = F_s^w, \\ = \sum_{e \in E} t_e(\bar{x}_e)\delta_{e,r_0^w}^w, & then\; 0 < P_s^w < F_s^w, \quad \forall w \in W, s \in S, \\ \geq \sum_{e \in E} t_e(\bar{x}_e)\delta_{e,r_0^w}^w, & then\; P_s^w = 0, \end{cases}$$

where r_0^w *is the shortest path in the public road network between OD-pair* w, *while* r_s^w *is the shortest path in the s-th subnetwork between OD-pair* w *under congestion* $\bar{x}$.

Proof. Since functions $t_e(x_e)$, $e \in E$, are strictly increasing, then the goal function (5) is convex. Hence, the optimization problem (5)–(10) has the unique solution [3]. In other words, if one applies mapping $\hat{\chi}(F)$ to a feasible travel demand pattern F then the unique equilibrium traffic assignment pattern x for a multi-subnet urban road network will be obtained. However, it is not clear if for any image of $\hat{\chi}(F)$ (feasible equilibrium traffic assignment pattern x) there exists the unique preimage (feasible travel demand pattern F).

Assume that there exist feasible travel demand patterns $\bar{F}$ and $\hat{F}$ such that $\bar{x} = \hat{\chi}(\bar{F}) = \hat{\chi}(\hat{F})$. According to (11), there also should exist route-flow assignment patterns $\bar{p}$, $\bar{f}$ and $\hat{p}$, $\hat{f}$ as well as matrix Δ, which satisfy the following matrix equations:

$$\bar{x} = \Delta \left(\frac{\bar{p}}{\bar{f}} \right) \quad \text{and} \quad \bar{x} = \Delta \left(\frac{\hat{p}}{\hat{f}} \right). \tag{15}$$

Moreover, since any feasible travel demand pattern satisfies $\sum_{r \in R_s^w} p_r^w = P_s^w$ and $\sum_{r \in R_0^w} f_r^w = F_0^w + \sum_{s \in S}(F_s^w - P_s^w)$ for any $w \in V \times V$, $s \in S$, then there exist vectors $\bar{P}$ and $\hat{P}$ as well as matrix A such that

$$\bar{P} = A \left(\frac{\bar{p}}{\bar{f}} \right) \quad \text{and} \quad \hat{P} = A \left(\frac{\hat{p}}{\hat{f}} \right). \tag{16}$$

Therefore, due to (15) and (16), the following matrix equations hold:

$$\begin{pmatrix} A \\ \Delta \end{pmatrix} \begin{pmatrix} \bar{p} \\ \bar{f} \end{pmatrix} = \begin{pmatrix} \bar{P} \\ \bar{x} \end{pmatrix} \quad \text{and} \quad \begin{pmatrix} A \\ \Delta \end{pmatrix} \begin{pmatrix} \hat{p} \\ \hat{f} \end{pmatrix} = \begin{pmatrix} \hat{P} \\ \bar{x} \end{pmatrix}.$$

Since $\bar{x}$ is actually the user-equilibrium assignment for $\bar{F}$ and $\hat{F}$, both systems have solutions, i.e. there exist nonzero vectors $\bar{p}$, $\bar{f}$ and nonzero vectors $\hat{p}$, $\hat{f}$ which solve these systems respectively. Let us subtract the first system from the second:

$$\begin{pmatrix} A \\ \Delta \end{pmatrix} \left[\begin{pmatrix} \bar{p} \\ \bar{f} \end{pmatrix} - \begin{pmatrix} \hat{p} \\ \hat{f} \end{pmatrix} \right] = \begin{pmatrix} \bar{P} - \hat{P} \\ \mathbb{O} \end{pmatrix}. \tag{17}$$

Thus, we obtain a non-zero route-flow assignment, which corresponds to the zero arc-flow assignment. For feasible (without negative components) route-flow assignment patterns $\bar{p}$, $\bar{f}$ and $\hat{p}$, $\hat{f}$, matrix equation (17) holds if and only if $\bar{p} = \hat{p}$ and $\bar{f} = \hat{f}$. Therefore, for any image of $\hat{\chi}(F)$ (feasible equilibrium traffic assignment pattern x) there exists the unique preimage (feasible travel demand pattern F). In other words, travel demand estimation problem (12)–(14) with mapping $\hat{\chi}(F)$ given by the optimization program (5)–(10) and definitional constraint (11) has the unique solution.

Let us consider the Lagrangian of the problem (5)–(10) with definitional constraint (11):

$$L = \sum_{e \in E} \int_0^{x_e} t_e(u)du + \sum_{w \in W} \left[\sum_{s \in S} t_s^w \left(P_s^w - \sum_{r \in R_s^w} p_r^w \right) \right.$$

$$+ t_0^w \left(F^w + \sum_{s \in S} \left(F_s^w - P_s^w \right) - \sum_{r \in R_0^w} f_r^w \right)$$

$$+ \sum_{s \in S} \sum_{r \in R_s^w} (-p_r^w) \eta_r^w + \sum_{r \in R_0^w} (-f_r^w) \xi_r^w + \sum_{s \in S} \left((-P_s^w) \gamma_s^w + (P_s^w - F_s^w) \zeta_s^w \right) \Bigg],$$

where t_0^w, t_s^w, $s \in S$, $\eta_r^w \geq 0$, $r \in R_s^w$ and $s \in S$, $\xi_r^w \geq 0$, $r \in R_0^w$, $\gamma_s^w \geq 0$, $\zeta_s^w \geq 0$, $s \in S$, for any $w \in W$ are Lagrangian multipliers. If we differentiate this Lagrangian and use Karush–Kuhn–Tucker conditions, then we obtain the following set of equalities and inequalities:

– for subnets:

$$\sum_{e \in E} t_e(x_e) \delta_{e,r}^w \begin{cases} = t_s^w \text{ for } p_r^w > 0, \\ \geq t_s^w \text{ for } p_r^w = 0, \end{cases} \quad \forall r \in R_s^w, s \in S, w \in W, \tag{18}$$

– for the public road network:

$$\sum_{e \in E} t_e(x_e) \delta_{e,r}^w \begin{cases} = t_0^w \text{ for } f_r^w > 0, \\ \geq t_0^w \text{ for } f_r^w = 0, \end{cases} \quad \forall r \in R_0^w, w \in W, \tag{19}$$

– for subnets within the public road network:

$$t_s^w \begin{cases} \leq t_0^w \text{ for } P_s^w = F_s^w, \\ = t_0^w \text{ for } 0 < P_s^w < F_s^w, \quad \forall s \in S, w \in W. \\ \geq t_0^w \text{ for } P_s^w = 0, \end{cases} \tag{20}$$

Therefore, the unique solution of the optimization problem (5)–(10) satisfies conditions (18)–(20). Thus, the equilibrium traffic assignment pattern $\bar{x}$ satisfies (18)–(20). No doubt that the shortest paths in the congested multi-subnet road network are active (non-zero route-flows) and so equilibrium travel times equal to corresponding travel times in shortest paths.   □

4 Multi-subnet Road Network with Disjoint Routes

Let us consider the particular case of a multi-subnet urban road network presented by the directed graph $G = (V, E)$. The set S is still the set of selected vehicle categories and $G = G_0 \cup \bigcup_{s \in S} G_s$, where $G_0 = (V_0, E_0)$ is the subgraph of public roads, which are open to public traffic, and $G_s = (V_s, E_s)$ is the subgraph of roads, which are open only for the s-th category of vehicles, $s \in S$. We also believe that there is only one OD-pair with non-zero travel demands, i.e., $|W| = 1$, $F_0 > 0$ and $F_s > 0$, $s \in S$. We assume that the topology of the graph G is such that any route initiating in the origin node of the OD-pair and terminating in its destination node has no common arcs with all other available routes between this OD-pair. The ordered sets of all possible routes between nodes of the single OD-pair we denote as R_0, $|R_0| = n_0$, for the subgraph G_0 and R_s,

$|R_s| = n_s$, for the subgraph G_s, $s \in S$. Demand $F_s > 0$ for any $s \in S$ is assigned between the available public routes R_0 and routes for vehicles of s-th category R_s. Thus, on the one hand, $\sum_{r \in R_s} p_r = P_s$, where p_r is the traffic flow of the s-th category vehicles through the route $r \in R_s$, while P_s is the overall traffic flow of the s-th category vehicles through the routes R_s. On the other hand, the difference $(F_s - P_s)$ is the traffic flow of the s-th category vehicles, which can be assigned between the available public routes R_0 for any $s \in S$, since P_s satisfies the following condition: $0 \le P_s \le F_s$ for any $s \in S$. Therefore, demand $F_0 > 0$ is assigned between the available public routes R_0 together with the traffic flow $\sum_{s \in S}(F_s - P_s)$: $\sum_{r \in R_0} f_r = F_0 + \sum_{s \in S}(F_s - P_s)$, where f_r is the traffic flow through the public route $r \in R_0$ between nodes of OD-pair $w \in W$.

Let us also introduce polynomial strictly increasing functions on the set of real numbers $t_r(\cdot)$, $r \in R_0 \cup \bigcup_{s \in S} R_s$, which are travel cost functions for the defined graph. We assume that

$$
\begin{aligned}
t_r(p_r) &= a_r^s + b_r^s (p_r)^{m_r^s}, \quad a_r^s \ge 0, \quad b_r^s > 0, \quad \forall r \in R_s, s \in S, \\
t_r(f_r) &= a_r^0 + b_r^0 (f_r)^{m_r^0}, \quad a_r^0 \ge 0, \quad b_r^0 > 0, \quad \forall r \in R_0,
\end{aligned}
\tag{21}
$$

where $m_r^s \ge 1$ for any $r \in R_s$, $s \in S$, and $m_r^0 \ge 1$ for any $r \in R_0$. Moreover, without loss of generality we believe that

$$
a_1^0 \le \ldots \le a_{n_0}^0 \quad \text{and} \quad a_1^s \le \ldots \le a_{n_s}^s \quad \forall s \in S. \tag{22}
$$

Fortunately, the travel demand pattern can be found explicitly for this particular case of multi-subnet road network.

Theorem 2. *Travel demand estimation problem* (12)–(14) *with mapping* $\hat{\chi}(F)$ *given by the optimization program* (5)–(10) *and definitional constraint* (11) *for the single-commodity multi-subnet urban road network with disjoint routes and polynomial cost functions* (21) *has the unique solution, which can be found explicitly as follows:*

$$
\text{if } a_1^s + b_1^s(p_1)^{m_1^s}
\begin{cases}
\le a_1^0 + b_1^0(f_1)^{m_1^s}, & \text{then } P_s = F_s, \\[2mm]
= a_1^0 + b_1^0(f_1)^{m_1^s}, & \text{then } 0 < P_s < F_s, \quad \forall s \in S, \\[2mm]
\ge a_1^0 + b_1^0(f_1)^{m_1^s}, & \text{then } P_s = 0,
\end{cases}
\tag{23}
$$

while

$$
P_s = \sum_{i=1}^{k_s} \sqrt[m_i^s]{\frac{a_1^s + b_1^s(p_1)^{m_1^s} - a_i^s}{b_i^s}}, \quad \forall s \in S, \tag{24}
$$

where $0 \le k_s \le n_s$, $s \in S$, *is such that*

$$
a_1^s \le \ldots \le a_{k_s}^s < a_1^s + b_1^s(p_1)^{m_1^s} \le a_{k_s+1}^s \le \ldots a_{n_s}^s \quad \forall s \in S, \tag{25}
$$

and

$$
F_0 + \sum_{s \in S}(F_s - P_s) = \sum_{i=1}^{k_0} \sqrt[m_i^0]{\frac{a_1^0 + b_1^0(f_1)^{m_1^0} - a_i^0}{b_i^0}}, \tag{26}
$$

where $0 \leq k_0 \leq n_0$ is such that

$$a_1^0 \leq \ldots \leq a_{k_0}^0 < a_1^0 + b_1^0(f_1)^{m_1^0} \leq a_{k_0+1}^0 \leq \ldots a_{n_0}^0. \tag{27}$$

Proof. According to the proof of the Theorem 1, the unique travel demand pattern in a multi-subnet urban road network has to satisfy conditions (18)–(20). For the single-commodity multi-subnet road network with disjoint routes conditions (18)–(20) have the following form:

$$t_r(p_r) \begin{cases} = t_s \text{ for } p_r > 0, \\ \geq t_s \text{ for } p_r = 0, \end{cases} \quad \forall r \in R_s, s \in S, \tag{28}$$

where $t_s = a_1^s + b_1^s(p_1)^{m_1^s}$ for $s \in S$, since, according to (22), the first route is the shortest one,

$$t_r(f_r) \begin{cases} = t_0 \text{ for } f_r > 0, \\ \geq t_0 \text{ for } f_r = 0, \end{cases} \quad \forall r \in R_0, \tag{29}$$

where $t_0 = a_1^0 + b_1^0(f_1)^{m_1^0}$, since, according to (22), the first route is the shortest one,

$$t_s \begin{cases} \leq t_0 \text{ for } P_s = F_s, \\ = t_0 \text{ for } 0 < P_s < F_s, \quad \forall s \in S. \\ \geq t_0 \text{ for } P_s = 0, \end{cases} \tag{30}$$

Therefore, (23) does hold.

Due to polynomial travel cost functions, expression (28) can be re-written as follows:

$$a_r^s + b_r^s(p_r)^{m_r^s} \begin{cases} = t_s \text{ for } p_r > 0, \\ \geq t_s \text{ for } p_r = 0, \end{cases} \quad \forall r \in R_s, s \in S,$$

or

$$p_r = \begin{cases} \sqrt[m_r^s]{\frac{t_s - a_r^s}{b_r^s}}, & \text{if } a_r^s \leq t_s, \\ 0, & \text{if } a_r^s > t_s, \end{cases} \quad \forall i \in R_s, s \in S.$$

Once condition (22) holds, then there exists k_s, $0 \leq k_s \leq n_s$, such that

$$\left. \begin{array}{l} \text{for } r \leq k_s, a_r^s < \\ \text{for } r > k_s, a_r^s \geq \end{array} \right\} t_s \quad \forall s \in S.$$

Hence, the following equalities hold:

$$\sum_{i=1}^{n_s} p_i = \sum_{i=1}^{k_s} p_i = \sum_{i=1}^{k_s} \sqrt[m_i^s]{\frac{t_s - a_i^s}{b_i^s}} = P_s \quad \forall s \in S,$$

thus

$$P_s = \sum_{i=1}^{k_s} \sqrt[m_i^s]{\frac{t_s - a_i^s}{b_i^s}} \quad \forall s \in S. \tag{31}$$

Therefore, conditions (24) and (25) do hold.

Secondly, due to polynomial travel cost functions, expression (29) can be re-written as follows:

$$a_r^0 + b_r^0 (f_r)^{m_r^0} \begin{cases} = t_0 \text{ for } f_r > 0, \\ \geq t_0 \text{ for } f_r = 0, \end{cases} \quad \forall r \in R_0,$$

or

$$f_r = \begin{cases} \sqrt[m_r^0]{\dfrac{t_0 - a_r^0}{b_r^0}} & \text{if } a_r^0 \leq t_0, \\ 0 & \text{for } a_r^0 > t_0, \end{cases} \quad \forall r \in R_s, s \in S. \tag{32}$$

Once condition (22) holds, then there exists k_0, $0 \leq k_0 \leq n_0$, such that

$$\left. \begin{array}{l} \text{for } r \leq k_0, a_r^0 < \\ \text{for } r > k_0, a_r^0 \geq \end{array} \right\} t_0.$$

Hence, the following equalities hold:

$$\sum_{i=1}^{n_0} f_i = \sum_{i=1}^{k_0} f_i = \sum_{i=1}^{k_0} \sqrt[m_i^0]{\frac{t_0 - a_i^0}{b_i^0}} = F_0 + \sum_{s \in S} (F_s - P_s) \quad \forall s \in S,$$

thus

$$F_0 + \sum_{s \in S} (F_s - P_s) = \sum_{i=1}^{k_0} \sqrt[m_i^0]{\frac{t_0 - a_i^0}{b_i^0}}. \tag{33}$$

Therefore, conditions (26) and (27) do hold.   □

Let us consider the travel demand estimation problem for the single-commodity multi-subnet urban road network with disjoint routes and linear cost functions in order to obtain important analytical relationships between travel demands.

Corollary 1. *Travel demand estimation problem* (12)–(14) *with mapping* $\hat{\chi}(F)$ *given by the optimization program* (5)–(10) *and definitional constraint* (11) *for the single-commodity multi-subnet urban road network with disjoint routes and linear cost functions* (21) *has the unique solution, which satisfies the following conditions:*

$$if \quad \frac{P_s + \sum\limits_{i=1}^{k_s} \dfrac{a_i^s}{b_i^s}}{\sum\limits_{i=1}^{k_s} \dfrac{1}{b_i^s}} \begin{cases} \leq \\ = \\ \geq \end{cases} \frac{F_0 + \sum\limits_{s \in S}(F_s - P_s) + \sum\limits_{i=1}^{k_0} \dfrac{a_i^0}{b_i^0}}{\sum\limits_{i=1}^{k_0} \dfrac{1}{b_i^0}} \quad \begin{array}{l} then\ P_s = F_s, \\[4pt] then\ 0 < P_s < F_s, \\[4pt] then\ P_s = 0, \end{array} \tag{34}$$

where $0 \leq k_s \leq n_s$, $s \in S$, *is such that*

$$a_1^s \leq \ldots \leq a_{k_s}^s < \frac{P_s + \sum\limits_{i=1}^{k_s} \dfrac{a_i^s}{b_i^s}}{\sum\limits_{i=1}^{k_s} \dfrac{1}{b_i^s}} \leq a_{k_s+1}^s \leq \ldots a_{n_s}^s \quad \forall s \in S, \tag{35}$$

and $0 \leq k_0 \leq n_0$ is such that

$$a_1^0 \leq \ldots \leq a_{k_0}^0 < \frac{F_0 + \sum\limits_{s \in S} (F_s - P_s) + \sum\limits_{i=1}^{k_0} \frac{a_i^0}{b_i^0}}{\sum\limits_{i=1}^{k_0} \frac{1}{b_i^0}} \leq a_{k_0+1}^0 \leq \ldots a_{n_0}^0. \quad (36)$$

Proof. Since travel cost functions are linear then t_s, $s \in S$, and t_0 can be obtained from (31) and (33) respectively as follows:

$$t_s = \frac{P_s + \sum\limits_{i=1}^{k_s} \frac{a_i^s}{b_i^s}}{\sum\limits_{i=1}^{k_s} \frac{1}{b_i^s}}, \quad s \in S, \quad \text{and} \quad t_0 = \frac{F_0 + \sum\limits_{s \in S} (F_s - P_s) + \sum\limits_{i=1}^{k_0} \frac{a_i^0}{b_i^0}}{\sum\limits_{i=1}^{k_0} \frac{1}{b_i^0}}.$$

□

The technique for the travel demand search in the multi-subnet road network with only disjoint routes and linear travel cost functions follows directly from the corollary. Let us consider its application to a simple topology network with a toll road.

5 Toll Road Counters for Travel Demand Estimation

In Fig. 1, we consider a simple topology network, which consists of 4 nodes and 4 arcs, and single OD-pair $(1, 3)$. We assume that the travel demand from origin 1 to destination 3 in the presented network includes drivers who are ready to pay fees for better passage conditions and drivers who are not ready to pay fees for passage. In other words, the overall travel demand from origin 1 to destination 3 is $F_0 + F_1$, where F_1 is drivers who are ready to pay fees for better travel conditions, while F_0 is drivers who are not ready to pay fees. The overall travel demand is assigned between the available disjoint public routes R_0, where R_0 consists of two routes: $1 \to 2 \to 3$ and $1 \to 4 \to 3$. We believe that travel time through both alternative routes is modeled by linear functions: $t_r(f_r) = a_r^0 + b_r^0 f_r$, $a_r^0 \geq 0$, $b_r^0 > 0$ for any $r \in R_0$, where f_r is the traffic flow through route r, $r \in R_0$.

Suppose that the road administration built a toll road from origin 1 to destination 3 and wants to estimate the total travel demand in the road network by toll road counters (Fig. 2). In other words, the number of drivers who chose the toll road for passage is counted by the fee payment system, and the road administration intends to use this data for the estimation of total travel demand from origin 1 to destination 3. Therefore, the road administration faces the multi-subnet urban road network with disjoint routes and one toll road subnet. Indeed, demand F_1 of drivers who are ready to pay fees for better passage conditions is assigned between the available disjoint public routes R_0 and routes for toll-paying drivers R_1, where R_1 in our example consists of a single route $1 \to 3$.

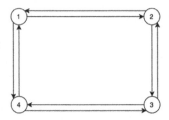

Fig. 1. Public road network

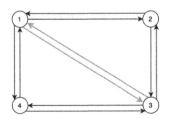

Fig. 2. Toll road subnetwork within the public road network

Thus, on the one hand, $\sum_{r \in R_1} p_r = P_1$, where p_r is the traffic flow of the toll-paying drivers through the route $r \in R_1$, while P_1 is the overall traffic flow of drivers through the routes R_1, counted by the fee payment system. On the other hand, the difference $(F_1 - P_1)$ is the traffic flow of the drivers who are ready to pay fees, but assigned between the available public routes R_0, since P_1 satisfies the following condition: $0 \leq P_1 \leq F_1$. Therefore, demand $F_0 > 0$ is assigned between the available public routes R_0 together with the traffic flow $(F_1 - P_1)$: $\sum_{r \in R_0} f_r = F_0 + (F_1 - P_1)$, where f_r is the traffic flow through the public route $r \in R_0$ from origin 1 to destination 3. We believe that travel time through the subnet routes is modeled by linear functions: $t_r(p_r) = a_r^1 + b_r^1 p_r$, $a_r^1 \geq 0$, $b_r^1 > 0$ for any $r \in R_1$, where p_r is the traffic flow through route r, $r \in R_1$. Moreover, we believe the ratio

$$\gamma = \frac{F_1}{F_0 + F_1} \tag{37}$$

is known. Actually, this ratio means how many percentages of drivers are ready to pay fees for passage and it can be evaluated due to the opinion poll of drivers.

According to the corollary, the travel demand estimation problem in the one-subnet urban road network with disjoint routes and linear cost functions has the unique solution, which satisfies (34), while actually used routes in toll road subnetwork within the public road network can be found due to (35), (36). Let us mention that for one-subnet urban road network with disjoint routes and linear travel time functions, the condition (34) can be relaxed. Indeed, if there exist k_0, $1 \leq k_0 \leq n_0$, and k_1, $1 \leq k_1 \leq n_1$, such that

$$\frac{P_1 + \sum_{i=1}^{k_1} \frac{a_i^1}{b_i^1}}{\sum_{i=1}^{k_1} \frac{1}{b_i^1}} \leq \frac{\left(\frac{1}{\gamma} - 1\right) P_1 + \sum_{i=1}^{k_0} \frac{a_i^0}{b_i^0}}{\sum_{i=1}^{k_0} \frac{1}{b_i^0}}, \tag{38}$$

then

$$F_1 = P_1 \quad \text{and} \quad F_0 = \left(\frac{1}{\gamma} - 1\right) P_1. \tag{39}$$

Actually, condition (38) means that the equilibrium travel time in public road network exceeds the equilibrium travel time through toll road subnetwork, i.e.

no one driver who ready to pay fees for passage can experience less travel time in public road network.

If condition (38) does not hold, then $0 < P_1 < F_1$ and there exist k_0, $1 \leq k_0 \leq n_0$, and k_1, $1 \leq k_1 \leq n_1$, such that

$$\frac{P_1 + \sum_{i=1}^{k_1} \frac{a_i^1}{b_i^1}}{\sum_{i=1}^{k_1} \frac{1}{b_i^1}} = \frac{F_0 + F_1 - P_1 + \sum_{i=1}^{k_0} \frac{a_i^0}{b_i^0}}{\sum_{i=1}^{k_0} \frac{1}{b_i^0}} \tag{40}$$

and, hence,

$$F_0 + F_1 = \left(\frac{P_1 + \sum_{i=1}^{k_1} \frac{a_i^1}{b_i^1}}{\sum_{i=1}^{k_1} \frac{1}{b_i^1}} + \frac{P_1 - \sum_{i=1}^{k_0} \frac{a_i^0}{b_i^0}}{\sum_{i=1}^{k_0} \frac{1}{b_i^0}} \right) \sum_{i=1}^{k_0} \frac{1}{b_i^0}, \tag{41}$$

while

$$F_1 = \gamma \left(F_0 + F_1 \right) \quad \text{and} \quad F_0 = \left(1 - \gamma \right) \left(F_0 + F_1 \right). \tag{42}$$

Actually, condition (40) means that the equilibrium travel time in toll road subnetwork is equal to the equilibrium travel time in public road network, i.e. the demand of drivers who are ready to pay toll for better passage conditions is not fully satisfied.

Therefore, obtained conditions allow the road administration to estimate travel demand in the multi-subnet urban road network only by toll road counters. Table 1 reflects available scenarios that can support decision-making.

Table 1. Scenarios for decision-making support.

Condition	Scenario	Estimation
Inequality (38) holds	The equilibrium travel time in toll road sub-network is less than equilibrium travel time in public road network, i.e. no one toll-paying driver can experience less travel time in public road network	Travel demand pattern can be obtained by (39)
Inequality (38) does not hold	The equilibrium travel time in toll road sub-network is equal to the equilibrium travel time in public road network, i.e. the demand of drivers who are ready to pay toll for better passage conditions is not fully satisfied	Travel demand pattern can be obtained by (42) under (41)

6   Conclusion

The present paper is devoted to the travel demand estimation problem in a multi-subnet urban road network. We formulated this problem as a bi-level optimization program and proved that it had the unique solution under quite natural assumption. Moreover, for the simple case of a road network topology with disjoint routes, we obtained important analytical results that allowed us to analyze different scenarios appearing within the travel demand estimation process in a multi-subnet urban road network. The findings of the paper contribute to the traffic theory and give fresh managerial insights for traffic engineers.

References

1. Bagloee, S., Ceder, A.: Transit-network design methodology for actual-size road networks. Transp. Res. Part B **45**, 1787–1804 (2011)
2. Bar-Gera, H.: Primal method for determining the most likely route flows in large road network. Transp. Sci. **40**(3), 269–286 (2006)
3. Bazaraa, M., Sherali, H., Shetty, C.: Nonlinear Programming: Theory and Algorithms, 2nd edn. Wiley, New York (1993)
4. Bell, M., Shield, C., Busch, F., Kruse, C.: A stochastic user equilibrium path flow estimator. Transp. Res. Part C **5**(3), 197–210 (1997)
5. Bierlaire, M.: The total demand scale: a new measure of quality for static and dynamic origin-destination trip tables. Transp. Res. Part B **36**, 837–850 (2002)
6. Fisk, C.: On combining maximum entropy trip matrix estimation with user optimal assignment. Transp. Res. Part B **22**(1), 69–73 (1988)
7. Frederix, R., Viti, F., Tampere, C.: Dynamic origin-destination estimation in congested networks: theoretical findings and implications in practice. Transp. A Transp. Sci. **9**(6), 494–513 (2013)
8. Hernandez, M., Valencia, L., Solis, Y.: Penalization and augmented Lagrangian for O-D demand matrix estimation from transit segment counts. Transp. A Transp. Sci. **15**(2), 915–943 (2019)
9. Heydecker, B., Lam, W., Zhang, N.: Use of travel demand satisfaction to assess road network reliability. Transportmetrica **3**(2), 139–171 (2007)
10. Kitamura, R., Susilo, Y.: Is travel demand instable? A study of changes in structural relationships underlying travel. Transportmetrica **1**(1), 23–45 (2005)
11. Krylatov, A., Raevskaya, A., Zakharov, V.: Travel demand estimation in urban road networks as inverse traffic assignment problem. Transp. Telecommun. **22**(2), 287–300 (2021)
12. Krylatov, A., Zakharov, V.: Competitive traffic assignment in a green transit network. Int. Game Theory Rev. **18**(2), 1640003 (2016)
13. Krylatov, A., Zakharov, V., Tuovinen, T.: Optimal transit network design. In: Optimization Models and Methods for Equilibrium Traffic Assignment. STTT, vol. 15, pp. 141–176. Springer, Cham (2020). https://doi.org/10.1007/978-3-030-34102-2_7
14. Krylatov, A., Zakharov, V., Tuovinen, T.: Principles of wardrop for traffic assignment in a road network. In: Optimization Models and Methods for Equilibrium Traffic Assignment. STTT, vol. 15, pp. 17–43. Springer, Cham (2020). https://doi.org/10.1007/978-3-030-34102-2_2

15. Lampkin, W., Saalmans, P.: The design of routes, service frequencies and schedules for a municipal bus undertaking: a case study. Oper. Res. Q. **18**, 375–397 (1967)
16. Lundgren, J., Peterson, A.: A heuristic for the bilevel origin-destination matrix estimation problem. Transp. Res. Part B **42**, 339–354 (2008)
17. Nguyen, S.: Estimating an OD matrix from network data: a network equilibrium approach. Publication 60, Centre de Recherche sur les Transports, Universitet de Motreal
18. Patriksson, M.: The Traffic Assignment Problem: Models and Methods. VSP, Utrecht (1994)
19. Sheffi, Y.: Urban Transportation Networks: Equilibrium Analysis with Mathematical Programming Methods. Prentice-Hall Inc., Englewood Cliffs (1985)
20. Shen, W., Wynter, L.: A new one-level convex optimization approach for estimating origin-destination demand. Transp. Res. Part B **46**, 1535–1555 (2012)
21. Sun, G., Bin, S.: Router-level internet topology evolution model based on multi-subnet composited complex network model. J. Internet Technol. **18**(6), 1275–1283 (2017)
22. Wardrop, J.G.: Some theoretical aspects of road traffic research. In: Proceedings of the Institution of Civil Engineers, vol. 2, pp. 325–378 (1952)
23. Wei, C., Asakura, Y.: A Bayesian approach to traffic estimation in stochastic user equilibrium networks. Transp. Res. Part C **36**, 446–459 (2013)
24. Xie, F., Levinson, D.: Modeling the growth of transportation networks: a comprehensive review. Netw. Spat. Econ. **9**, 291–307 (2009)
25. Yang, H., An, S.: Robustness evaluation for multi-subnet composited complex network of urban public transport. Alex. Eng. J. **60**, 2065–2074 (2021)
26. Yang, H., Sasaki, T., Iida, Y., Asakura, Y.: Estimation of origin-destination matrices from link traffic counts on congested networks. Transp. Res. Part B **26**(6), 417–434 (1992)
27. Zakharov, V.V., Krylatov, A.Y.: Transit network design for green vehicles routing. Adv. Intell. Syst. Comput. **360**, 449–458 (2015)
28. Zakharov, V., Krylatov, A., Ivanov, D.: Equilibrium traffic flow assignment in case of two navigation providers. In: Camarinha-Matos, L.M., Scherer, R.J. (eds.) PRO-VE 2013. IAICT, vol. 408, pp. 156–163. Springer, Heidelberg (2013). https://doi.org/10.1007/978-3-642-40543-3_17
29. Zhao, F., Zeng, X.: Optimization of transit route network, vehicle headways, and timetables for large-scale transit networks. Eur. J. Oper. Res. **186**, 841–855 (2008)

The Shortest Simple Path Problem with a Fixed Number of Must-Pass Nodes: A Problem-Specific Branch-and-Bound Algorithm

Andrei Kudriavtsev[1] , Daniel Khachay[1] , Yuri Ogorodnikov[1] , Jie Ren[2],
Sheng Cheng Shao[2], Dong Zhang[2], and Michael Khachay[1(✉)]

[1] Krasovsky Institute of Mathematics and Mechanics, Yekaterinburg, Russia
{kudriavtsev,dmx,yogorodnikov,mkhachay}@imm.uran.ru
[2] Huawei Technologies Co. Ltd., Shenzhen, China
{renjie21,shaoshengcheng,zhangdong48}@huawei.com

Abstract. The Shortest Simple Path Problem with Must-Pass Nodes
is the well-known combinatorial optimization problem having numerous applications in operations research. In this paper, we show, that
this problem remains intractable even for any fixed number of must-
pass nodes. In addition, we propose a novel problem-specific branch-and-
bound algorithm for this problem and prove its high performance by a
numerical evaluation. The experiments are carried out on the real-life
benchmark dataset 'Rome99' taken from the 9th DIMACS Implementation Challenge. The results show that the proposed algorithm outperforms the well-known solver Gurobi equipped with the best known MILP
models both in obtained accuracy and execution time.

Keywords: Shortest Simple Path Problem with Must-Pass Nodes ·
Branch-and-Bound algorithm · Milp models

1 Introduction

The Shortest Simple Path Problem with Must-Pass Nodes (SSPP-MPN) is the
well-known combinatorial optimization problem studied from the middle of 1960s
(see, e.g. [7,15]). The problem has numerous applications in operations research,
e.g. in the field of optimal routing in communication networks [13,14,16].

As for the classic Shortest Path Problem (SPP), in SSPP-MPN, the goal is to
find a path connecting given source and destination nodes in a weighted network
by a path of the smallest total cost. The only difference is that this path should
visit (*pass*) a given subset of dedicated nodes.

This seemingly minor detail makes a tremendous impact on the computational complexity of the problem in question. Indeed, without loss of generality,
assume the cost function to be non-negative. Then, the SPP can be solved to
optimality in quadratic time by the seminal Dijkstra's algorithm [5]. Meanwhile,

© Springer Nature Switzerland AG 2021
D. E. Simos et al. (Eds.): LION 2021, LNCS 12931, pp. 198–210, 2021.
https://doi.org/10.1007/978-3-030-92121-7_17

the SSPP-MPN is strongly NP-hard enclosing the classic Traveling Salesman Problem (TSP).

Furthermore, under our assumption, the simple-path constraint, optional for the SPP, is obligatory for the SSPP-MPN. Indeed, relaxation of this constraint immediately implies tractability of the SSPP-k-MPN for any fixed number k of must-pass nodes, whilst otherwise, as we show in this paper, this problem remains NP-hard even for $k = 1$.

Related Work. To the best of our knowledge, the SSPP-MPN was introduced in [15], where the first simple but erroneous algorithm (as was shown in [7]) was proposed. In [11], dynamic programming scheme and the first branch-and-bound algorithm based on the flow MILP-model for the classic Shortest Path Problem were developed. Although these algorithms are correct, they can hardly be applied to solving even moderate-size instances of the problem, due to their high computational complexity.

Recently, a substantial progress in the design of algorithms for the SSPP-MPN and related problems was achieved. For instance, in the paper [1], a number of compact MILP-models were proposed that gives us an opportunity for solving the problem using the well-known MIP-solvers. In [16] an efficient multi-stage meta-heuristic algorithm were proposed.

Numerous promising algorithmic results were obtained for the well-known shortest k-Vertex-Disjoint Paths Problem (or just k-DPP), which appears to be close to the SSPP-MPN. In k-DPP, we are given by an edge-weighted graph $G = (V, E, c)$ and a collection of source-destination pairs $C = \{(s_i, t_i) \in V^2\}$. The goal is to to find k vertex-disjoint simple paths of the smallest total cost, such that each ith path connects s_i with t_i.

As known [12], the DPP is NP-hard if k belongs to the instance. Furthermore, for the digraphs, the k-DPP is NP-hard for any fixed $k \geq 2$ as well [8].

Nevertheless, many efficient algorithms were proposed for the case of planar graphs. Thus, the authors of [17] proved that the planar k-DPP can be solved to optimality in $O(kn \log n)$ time, if all the sources belong to one face, while all the destinations to another one. In [4], $O(4^k n^{\omega/2+2})$ time FPT algorithm was proposed for the case of the planar k-DPP, where all sources and destinations are located at the boundary of a common face.

Finally, we should notice the recent breakthrough result obtained by Börklund and Husfeldt [2], who proved that k-DPP is polynomially solvable for $k = 2$ in the class of undirected graphs.

Unlike to the aforementioned problems, the SSPP-MPN with a fixed number of must-pass nodes is rather weakly explored in terms of algorithm design. Even the question on polynomial solvability/intractability of the problem (becoming quite topical in the light of [2]) still remains open.

Apparently, few elegant heuristics proposed in [9,13] together with their numerical evaluation on instances of small and moderate size are the only algorithmic results for this problem. In this paper, we try to bridge this gap.

Our Contribution is two-fold:

(i) we prove that the SSPP-k-MPN is NP-hard for an arbitrary $k \geq 1$;
(ii) we propose a novel problem-specific branch-and-bound algorithm and prove its efficiency numerically on a real-life dataset from the 9th DIMACS Implementation Challenge – Shortest Paths [6].

The rest of the paper is structured as follows. In Sect. 2, we remind formulation of the SSPP-MPN and MILP-models, which afterwards will be used in the numerical experiments as baselines. In Sect. 3, we provide a short explanation of the mutual polynomial-time cost-preserving reduction of the SSPP-k-MPN and $(k+1)$-DPP and obtain the NP-hardness of the former problem as a simple consequence.

Further, in Sect. 4, we discuss the main idea of the proposed algorithm. As in Sect. 3, we keep the description short postponing the complete version to the forthcoming paper. Section 5 provides numerical evaluation carried out on the DIMACS [6] dataset 'Rome99'. Finally, in Sect. 6 we summarize our results and discuss some questions that remain open.

2   Problem Statement

An instance of the SSPP-MPN is given by a weighted directed graph $G = (V, A, c)$, where $c\colon A \to \mathbb{R}_+$ specifies direct transportation costs, an ordered pair $(s, t) \in V^2$, where s and t are called *a source* and *destination*, respectively, and a finite subset $F \subset V$ of *must-pass* nodes, each of them should be visited by any feasible path. The goal is to construct a shortest *simple* (or *elementary*) feasible path departing from s and arriving to t.

For decades, a number of MILP models were proposed for the SSPP-MPN. The first of them having the following form:

$$(DFJ)\colon \ \min \sum_{(i,j)\in A} c_{ij}x_{ij} \tag{1}$$

$$s.t. \ \sum_{(i,j)\in A} x_{ij} - \sum_{(j,i)\in A} x_{ji} = \begin{cases} -1, & i = t, \\ 1, & i = s, \\ 0, & \text{otherwise} \end{cases} \quad (j \in V) \tag{2}$$

$$\sum_{(i,j)\in A} x_{ij} = \sum_{(j,i)\in A} x_{ji} = 1 \quad (i \in F) \tag{3}$$

$$\sum_{i,j\in S} x_{ij} \leq |S| - 1, \quad (\varnothing \neq S \subset V) \tag{4}$$

$$x_{ij} \in \{0, 1\}, \quad (i, j \in V) \tag{5}$$

appears to be a straightforward adaptation of the classic Dantzig-Fulkerson–Johnson model [3] for the Asymmetric Traveling Salesman Problem (ATSP).

Hereinafter, each variable x_{ij} indicates an assignment of the arc (i, j) to the s-t-path to be constructed. The objective function minimized in equation

(1) denotes the total transportation cost. Equations (2) and (4) represent the flow conservation and subtour elimination constraints, respectively, whilst eq. (3) ensures that each must-pass node is visited exactly once.

As for the ATSP, model DFJ appears to be the most simple and convenient for the theoretical treatment of the problem. Unfortunatelly, its application for solving large-scale instances of the problem using one of the well-known MIP-solvers, e.g. CPLEX or Gurobi, is difficult due to the exponential number of constraints.

Recently [1], several novel approaches for developing polynomial-size MILP models for the SSPP-MPN were intoduced and numerically compared. In this paper, for numerical comparison with our Branch-and-Bound algorithm, we take two best-performers. We call these models A_1 and A_2, respectively.

To ensure subtour elimination, model A_1 incorporates dual variables π_i and penalties with a big $\mathcal{M}$ constant, as follows:

$$(A_1) : \min \sum_{(i,j)\in A} c_{ij}x_{ij} \tag{6}$$

$$s.t. \quad (2), (3), (5)$$

$$\sum_{(i,s)\in A} x_{is} = \sum_{(j,t)\in A} x_{tj} = 0 \tag{7}$$

$$\pi_j - \pi_i \leq c_{ij} + \mathcal{M} \cdot (1 - x_{ij}) \quad ((i,j) \in A) \tag{8}$$

$$\pi_j - \pi_i \geq c_{ij} - \mathcal{M} \cdot (1 - x_{ij}) \quad ((i,j) \in A) \tag{9}$$

$$\pi_s = 0 \tag{10}$$

$$\pi_i \geq 0 \quad (i \in V). \tag{11}$$

On the other hand, in A_2, the same affect is obtained by inclusion of the auxiliary continuous flow variables:

$$(A_2) : \min \sum_{(i,j)\in A} c_{ij}x_{ij} \tag{12}$$

$$s.t. \quad (2), (3), (5), (7)$$

$$\sum_{(ij)\in A} x_{ij} \leq 1 \quad (j \in V \setminus (F \cup \{s,t\})) \tag{13}$$

$$\sum_{(i,j)\in A} f_{ij} - \sum_{(j,i)\in A} f_{ji} = \begin{cases} |F| + 1, & i = s, \\ -1, & i \in F \cup t, \quad (i \in V) \\ 0, & \text{otherwise} \end{cases} \tag{14}$$

$$x_{ij} \leq f_{ij} \leq (|F| + 1) \cdot x_{ij} \quad ((i,j) \in A). \tag{15}$$

As it shown in [1], in both models, the number of decision variables and constraints depends linearly on $|V| + |A|$.

Hereinafter, we consider the SSPP-MPN setting, where the number of must-pass nodes $|F| = k$ for some $k \in \mathbb{N}$. We call this problem SSPP-*k*-MPN.

3   Computational Complexity

In this section, we briefly discuss the polynomial-time cost-preserving reduction between the SSPP-k-MPN and $(k+1)$-DPP for any fixed k, which leads us to the proof of the intractabllity of the problem in question.

Theorem 1. *For any fixed $k \in \mathbb{N}$, SSPP-k-MPN is polynomially equivalent to $(k+1)$-DPP.*

Consider a short sketch of the proof.

(i). Indeed, to prove the cost-preserving reduction of $(k+1)$-DPP to SSPP-k-MPN, suppose that an instance of the former problem is specified by an edge-weighted digraph $G = (V, A, c)$ and a finite collection of source-destination pairs $C = \{(s_i, t_i) \colon i \in \{1, \ldots, k+1\}\}$. Augment the graph G with new vertices $W_1, \ldots, W_k$ and zero-cost arcs (t_i, W_i) and (W_i, s_{i+1}) and consider the SSPP-k-MPN instance defined the resulting digraph, the source s_1, the destination t_k, and the set of must-pass nodes $F = \{W_1, \ldots, W_k\}$ (Fig. 1 illustrates the case of $k = 1$). By construction, each feasible path in the constructed instance induces a feasible solution of the initial $(k+1)$-DPP instance of the same cost.

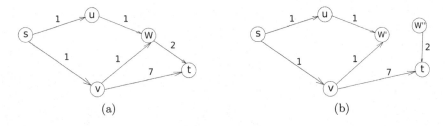

Fig. 1. Example of reducing 2-DPP to SSPP-1-MPN: (a) the initial 2-DPP instance and (b) the corresponding SSPP-1-MPN instance with a single must-pass node W

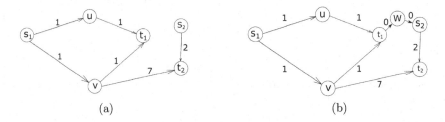

Fig. 2. Example of the backward reduction: (a) the initial SSPP-1-MPN instance with the must-pass node W and (b) the corresponding instance of the 2-DPP (with OD-pairs (s, W') and (W'', t))

(ii). On the other hand, consider an instance of the SSPP-k-MPN defined by a digraph $G = (V, A, c)$, a source s, destination t, and a set of must-pass nodes $F = \{W_1, \ldots, W_k\}$, for some fixed k. Split each must-pass node W_i into W_i' and W_i'', such that each inbound arc of W_i become incident to W_i' and each outbound arc incident to W_i''. Then, to an arbitrary ordering $W_{i_1}, \ldots, W_{i_k}$ of the set F, assign the appropriate instance of the $(k+1)$-DPP specified by the graph G with splitted nodes W_i and the collection of source-destination pairs $C = \{(s, W_{i_1}'), (W_{i_1}'', W_{i_2}'), \ldots, (W_{i_k}'', t)\}$ (Fig. 2). Again, any feasible solution of each among $k!$ produced $(k+1)$-DPP instances induces a same-cost feasible solution of the initial instance of the SSPP-k-MPN. Therefore, we are done with the reverse reduction.

As a simple consequence of Theorem 1 and complexity results obtained in [8] for the k-DPP, we obtain

Theorem 2

(i) SSPP-k-MPN on digraphs is strongly NP-hard for any fixed $k \geq 1$
(ii) for any fixed k, an arbitrary polynomial time algorithm for the k-DPP induces the appropriate algorithm for the SSPP-k-MPN

4   Branch-and-Bound Algorithm

Our algorithm extends the branching framework proposed in [11].

Main Idea. At any node of the branching tree, we introduce an auxiliary weighted digraph $G' = (Z, A')$, where

(i) $Z = \{s, t\} \cup F$;
(ii) for any nodes $z' \neq z''$, $(z', z'') \in A'$ if and only if, in the graph G, there exists a path $P = z', z_1, \ldots, z_r, z''$ for some r, such that $\{z_1, \ldots, z_r\} \cap Z = \varnothing$;
(iii) we weight any arc $(z', z'') \in A'$ with a length of the shortest such path P.

Then, we find[1] the shortest Hamiltonian s-t path P' in the graph G' and transform it to the corresponding path P in the graph G.

If the obtained P is simple, then a feasible solution of the SSPP-MPN is found. In this case, we update the record and branch cut the branch off. Otherwise, we proceed with branching at the current node by removing (from the graph G) the edges incident to a node visited by the path P more than once.

A branch is pruned any time when one of the following conditions is met:

(i) a feasible solution found, in this case we check and possibly update the current upper bound
(ii) the auxiliary graph G' has no Hamiltonian paths
(iii) a non-simple path P is obtained, whose cost is greater than the current upper bound.

---

[1] Or show that G' has no Hamiltonian paths. Since $|F|$ is fixed, this task can be solved in a constant time.

Branching. Our branching strategy is based on arc exclusion for any vertex of the graph G visited by the appropriate relaxed solution more than once. Suppose, at the current tree-node, we found a not-simple path (see Fig. 3). Eliminating separately each arc of this path incident to a more-than-once-passed vertex[2], we obtain child nodes in the branching tree. In this paper, we use Breadth-first search strategy for exploration of the branching tree.

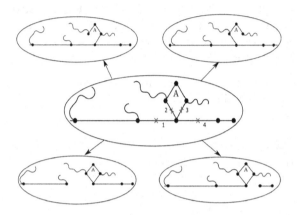

Fig. 3. Non-simple path and the appropriate child nodes in a search tree

Stopping Criteria in our algorithm are as follows:

(i) the instance is infeasible (there exists a must-pass node, which cannot be visited by a simple path)
(ii) an exact solution is found (at root relaxation)
(iii) an admissible value of the gap is achieved (at an arbitrary node of the search tree)
(iv) the entire search tree is explored
(v) established calculation time limit is reached

Lower Bounds. We use a standard promotion strategy for the lower bounds. Each not-a-leaf node of the search tree contains its own lower bound, called *local LB*, while the bound belonging to the root node is called *global LB* (or just LB). For an arbitrary not-a-leaf node of the branching tree,

(i) we initialize the local LB with a cost of the found relaxed solution,
(ii) update the local LB for the first time, when all its child nodes initialize their local LBs,
(iii) after that, we recalculate the local LB any time, when an arbitrary child node updates its own local LB.

---

[2] Highlighted by red crosses.

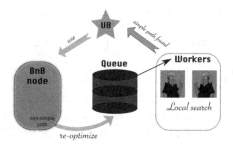

Fig. 4. Local-search heuristic

Local Search. In parallel with the main branching procedure, we use a simple local search heuristic. Any time, when in a tree node, we obtain an enough short but not a simple path, we provide it to the pool of parallel workers with intension to make it simple (Fig. 4). Our experiments show that this simple heuristic significantly speeds up the overall computation process.

Implementation. Our algorithm is implemented on *Python* 3.8 including *networkx* and *multiprocessing* packages, without any outer or non-standard dependecies.

5   Numerical Evaluation

To evaluate performance of the proposed algorithm, we carry out the following numerical experiment.

Experimental Setup. As a test benchmark, we take '*Rome99*' dataset from the 9th Implementation Challenge—Shortest Paths [6] that provides a large portion of the directed road network of the city of Rome, Italy, from 1999. This network is represented by a weighted digraph G of 3353 vertices and 8870 arcs.

For each $k \in \{2, 4, 6, 8\}$, we generate 500 instances on the same graph G with k must-pass nodes exactly, where the source s and destination d vertices and the subset F are sampled at random.

As a baseline, we use the well-known MIP-optimizer Gurobi [10] applied to the aforementioned models A_1 and A_2. We establish 1% gap tolerance for all the algorithms. Time limits are 300 s for our algorithm vs. 3600 s for Gurobi. Computational platform is Intel(R) Xeon(R) CPU 4×2.60 GHz 8 GB RAM with Centos 8 Linux OS.

Results. In Table 1 and Fig. 5a–5c one can find the short summary of the obtained results in terms of the compared algorithms: the proposed BnB and the Gurobi optimizer applied to the MILP models A_1 and A_2. First group of columns presents the ratio of obtained feasible solutions (see Fig. 5a). Second

Table 1. Evaluation summary: best values are highlighted

	Feas. sol. ratio (%)			Avg. gap (%)			Avg. time (sec)		
# MPN	A_1	A_2	BnB	A_1	A_2	BnB	A_1	A_2	BnB
2	97.4	100.0	100.0	26.15	4.47	**3.23**	2842.56	1596.08	**29.72**
4	89.6	99.8	99.8	26.15	13.36	**3.70**	3398.52	3021.57	**47.98**
6	32.4	99.4	99.4	32.19	15.59	**4.55**	3580.81	3365.34	**70.85**
8	16.6	99.0	99.4	33.85	15.38	**4.89**	3594.15	3477.99	**241.15**

group reflects the average gaps for the aforementioned methods calculated by the formula (UB − LB)/UB. Here UB is the objective value of the best found solution and LB is the best lower bound, respectively. Finally, last group shows the average computation time of the proposed methods (see Fig. 5c). The time limit for BnB was set to 300 s, while for A_1 and A_2 computation time was limited to 1 h.

As it follows from Table 1, the proposed algorithm finds feasible solution in 99% of instances, even for 8 must-pass nodes. It also provides a significantly better approximation ratio in comparison to the solutions obtained by Gurobi with significantly less computational time consumption.

Table 2. Gaps probability distributions

	Gap percentiles (%)							
	# of must-pass nodes							
	2		4		6		8	
α level (%)	BnB	A_2	BnB	A_2	BnB	A_2	BnB	A_2
10	0.0	0.0	0.2	0.6	0.6	1.1	0.7	5.7
20	0.1	0.0	0.7	1.0	0.9	7.3	1.2	8.9
30	0.4	0.2	1.0	5.6	1.6	10.4	2.0	11.1
40	0.8	0.7	1.6	9.5	2.2	13.8	2.9	13.0
50	1.5	0.9	2.4	13.2	3.1	16.8	3.7	15.4
60	2.5	1.0	3.4	15.7	4.3	18.0	4.9	17.5
70	4.2	5.4	4.3	19.0	5.5	20.2	6.2	19.8
80	5.8	8.9	5.9	23.1	7.1	23.0	7.8	22.3
90	8.7	16.2	9.6	27.4	10.9	27.4	10.3	24.8
100	33.1	34.4	28.7	40.8	29.3	41.6	33.4	34.4

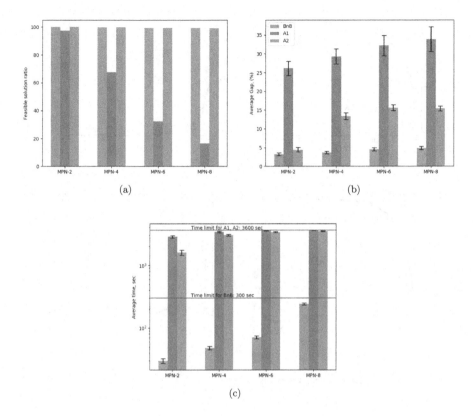

Fig. 5. (a) obtained ratio of feasible solutions; (b) average gap with 95% confidence bounds; (c) average time complexity with 95% confidence bounds

Sufficiently large amount of the considered empirical data allows us not to restrict ourselves to simple averages and leads to more complicated statistical findings. In the sequel, we present empirical estimations for probabilistic distributions of the observed gaps. We exclude A_1 from this study because this algorithm gives too small ratio of feasible solutions (see Table 1 and Fig. 5a). We present the estimated distributions in Table 2 and illustrate them together with the appropriate densities in Fig. 6a–d respectively.

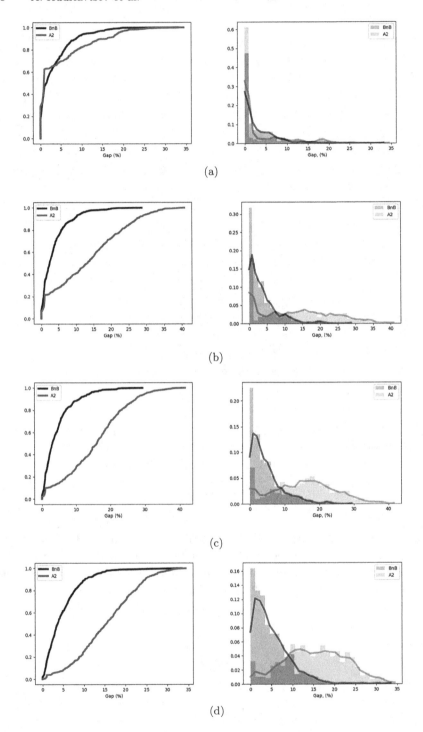

Fig. 6. Empirical distribution functions and densities for random gaps obtained by BnB and A_2 for: (a) two, (b) four, (c) six, and (d) eight must-pass nodes

6   Conclusion

In this paper, we proposed a novel branch-and-bound algorithm for the Shortest Simple Path Problem with a fixed number of Must-Pass Nodes. Results of the numerical evaluation show that the proposed technique outperforms well-known Gurobi solver applied to the best-known models for SSPP-MPN to date. As a future work, we plan to extend our algorithm to a number of related problems and test it on data sets of a larger scale.

Acknowledgements. This research was performed as a part of research carried out in the Ural Mathematical Center with the financial support of the Ministry of Science and Higher Education of the Russian Federation (Agreement number 075-02-2021-1383) and partially funded by Contract no. YBN2019125124. All computations were performed on supercomputer 'Uran' at Krasovsky Institute of Mathematcs and Mechanics.

References

1. Castro de Andrade, R.: New formulations for the elementary shortest-path problem visiting a given set of nodes. Eur. J. Oper. Res. **254**(3), 755–768 (2016). https://doi.org/10.1016/j.ejor.2016.05.008
2. Björklund, A., Husfeldt, T.: Shortest two disjoint paths in polynomial time. SIAM J. Comput. **48**(6), 1698–1710 (2019). https://doi.org/10.1137/18M1223034
3. Dantzig, G., Fulkerson, R., Johnson, S.: Solution of a large-scale traveling salesman problem. J. Oper. Res. Soc. Am. **2**(4), 393–410 (1954). https://doi.org/10.1287/opre.2.4.393
4. Datta, S., Iyer, S., Kulkarni, R., Mukherjee, A.: Shortest k-disjoint paths via determinants, February 2018
5. Dijkstra, E.W.: A note on two problems in connexion with graphs. Numer. Math. **1**(1), 269–271 (1959). https://doi.org/10.1007/BF01386390
6. DIMACS: 9th DIMACS Implementation Challenge - Shortest Paths (2006). http://users.diag.uniroma1.it/challenge9/download.shtml. Accessed 4 Feb 2021
7. Dreyfus, S.: An appraisal of some shortest path algorithm. Oper. Res. **17**, 395–412 (1969). https://doi.org/10.1287/opre.17.3.395
8. Fortune, S., Hopcroft, J., Wyllie, J.: The directed subgraph homeomorphism problem. Theor. Comput. Sci. **10**(2), 111–121 (1980). https://doi.org/10.1016/0304-3975(80)90009-2
9. Gomes, T., Martins, L., Ferreira, S., Pascoal, M., Tipper, D.: Algorithms for determining a node-disjoint path pair visiting specified nodes. Opt. Switch. Netw. **23** (2017). https://doi.org/10.1016/j.osn.2016.05.002
10. Gurobi Optimization LLC.: Gurobi optimizer reference manual (2020). http://www.gurobi.com
11. Ibaraki, T.: Algorithms for obtaining shortest paths visiting specified nodes. SIAM Rev. **15**(2), 309–317 (1973). http://www.jstor.org/stable/2028603
12. Karp, R.: Reducibility among combinatorial problems. vol. 40, pp. 85–103 (01 1972). https://doi.org/10.1007/978-3-540-68279-0_8
13. Martins, L., Gomes, T., Tipper, D.: Efficient heuristics for determining node-disjoint path pairs visiting specified nodes. Networks **70**(4), 292–307 (2017) https://doi.org/10.1002/net.21778

14. Rak, J.: Resilient Routing in Communication Networks. Computer Communications and Networks, Springer (2015)
15. Saksena, J.P., Kumar, S.: The routing problem with 'K' specified nodes. Oper. Res. **14**(5), 909–913 (1966). http://www.jstor.org/stable/168788
16. Su, Z., Zhang, J., Lu, Z.: A multi-stage metaheuristic algorithm for shortest simple path problem with must-pass nodes. IEEE Access **7**, 52142–52154 (2019). https://doi.org/10.1109/ACCESS.2019.2908011
17. Verdière, E.C.D., Schrijver, A.: Shortest vertex-disjoint two-face paths in planar graphs ACM Trans. Algor. **7**(2), 1–12 (2011). https://doi.org/10.1145/1921659.1921665

Medical Staff Scheduling Problem in Chinese Mobile Cabin Hospitals During Covid-19 Outbreak

Shaowen Lan[1,2], Wenjuan Fan[1,2(✉)], Kaining Shao[1,2], Shanlin Yang[1,2],
and Panos M. Pardalos[3]

[1] School of Management, Hefei University of Technology, Hefei 230009, China
fanwenjuan@hfut.edu.cn
[2] Key Laboratory of Process Optimization and Intelligent Decision-Making
of Ministry of Education, Hefei 230009, China
[3] Center for Applied Optimization, Department of Industrial and Systems
Engineering, University of Florida, Gainesville, FL 32611-6595, USA

Abstract. In this paper, we discuss the medical staff scheduling problem in the Mobile Cabin Hospital (MCH) during the pandemic outbreaks. We investigate the working contents and patterns of the medical staff in the MCH of Wuhan during the outbreak of Covid-19. Two types of medical staff are considered in the paper, i.e., physicians and nurses. Besides, two different types of physicians are considered, i.e., the expert physician and general physician, and the duties vary among different types of physicians. The objective of the studied problem is to get the minimized number of medical staff required to accomplish all the duties in the MCH during the planning horizon. To solve the studied problem, a general Variable Neighborhood Search (general VNS) is proposed, involving the initialization, the correction strategy, the neighborhood structure, the shaking procedure, the local search procedure, and the move or not procedure. The mutation operation is adopted in the shaking procedure to make sure the diversity of the solution and three neighborhood structure operations are applied in the local search procedure to improve the quality of the solution.

Keywords: Medical staff scheduling · Mobile Cabin Hospital · Covid-19 pandemic · Variable neighborhood search

1 Introduction

In the past years, there are several times of pandemic outbreaks all over the world, each bringing huge threats to human health and society operation. Take the Covid-19 pandemic as an example, it is a new infectious disease and spreads world-wide quickly. Patients suffer from the pandemic with lower respiratory tract infection and the symptoms including fever, dry cough, and even dyspnea [5]. The disease caused by Covid-19 has an incubation period, and the

© Springer Nature Switzerland AG 2021
D. E. Simos et al. (Eds.): LION 2021, LNCS 12931, pp. 211–218, 2021.
https://doi.org/10.1007/978-3-030-92121-7_18

patients in the incubation period usually do not have obvious symptoms, such as cough and fever, hence many of them are infected without being aware of it, and they may continue to infect others. Therefore, once an epidemic appears and it is not contained in time, it will spread quickly in a short time and may become a pandemic.

The outbreaks of pandemic bring huge pressure on the local hospitals due to the strong infectiousness of the disease and the explosive growth of infected patients. The inefficiency isolation, treatment, and management measures of the hospitals will exacerbate the spread of the pandemic, especially the nosocomial infection. The mobile cabin hospitals (MCHs) are a kind of modular health equipment with emergency treatment, which can relieve the shortage of medical sources [6]. MCH is used to quarantine and diagnose the mild-moderate patients, which plays an important role in controlling the Covid-19 pandemic in Wuhan, China [4]. In Wuhan city, the MCHs are composed of different functions units, such as the ward unit, severe disease observation and treatment unit, medical imaging examination unit, clinic testing unit, nucleic acid test unit, etc.

Patients in a MCH are with similar symptoms and degrees of illness, therefore, the treatment plans for these patients are very similar. It can help medical staff develop a relatively standardized work model and improve the efficiency of the diagnosis. MCH is composed of many ward units, and the clinical treatment and medical observation of patients are provided in the ward unit. Medical staff in the MCHs include physicians and nurses, who need to undertake different duties in various units. Physicians, in the ward units, are supposed to diagnose and provide the treatment plan for the patients, and nurses in the units need to observe the vital signs of patients, dispense medicine for patients, etc. The times of doing the duties, i.e., diagnosis and the medical observation for each patient, are quite different in a day. Therefore, the required number of physicians and nurses in the same unit at a time is different.

Considering the above background, a novel medical staff scheduling problem in MCH during a pandemic outbreak is investigated in this paper. It takes into account the infectivity of the pandemic and the MCH can effectively prevent the spread of the pandemic. In the following, the details of the studied problem are illustrated in Sect. 2. The proposed general VNS is introduced in Sect. 3. We summarize our work in Sect. 4. The experimental results are provided in Sect. 5.

2 Problem Description

In this paper, medical staff denote the group of physicians and nurses, where two types of physicians are considered, i.e., the general physicians and the expert physicians. Generally, the number of physicians is very limited, especially the expert physicians. Five types of duties are considered, i.e., expert physician consultation, general diagnosis, medical observation, medicine distribution, and nucleic acid test. It is worth noting that all patients are with mild symptoms when they enter the MCH and part of the patients may deteriorate and need to be transferred to large hospitals. The expert physician consultation, fulfilled by the expert physicians, is usually provided to patients with deteriorating symptoms or who may satisfy the criteria of recovery. The general diagnosis provided by the general physicians is to make the treatment plan for patients with mild symptoms. Medical observation, medicine distribution, and nucleic acid test need to be accomplished by different nurses in a shift. Variances in the working contents and the resource limitations of medical staff require that the specific schedules should be formulated for different types of medical staff. Nurses are operated round on three shifts in a day and three shifts are SN1 (8 AM–4 PM), SN2 (4 PM–0 AM), SN3 (0 AM–8 AM).

The purpose of the paper is to get the minimum number of medical staff required to accomplish all the duties in the MCH during the planning horizon, generally two months. For each physician or each nurse, the total working time in the planning horizon cannot exceed W^I or W^N, respectively. Besides, for each physician or nurse, the maximum number of days of them in the MCH is denoted by D^I or D^N, respectively, because that the medical staff in MCH is dispatched from different hospitals and they also need to go back to work. The arriving day for each physician or nurse is denoted by $A_i^I, i = 1, 2, \ldots, I$, where I is the total number of physicians, or $A_r^N, r = 1, 2, \ldots, R$, where R is the total number of nurses. For each medical staff, s/he cannot work in two successive shifts. In a week, each medical staff has a day off. In a shift, the minimum number of physicians for a patient is $I_p, p = 1, 2, \ldots, P$, where P is the total number of patients. For each physician, the maximum number of patients s/he can diagnose is $N_i^{IP}, i = 1, 2, \ldots, I$, in a shift. For each nurse, the maximum number of patients s/he can serve is $N_r^{NP}, r = 1, 2, \ldots, R$, in a shift.

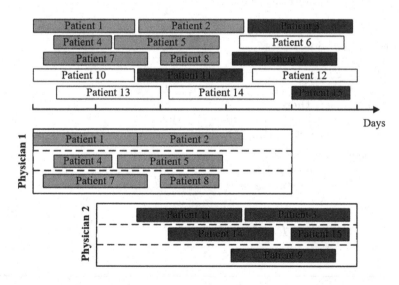

Fig. 1. An example of the assignment between physicians and patients.

The length of patients' stay in the MCH is different. For each patient, the length of stay is denoted by L_p, $p = 1, 2, \ldots, P$. For each patient, s/he is under the charge of a physician, which means the physician needs to diagnose the patient until s/he leaves the MCH. There is an assignment between physician and patient, and when the physician is on duty, the diagnosis of the assigned patient should be fulfilled by the physician. An example of the assignment between physicians and patients is provided in Fig. 1. In the example, each physician cannot be assigned more than three patients at the same time. Besides, the arriving day of the assigned patients should be later than the arriving day of the physician and the leaving time of the assigned patients should be earlier than the leaving day of the physician.

3   The Proposed VNS

To solve the studied medical staff scheduling problem in the MCH, a general VNS [1,2] is proposed, which involves the initialization, the correction strategy, the neighborhood structure, the shaking procedure, the local search procedure, and the move or not procedure. The processes of the proposed general VNS are provided in Algorithm 1.

Algorithm 1: Proposed general VNS

Input: neighborhood structure n_k, $k = 1, 2, \ldots, k_{max}$, k_{max} denotes the
maximum number of neighborhood structures, the time limitation T_{max}

Output: solution x and fitness f_x

1 $x, f_x \leftarrow Initialization function$;
2 $t \leftarrow 0$;
3 $f^{best} \leftarrow f_x$;
4 **while** $t \leq T_{max}$ **do**
5     $k \leftarrow 0$;
6   **while** $k \leq k_{max}$ **do**
7        $x', f_{x'} \leftarrow Local\ search\ function(x, f_x, k)$;
8   **if** $f_{x'} \leq f_x$ **then**
9           $x, f_x \leftarrow Shaking\ function(x', f_{x'})$;
10          $k \leftarrow 0\ f^{best} \leftarrow min f_x, f_{x'}$;
11   **else**
12          $k^s \leftarrow Random(1, k_{max})$;
13          $x, f_x \leftarrow Shaking\ function(x', f_{x'}, k^s)$;
14          $k \leftarrow k + 1$;
15   **if** $f_x < f^{best}$ **then**
16             $f^{best} \leftarrow f_x$

17 return x, f_x;

For the general VNS, two encoding schemes are applied. One of them is for the physician scheduling and another one is for the nurse scheduling. An example of solution for the proposed VNS is provided in Fig. 2. Three parts are involved in the solution. The first part is for the physician scheduling, each element of the solution means that the index of physician whom the specific patient is assigned to. In the figure, each patient is assigned to a physician, for example, patients P_1 and P_4 are assigned to physician I_1, and patients P_2 is assigned to physician I_2. An element with "zero" value is the second part of the solution, which means the encoding for physician scheduling is end and the encoding for nurse scheduling will be start. Therefore, the last part is for the nurse scheduling, each element of the solution means the index of nurse whom the duty is assigned to. Duties are grouped by the shift and each duty is assigned to a nurse. Besides, a nurse cannot be assigned to more than one duty in a shift.

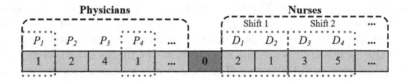

Fig. 2. An example of solution for the proposed VNS.

Initialization: First, get the range of each element in the solution. Then, randomly generate a value in the range of each element in the solution. In this way, generate several feasible solutions for the problem, and choose the best of them as the initial solution.

Correction Strategy: The aim of the correction strategy is to ensure the feasibility of the solution, by correcting the elements out of range in the solution. Besides, some constraints for the studied problem should be verified in the correction strategies. For example, the maximum number of patients can be assigned to a physician at the same time, the arriving day of patients cannot be earlier than the assigned physician, and the leaving day of patients cannot be later than the assigned physicians.

Neighborhood Structure: There are three neighborhood structures applied in the paper, i.e., swapping, insertion, and 2-opt. The swapping neighborhood is to randomly select two elements and swap the values between them in the physician scheduling part and nurse scheduling part of the solution, respectively. The insertion neighborhood is to randomly select one element in the solution, and then insert an element before the selected element and delete the final element of the solution. Besides, if the selected element is in the physician scheduling part, delete the final element of the physician scheduling part. In 2-opt neighborhood, it needs to choose the operation is adopted in the physician scheduling part or nurse scheduling part at first. Then, select two elements of the solution in the specific range, and reverse the values between these two elements.

Shaking Procedure: The shaking procedure is to make sure the diversity of the solution to avoid trapping into local optimal. In this paper, if the solution is improved by the local search procedure, the shaking procedure adopts the mutation operation in the Genetic Algorithm [3]. In detail, randomly choose an element in the solution, which are from the physician scheduling part and nurse scheduling part of the solution, respectively. Then, in the physician scheduling part, randomly select a value a physician who is scheduled in the current solution and replace the value of the element with the index of the selected physician. In the nurse scheduling part, select a nurse who is not in the shift of the chosen duty and replace the value of the element with the new one. Otherwise, randomly select a neighborhood structure and perturb the solution in the selected neighborhood.

Local Search Procedure: The local search procedure in the paper is to search the local optimal in the special neighborhood structure. There are three neighborhood structure operations applied, i.e., swapping, insertion, and 2-opt. The input of the local search procedure is a solution, the fitness of the solution, and the special neighborhood structure. The local search procedure will return a new solution and its fitness. The local search procedure is provided in Algorithm 2.

Algorithm 2: Local search procedure

Input: neighborhood structure $n_k, k = 1, 2, \ldots, k_{max}$, solution x and fitness f_x
Output: new solution x and fitness f_x

1 **while** $k \leq k_{max}$ **do**
2     $x', f_{x'} \leftarrow$ *Search solution in neighborhood* $k(x, f_x, k)$;
3   **if** $f_{x'} \leq f_x$ **then**
4        $x, f_x \leftarrow x', f_{x'}$;
5        $k \leftarrow 0$
6   **else**
7        $k \leftarrow k + 1$;

8 **return** x, f_x;

Move or Not: If the solution provided by the local search procedure is better than the current solution, the current solution and its fitness will be replaced by the better ones and the neighborhood structure is not changed. Otherwise, the current solution remains and the neighborhood structure is changed to the next one.

4 Experiments

In this section, the proposed approach is evaluated. All experiments are carried out in Python running on Windows 10 with an Intel (R) Core^{TM} i5-8400

CPU @2.81 GHz and 16 GB RAM. Ten instances with different scales have been tested where the number of patients is from 100 to 1000 with increments of 100. Besides, the proposed general VNS is compared with four variants of it, i.e., VNS1, VNS2, VNS3, and VNS4. VNS1, VNS2, and VNS3 are only use two neighborhood structures in the local search procedure. In detail, VNS1 does not apply the 2-opt neighborhood, VNS2 does not apply the swapping neighborhood, and VNS3 does not apply the insertion neighborhood. VNS4 does not adopt the mutation operation in the shaking procedure when the solution is improved by the local search procedure.

For each instance, each algorithm is executed 10 times and the maximum iteration is set as 200. The Relative Percentage Deviation (RPD) is used to measure the performance of the algorithms which is defined by $RPD_a = \frac{(AveObj_a - Min)}{Min} * 100$. Where RPD_a and $AveObj_a$ represent the RPD value and the average objective value in 10 runs of the algorithm a, respectively. Min is the best-found objective value of the instance. The results of these compared algorithms can be seen in Table 1 and the average objective value of 10 runs and the RPD value are denoted as Ave and RPD, respectively. The experimental results show that the proposed general VNS is better than other competitors which verify the effectiveness of the designed neighborhood structures and the proposed shaking procedure.

Table 1. The experimental results between all compared algorithms.

Patient number	Proposed VNS		VNS1		VNS2		VNS3		VNS4	
	Ave	RPD	Ave	RPD	Ave	RPD	Ave	RPD	Ave	RPD
100	35.6	22.76	37.1	27.93	38.0	31.03	38.3	32.07	37.6	29.66
200	52.3	8.96	54.5	13.54	53.5	11.46	53.5	11.46	53.1	10.63
300	101.1	11.1	102.5	12.64	101.8	11.87	101.4	11.43	105.4	15.82
400	133.2	8.29	137.2	11.54	140.4	14.15	139.3	13.25	141.2	14.8
500	184.6	5.49	187	6.86	188.7	7.83	187	6.86	188.9	7.94
600	220.7	7.14	224.7	9.08	223.4	8.45	226.9	10.15	225.5	9.47
700	274.3	5.1	278.4	6.67	276.4	5.9	280.2	7.36	279.2	6.97
800	299.6	2.6	305.9	4.76	303.4	3.9	305.7	4.69	300.4	2.88
900	361.3	4.12	368.9	6.31	370.5	6.77	372.7	7.41	367.9	6.02
1000	389.6	2.26	395	3.67	395.5	3.81	389.7	2.28	392.7	3.07

5 Conclusions

The problem studied in the paper is inspired by the real-world pandemic outbreak, i.e., the Covid-19 pandemic. The MCH can deal with the shortage of medical sources during the pandemic outbreaks, which can release the burden of the local hospitals and effectively slow down the spread of the pandemic. In this paper, we investigate the medical staff scheduling problem in the MCH by considering multiple specific features, for example, the different arriving days and leaving days of medical staff, the duty assignment rules between physicians and patients, and the various duties of physicians and nurses. Such features increase the difficulty of the problem and make our studied problem closer to the reality. To solve the studied problem, a general VNS is proposed with the specific encoding strategy for the problem and the shaking procedure and local search procedure.

Acknowledgments. This work is supported by the National Natural Science Foundation of China (Nos. 72071057 and 71922009), the Basic scientific research Projects in central colleges and Universities (JZ2018HGTB0232), and Innovative Research Groups of the National Natural Science Foundation of China (71521001).

References

1. Hansen, P., Mladenović, N., Moreno Pérez, J.A., Moreno Pérez, J.A.: Variable neighbourhood search: methods and applications. Ann. Oper. Res. **175**(1), 367–407 (2010). https://doi.org/10.1007/s10479-009-0657-6
2. Mladenović, N., Hansen, P.: Variable neighborhood search. Comput. Oper. Res. **24**(11), 1097–1100 (1997). https://doi.org/10.1016/S0305-0548(97)00031-2
3. Pan, J.C.H., Shih, P.H., Wu, M.H., Lin, J.H.: A storage assignment heuristic method based on genetic algorithm for a pick-and-pass warehousing system. Comput. Ind. Eng. **81**, 1–13 (2015). https://doi.org/10.1016/j.cie.2014.12.010
4. Wang, B., et al.: Epidemiological and clinical course of 483 patients with COVID-19 in Wuhan, China: a single-center, retrospective study from the mobile cabin hospital. Eur. J. Clin. Microbiol. Infect. Dis. **39**(12), 2309–2315 (2020). https://doi.org/10.1007/s10096-020-03927-3
5. Yuki, K., Fujiogi, M., Koutsogiannaki, S.: COVID-19 pathophysiology: a review. Clin. Immunol. **215**, 108427 (2020). https://doi.org/10.1016/j.clim.2020.108427
6. Zhang, J., et al.: The clinical characteristics and prognosis factors of mild-moderate patients with COVID-19 in a mobile cabin hospital: a retrospective, single-center study. Front. Public Health **8**, 1–11 (2020). https://doi.org/10.3389/fpubh.2020.00264

Performance Evaluation of Adversarial Attacks on Whole-Graph Embedding Models

Mario Manzo[1], Maurizio Giordano[2], Lucia Maddalena[2],
and Mario R. Guarracino[3,4(✉)]

[1] University of Naples "L'Orientale", Naples, Italy
[2] National Research Council, Rome, Italy
[3] University of Cassino and Southern Lazio, Cassino, Italy
mario.guarracino@unicas.it
[4] National Research University Higher School of Economics, Moscow, Russia

Abstract. Graph embedding techniques are becoming increasingly common in many fields ranging from scientific computing to biomedical applications and finance. These techniques aim to automatically learn low-dimensional representations for a variety of network analysis tasks. In literature, several methods (e.g., random walk-based, factorization-based, and neural network-based) show very promising results in terms of their usability and potential. Despite their spreading diffusion, little is known about their reliability and robustness, particularly when applied to the real world of data, where adversaries or malfunctioning/noisy data sources may supply deceptive data. The vulnerability emerges mainly by inserting limited perturbations in the input data when these lead to a dramatic deterioration in performance. In this work, we propose an analysis of different adversarial attacks in the context of whole-graph embedding. The attack strategies involve a limited number of nodes based on the role they play in the graph. The study aims to measure the robustness of different whole-graph embedding approaches to those types of attacks, when the network analysis task consists in the supervised classification of whole-graphs. Extensive experiments carried out on synthetic and real data provide empirical insights on the vulnerability of whole-graph embedding models to node-level attacks in supervised classification tasks.

Keywords: Whole-graph embedding · Adversarial attacks · Graph classification

1 Introduction

Graph structure plays an important role in many real-world applications. Representation learning on structured data with machine and deep learning methods has shown promising results in various applications, including drug screening [46], protein analysis [41], and knowledge graph completion [27].

© Springer Nature Switzerland AG 2021
D. E. Simos et al. (Eds.): LION 2021, LNCS 12931, pp. 219–236, 2021.
https://doi.org/10.1007/978-3-030-92121-7_19

Many graph embedding methods have been developed aiming at mapping graph data into a vector space [5]. The result is a low-dimensional feature representation for each node in the graph where the distance between the nodes in the destination space is preserved as much as possible. Actually, working on embedded data turns out to be easier and faster than on the original data. Furthermore, the resulting vectors in the transformed space can be adopted for downstream analysis, either by analyzing the target space or by applying machine learning (ML) techniques to the vector space. Indeed, by maintaining the topological information of the graph, low-dimensional representations can be adopted as features for different tasks such as graphs/nodes classification or clustering.

Despite the remarkable success, the lack of interpretability and robustness of these models makes them highly risky in fields like biomedicine, finance, and security, to name a few. Typically, sensitive information concerns the user-user relationship within the graph. A user who connects with many users with sensitive information may have sensitive information as well. As heuristics learned from graph-based methods often produce good predictions, they could also jeopardize the model. For example, an ill-intentioned person could disguise himself by connecting to other people on a social network. Such an "attack" on the model is simple enough but could lead to severe consequences [11]. Due to a large number of daily interactions, even if only a few of them are fraudulent, the ill-intentioned could gain enormous benefits.

The concept of graph robustness was first introduced in the 1970s [10] and is certainly interdisciplinary. This aspect has generated a variety of points of view, opening up to challenging and implicit problems with the aim of providing fundamental knowledge.

Robustness in networked systems is commonly defined as a measure of their ability to continue operating when one or more of their parts are naturally damaged or targeted by an attack [4]. The study of network robustness concerns the understanding of interconnected complex systems. For example, consider a network that is prone to natural failures or targeted attacks. A natural failure occurs when a single element fails due to natural causes such as obsolescence. The consequence is an additional load of the whole remaining network, causing a series of cascading faults. Not all failures come from natural causes; some may be induced by targeted attacks, penetrating the network and sabotaging an important part of it. The antonym of network robustness is *vulnerability* [42], defined as a measure of a network's susceptibility to the spread of perturbations across the network. The concepts of *robustness* and *vulnerability* can be extended to different types of networks, such as biological ones. Also in this case, they are two important indicators to verify the possible fault of a part of the network or any criticalities that can compromise the general functions with irreversible impact.

Robustness and vulnerability analysis is a crucial problem for today's research focusing on machine learning, deep learning, and AI algorithms operating on networked data in several domains, from cybersecurity to online financial trading, from social media to big-data analytics. In these contexts, while the networked systems (i.e., the graph-structured data) are the target of the attacks or perturbations, the real goal is to cause either the malfunctioning (intentionally or not) or an induced fraudulent behavior of the algorithms which operate on the modified data.

According to this interpretation, adversarial machine learning [23] is the area of research in which ML models vulnerability is studied under adversarial manipulation of their input intended to cause incorrect classification [12]. Neural networks and many other machine learning models are highly vulnerable to adversarial perturbations of the input to the model either at train or at test time, or both.

Several works on adversarial machine learning in the literature focus on the computer vision domain [2,34] with application to image recognition. Specifically, they address the problem of studying and improving the robustness of classification methods when adversarial images are present in the training and/or testing stages. More recently, adversarial ML has been increasingly utilized in other domains, such as natural language processing (NLP) [16] and cybersecurity [36]. Examples of applications in computer vision and NLP domains include handling autonomous cars' systems vulnerability, fake news, and financial fraud detection algorithms. In the cybersecurity domain, adversaries can be terrorists and fraudulent attackers. Examples of AI cyber systems that can be vulnerable to adversarial attacks are detection algorithms of malware stealing user information and/or collecting money and network worms causing network damages and malicious functionality.

In this work, we focus on adversarial ML techniques and approaches in the domain of machine learning models applied to the classification of biological networks. In this domain, we do not think of a scenario in which a "real adversary" intentionally introduces malicious perturbations in the input of learning models. In our interpretation of "adversarial attacks" within the realm of biological networks, we mean any type of perturbation to the graph structure, either due to noise introduced by the experimental environment from where the biological data is extracted or to the lack of information due to corrupted sources or incomplete pre-processing of raw data.

We propose a broad experimentation phase to address the various aspects mentioned above, using several methods and datasets. To the best of our knowledge, a performance analysis of whole-graph embedding methods under conditions of adversarial attacks has never been carried out.

The paper is structured as follows. Section 2 provides an overview of the state-of-art about adversarial attacks on whole-graph embedding models. Section 3 gives details about the problem statement. Section 4 provides a comprehensive experimental phase, while Sect. 5 concludes the paper.

2   Related Work

The literature concerning adversarial attacks for graph data is very recent and often aimed at node-level or link-level applications [8,39]. Here, we focus on graph-level applications, and specifically on adversarial attacks on whole-graph embedding methods, for which few recent papers (mainly preprints) are available.

In [40], Tang et al. design a surrogate model that consists of convolutional and pooling operators to generate adversarial samples to fool the hierarchical Graph Neural Networks (GNN)-based graph classification models. Nodes preserved by the pooling operator are set as attack targets. Then the attack targets are perturbed slightly to trick the pooling operator in hierarchical GNNs into selecting the wrong nodes to preserve. Furthermore, a robust training on the target models is performed to demonstrate that the retrained graph classification models can better defend against the attack from the adversarial samples.

Chen et al. [7] propose a graph attack framework named GraphAttacker that works to adjust the structures and to provide the attack strategies according to the graph analysis tasks. It generates adversarial samples based on the Generative Adversarial Network (GAN) through alternate training on three key components: the Multi-strategy Attack Generator, the Similarity Discriminator, and the Attack Discriminator. Furthermore, to achieve attacks within perturbation budget, a novel Similarity Modification Rate to quantify the similarity between nodes and thus to constrain the attack budget is introduced.

Another graph attack framework, named Graph Backdoor, is presented by Xi et al. [48]. It can be applied readily without knowing data models or tuning strategies to optimize both attack effectiveness and evasiveness. It works in different ways: i) graph-oriented – it defines triggers as specific subgraphs, including topological structures and descriptive features, entailing a large design spectrum for the adversary; ii) input-tailored – it dynamically adapts triggers to individual graphs; and iii) attack-extensible – it can be instantiated for both transductive and inductive tasks.

The vulnerability of Graph Convolutional Networks (GCNs) to adversarial attacks has been debated in the literature. In [24], Jin et al. introduce a robustness certificate for graph classification using GCNs under structural attacks. The method is based on Lagrange dualization and convex envelope, which result in tight approximation bounds computable by dynamic programming. Applied in conjunction with robust training, it allows an increased number of graphs to be certified as robust.

Faber et al. [14] discuss the particularities of explaining GNN predictions. In graphs, the structure is fundamental, and a slight modification can lead to little knowledge of the data. Therefore, the explanation is reflected in adversarial attacks. The authors argue that the explanation methods should stay with the training data distribution and produce Distribution Compliant Explanation (DCE). To this end, they propose a novel explanation method, Contrastive GNN Explanation, for graph classification that adheres to DCE.

You et al. [49] propose a graph contrastive learning (GraphCL) framework for learning unsupervised representations of graph data. The impact of various combinations of graph augmentations in different tasks (semi-supervised, unsupervised, transfer learning, and adversarial attacks) is explored. The proposed framework can produce graph representations of similar or better generalizability, transferability, and robustness than state-of-the-art methods.

In [9], Chung et al. present a framework named Graph Augmentations with Bi-level Optimization (GABO). It is built to provide a graph augmentation approach based on bi-level optimization to investigate the effects on graph classification performance. The augmentation procedure can be applied without a priori domain knowledge about the task. Indeed, the framework combines a Graph Isomorphism Network (GIN) layer augmentation generator with a bias transformation.

All the above described approaches propose different types of adversarial attacks. However, none of them shares our aim, i.e., to compare the robustness to adversarial attacks of different whole-graph embedding methods.

3   Background

In this section, we introduce the formalization of a graph adversarial attack for the graph classification task. We will first give some preliminary notions about graphs and the whole-graph embedding problem. Then, we introduce the graph adversarial attack and related strategies for graph classification.

3.1   Whole-Graph Embedding

A graph $G = (V, E)$ is represented by a pair of sets: $V = \{v_i\}_{i=1}^N$ is the set of nodes, and $E \subseteq V \times V$ is the set of edges, each one represented by a pair of nodes (v_i, v_j), where v_i is the source node and v_j the target node. This definition holds for unweighted graphs, which means graphs whose vertices relation is simply represented by a connection between them. Let W be a set of real numbers, called weights, such that for each $(v_i, v_j) \in E$ there exists a weight $w_{i,j} \in W$ associated to the edge; then $G(V, E, W)$ is called a weighted graph. An alternative representation of a weighted graph is through its adjacency matrix $A = \{A_{i,j}\}$, whose elements are:

$$A_{i,j} = \begin{cases} w_{i,j} & \text{if } (v_i, v_j) \in E \\ 0 & \text{otherwise} \end{cases}$$

For unweighted graphs, a unitary weight is considered for each edge to obtain the adjacency matrix. In general, the adjacency matrix A is not symmetric, since the occurrence of an edge from node v to node u does not imply the existence of the edge (u, v). This is only the case of undirected graphs, in which the connection between two nodes u and v has no direction, thus both $(u, v) \in E$ and $(v, u) \in E$ and A is symmetric. In the following, we will refer to generic graphs $G = (V, E)$, specifying their weights or their directionality only if needed.

In a very general definition, graph embedding learns a mapping from a graph to a vector space with the purpose of preserving main graph properties.

Definition 1. *Given a graph $G = (V, E)$, a graph embedding (or node-level graph embedding) is a mapping $\phi: v_i \in V \to \mathbf{y}_i \in \mathbb{R}^d, i = 1, \ldots, N, d \in \mathbb{N}$, such that the function ϕ preserves some proximity measure defined on graph G.*

Specifically, it is a space reduction that maps the nodes of a graph into a d-dimensional feature vector space, also known as *latent space*, trying to maintain structural information in terms of connections between vertices. The goal of keeping as much information as possible about the graph space in the transformation is linked to the choice of node/edge properties for the initial representation of the graph. The criticality concerns the final latent space that expresses valuable information, for applications such as classification or grouping, despite being in a lower-dimensional search space.

The concept of graph embedding refers to node-level since it maps each node in a graph into a vector, preserving node-node proximity (similarity/distance).

Definition 2. *Given a set of graphs $\mathcal{G} = \{G_i\}_{i=1}^M$ with the same set of vertices V, a whole-graph embedding is a mapping $\psi : G_i \rightarrow \mathbf{y}_i \in \mathbb{R}^d, i = 1,\ldots,M$, $d \in \mathbb{N}$, such that the function ψ preserves some proximity measure defined on $\mathcal{G}$.*

In this context, the fundamental condition is that the nodes of the graphs represent the same information. This requires an alignment procedure that verifies this property to provide compliant embedding.

Unlike graph embedding, which is adopted in applications such as link prediction and node label predictions, whole-graph embedding is more suited to graph classification, graph similarity ranking, and graph visualization.

3.2   Graph Adversarial Attacks

Generally, a network can become damaged through two primary ways: natural failure and targeted attack. Natural failures typically occur when a part fails due to natural causes. This results in the malfunction or elimination of a node or edge in the graph. Despite random network failures are much more common, they are less harmful than targeted attacks. This phenomenon has been verified across a range of graph structures [4]. Otherwise, targeted attacks carefully and through precise rules select the nodes and edges of the network for removal to maximally disrupt network functionality.

Our attention is focused on the modifications to the discrete structures and different attack strategies. Generally, the attacker tries to add or delete edges from G to create the new graph. These kinds of actions are varied since adding or deleting nodes can be performed by a series of modifications to the edges. Editing edges requires more effort than editing nodes. Indeed choosing a node only requires $O(|V|)$ complexity, while choosing an edge requires $O(|V|^2)$. In our experiments, we consider two attack strategies

- Degree-based Attack (DA): a percentage p of graph nodes having the highest degree is removed. The degree (or connectivity) δ_{v_i} of node v_i is the number of edges connected to it and can be computed using the graph adjacency matrix $A = \{A_{i,j}\}$ as

$$\delta_{v_i} = \sum_{j \neq i} A_{i,j}.$$

The effect of a DA is to reduce the total number of edges in the network as fast as possible [22]. It only takes into account the neighbors of the target node v when making a decision and can be considered a local attack. It is performed with a low computational overhead.

- Betweenness-based Attack (BA): a percentage p of graph nodes having the highest betweenness centrality is removed. The betweenness centrality for a node v_i is defined as

$$b_{v_i} = \sum_{j,k \neq i} \frac{\sigma_{j,k}(v_i)}{\sigma_{j,k}},$$

where $\sigma_{j,k}$ is the total number of shortest paths from node v_j to node v_k and $\sigma_{j,k}(v_i)$ is the number of those paths that pass through the target node v_i. The effect of a BA is to destroy as many paths as possible [22]. It is considered a global attack strategy due to the path information is aggregated from the whole network. Clearly, global information carries significant computational overhead compared to local attacks.

The robustness of the whole-graph embedding methods to adversarial attacks will be evaluated in terms of their performance for the task of graph classification on the attacked data.

4   Experiments

In our experiments, we analyze and compare the behavior of some whole-graph embedding methods under attack conditions for the task of graph classification. There are different challenges in this topic [40]

- *Selection of the target nodes and edges for the attack.* Suppose one or a few nodes or edges are perturbed at random. In that case, the graph classification results may not change because such a perturbation may not affect or destroy the intrinsic characteristics of graphs discriminating for the classification. In this regard, node selection strategies have been chosen as illustrated in Sect. 3.2.
- *Parameters setting to generate effective results.* The choice is undoubtedly difficult as the starting graphs are perturbed. A consequence could also fall on the computational costs during classification. As is well known, optimizing the parameters is a crucial aspect for obtaining the best performance. Concerning this point, we explored the parameter space to choose those that lead to the best results.
- *Robustness* is always an essential factor in evaluating the performance of the models. In the scenario of adversarial attacks, how to improve the robustness of the classification models? This is one of the two crucial points on which the paper was founded. In fact, as it is possible to observe through provided results, it is not certain that, by weakening the structure of the graphs, the transformation into a vector space, through the embedding phase, necessarily

produces an unrepresentative features vector, affecting the classification. We will see how some methods adapt even when the graph structures are less dense and informative.

- *Vulnerability*, in the same way, is always an essential factor in evaluating the performance of the models. In the scenario of adversarial attacks, how to identify the vulnerability of the classification models? It is the second crucial node on which the paper was founded. As it is possible to observe through the provided results, also in this case, by weakening the structure of the graphs, the transformation into a vector space, through the embedding phase, could produce an unrepresentative feature vector, affecting the classification. We will see how some methods do not fit when the graph structures are less dense and informative.

- *Data-driven selection*. The choice of data is driven by the characteristics of the graphs. In this way, models can show robustness or highlight critical issues when a variation of the data occurs. We decided to stress the various methods chosen for the evaluation based on different characteristics related to data. As we can see from Table 1, for example, three of the five datasets are unweighted. This detail is fundamental for calculating the centrality measures and, therefore, for selecting the nodes to be attacked.

4.1   Datasets

Table 1 illustrates the main properties of the datasets adopted in the experiments and includes synthetic and real network datasets, concerning some case studies of our current research on graph classification and clustering [17,29,30].

LFR is a synthetic dataset introduced in [21] based on the Lancichinetti–Fortunato–Radicchi (LFR) method [25]. As described in [29], we generated two classes of graphs containing 81 nodes, constructed using two different values of the parameter μ (expected proportion of edges having a vertex in one community and the other vertex in another community): 600 graphs with 0.1 μ and 1000 with 0.5 μ. Therefore, this dataset includes many small and unweighted graphs, subdivided into classes differing by well-defined community properties.

The MREG model [47] is adopted to generate the synthetic Multiple Random Eigen Graphs (MREG) dataset. Settings for MREG parameters, chosen based on the authors suggestions and our previous choices [29], are: $d = 2$ (model dimension), $n = 100$ (number of nodes), $h_1, h_2 \in \mathbb{R}^n$, where $h_1(i) = 0.1, \forall i$, $h_2(i) = -0.1, i = 1, \ldots, 50$, $h_2(i) = 0.1, i = 51, \ldots, 100$. The total number of unweighted graphs is 300, each composed of 100 nodes each, equally subdivided into 3 classes using $\lambda = [24.5, 4.75]$ for class c1, $\lambda = [20.75, 2.25]$ for class c2, and $\lambda = [24.5, 2.25]$ for class c3. In [47], further parameters' details are given.

The Brain fMRI dataset contains real networks built in [3] from functional magnetic resonance imaging (fMRI) time-series data [1] from the Center for Biomedical Research Excellence (COBRE) dataset. It is composed of 54 graphs from Schizophrenia subjects and 70 graphs from healthy controls. Each graph includes 263 nodes corresponding to different brain regions. The edges weights represent the Fisher-transformed correlation between the fMRI time-series of the nodes after ranking [3], and we only kept the weights of the positively correlated edges. The dataset ends up including dense weighted graphs with a high average degree but a small diameter.

The Kidney dataset describes real metabolic networks created for validating related research [18, 20, 30]. It contains networks derived from data of 299 patients divided into three classes: 159 clear cell Renal Cell Carcinoma (KIRC), 90 Papillary Renal Cell Carcinoma (KIRP), and 50 Solid Tissue Normal samples (STN). We obtained the networks by mapping gene expression data coming from the Genomic Data Commons (GDC, https://portal.gdc.cancer.gov) portal (Projects TCGA-KIRC and TCGA-KIRP) on the biochemical reactions extracted from the kidney tissue metabolic model [44] (https://metabolicatlas.org). Specifically, given the stoichiometric matrix of the metabolic model, the graph nodes represent the metabolites, and the edges connect reagent and product metabolites in the same reaction, weighted by the average of the expression values of the genes/enzymes catalyzing that reaction [20]. Different reactions represented by multiple edges connecting two metabolites were fused in a single edge, where the weight includes the sum of the weights of the fused edges. Disconnected nodes, due to reactions not catalyzed by an enzyme, and recurrent metabolites, were not included [20]. The simplification procedure described in [18] is applied to reduce the complexity of the network, leading to reduce the number of nodes from 4022 to 1034. Overall, the dataset includes sparse weighted graphs with a small average degree but wide diameter.

MUTAG [13] is a popular benchmark dataset and is composed of networks of 188 mutagenic aromatic and heteroaromatic nitro compounds. The nodes represent atoms, while the edges represent chemical bonds between them. The graphs contain both vertex and edge labels. The two classes indicate whether or not the compound has mutagenic effects on a bacterium. Contrary to the other datasets, the nodes are not perfectly aligned. Indeed, the MUTAG networks have an average of eighteen vertices, but the labels are only seven.

4.2   Compared Methods

In the experiments, we compared the classification results obtained using the network embeddings produced by seven whole-graph embedding methods, briefly described in the following

- GL2vec [6]. It is an extended version of Graph2vec. The method is named Graph and Line graph to vector (GL2vec) because it concatenates the embedding of an original graph to that of the corresponding line graph. The line graph is an edge-to-vertex dual graph of the original graph. Specifically, GL2vec integrates either the edge label information or the structural information, which Graph2vec misses with the embeddings of the line graph.

Table 1. Main properties of the adopted datasets

	LFR	MREG	Kidney	Brain fMRI	MUTAG
Graphs	1600	300	299	124	188
Classes	2	3	3	2	2
Samples per class	600/1000	100/100/100	159/90/50	70/54	125/63
Vertices	82	100	1034	263	17.93
Average edges	844.45	1151.71	3226.00	19748.88	39.59
Average edge density	0.13	0.23	0.01	0.57	0.138454
Distinct vertex labels	82	100	1034	263	7
Edge weights	✗	✗	✓	✓	✗
Minimum diameter	3	2	126	0.03	5
Maximum diameter	7	3	455.36	0.07	15
Average degree	20.60	23.03	6.24	150.18	2.19

- Graph2vec [33]. It provides a Skip-Gram neural network model, typically adopted in the NLP domain. It learns data-driven distributed representations of arbitrarily sized graphs. The resulting embeddings are learned in an unsupervised manner and are task-unaware.
- IGE [15]. It extracts handcrafted invariant features based on graph spectral decomposition. These features are easy to compute, permutation-invariant, and include sufficient information on the graph's structure.
- NetLSD [43]. It computes a compact graph signature derived from the solution of the heat equation involving the normalized Laplacian matrix. It is permutation and size-invariant, scale-adaptive, and computationally efficient.
- FGSD [45]. It provides a graph representation based on a family of graph spectral distances with uniqueness, stability, sparsity, and computational efficiency properties.
- FeatherGraph [38]. It adopts characteristic functions defined on graph vertices to describe the distribution of node features at multiple scales. The probability weights of the characteristic function are defined as the transition probabilities of random walks. The node-level features are combined by mean pooling to create graph-level statistics.
- Netpro2vec [31]. It is a neural-network method that produces embedding of whole-graphs which are independent from the task and nature of the data. It first represents graphs as textual documents whose words are formed by processing probability distributions of graph node paths/distances (e.g., the Transition Matrix, TM, or the Node Distance Distribution, NDD). Then, it embeds graph documents by using the doc2vec method [26].

4.3   Implementation Details

For the first six whole-graph embedding methods (GL2vec, Graph2vec, IGE, NetLSD, FGSD, and FeatherGraph), we used their implementation provided in the Karate Club software [37]. Our Netpro2vec framework, implemented in Python, is publicly available (https://github.com/cds-group/Netpro2vec). It also includes the code for extracting the NDD and TM distribution matrices, based on the GraphDistances R package [19], and the doc2vec embedding is performed using the gensim Natural Language Processing (NLP) library [35]. Even though the method can exploit different distribution distances as well their combinations, in the experimental results, we only report the two obtained using NDD (Netpro2vec^{ndd}) and the combination of NDD with TM1 (Netpro2vec$^{ndd+tm1}$), which lead to the best performance results.

The dimension d of the latent feature space for GL2Vec, Graph2Vec, FGSD, and Netpro2vec was set to 512; this value has been experimentally chosen so as to maximize accuracy. Instead, for IGE, FeatherGraph, and NetLSD, the output dimension cannot be specified as an input parameter.

For classification, we adopted an SVM model with a linear kernel (https://scikit-learn.org/stable/modules/generated/sklearn.svm.SVC.html). To avoid hiding the effect of adversarial attacks, we did not apply any feature selection, even though it could certainly provide higher performance results for any of the considered methods. We validated the developed models through ten-fold stratified cross-validation iterated ten times, measuring the mean and standard deviation of classification accuracy and Matthews correlation coefficient (MCC) [32].

All the experiments were run on Google Colab Machine, which provides by default a virtual machine based on a bi-processor with two CPUs @ 2.30 GHz Intel(R) Xeon(R), 13 GB RAM and 33 GB HDD.

4.4   Performance Evaluation

Performance results obtained using the seven whole-graph embedding methods described in Sect. 4.2 on the five datasets detailed in Sect. 4.1 are reported in the bar plots of Fig. 1. Detailed numerical results are given in Tables 2 and 3. Here, we consider the results achieved using the original network data (Unattacked), as well as those using data that underwent the removal of the 30% and 50% of the nodes having the highest betweenness centrality (BA) or the highest degree (DA), respectively. The choice of these percentages p of nodes to be removed aims at investigating the effects of both *moderate* (30%) and *strong* (50%) adversarial attacks. The performance is evaluated in terms of the Accuracy and MCC values, defined as

$$\text{Accuracy} = \frac{\text{TP} + \text{TN}}{\text{TP} + \text{FN} + \text{FP} + \text{TN}},$$

$$\text{MCC} = \frac{\text{TP} \cdot \text{TN} - \text{FP} \cdot \text{FN}}{\sqrt{(\text{TP} + \text{FP})(\text{TP} + \text{FN})(\text{TN} + \text{FP})(\text{TN} + \text{FN})}},$$

respectively. Here, TP, TN, FP, and FN indicate the number of true positives, true negatives, false positives, and false negatives. While the Accuracy provides the percentage of correctly classified samples, MCC gives the correlation coefficient between observed and predicted binary classifications.

Table 2. Accuracy (%) and MCC (mean and std over ten iterations) of adversarial attacks on whole-graph embedding models (30% of attacked nodes). In boldface the best results for each dataset and each attack.

Dataset	Method	Accuracy			MCC		
		Unattacked	BA	DA	Unattacked	BA	DA
LFR	GL2Vec	94.59 ± 1.75	84.66 ± 2.45	87.66 ± 2.12	0.88 ± 0.03	0.66 ± 0.05	0.74 ± 0.04
	Graph2vec	91.94 ± 2.04	84.41 ± 2.74	89.44 ± 2.13	0.82 ± 0.04	0.66 ± 0.05	0.77 ± 0.04
	IGE	100.00 ± 0.00	97.06 ± 1.34	97.17 ± 1.17	1.00 ± 0.00	0.93 ± 0.02	0.93 ± 0.02
	NetLSD	100.00 ± 0.00	$\mathbf{99.09 \pm 0.73}$	99.04 ± 0.71	1.00 ± 0.00	$\mathbf{0.98 \pm 0.01}$	0.97 ± 0.01
	FGSD	100.00 ± 0.00	97.97 ± 0.99	$\mathbf{99.15 \pm 0.68}$	1.00 ± 0.00	0.95 ± 0.02	$\mathbf{0.98 \pm 0.01}$
	FeatherGraph	100.00 ± 0.00	98.99 ± 0.69	99.00 ± 0.74	1.00 ± 0.00	0.97 ± 0.01	0.97 ± 0.01
	Netpro2vec^{ndd}	100.00 ± 0.00	98.41 ± 0.96	97.40 ± 1.16	1.00 ± 0.00	0.96 ± 0.01	0.94 ± 0.02
	Netpro2vec$^{ndd+tm1}$	100.00 ± 0.00	95.26 ± 1.16	72.13 ± 2.99	1.00 ± 0.00	0.89 ± 0.03	0.38 ± 0.07
MREG	GL2Vec	66.83 ± 7.14	$\mathbf{65.23 \pm 7.63}$	64.23 ± 8.09	0.44 ± 0.10	$\mathbf{0.48 \pm 0.11}$	0.47 ± 0.12
	Graph2vec	38.70 ± 8.46	40.60 ± 9.46	46.27 ± 9.30	0.08 ± 0.13	0.11 ± 0.14	0.19 ± 0.14
	IGE	65.80 ± 8.60	62.67 ± 7.53	65.70 ± 7.55	0.49 ± 0.12	0.44 ± 0.11	0.49 ± 0.11
	NetLSD	$\mathbf{71.57 \pm 7.05}$	58.03 ± 8.38	58.30 ± 8.29	$\mathbf{0.58 \pm 0.10}$	0.37 ± 0.12	0.38 ± 0.12
	FGSD	59.60 ± 6.08	59.73 ± 7.71	65.93 ± 7.97	0.40 ± 0.09	0.40 ± 0.11	$\mathbf{0.49 \pm 0.12}$
	FeatherGraph	59.60 ± 6.08	62.80 ± 7.77	$\mathbf{66.10 \pm 6.77}$	0.40 ± 0.09	0.44 ± 0.11	0.49 ± 0.10
	Netpro2vec^{ndd}	63.07 ± 7.60	42.70 ± 8.73	55.33 ± 9.44	0.45 ± 0.11	0.14 ± 0.13	0.33 ± 0.14
	Netpro2vec$^{ndd+tm1}$	36.80 ± 8.48	34.77 ± 7.95	30.87 ± 8.47	0.05 ± 0.12	0.02 ± 0.12	0.03 ± 0.13
Brain fMRI COBRE	GL2Vec	No conv	No conv	No conv	No conv	No conv	No conv
	Graph2vec	43.85 ± 11.27	46.29 ± 13.43	42.58 ± 12.38	-0.18 ± 0.24	-0.09 ± 0.27	-0.18 ± 0.25
	IGE	44.88 ± 14.70	48.99 ± 13.33	53.69 ± 14.64	-0.11 ± 0.30	-0.03 ± 0.28	0.05 ± 0.30
	NetLSD	56.12 ± 6.59	55.98 ± 12.10	$\mathbf{56.01 \pm 8.70}$	0.01 ± 0.18	0.09 ± 0.26	0.03 ± 0.21
	FGSD	56.54 ± 2.20	54.68 ± 12.75	48.31 ± 13.90	0.00 ± 0.00	0.07 ± 0.26	-0.06 ± 0.29
	FeatherGraph	53.77 ± 5.89	52.65 ± 7.77	53.77 ± 12.30	-0.06 ± 0.16	-0.08 ± 0.17	0.01 ± 0.28
	Netpro2vec^{ndd}	$\mathbf{56.58 \pm 12.74}$	58.58 ± 10.20	52.97 ± 11.36	$\mathbf{0.11 \pm 0.27}$	0.14 ± 0.23	-0.00 ± 0.27
	Netpro2vec$^{ndd+tm1}$	$\mathbf{56.58 \pm 12.74}$	$\mathbf{59.18 \pm 13.32}$	53.30 ± 12.70	$\mathbf{0.11 \pm 0.27}$	$\mathbf{0.17 \pm 0.28}$	0.05 ± 0.27
Kidney RNASeq	GL2Vec	90.09 ± 4.74	82.58 ± 6.73	59.83 ± 6.05	0.83 ± 0.08	0.71 ± 0.11	0.25 ± 0.16
	Graph2vec	90.79 ± 5.11	79.87 ± 7.05	58.08 ± 5.94	0.83 ± 0.08	0.66 ± 0.12	0.21 ± 0.17
	IGE	No conv	No conv	No conv	No conv	No conv	No conv
	NetLSD	53.46 ± 7.02	59.07 ± 7.14	62.23 ± 8.68	0.11 ± 0.16	0.25 ± 0.15	0.36 ± 0.15
	FGSD	No conv	No conv	No conv	No conv	No conv	No conv
	FeatherGraph	81.51 ± 7.96	81.67 ± 6.44	84.36 ± 6.64	0.68 ± 0.13	0.69 ± 0.10	0.74 ± 0.11
	Netpro2vec^{ndd}	83.53 ± 6.42	87.22 ± 6.17	85.83 ± 6.19	0.71 ± 0.11	$\mathbf{0.79 \pm 0.10}$	0.76 ± 0.10
	Netpro2vec$^{ndd+tm1}$	$\mathbf{91.27 \pm 4.45}$	$\mathbf{87.33 \pm 5.86}$	$\mathbf{90.91 \pm 5.60}$	$\mathbf{0.86 \pm 0.07}$	0.79 ± 0.09	$\mathbf{0.85 \pm 0.09}$
MUTAG	GL2Vec	76.11 ± 8.48	59.09 ± 10.57	67.82 ± 9.44	0.31 ± 0.24	0.05 ± 0.24	0.27 ± 0.21
	Graph2vec	66.32 ± 9.72	64.07 ± 9.79	63.63 ± 10.02	0.39 ± 0.24	0.15 ± 0.24	0.15 ± 0.24
	IGE	83.72 ± 7.92	83.22 ± 8.13	$\mathbf{84.93 \pm 7.45}$	0.61 ± 0.16	0.63 ± 0.17	$\mathbf{0.67 \pm 0.16}$
	NetLSD	$\mathbf{86.23 \pm 7.68}$	$\mathbf{83.71 \pm 8.03}$	83.38 ± 7.54	0.69 ± 0.16	$\mathbf{0.63 \pm 0.18}$	0.62 ± 0.17
	FGSD	86.01 ± 7.77	78.68 ± 9.57	81.21 ± 8.79	$\mathbf{0.70 \pm 0.16}$	0.55 ± 0.19	0.60 ± 0.18
	FeatherGraph	82.40 ± 8.24	69.72 ± 7.22	68.23 ± 7.20	0.60 ± 0.17	0.23 ± 0.22	0.17 ± 0.23
	Netpro2vec^{ndd}	71.42 ± 9.19	61.75 ± 8.78	62.34 ± 0.06	0.35 ± 0.21	0.00 ± 0.22	0.12 ± 0.23
	Netpro2vec$^{ndd+tm1}$	72.06 ± 9.64	73.95 ± 8.60	76.51 ± 9.27	0.34 ± 0.19	0.41 ± 0.20	0.47 ± 0.21

For the LFR dataset, the performance on unattacked graphs is high for all methods. Indeed, as already shown in [31], most of the considered methods succeed in producing whole-graph embeddings that lead to an almost perfect linear separation between the two LFR classes. In the case of moderate attacks, NetLSD and FGSD (but also FeatherGraph and Netpro2vec with NDD) respond better to both the types of adversarial attack, showing a lower reduction in Accuracy and MCC values as compared to the other methods. For stronger attacks, FeatherGraph reveals the most robust method, experiencing only a slight performance decrease.

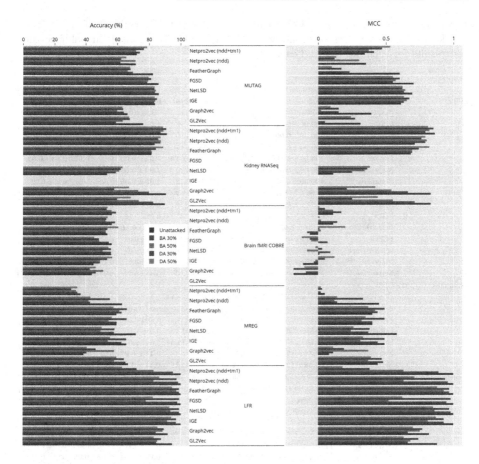

Fig. 1. Bar plots of the accuracy (%) and MCC of adversarial attacks on whole-graph embedding models

In the case of the MREG dataset, discordant results can be observed. Indeed, the best-performing method using unattacked data (NetLSD) experiences a dramatic performance decrease when handling both types of attacks. In contrast, the second-best method (GL2vec) is subject to a much lower performance decrease under attack, and this behavior holds whichever the strength of the attack. On the other side, the attacks can even be beneficial for classification performance, as for FeatherGraph that improves its performance under the moderate BA and the stronger DA.

For both the fMRI and Kidney datasets, Netpro2vec, mainly when based on NDD+TM1, appears to be the method that best exploits the network edges' weights. At the same time, it proves to be quite robust to adversarial attacks, experiencing slightly decreased performance for both moderate and strong attacks. Instead, NetLSD improves its performance when handling moderate DAs, showing the best performance among all the compared methods, and

Table 3. Accuracy (%) and MCC (mean and std over ten iterations) of adversarial attacks on whole-graph embedding models (50% of attacked nodes). In boldface the best results for each dataset and each attack.

Dataset	Method	Accuracy			MCC		
		Unattacked	BA	DA	Unattacked	BA	DA
LFR	GL2Vec	94.59 ± 1.75	85.36 ± 2.65	83.04 ± 2.56	0.88 ± 0.03	0.68 ± 0.05	0.63 ± 0.05
	Graph2vec	91.94 ± 2.04	88.74 ± 2.36	85.51 ± 2.69	0.82 ± 0.04	0.75 ± 0.05	0.70 ± 0.05
	IGE	100.00 ± 0.00	91.46 ± 2.10	94.17 ± 1.85	1.00 ± 0.00	0.81 ± 0.04	0.87 ± 0.03
	NetLSD	100.00 ± 0.00	93.60 ± 1.93	92.97 ± 1.99	1.00 ± 0.00	0.86 ± 0.04	0.85 ± 0.04
	FGSD	100.00 ± 0.00	77.96 ± 2.54	82.58 ± 2.97	1.00 ± 0.00	0.52 ± 0.05	0.62 ± 0.06
	FeatherGraph	100.00 ± 0.00	$\mathbf{97.17 \pm 1.19}$	$\mathbf{94.62 \pm 1.79}$	1.00 ± 0.00	$\mathbf{0.93 \pm 0.02}$	$\mathbf{0.88 \pm 0.03}$
	Netpro2vec^{ndd}	100.00 ± 0.00	82.99 ± 2.55	86.67 ± 2.40	1.00 ± 0.00	0.63 ± 0.05	0.71 ± 0.05
	Netpro2vec$^{ndd+tm1}$	100.00 ± 0.00	82.99 ± 2.55	62.99 ± 3.73	1.00 ± 0.00	0.63 ± 0.05	0.18 ± 0.08
MREG	GL2Vec	66.83 ± 7.14	58.07 ± 8.46	$\mathbf{59.57 \pm 8.28}$	0.44 ± 0.10	0.37 ± 0.12	$\mathbf{0.39 \pm 0.12}$
	Graph2vec	38.70 ± 8.46	58.07 ± 8.46	40.77 ± 8.18	0.08 ± 0.13	0.37 ± 0.12	0.11 ± 0.12
	IGE	65.80 ± 8.60	54.13 ± 6.73	55.53 ± 7.47	0.49 ± 0.12	0.31 ± 0.10	0.33 ± 0.11
	NetLSD	$\mathbf{71.57 \pm 7.05}$	49.37 ± 7.78	50.20 ± 7.48	$\mathbf{0.58 \pm 0.10}$	0.24 ± 0.11	0.26 ± 0.11
	FGSD	59.60 ± 6.08	54.60 ± 7.98	56.63 ± 8.13	0.40 ± 0.09	0.32 ± 0.12	0.35 ± 0.12
	FeatherGraph	59.60 ± 6.08	$\mathbf{61.37 \pm 8.02}$	57.73 ± 8.18	0.40 ± 0.09	$\mathbf{0.42 \pm 0.12}$	0.37 ± 0.12
	Netpro2vec^{ndd}	63.07 ± 7.60	41.83 ± 9.18	42.93 ± 7.92	0.45 ± 0.11	0.12 ± 0.14	0.14 ± 0.12
	Netpro2vec$^{ndd+tm1}$	36.80 ± 8.48	33.87 ± 8.71	34.70 ± 7.93	0.05 ± 0.12	0.00 ± 0.13	0.02 ± 0.12
Brain fMRI COBRE	GL2Vec	No conv	No conv	No conv	No conv	No conv	No conv
	Graph2vec	43.85 ± 11.27	51.13 ± 11.93	46.27 ± 13.14	-0.18 ± 0.24	-0.00 ± 0.25	-0.10 ± 0.27
	IGE	44.88 ± 14.70	48.96 ± 13.21	56.83 ± 13.16	-0.11 ± 0.30	-0.02 ± 0.27	0.12 ± 0.27
	NetLSD	56.12 ± 6.59	50.68 ± 10.30	54.41 ± 6.25	0.01 ± 0.18	-0.08 ± 0.20	-0.02 ± 0.13
	FGSD	56.54 ± 2.20	48.49 ± 13.51	45.22 ± 11.42	0.00 ± 0.00	-0.05 ± 0.28	-0.12 ± 0.23
	FeatherGraph	53.77 ± 5.89	52.63 ± 12.04	$\mathbf{60.65 \pm 13.51}$	-0.06 ± 0.16	0.02 ± 0.26	$\mathbf{0.20 \pm 0.28}$
	Netpro2vec^{ndd}	$\mathbf{56.58 \pm 12.74}$	52.46 ± 13.24	53.83 ± 11.31	$\mathbf{0.11 \pm 0.27}$	0.02 ± 0.28	0.01 ± 0.26
	Netpro2vec$^{ndd+tm1}$	$\mathbf{56.58 \pm 12.74}$	$\mathbf{56.35 \pm 12.46}$	53.83 ± 11.31	$\mathbf{0.11 \pm 0.27}$	$\mathbf{0.11 \pm 0.25}$	0.01 ± 0.26
Kidney RNASeq	GL2Vec	90.09 ± 4.74	73.39 ± 7.56	68.49 ± 7.34	0.83 ± 0.08	0.55 ± 0.12	0.44 ± 0.14
	Graph2vec	90.79 ± 5.11	73.02 ± 7.30	67.42 ± 7.44	0.83 ± 0.08	0.54 ± 0.12	0.42 ± 0.14
	IGE	No conv	No conv	No conv	No conv	No conv	No conv
	NetLSD	53.46 ± 7.02	61.13 ± 8.00	63.27 ± 7.76	0.11 ± 0.16	0.34 ± 0.14	0.38 ± 0.13
	FGSD	No conv	No conv	No conv	No conv	No conv	No conv
	FeatherGraph	81.51 ± 7.96	81.37 ± 6.83	$\mathbf{89.00 \pm 4.79}$	0.68 ± 0.13	0.69 ± 0.11	$\mathbf{0.82 \pm 0.07}$
	Netpro2vec^{ndd}	83.53 ± 6.42	87.52 ± 5.66	87.35 ± 5.30	0.71 ± 0.11	0.80 ± 0.10	0.79 ± 0.08
	Netpro2vec$^{ndd+tm1}$	$\mathbf{91.27 \pm 4.45}$	$\mathbf{89.20 \pm 5.36}$	88.87 ± 5.68	$\mathbf{0.86 \pm 0.07}$	$\mathbf{0.82 \pm 0.09}$	0.81 ± 0.09
MUTAG	GL2Vec	76.11 ± 8.48	63.37 ± 9.25	67.39 ± 10.10	0.31 ± 0.24	0.00 ± 0.22	0.24 ± 0.24
	Graph2vec	66.32 ± 9.72	59.86 ± 9.32	63.10 ± 8.28	0.39 ± 0.24	0.02 ± 0.24	0.09 ± 0.21
	IGE	83.72 ± 7.92	83.93 ± 7.85	83.56 ± 7.47	0.61 ± 0.16	$\mathbf{0.65 \pm 0.17}$	$\mathbf{0.63 \pm 0.17}$
	NetLSD	$\mathbf{86.23 \pm 7.68}$	$\mathbf{84.25 \pm 7.61}$	83.48 ± 7.77	0.69 ± 0.16	0.64 ± 0.16	0.62 ± 0.17
	FGSD	86.01 ± 7.77	79.27 ± 8.80	$\mathbf{79.54 \pm 9.17}$	$\mathbf{0.70 \pm 0.16}$	0.57 ± 0.17	0.55 ± 0.20
	FeatherGraph	82.40 ± 8.24	66.44 ± 2.30	67.82 ± 4.63	0.60 ± 0.17	0.00 ± 0.00	0.10 ± 0.18
	Netpro2vec^{ndd}	71.42 ± 9.19	71.22 ± 8.09	65.55 ± 8.54	0.35 ± 0.21	0.30 ± 0.22	0.13 ± 0.23
	Netpro2vec$^{ndd+tm1}$	72.06 ± 9.64	71.60 ± 11.32	78.52 ± 8.30	0.34 ± 0.19	0.37 ± 0.20	0.53 ± 0.18

the same can be said for FeatherGraph under strong attack. Other methods, such as GL2vec on fMRI or IGE and FGSD on Kidney, fail to reach convergence in all the unattacked and attacked cases, yielding no classification model.

On the MUTAG dataset, the best-performing method (NetLSD) experiences a tiny performance decrease in handling moderate and strong adversarial attacks. The same can also be said for IGE and Netpro2vec, which maintain similar performance, if not better, regardless of the attacks.

Overall, we can conclude that FeatherGraph, NetLSD, Netpro2vec, and IGE appear to be more robust than the other methods under both moderate and strong adversarial attacks. We also observed unexpected behaviors in some cases, where an improvement in performance rather than a degradation occurs. Indeed, removing central (important) nodes does not always weaken the significance of the graph description. Therefore, these nodes can be considered important but not fundamental for the transformation from a graph to a vector space. This point deserves further attention in future experiments.

We wish to emphasize that the above analysis is primarily intended to investigate the robustness to adversarial attacks of the considered methods rather than their performance for the classification task. Indeed, further optimization steps, such as feature selection or class balancing, have been purposely omitted for all the methods, which would certainly help in achieving better classification performance.

5   Conclusions and Future Work

We have analyzed and compared different whole-graph embedding methods to understand their behavior under adversarial attacks better. We have performed attacks on graphs supposing the subsequent data analysis task is supervised classification. During the attacks, we have analyzed the unique features of each embedding method to highlight strengths and weaknesses, varying the type of attack and dataset. In this regard, the robustness of the graph analysis task model is an important issue. Future works concern many directions. First, extending the analysis on different types of datasets and attacks to propose defense mechanisms that can partially or entirely erase the highlighted limits of existing solutions. Besides, it would be interesting to analyze the embedding features that the methods create for the classification task. Methods like SHapley Additive exPlanations (SHAP) [28] could be applied to learn feature importance and explain the model output.

Acknowledgments. This work has been partially funded by BiBiNet project (H35F21000430002) within POR-Lazio FESR 2014–2020. It was carried out also within the activities of the authors as members of the ICAR-CNR INdAM Research Unit and partially supported by the INdAM research project "Computational Intelligence methods for Digital Health". Mario Manzo thanks Prof. Alfredo Petrosino for the guidance and supervision during the years of working together. The work of Mario R. Guarracino was conducted within the framework of the Basic Research Program at the National Research University Higher School of Economics (HSE).

References

1. Aine, C.J., Jeremy Bockholt, H., Bustillo, J.R., et al.: Multimodal neuroimaging in schizophrenia: description and dissemination. Neuroinformatics **15**(4), 343–364 (2017)
2. Akhtar, N., Mian, A.: Threat of adversarial attacks on deep learning in computer vision: a survey. IEEE Access **6**, 14410–14430 (2018). https://doi.org/10.1109/ACCESS.2018.2807385

3. Arroyo-Relión, J.D., et al.: Network classification with applications to brain connectomics [Internet]. Ann. Appl. Stat. **13**(3), 1648 (2019)
4. Beygelzimer, A., et al.: Improving network robustness by edge modification. Physica A Stat. Mech. Appl. **357**(3–4), 593–612 (2005)
5. Cai, H., Zheng, V.W., Chang, K.C.-C.: A comprehensive survey of graph embedding: problems, techniques, and applications. IEEE Trans. Knowl. Data Eng. **30**(9), 1616–1637 (2018)
6. Chen, H., Koga, H.: GL2vec: graph embedding enriched by line graphs with edge features. In: Gedeon, T., Wong, K.W., Lee, M. (eds.) ICONIP 2019. LNCS, vol. 11955, pp. 3–14. Springer, Cham (2019). https://doi.org/10.1007/978-3-030-36718-3_1
7. Chen, J., et al.: GraphAttacker: a general multi-task graphattack framework. In: arXiv preprint arXiv:2101.06855 (2021)
8. Chen, L., et al.: A survey of adversarial learning on graphs. In: CoRR abs/2003.05730 (2020). arXiv: 2003.05730
9. Chung, H.W., Datta, A., Waites, C.: GABO: graph augmentations with bi-level optimization. In: arXiv preprint arXiv:2104.00722 (2021)
10. Chvátal, V.: Tough graphs and Hamiltonian circuits. Discret. Math. **5**(3), 215–228 (1973)
11. Dai, H., et al.: Adversarial attack on graph structured data. In: CoRR abs/1806.02371 (2018). arXiv: 1806.02371
12. Dalvi, N., et al.: Adversarial classification. In: Proceedings of the Tenth ACM SIGKDD International Conference on Knowledge Discovery and Data Mining, KDD 2004, pp. 99–108. Association for Computing Machinery, Seattle (2004). https://doi.org/10.1145/1014052.1014066. ISBN 1581138881
13. Debnath, A.K., et al.: Structure-activity relationship of mutagenic aromatic and heteroaromatic nitro compounds. Correlation with molecular orbital energies and hydrophobicity. J. Med. Chem. **34** (1991). https://doi.org/10.1021/jm00106a046
14. Faber, L., Moghaddam, A.K., Wattenhofer, R.: Contrastive graph neural network explanation. In: arXiv preprint arXiv:2010.13663 (2020)
15. Galland, A., Lelarge, M.: Invariant embedding for graph classification. In: ICML 2019 Workshop on Learning and Reasoning with Graph-Structured Representations (2019)
16. Gao, J., et al.: Black-box generation of adversarial text sequences to evade deep learning classifiers. In: 2018 IEEE Security and Privacy Workshops (SPW), pp. 50–56 (2018). https://doi.org/10.1109/SPW.2018.00016
17. Granata, I., et al.: A short journey through whole graph embedding techniques. In: International Conference on Network Analysis (NET 2020) (2020)
18. Granata, I., et al.: Model simplification for supervised classification of metabolic networks. Ann. Math. Artif. Intell. **88**(1), 91–104 (2020)
19. Granata, I., Guarracino, M.R., Maddalena, L., Manipur, I.: Network distances for weighted digraphs. In: Kochetov, Y., Bykadorov, I., Gruzdeva, T. (eds.) MOTOR 2020. CCIS, vol. 1275, pp. 389–408. Springer, Cham (2020). https://doi.org/10.1007/978-3-030-58657-7_31 ISBN 978-3-030-58657-7
20. Granata, I., et al.: Supervised classification of metabolic networks. In: 2018 IEEE International Conference on Bioinformatics and Biomedicine (BIBM), pp. 2688–2693. IEEE (2018)
21. Gutiérrez-Gómez, L., Delvenne, J.-C.: Unsupervised network embeddings with node identity awareness. Appl. Netw. Sci. **4**(1), 82 (2019)
22. Holme, P., et al.: Attack vulnerability of complex networks. Phys. Rev. E **65**(5), 056109 (2002)

23. Huang, L., et al.: Adversarial machine learning. In: Proceedings of the 4th ACM Workshop on Security and Artificial Intelligence, AISec 2011, pp. 43–58. Association for Computing Machinery, Chicago (2011). https://doi.org/10.1145/2046684.2046692. ISBN 9781450310031
24. Jin, H., et al.: Certified robustness of graph convolution networks for graph classification under topological attacks. In: Advances in Neural Information Processing Systems, vol. 33 (2020)
25. Lancichinetti, A., Fortunato, S., Radicchi, F.: Benchmark graphs for testing community detection algorithms. Phys. Rev. E **78**(4), 046110 (2008)
26. Le, Q., Mikolov, T.: Distributed representations of sentences and documents. In: International Conference on Machine Learning, pp. 1188–1196 (2014)
27. Lin, Y., et al.: Learning entity and relation embeddings for knowledge graph completion. In: Proceedings of the AAAI Conference on Artificial Intelligence. vol. 29, January 2015
28. Lundberg, S.M., Lee, S.-I.: A unified approach to interpreting model predictions. In: Advances in Neural Information Processing Systems, vol. 30, pp. 4765–4774 (2017)
29. Maddalena, L., et al.: On whole graph embedding techniques. In: International Symposium on Mathematical and Computational Biology (BIOMAT 2020), November 2020
30. Manipur, I., et al.: Clustering analysis of tumor metabolic networks. BMC Bioinform. (2020). https://doi.org/10.1186/s12859-020-03564-9. ISSN 1471–2105
31. Manipur, I., et al.: Netpro2vec: a graph embedding framework for biomedical applications. IEEE/ACM Trans. Comput. Biol. Bioinform. (2021)
32. Matthews, B.W.: Comparison of the predicted and observed secondary structure of T4 phage lysozyme. Biochimica et Biophysica Acta (BBA) - Protein Struct. **405**(2), 442–451 (1975). https://doi.org/10.1016/0005-2795(75)90109-9. ISSN 0005–2795
33. Narayanan, A., et al.: graph2vec: learning distributed representations of graphs. In: ArXiv abs/1707.05005 (2017)
34. Qiu, S., et al.: Review of artificial intelligence adversarial attack and defense technologies. Appl. Sci. **9**(5) (2019). https://doi.org/10.3390/app9050909. ISSN 2076–3417
35. Řehůřek, R., Sojka, P.: Software framework for topic modelling with large corpora. English. In: Proceedings of the LREC 2010 Workshop on New Challenges for NLP Frameworks, pp. 45–50. ELRA, Valletta, May 2010
36. Rosenberg, I., Shabtai, A., Rokach, L., Elovici, Y.: Generic black-box end-to-end attack against state of the art API call based malware classifiers. In: Bailey, M., Holz, T., Stamatogiannakis, M., Ioannidis, S. (eds.) RAID 2018. LNCS, vol. 11050, pp. 490–510. Springer, Cham (2018). https://doi.org/10.1007/978-3-030-00470-5_23 ISBN 978-3-030-00470-5
37. Rozemberczki, B., Kiss, O., Sarkar, R.: Karate Club: an API oriented open-source Python framework for unsupervised learning on graphs. In: Proceedings of the 29th ACM International Conference on Information and Knowledge Management (CIKM 2020). ACM (2020)
38. Rozemberczki, B., Sarkar, R.: Characteristic functions on graphs: birds of a feather, from statistical descriptors to parametric models. In: Proceedings of the 29th ACM International Conference on Information & Knowledge Management, pp. 1325–1334 (2020)
39. Sun, L., et al.: Adversarial attack and defense on graph data: a survey. In: CoRR abs/1812.10528 (2018). arXiv: 1812.10528

40. Tang, H., et al.: Adversarial attack on hierarchical graph pooling neural networks. In: arXiv preprint arXiv:2005.11560 (2020)
41. Thorne, T., Stumpf, M.P.H.: Graph spectral analysis of protein interaction network evolution. J. Royal Soc. Interface **9**(75), 2653–2666 (2012)
42. Tong, H., et al.: On the vulnerability of large graphs. In: 2010 IEEE International Conference on Data Mining, pp. 1091–1096. IEEE (2010)
43. Tsitsulin, A., et al.: NetLSD: hearing the shape of a graph. In: Proceedings of the 24th ACM SIGKDD International Conference on Knowledge Discovery & Data Mining, pp. 2347–2356 (2018)
44. Uhlén, M., et al.: Tissue-based map of the human proteome. Science **347**(6220) (2015)
45. Verma, S., Zhang, Z.-L.: Hunt for the unique, stable, sparse and fast feature learning on graphs. In: Advances in Neural Information Processing Systems, pp. 88–98 (2017)
46. Vlietstra, W.J., et al.: Using predicate and provenance information from a knowledge graph for drug efficacy screening. J. Biomed. Semant. **9**(1), 1–10 (2018)
47. Wang, S., et al.: Joint embedding of graphs. IEEE Trans. Pattern Anal. Mach. Intell. **43**(4), 1324–1336 (2021). https://doi.org/10.1109/TPAMI.2019.2948619
48. Xi, Z., et al.: Graph backdoor. In: 30th USENIX Security Symposium (USENIX Security 2021) (2021)
49. You, Y., et al.: Graph contrastive learning with augmentations. In: Advances in Neural Information Processing Systems, vol. 33 (2020)

Algorithm Selection on Adaptive Operator Selection: A Case Study on Genetic Algorithms

Mustafa Mısır[1,2(✉)]

[1] Istinye University, Istanbul, Turkey
mustafa.misir@istinye.edu.tr
[2] Duke Kunshan University, Kunshan, China

Abstract. The present study applies Algorithm Selection (AS) to Adaptive Operator Selection (AOS) for further improving the performance of the AOS methods. AOS aims at delivering high performance in solving a given problem through combining the strengths of multiple operators. Although the AOS methods are expected to outperform running each operator separately, there is no one AOS method can consistently perform the best. Thus, there is still room for improvement which can be provided by using the best AOS method for each problem instance being solved. For this purpose, the AS problem on AOS is investigated. The underlying AOS methods are applied to choose the crossover operator for a Genetic Algorithm (GA). The Quadratic Assignment Problem (QAP) is used as the target problem domain. For carrying out AS, a suite of simple and easy-to-calculate features characterizing the QAP instances is introduced. The corresponding empirical analysis revealed that AS offers improved performance and robustness by utilizing the strenghts of different AOS approaches.

1 Introduction

Various algorithms, each dedicated to solve a particular problem, have been developed to tackle the optimization and decision problems. The studies on different problems, however, have revealed that each algorithm has its own strengths and weaknesses. Thus, an algorithm that performs well on a certain set of problem instances, delivers poor performance on other instances of the same problem. In other words, there is no one algorithm performs well on all the instance of a particular problem. This claim is also supported by the No Free Lunch theorem [1]. From this perspective, one way to outperform such algorithms is to develop a strategy to determine which algorithm is likely to perform the best on a given problem instance. This idea has been studied under the Algorithm Selection Problem (ASP) [2]. The traditional approach to handle the ASP is generating performance prediction models. Such a model can forecast the performance of the existing algorithms on a new problem instance. Then, the algorithm(s) which is expected to perform the best can be applied to solve this particular instance.

© Springer Nature Switzerland AG 2021
D. E. Simos et al. (Eds.): LION 2021, LNCS 12931, pp. 237–251, 2021.
https://doi.org/10.1007/978-3-030-92121-7_20

Further, fine-tuned, selection among algorithms can be done through Adaptive Operator Selection (AOS) [3], a.k.a. adaptive operator allocation [4], similarly to Selection Hyper-heuristics (SHHs) [5]. AOS can be referred as Online Algorithm Selection (AS) while the aforementioned ASP is mainly concerned with Offline AS. Online AS has more potential compared to Offline AS considering that Offline AS is a special case of Online AS. This potential can be justified by referring to the condition that the performance of algorithms can differ on different parts of the search space [6]. Determining the best algorithm on-the-fly while a problem instance solved can help to deliver better performance than the traditional Offline AS. The possible performance gain of combining multiple algorithms in an AOS setting was also shown from a theoretical perspective [7]. It should be noted that algorithms used under AOS are usually expected to be used to perturb a given solution. Thus, unlike the traditional algorithm selection strategies, the AOS methods can take advantage of the algorithms' strengths at the solution level. However, the algorithm base used under AOS is mostly composed of simple move operators that make rather small changes on a solution. Yet, there exists studies [8] utilizing complete algorithms as well.

This work focuses on using AS without loosing the advantages of AOS so that the performance beyond the independent performance of Offline AS and AOS alone can be reached or exceeded. An existing AS system, i.e. ALORS [9], is accommodated for performing AS. The task of AOS stays the same while AS is employed to choose the best AOS method. In other words, AS performs selection decision at a higher level. In order to experimentally analyze the proposed idea, a suite of well-known AOS techniques are employed. The goal of AOS here is to pick the best crossover operator at each generation for a Genetic Algorithm (GA). The GA is applied to the widely-used benchmarks of the Quadratic Assignment problem.

In the remainder of the paper, Sect. 2 gives an overview both on AS and AOS. Section 3 introduces the proposed approach. Then, a detailed experimental analysis is presented in Sect. 4. The paper is finalized with a summary and discussed in Sect. 5.

2   Background

2.1   Adaptive Operator Selection

Adaptive Operator Selection (AOS) [3,10] has been mainly studied to choose the right operator on-the-fly for evolutionary algorithms (EAs). AOS is composed of two sub-mechanisms including *Operator Selection* and *Credit Assignment*. Operator Selection is the main component that decides which operator to apply next. Next can mean next generation for evolutionary algorithms while it can mean next iteration for other type of algorithms. This decision usually depends on a measure which reflects the performance or behavior of each operator. Credit Assignment is responsible to take care of this evaluation step of operators. While these terminologies belong to AOS, they have been widely used in a broder perspective in Selection Hyper-heuristics[1] [11,12].

From the operator selection perspective, the idea is to efficiently explore the search space while exploiting good search regions. Exploration and exploitation dilemma has been the central subject of reinforcement learning (RL) [13]. In terms of AOS, multi-armed bandits (MABs) [14] have been widely used to address this dilemma. A MAB can be considered a single state RL[2]. Prior to utilizing MAB for AOS, Probability Matching (PM) [15] has been used to perform basic AOS through roulette wheel selection referring to the current credits of each operator. The shortcomings of PM have been tried to addressed by Adaptive Pursuit (AP) [4] which comes from Learning Automata (LA) [16], i.e. stateless RL. Pure LA has been also used in SHHs [17,82,83]. An extreme value based credit assignment strategy was delivered for AOS as ExAOS [18] by integrating it to some existing selection approaches like PM and AP. A comparative analysis of AOS with PM and AP using varying credit assignment schemes for multi-objective optimization was reported in [19]. Both PM and AP were also integrated to NSGA-III in [20]. LSAOS-DE [21] was offered as a landscape-based AOS for differential evolution. In [22], Hybrid Non-dominated Sorting Genetic Algorithm (HNSGA) with AOS was proposed as a variant of NSGA-II [23], using a credit assignment step depends on the survival of the solutions generated by each operator. Evolvability metrics were accommodated for credit assignment in [24]. The speed up achieved for the sake of expensive calculation of the evolvability metrics, in [25]. Self-organizing Neural Networks [26] was proposed, requiring an offline training phase for a continous state-discrete action markov decision process (MDP) scenario. In [27–29], AOS was emulated in distributed settings.

Regarding MAB, the objective is to minimize the regret which comes from the total reward achieved compared to the maximum reward could have been collected. As a MAB algorithm, Upper Confidence Bound (UCB) [30] has been widely referred for AOS [3] mainly due to its theoretical guarantees. In order to revise this idea considering the changing search requirements, Dynamic Multi-Armed Bandit (DMAB) [3] was proposed. DMAB essentially employs the Page-Hinkley (PH) statistical test [31] to determine the drastic changes in the reward distribution, resulting in restarting the whole AOS learning process. It is then extend with ExAOS as Ex-DMAB [32]. Sliding Multi-Armed Bandit (SLMAB) [33] was developed, limiting the reward update scheme. Two credit-assignment, reward schemes were incorporated as SR-B and AUC-B in [34]. While SR-B utilized a rank-based reward scheme, AUC-B basically uses Area Under Curve (AUC) [35] from machine learning. Fitness-Rate-Ranking based Multi-armed Bandit (FRRMAB) [36] was developed to perform AOS relying on instance fitness improvement, as also in [37], and applied in a multi-objective scenario [38]. Fitness-Rate-Average based Multi-armed Bandit (FRAMAB) [39] borrows the basic idea of FRRMAB while incorporating average fitness improvement, as a hyper-heuristic. In [40], FRRMAB was also used as a hyper-heuristic component

---

[1] a hyper-heuristic bibliography: https://mustafamisir.github.io/hh.html.

[2] http://www.yisongyue.com/courses/cs159/lectures/rl-notes.pdf.

of MOEAs for Search-based Software Engineering (SBSE) [41]. Self-organizing Neural Network (SONN) was applied as a AOS for crossover selection of a GA in [26].

2.2 Algorithm Selection

Algorithm Selection (AS) is about predicting the high performing algorithm, hopefully the best, among a set of available algorithms for a particular problem instance. In order to perform such prediction, an AS model is generated to map a group of instance features that are able to reflect the differences of the instance of the corresponding problem to the algorithms' performance. SATZilla [42] is one of the well known AS approaches due to its success in the SAT competitions[3]. It accommodates pure AS as well as a number of supportive components such as pre-solvers and back-up solvers. Its recent version [43] additionally incorporates the cost-sensitive classification models, as in [44] offering an AS system based on cost-sensitive hierarchical clustering (CSHC). ArgoSmArT[4] [45] was introduced as another AS system yet considering combination of different policies used in the SAT solvers. It essentially seeks the best combination of policies, in particular for variable selection, polarity selection and restart policies. ArgoSmArT k-NN[5] [46] is a general AS system operating based on k-nearest neighbors. SNAPP[6] [47] also utilizes k-NN for AS. Although all these AS approaches happen to provide effective solutions for AS, they can also suffer for providing the best AS decisions at a high level. AutoFolio [48] seeks the best possible AS setting through algorithm configuration/parameter tuning [49]. It essentially search across the space of possible components to be used in an AS method.

Besides utilizing AS for choosing an algorithm for a given instance, it is also possible to derive effective algorithm schedules, used in an AS approach called 3S[7], [50]. An algorithm schedule is all about generating a schedule that assigns a time budget to run each selected algorithm in an interleaving manner. While it a general algorithm schedule can be derived for a particular problem as static schedules, a different schedule can be used depending on the instance to be solved as dynamic schedules [51] – CPHydra[8]. SUNNY [52] also falls into this category as a schedule based portfolio system without any offline training, targeting the Constraint Satisfaction Problems (CSPs) [53]. Its generic version[9] is also available to be able to show its capabilities beyond the CSPs. T-SUNNY [54] was introduced as an extension by additionally performing a training phase to deliver better algorithm schedules. These schedules are primarily expected to be run on a single CPU (core). In [55], 3S was extended as p3S such that parallel

---

[3] http://www.satcompetition.org/.

[4] http://argo.matf.bg.ac.rs/downloads/software/ArgoSmart.zip.

[5] http://argo.matf.bg.ac.rs/downloads/software/ArgoSmartkNN.zip.

[6] https://sites.google.com/site/yurimalitsky/downloads/SNNAP_ver1.5.zip.

[7] https://sites.google.com/site/yurimalitsky/downloads/3S-2011.tar.

[8] http://homepages.laas.fr/ehebrard/cphydra.html.

[9] https://github.com/CP-Unibo/sunny-as.

schedules can be generated. aspeed [56] was developed to deliver both sequential and parallel solver schedules with the help of answer set programming.

The performance of AS heavily depends on the underlying algorithm set. Algorithm portfolios [57] is a tightly related concept to AS, aiming at determining an algorithm set mainly involving diverse algorithms. Diversity here refers to the capability of solving different types of instances so that the algorithm base can help to handle a varying set of instances. The main purpose of algorithm portfolios is to perform the constituent algorithms in parallel so that at least one of the algorithms can solve the target problem instance. ppfolio[10] [58] offers such a basic setting. Yet, they can also be used to come up with an efficient portfolio to be used for algorithm selection.

Algorithms portfolios can be automatically generated at some extend. ISAC [59] return different configurations for a given algorithm such that a diverse portfolio is achieved. Evolving ISAC (EISAC) [60,61] updates such configuration based portfolios over-time with new instances. ISAC+ [62] integrates multiple algorithms while utilizing CSHC [44] as a method to choose which algorithm-configuration pairs to keep. Hydra[11] [63] was introduced as as another configuration-based portfolio tool. Its version of directly targeting mixed-integer programming was also offered as a separate tool, i.e. Hydra-MIP [64]. OSCAR [65] provides a portfolio generation system to be used for Online AS by choosing from available design choices of a given algorithm. A similar idea has been followed in [66] by using Design of Experiments (DOE) [67] to perform cheap algorithm configuration which can drastically narrow down the configuration space of a given algorithm. ADVISER[12] [68] offers a rather basic algorithm configuration based portfolio system yet operating as-a-service. Any portfolio generation task, concerning different both parametric and non-parametric algorithms, was handled on a remote machine after submitting it through a web interface. ADVISER+[13] [69] extends ADVISER by providing new functionalities on the front-end, helping better user experience and the back-end, by running the submitted portfolio generation tasks in parallel for better response time.

One of the issue with the algorithm selection techniques is their need for the complete performance data. Depending on the algorithms and the problem type, it can be quite costly mainly to generate such data. ALORS [9] was introduced to address this issue by generating algorithm selection models with incomplete/sparse data. Matrix factorization (MF) [70] was incorporated for this purpose since it is able to extract useful features both for instances and algorithms [71]. In case AS will be applied to a totally new problem, this incomplete data can be chosen such that it will reveal more information at a cheaper cost than randomly picking them [72].

---

[10] http://www.cril.univ-artois.fr/~roussel/ppfolio/.

[11] http://www.cs.ubc.ca/labs/beta/Projects/Hydra/ – unrelated to the aforementioned CPHydra.

[12] http://research.larc.smu.edu.sg/adviser/.

[13] http://research.larc.smu.edu.sg/adviserplus/.

In order to provide a unified testbed for AS, ASlib[14] [73] was released. ASlib provides various AS benchmarks from different problem domains. Each benchmark is mainly composed of a set of algorithms, a group of problem instances and their features.

3 Algorithm Selection for AOS

Algorithm Selection (AS) is performed to build a rank-prediction model that can forecast the Adaptive Operator Selection (AOS) methods' performance in a per-instance basis. The predictions lead to determine the expected best AOS method for each selection task. AOS, here, is utilized to choose among the crossover operators of a Genetic Algorithm (GA) for the Quadratic Assignment Problem (QAP). In that respect, AS performs the selection of selectors, similarly to a work using AS on Selection Hyper-heuristics (SHHs) [74]. ALORS is used as the AS technique. Unlike the traditional AS techniques, ALORS performs selection through a feature-to-feature mapping. The source features are the ones, directly derived from the problem instances of a domain. The target features are automatically extracted ones through Matrix Factorization (MF). For MF, Singular Value Decomposition (SVD) [75] is employed. SVD is applied to the performance data for deriving matrices of rank 3 which is the dimension of the resulting matrices. The aforementioned mapping is achieved by Random Forest (RF) [76].

3.1 Instance Features

The present study targets the Quadratic Assignment Problem (QAP) [77,78]. The QAP is concerned with matching n facilities to n locations such that the cost incurred due to the movements between these facilities is minimized. The cost is calculated through a weight or flow element and distance as shown in Eq. 1. $f_{\pi_i \pi_j}$ represents the flow between the facilities π_i and π_j in the solution π. d_{ij} shows the distance between the locations i and j.

$$min \sum_{i}^{n} \sum_{j}^{n} f_{\pi_i \pi_j} d_{ij} \qquad (1)$$

For representing the QAP instances, 18 simple features are introduced as listed in Table 1. n as the size of the given QAP instance is used as the first feature. The remaining features are directly extracted from the flow and distance matrices. Among the latter group, there are also basic features considering both flow and distance related measures.

---

[14] http://aslib.net.

Table 1. The QAP instance features (std: standard deviaton; coeff_var: coefficient of variation; non_zero_ratio: ratio of the number of non-zero entries)

Source	Features
Basic (1)	size (n)
Distance matrix (6)	mean_cost, min_cost, max_cost, std_cost, coeff_var_cost, non_zero_ratio_cost
Flow matrix (6)	mean_flow, min_flow, max_flow, std_flow, coeff_var_flow, non_zero_ratio_flow
Joint (5)	mean_cost_×_mean_flow, max_cost_×_max_flow, max_cost_×_min_flow, min_cost_×_max_flow, min_cost_×_min_flow

4   Computational Analysis

The Algorithm Selection (AS) dataset is composed of 6 AOS methods including Self-Organizing Neural Network (SONN) [26], Probability Matching (PM) [15], Adaptive Pursuits (AP) [4], Multi-Armed Bandit (MAB) [32], Reinforcement Learning (RL) [79] and Random. Random is essentially a uniform random selection strategy. For the target problem, 134 Quadratic Assignment Problem (QAP) instances from QAPLIB[15] [80] are accommodated. As the candidate set for the AOS methods, the crossover operators used in the Genetic Algorithm (GA) consists of cycle crossover (CX), distance-preserving crossover (DPX), partially-mapped crossover (PMX) and order crossover (OX), as in [81]. The AOS × QAP instance performance data was directly taken from [26] which also took it from [81]. Thus, the settings used to generate were the same.

Table 2 reports the average rank across all the instances both for each each AOS method and ALORS. The results illustrate that ALORS outperforms all the constituent AOS approaches. Besides that the predictions are more robust than the standalone use of AOS, referring to the standard deviations. Among the AOS techniques, SONN delivers the best performance while the worst performance comes by PM comes with the worst performance, even falling behind Random. Table 3 gives the further details on each instance group. Although the instance group sizes are unbalanced, their performance in terms of average ranks still reveals the success of ALORS. Being said that in this evaluation SONN falls behind RL.

Figure 1 shows the selection distribution concerning each AOS method. AS' decisions are somewhat aligned with the performance of the AOS methods (Table 2). SONN as the best performing AOS method is selected most frequently. A similar trend is realized for AP, especially for the cases SONN delivers relatively poor performance. Random selection seems to be helpful in particular for the instances that the AOS strategies offer similar or the same performance in terms of the solution quality.

---

[15] http://anjos.mgi.polymtl.ca/qaplib/.

Table 2. The average rank performance of AS, i.e. ALORS, besides the constituent AOS methods

Method	Avg. Rank (Std.)
Random	4.22 ± 1.37
PM	4.50 ± 1.32
AP	4.07 ± 1.36
MAB	4.00 ± 1.18
RL	3.87 ± 1.40
SONN	3.71 ± 1.40
ALORS	**3.62 ± 1.19**

Table 3. The rank performance of AS on each instance group (the size of each group is shown in parenthesis)

	Random	PM	AP	MAB	RL	SONN	ALORS
bur (8)	4.38	**3.94**	**3.94**	**3.94**	**3.94**	**3.94**	**3.94**
chr (14)	5.11	5.04	4.82	3.71	3.75	**2.79**	**2.79**
els (1)	4	4	4	4	4	4	4
esc (20)	4	4.15	4	**3.88**	4.1	**3.88**	4
kra (3)	4.83	4.17	6.5	5.5	4	**1.5**	**1.5**
had (5)	4	4	4	4	4	4	4
lipa (16)	4.69	4.69	3.91	4.19	3.78	**3.22**	3.53
nug (15)	3.83	4	4.23	3.63	**3.57**	4.4	4.33
rou (3)	5	4	4.67	**3**	4.33	3.5	3.5
scr (3)	**3.83**	**3.83**	5	**3.83**	**3.83**	**3.83**	**3.83**
sko (13)	**3.27**	6	4.04	4.08	3.81	3.35	3.46
ste (3)	3	5.33	**2.83**	5.33	4	4.17	3.33
tai (26)	4.38	4.29	**3.62**	4.12	4.02	3.96	**3.62**
tho (2)	6.5	**2.5**	3.75	5	**2.5**	4.75	3
wil (2)	**1.75**	6.5	2.75	3.5	3.5	6.5	3.5
AVG	4.17	4.43	4.14	4.11	3.81	3.85	*3.49*
Std.	1.07	0.97	0.90	0.68	0.42	1.06	*0.68*

Figure 2 reports the contribution of each QAP instance feature on the AS model derived by ALORS. The contribution is identified in the form of Gini importance provided by Random Forest (RF). RF is the method used for feature-to-feature mapping of ALORS. The outcome shows that non_zero_ratio_flow happens to be the most significant feature among the feature set. There are other features like mean_flow, coeff_var_cost and std_cost, also contributing to the selection model. The features utilizing min_cost or min_flow are the least critical features as their importance measures are mostly zero.

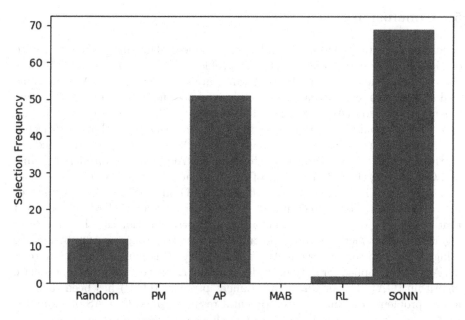

Fig. 1. AOS selection frequencies

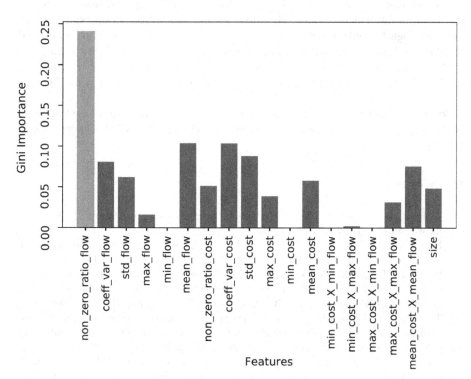

Fig. 2. Feature importance analysis

5   Conclusion

Algorithm Selection (AS) has been known as a meta-algorithm managing a suite of algorithms aiming to solve a target problem. The idea is to pick the best algorithm(s) when a new problem instance arrives. The existing AS studies as well as SAT competitions revealed the success and effectiveness of AS. Adaptive Operator Selection (AOS) is a field also relates to AS. AS as a term is mainly used to denote offline selection procedures, meaning that algorithms are selected before hand a target instance is being solved. AOS, however, performs selection on-the-fly, in an online manner. In that respect, AOS can be considered Online AS. Considering the Offline aspect of AS, the traditional AS can be categorized as Offline AS. This study applies Offline AS to Online AS, i.e. AOS, for better performance than the direct use of AOS. Existing empirical results revealed that there is no one AOS methods always perform superior as other, low-level algorithms. This fact gives an opportunity to take advantage of AOS by placing a higher level on top of various AOS methods. For this purpose, an AS system, i.e. ALORS is applied for automatically choosing crossover operators in a genetic algorithm (GA). The quadratic assignment problem (QAP) has been targeted as the problem domain. The experimental results showed that AS offers better performance with higher robustness than the constituent AOS methods.

For enhancing the reported contributions, as follow-up research, the application domains will be expanded, going beyond the QAP. Other existing AOS methods together with heuristic selection strategies will be used to further extend the dataset. The portfolio aspects of AOS will be investigated while offering parallelization of the constituent AOS methods. Problem independent features like in [65, 74] will be incorporated for cross-domain algorithm selection for AOS.

Acknowledgement. This study was supported by the 2232 Reintegration Grant from Scientific and Technological Research Council of Turkey (TUBITAK) under Project 119C013.

References

1. Wolpert, D., Macready, W.: No free lunch theorems for optimization. IEEE Trans. Evol. Comput. **1**, 67–82 (1997)
2. Kerschke, P., Hoos, H.H., Neumann, F., Trautmann, H.: Automated algorithm selection: survey and perspectives. Evol. Comput. **27**(1), 3–45 (2019)
3. Da Costa, L., Fialho, A., Schoenauer, M., Sebag, M.: Adaptive operator selection with dynamic multi-armed bandits. In: Proceedings of Genetic and Evolutionary Computation Conference (GECCO), Atlanta, GA, USA, pp. 913–920 (2008)
4. Thierens, D.: An adaptive pursuit strategy for allocating operator probabilities. In: Proceedings of the 7th International Conference on Genetic and Evolutionary Computation (GECCO), pp. ACM. 1539–1546 (2005)
5. Mısır, M.: Hyper-heuristics: autonomous problem solvers. In: Pillay, N., Qu, R. (eds.) Automated Design of Machine Learning and Search Algorithms. NCS, pp. 109–131. Springer, Cham (2021). https://doi.org/10.1007/978-3-030-72069-8_7

6. Davis, L.: Adapting operator probabilities in genetic algorithms. In: Proceedings of the 3rd International Conference on Genetic Algorithms (ICGA). pp. 61–69 (1989)
7. He, J., He, F., Dong, H.: Pure strategy or mixed strategy? In: Hao, J.-K., Middendorf, M. (eds.) EvoCOP 2012. LNCS, vol. 7245, pp. 218–229. Springer, Heidelberg (2012). https://doi.org/10.1007/978-3-642-29124-1_19
8. Grobler, J., Engelbrecht, A., Kendall, G., Yadavalli, S.: Alternative hyper-heuristic strategies for multi-method global optimization. In: Proceedings of the IEEE Congress on Evolutionary Computation (CEC), Barcelona, Spain, pp. 826–833, 18–23 July 2010
9. Mısır, M., Sebag, M.: ALORS: an algorithm recommender system. Artif. Intell. **244**, 291–314 (2017)
10. Thierens, D.: Adaptive strategies for operator allocation. Paramet. Sett. Evol. Algor. **54**, 77–90 (2007)
11. Mısır, M.: Intelligent hyper-heuristics: a tool for solving generic optimisation problems. PhD thesis, Department of Computer Science, KU Leuven (2012)
12. Burke, E.K., et al.: Hyper-heuristics: a survey of the state of the art. J. Oper. Res. Soc. **64**(12), 1695–1724 (2013)
13. Kaelbling, L.P., Littman, M.L., Moore, A.W.: Reinforcement learning: a survey. J. Artif. Intell. Res. **4**, 237–285 (1996)
14. Lai, T.L., Robbins, H.: Asymptotically efficient adaptive allocation rules. Adv. Appl. Math. **6**(1), 4–22 (1985)
15. Goldberg, D.: Probability matching, the magnitude of reinforcement, and classifier system bidding. Mach. Learn. **5**(4), 407–425 (1990)
16. Thathachar, M., Sastry, P.: Networks of Learning Automata: Techniques for Online Stochastic Optimization. Kluwer Academic Publishers, Boston (2004). https://doi.org/10.1007/978-1-4419-9052-5
17. Mısır, M., Verbeeck, K., De Causmaecker, P., Vanden Berghe, G.: An intelligent hyper-heuristic framework for CHeSC 2011. In: Hamadi, Y., Schoenauer, M. (eds.) LION 2012. LNCS, pp. 461–466. Springer, Heidelberg (2012). https://doi.org/10.1007/978-3-642-34413-8_45
18. Fialho, Á., Da Costa, L., Schoenauer, M., Sebag, M.: Extreme value based adaptive operator selection. In: Rudolph, G., Jansen, T., Beume, N., Lucas, S., Poloni, C. (eds.) PPSN 2008. LNCS, vol. 5199, pp. 175–184. Springer, Heidelberg (2008). https://doi.org/10.1007/978-3-540-87700-4_18
19. Hitomi, N., Selva, D.: A classification and comparison of credit assignment strategies in multiobjective adaptive operator selection. IEEE Trans. Evol. Comput. **21**, 294–314 (2016)
20. Gonçalves, R.A., Pavelski, L.M., de Almeida, C.P., Kuk, J.N., Venske, S.M., Delgado, M.R.: Adaptive operator selection for many-objective optimization with NSGA-III. In: Trautmann, H., et al. (eds.) EMO 2017. LNCS, vol. 10173, pp. 267–281. Springer, Cham (2017). https://doi.org/10.1007/978-3-319-54157-0_19
21. Sallam, K.M., Elsayed, S.M., Sarker, R.A., Essam, D.L.: Landscape-based adaptive operator selection mechanism for differential evolution. Inf. Sci. **418**, 383–404 (2017)
22. Mashwani, W.K., Salhi, A., Yeniay, O., Jan, M.A., Khanum, R.A.: Hybrid adaptive evolutionary algorithm based on decomposition. Appl. Soft Comput. **57**, 363–378 (2017)
23. Deb, K., Pratap, A., Agarwal, S., Meyarivan, T.: A fast and elitist multiobjective genetic algorithm: NSGA-II. IEEE Trans. Evol. Comput. **6**(2), 182–197 (2002)

24. Soria Alcaraz, J.A., Ochoa, G., Carpio, M., Puga, H.: Evolvability metrics in adaptive operator selection. In: Proceedings of the Annual Conference on Genetic and Evolutionary Computation (GECCO), pp. 1327–1334. ACM (2014)
25. Soria-Alcaraz, J.A., Espinal, A., Sotelo-Figueroa, M.A.: Evolvability metric estimation by a parallel perceptron for on-line selection hyper-heuristics. IEEE Access **5**, 7055–7063 (2017)
26. Teng, T.H., Handoko, S.D., Lau, H.C.: Self-organizing neural network for adaptive operator selection in evolutionary search. In: Proceedings of the 10th Learning and Intelligent OptimizatioN Conference (LION). LNCS, Naples, Italy (2016)
27. Candan, C., Goeffon, A., Lardeux, F., Saubion, F.: A dynamic island model for adaptive operator selection. In: Proceedings of the 14th International Conference on Genetic and Evolutionary Computation Conference (GECCO), pp. 1253–1260. ACM (2012)
28. Candan, C., Goëffon, A., Lardeux, F., Saubion, F.: Non stationary operator selection with island models. In: Proceedings of the 15th Annual Conference on Genetic and Evolutionary Computation, pp. 1509–1516. ACM (2013)
29. Goëffon, A., Lardeux, F., Saubion, F.: Simulating non-stationary operators in search algorithms. Appl. Soft Comput. **38**, 257–268 (2016)
30. Auer, P., Cesa-Bianchi, N., Fischer, P.: Finite-time analysis of the multiarmed bandit problem. Mach. Learn. **47**(2), 235–256 (2002)
31. Page, E.: Continuous inspection schemes. Biometrika **41**(1/2), 100–115 (1954)
32. Fialho, Á., Da Costa, L., Schoenauer, M., Sebag, M.: Dynamic multi-armed bandits and extreme value-based rewards for adaptive operator selection in evolutionary algorithms. In: Stützle, T. (ed.) LION 2009. LNCS, vol. 5851, pp. 176–190. Springer, Heidelberg (2009). https://doi.org/10.1007/978-3-642-11169-3_13
33. Fialho, Á., Da Costa, L., Schoenauer, M., Sebag, M.: Analyzing bandit-based adaptive operator selection mechanisms. Ann. Math. Artif. Intell. **60**(1), 25–64 (2010)
34. Fialho, Á., Schoenauer, M., Sebag, M.: Toward comparison-based adaptive operator selection. In: Proceedings of the 12th Annual Conference on Genetic and Evolutionary Computation (GECCO), 767–774. ACM (2010)
35. Bradley, A.P.: The use of the area under the ROC curve in the evaluation of machine learning algorithms. Patt. Recogn. **30**(7), 1145–1159 (1997)
36. Li, K., Fialho, A., Kwong, S., Zhang, Q.: Adaptive operator selection with bandits for a multiobjective evolutionary algorithm based on decomposition. IEEE Trans. Evol. Comput. **18**(1), 114–130 (2014)
37. Mashwani, W.K., Salhi, A., Yeniay, O., Hussian, H., Jan, M.: Hybrid non-dominated sorting genetic algorithm with adaptive operators selection. Appl. Soft Comput. **56**, 1–18 (2017)
38. Zhang, Q., Liu, W., Li, H.: The performance of a new version of moea/d on cec09 unconstrained mop test instances. In: IEEE Congress on Evolutionary Computation (CEC), pp. 203–208. IEEE (2009)
39. Ferreira, A.S., Gonçalves, R.A., Pozo, A.: A multi-armed bandit selection strategy for hyper-heuristics. In: IEEE Congress on Evolutionary Computation (CEC), pp. 525–532. IEEE (2017)
40. Strickler, A., Lima, J.A.P., Vergilio, S.R., Pozo, A.T.: Deriving products for variability test of feature models with a hyper-heuristic approach. Appl. Soft Comput. **49**, 1232–1242 (2016)
41. Harman, M., Mansouri, S.A., Zhang, Y.: Search-based software engineering: trends, techniques and applications. ACM Comput. Surv. (CSUR) **45**(1), 11 (2012)
42. Xu, L., Hutter, F., Hoos, H., Leyton-Brown, K.: SATzilla: portfolio-based algorithm selection for SAT. J. Artif. Intell. Res. **32**(1), 565–606 (2008)

43. Xu, L., Hutter, F., Shen, J., Hoos, H., Leyton-Brown, K.: Satzilla 2012: Improved algorithm selection based on cost-sensitive classification models. In: Proceedings of SAT Challenge 2012: Solver and Benchmark Descriptions, pp. 57–58 (2012)
44. Malitsky, Y., Sabharwal, A., Samulowitz, H., Sellmann, M.: Algorithm portfolios based on cost-sensitive hierarchical clustering. In: Proceedings of the 23rd International Joint Conference on Artifical Intelligence (IJCAI). pp. 608–614 (2013)
45. Nikolić, M., Marić, F., Janičić, P.: Instance-based selection of policies for SAT solvers. In: Kullmann, O. (ed.) SAT 2009. LNCS, vol. 5584, pp. 326–340. Springer, Heidelberg (2009). https://doi.org/10.1007/978-3-642-02777-2_31
46. Nikolić, M., Marić, F., Janičić, P.: Simple algorithm portfolio for SAT. Arti. Intell. Rev. 40(4), 457–465 (2011). https://doi.org/10.1007/s10462-011-9290-2
47. Collautti, M., Malitsky, Y., Mehta, D., O'Sullivan, B.: SNNAP: solver-based nearest neighbor for algorithm portfolios. In: Blockeel, H., Kersting, K., Nijssen, S., Železný, F. (eds.) ECML PKDD 2013. LNCS (LNAI), vol. 8190, pp. 435–450. Springer, Heidelberg (2013). https://doi.org/10.1007/978-3-642-40994-3_28
48. Lindauer, M., Hoos, H.H., Hutter, F., Schaub, T.: AutoFolio: an automatically configured algorithm selector. J. Artif. Intell. Res. 53, 745–778 (2015)
49. Hutter, F., Hoos, H., Stutzle, T.: Automatic algorithm configuration based on local search. In: Proceedings of the National Conference on Artificial Intelligence, vol. 22, 1152p. Menlo Park, CA, AAAI Press; MIT Press; Cambridge, MA; London (2007)
50. Kadioglu, S., Malitsky, Y., Sabharwal, A., Samulowitz, H., Sellmann, M.: Algorithm selection and scheduling. In: Lee, J. (ed.) CP 2011. LNCS, vol. 6876, pp. 454–469. Springer, Heidelberg (2011). https://doi.org/10.1007/978-3-642-23786-7_35
51. O'Mahony, E., Hebrard, E., Holland, A., Nugent, C., O'Sullivan, B.: Using case-based reasoning in an algorithm portfolio for constraint solving. In: Irish Conference on Artificial Intelligence and Cognitive Science (2008)
52. Amadini, R., Gabbrielli, M., Mauro, J.: Sunny: a lazy portfolio approach for constraint solving. Theory Pract. Logic Program. 14, 509–524 (2014)
53. Kumar, V.: Algorithms for constraint-satisfaction problems: a survey. AI Mag. 13(1), 32 (1992)
54. Lindauer, M., Bergdoll, R.-D., Hutter, F.: An empirical study of per-instance algorithm scheduling. In: Festa, P., Sellmann, M., Vanschoren, J. (eds.) LION 2016. LNCS, vol. 10079, pp. 253–259. Springer, Cham (2016). https://doi.org/10.1007/978-3-319-50349-3_20
55. Malitsky, Y., Sabharwal, A., Samulowitz, H., Sellmann, M.: Parallel SAT solver selection and scheduling. In: Milano, M. (ed.) CP 2012. LNCS, pp. 512–526. Springer, Heidelberg (2012). https://doi.org/10.1007/978-3-642-33558-7_38
56. Hoos, H., Kaminski, R., Lindauer, M., Schaub, T.: aspeed: Solver scheduling via answer set programming. Theory Pract. Logic Program. 1–26 (2014)
57. Gomes, C., Selman, B.: Algorithm portfolios. Artif. Intell. 126(1), 43–62 (2001)
58. Roussel, O.: Description of ppfolio 2012. In: Proceedings of SAT Challenge, 46 p (2012)
59. Kadioglu, S., Malitsky, Y., Sellmann, M., Tierney, K.: ISAC-instance-specific algorithm configuration. In: Proceedings of the 19th European Conference on Artificial Intelligence (ECAI'10), pp. 751–756 (2010)
60. Malitsky, Y., Mehta, D., O'Sullivan, B.: Evolving instance specific algorithm configuration. In: Proceedings of the 6th International Symposium on Combinatorial Search (SoCS) (2013)

61. Malitsky, Y.: Evolving instance-specific algorithm configuration. In: Instance-Specific Algorithm Configuration, pp. 93–105. Springer, Cham (2014). https://doi.org/10.1007/978-3-319-11230-5_9

62. Ansótegui, C., Malitsky, Y., Sellmann, M.: MaxSAT by improved instance-specific algorithm configuration. In: Proceedings of the 28th AAAI Conference on Artificial Intelligence (AAAI) (2014)

63. Xu, L., Hoos, H., Leyton-Brown, K.: Hydra: Automatically configuring algorithms for portfolio-based selection. In: Proceedings of the 24th AAAI Conference on Artificial Intelligence (AAAI), pp. 210–216 (2010)

64. Xu, L., Hutter, F., Hoos, H., Leyton-Brown, K.: Hydra-MIP: automated algorithm configuration and selection for mixed integer programming. In: Proceedings of the 18th RCRA International Workshop on Experimental Evaluation of Algorithms for Solving Problems with Combinatorial Explosion (2011)

65. Mısır, M., Handoko, S.D., Lau, H.C.: OSCAR: online selection of algorithm portfolios with case study on memetic algorithms. In: Dhaenens, C., Jourdan, L., Marmion, M.-E. (eds.) LION 2015. LNCS, vol. 8994, pp. 59–73. Springer, Cham (2015). https://doi.org/10.1007/978-3-319-19084-6_6

66. Gunawan, A., Lau, H.C., Mısır, M.: Designing a portfolio of parameter configurations for online algorithm selection. In: the 29th AAAI Conference on Artificial Intelligence: Workshop on Algorithm Configuration (AlgoConf), Austin/Texas, USA (2015)

67. Montgomery, D.C.: Design and Analysis of Experiments, John Wiley & Sons, Hoboken (2017)

68. Mısır, M., Handoko, S.D., Lau, H.C.: ADVISER: a web-based algorithm portfolio deviser. In: Dhaenens, C., Jourdan, L., Marmion, M.-E. (eds.) LION 2015. LNCS, vol. 8994, pp. 23–28. Springer, Cham (2015). https://doi.org/10.1007/978-3-319-19084-6_3

69. Lau, H., Mısır, M., Xiang, L., Lingxiao, J.: ADVISER+: toward a usable web-based algorithm portfolio deviser. In: Proceedings of the 12th Metaheuristics International Conference (MIC), Barcelona, Spain, pp. 592–599 (2017)

70. Koren, Y., Bell, R., Volinsky, C.: Matrix factorization techniques for recommender systems. Computer **42**(8), 30–37 (2009)

71. Mısır, M.: Matrix factorization based benchmark set analysis: a case study on HyFlex. In: Shi, Y., et al. (eds.) SEAL 2017. LNCS, vol. 10593, pp. 184–195. Springer, Cham (2017). https://doi.org/10.1007/978-3-319-68759-9_16

72. Mısır, M.: Data sampling through collaborative filtering for algorithm selection. In: the 16th IEEE Congress on Evolutionary Computation (CEC), pp. 2494–2501. IEEE (2017)

73. Bischl, B., et al.: ASlib: a benchmark library for algorithm selection. Artif. Intell. **237**, 41–58 (2017)

74. Mısır, M.: Algorithm selection across selection hyper-heuristics. In: the Data Science for Optimization (DSO) @ IJCAI 2020 workshop at the 29th International Joint Conference on Artificial Intelligence (IJCAI). (2021)

75. Golub, G.H., Reinsch, C.: Singular value decomposition and least squares solutions. Numer. Math. **14**(5), 403–420 (1970)

76. Breiman, L.: Random forests. Mach. Learn. **45**(1), 5–32 (2001)

77. Lawler, E.L.: The quadratic assignment problem. Manag. Sci. **9**(4), 586–599 (1963)

78. Burkard, R.E., Cela, E., Pardalos, P.M., Pitsoulis, L.S.: The quadratic assignment problem. In: Du, D.Z., Pardalos, P.M. (eds.) Handbook of Combinatorial Optimization, pp. 1713–1809. Springer, Boston (1998). https://doi.org/10.1007/978-1-4613-0303-9_27

79. Handoko, S.D., Nguyen, D.T., Yuan, Z., Lau, H.C.: Reinforcement learning for adaptive operator selection in memetic search applied to quadratic assignment problem. In: Proceedings of the Companion Publication of the 2014 Annual Conference on Genetic and Evolutionary Computation, pp. 193–194. ACM (2014)

80. Burkard, R.E., Karisch, S.E., Rendl, F.: QAPLIB-a quadratic assignment problem library. J. Glob. Optim. **10**(4), 391–403 (1997)

81. Francesca, G., Pellegrini, P., Stützle, T., Birattari, M.: Off-line and on-line tuning: a study on operator selection for a memetic algorithm applied to the QAP. In: Merz, P., Hao, J.-K. (eds.) EvoCOP 2011. LNCS, vol. 6622, pp. 203–214. Springer, Heidelberg (2011). https://doi.org/10.1007/978-3-642-20364-0_18

82. Mısır, M., Wauters, T., Verbeeck, K., Vanden Berghe, G.: A new learning hyper-heuristic for the traveling tournament problem. In: Proceedings of the 8th Meta-heuristic International Conference (MIC) (2009)

83. Mısır, M., Verbeeck, K., De Causmaecker, P., Vanden Berghe, G.: A new hyper-heuristic as a general problem solver: an implementation in HyFlex. J. Sched. **16**(3), 291–311 (2013)

Inverse Free Universum Twin Support Vector Machine

Hossein Moosaei[1,2]([ICON]) [ORCID] and Milan Hladík[2] [ORCID]

[1] Department of Mathematics, Faculty of Science, University of Bojnord,
Bojnord, Iran
moosaei@ub.ac.ir
[2] Department of Applied Mathematics, Faculty of Mathematics and Physics,
Charles University, Prague, Czech Republic
{hmoosaei,hladik}@kam.mff.cuni.cz

Abstract. Universum twin support vector machine (ᵁ-TSVM) is an efficient method for binary classification problems . In this paper, we improve the ᵁ-TSVM algorithm and propose an improved Universum twin bounded support vector machine (named as IUTBSVM) . Indeed, by introducing a different Lagrangian function for the primal problems, we obtain new dual formulations so that we do not need to compute inverse matrices. Also to reduce the computational time of the proposed method, we suggest smaller size of the rectangular kernel matrices than the other methods. Numerical experiments on several UCI benchmark data sets indicate that the IUTBSVM is more efficient than the other three algorithms, namely ᵁ-SVM, TSVM, and ᵁ-TSVM in sense of the classification accuracy.

Keywords: Support vector machine · Twin SVM · Universum data · ᵁ-SVM, ᵁ-TSVM

1 Introduction

Support vector machine (SVM) [21,22] emerges from the idea of structural risk minimization (SRM) and has been used in many applications such as heart disease, text categorization, computational biology, bioinformatics, image classification, real-life data sets, lung cancer, colon tumor, etc. [1–3,8,12,20,23]. The standard SVM finds two parallel support hyperplanes with maximum margin between them by solving a constrained quadratic programming problem (QPP).

The generalized eigenvalue proximal SVMs (GEPSVMs) [10] were proposed to reduce the computational cost of SVM. In this method, two nonparallel hyperplanes are generated such that each of them is closest to one of the two classes and as far as possible from the other class. Motivated by GEPSVMs, Jayadeva et al. [7] proposed twin support vector machine (TSVM) so that the main idea of the both methods is the same, but formulations are different from that of

---

The authors were supported by the Czech Science Foundation Grant P403-18-04735S.

D. E. Simos et al. (Eds.): LION 2021, LNCS 12931, pp. 252–264, 2021.
https://doi.org/10.1007/978-3-030-92121-7_21

GEPSVMs and are similar to SVM. Indeed, the TSVM generates two nonparallel hyperplanes such that each plane is close enough to its own class and as far as possible from the other class by solving two smaller size of QPPs rather than a single large QPP in SVM; this makes it faster than classical SVM [7]. The SVM-based methods were noticed very quickly so far and researchers proposed variants of SVM and TSVM to improve the performance of classification [11,16–19].

In supervised learning algorithms, we face labeled data and consider them for classifying data and ignore any other information which is related to the problem and data. In order to overcome this disadvantage, a new research line is opened. It allows us to incorporate the prior knowledge embedded in the unlabeled samples into the supervised learning problem. Indeed, the new unlabeled data, which belongs to no class, is used in model and these new data are named as Universum. The Universum learning, which has been proved to be useful to the supervised and semi-supervised learning, was also applied to increase the performance of various interesting applications, such as text classification [9], electroencephalogram (EEG) signals for the diagnosis of neurological disorders [14], and so on. However, constructing an impressive Universum classifier is still a challenging topic and it would be the main concentrate of this study. Weston et al. [24] used of prior information of data in the SVM model and termed their model as Universum SVM ($\mathfrak{U}$-SVM). They shown that the $\mathfrak{U}$-SVM performs better generalization performance than standard SVM. Sinz et al. [15] analyzed $\mathfrak{U}$-SVM and presented a least-squares version of the $\mathfrak{U}$-SVM algorithm. They also concluded that the use of Universum data is very useful. To improve the $\mathfrak{U}$-SVM, Qi et al. [13] firstly incorporated Universum data into TSVM and proposed twin support vector machines with Universum data ($\mathfrak{U}$-TSVM). Their experiments on various data sets showed that the classification accuracy of $\mathfrak{U}$-TSVM is better than the classical TSVM algorithms that do not use Universum data.

In this paper, we propose an improved version of twin bounded support vector machine with Universum data and we name it IUTBSVM. In fact, one of the common drawbacks, which is inevitable in many methods, is computing the inverse matrices. In contrast to these methods, we introduce a different Lagrangian function for the primal problems so that we get different dual problems, which overcomes this drawback, i.e., we do not need to compute inverse matrices for solving the dual problems. In IUTBSVM, we will also use the reduced kernel, which has not been utilized for $\mathfrak{U}$-TSVM so far. This reduces the computational cost of nonlinear problems compared to other methods.

To verify the effectiveness of our proposed method, we conduct experiments on several UCI benchmark data sets. The experiments demonstrate efficiency of our proposed algorithm over three other algorithms, $\mathfrak{U}$-SVM, TSVM and $\mathfrak{U}$-TSVM, in sense of the classification accuracy.

2   Related Works

In this section, we briefly review the $\mathfrak{U}$-SVM, TSVM, and $\mathfrak{U}$-TSVM formulations.

The following notation is used throughout this paper. Suppose that a data set (x_i, y_i) is given for training with the input $x_i \in \mathbb{R}^n$ and the corresponding target value or label $y_i = 1$ or -1, i.e.,

$$\{(x_1, y_1), \ldots, (x_m, y_m)\} \in (\mathbb{R}^n \times \{\pm 1\})^m. \tag{1}$$

We will denote the Universum data by

$$\mathfrak{U} = \{x_1^*, \ldots, x_r^*\},$$

which includes r samples. The data points belonging to positive and negative classes are represented by matrices A and B, respectively. Also the data points in $\mathfrak{U}$, represented by matrix U so that each row of U represents a Universum sample. The function $\phi \colon \mathbb{R}^l \to \mathbb{R}^p$ is a nonlinear map to higher dimension ($p > l$).

2.1   Universum Support Vector Machine

The Universum support vector machines ($\mathfrak{U}$-SVM) was proposed by Weston et al. [24]. We associate the Universum data with both classes so that we have

$$(x_1^*, 1), \ldots, (x_r^*, 1), (x_1^*, -1), \ldots, (x_r^*, -1).$$

That is why we extend the notation as follows:

$$
\begin{aligned}
x_{m+j} = x_j^*, \;\; y_{m+j} = 1, && \text{for } j = 1, \ldots, r, \\
x_{m+r+j} = x_j^*, \;\; y_{m+r+j} = -1, && \text{for } j = 1, \ldots, r, \\
\psi_{m+j} = \psi_j, && \text{for } j = 1, \ldots, 2r.
\end{aligned}
$$

The $\mathfrak{U}$-SVM formulation reads as follows:

$$
\begin{aligned}
\min_{w,b,\xi,\psi} \;\; & \frac{1}{2}\|w\|^2 + C\sum_{i=1}^{m}\xi_i + C_u\sum_{j=1}^{2r}\psi_j \\
s.t. \;\; & y_i\left(w^\top\phi(x_i) + b\right) \geq 1 - \xi_i, \\
& y_j\left(w^\top\phi(x_j) + b\right) \geq -\varepsilon - \psi_j, \\
& \xi_i \geq 0, \;\; \psi_j \geq 0, \quad i = 1, \ldots, m, \;\; j = m+1, \ldots, m+2r,
\end{aligned} \tag{2}
$$

where parameters C and C_u, which control the trade-off between the minimization of training errors and the maximization of the number of Universum samples, are positive real numbers, and $\xi_i, \psi_j \in \mathbb{R}$, are slack variables. Next, $\epsilon > 0$ is a parameter for the ϵ-insensitive tube. For the case of $C_u = 0$, this formulation is reduced to a standard SVM classifier [4]. In general, the parameter ϵ can be set to zero or a small positive value.

The Wolfe dual to problem (2) can be described as follows:

$$
\min_{\alpha} \;\; \frac{1}{2}\sum_{i=1}^{m+2r}\sum_{j=1}^{m+2r}\alpha_i\alpha_j y_i y_j \phi(x_i)^T\phi(x_j) - \sum_{i=1}^{m+2r}\mu_i\alpha_i
$$

$$
\text{subject to} \;\; \sum_{i=1}^{m+2r} y_i\alpha_i = 0, \tag{3}
$$

$$
\begin{aligned}
& 0 \leq \alpha_i \leq C, \;\; \mu_i = 1, \quad i = 1, \ldots, m, \\
& 0 \leq \alpha_i \leq C_u, \;\; \mu_i = -\epsilon, \quad i = m+1, \ldots, m+2r,
\end{aligned}
$$

where α_i is the Lagrange multiplier. By solving the above quadratic programming problem, the decision function of $\mathfrak{U}$-SVM can be obtained as folows:

$$f(x) = \text{sgn}(w^T \phi(x) + b) = \text{sgn}(\sum_{i=1}^{m+2r} \alpha_i y_i \phi(x_i)^T \phi(x) + b).$$

As in SVM, a new sample is classified as $+1$ or -1, by using of the above decision function.

2.2 Universum Twin Support Vector Machine

Universum twin support vector machine, which is called $\mathfrak{U}$-TSVM, was proposed by Qi et al. [13]. The nonlinear $\mathfrak{U}$-TSVM solves the following pair of QPPs:

$$\min_{w_1, b_1, q_1, \phi_1} \|K(A, D^T)w_1 + e_1^T b_1\|^2 + c_1 e_2^T q_1 + c_u e_u^T \phi_1,$$

$$\text{subject to} \quad -(K(B, D^T)w_1 + e_2 b_1) + q_1 \geq e_2, \tag{4}$$
$$(K(U, D^T)w_1 + e_u b_1) + \phi_1 \geq (-1 + \epsilon)e_u,$$
$$q_1 \geq 0, \phi_1 \geq 0,$$

and

$$\min_{w_2, b_2, q_2, \phi_2} \|K(B, D^T)w_2 + e_2^T b_2\|^2 + c_2 e_1^T q_2 + c_u e_u^T \phi_2,$$

$$\text{subject to} \quad (K(A, D^T)w_2 + e_1 b_2) + q_2 \geq e_1, \tag{5}$$
$$-(K(U, D^T)w_2 + e_u b_2) + \phi_2 \geq (-1 + \epsilon)e_u,$$
$$q_2 \geq 0, \phi_2 \geq 0,$$

where c_1, c_2, c_u are positive parameters, e_1, e_2, e_u are vectors of ones of appropriate dimensions, q_1, q_2, ϕ_1 and ϕ_2 are slack vectors, $D = [A^T \ B^T]^T$, and K is an appropriately chosen kernel.

By applying the KKT conditions, the Wolfe dual formulations of (4) and (5) can be obtained, respectively as follows:

$$\max_{\alpha_1, \mu_1} e_2^T \alpha_1 - \frac{1}{2}(\alpha_1^T G - \mu_1^T O)(H^T H)^{-1}(G^T \alpha_1 - O^T \mu_1) + (\epsilon - 1)e_u^T \mu_1$$

$$\text{subject to} \quad 0 \leq \alpha_1 \leq c_1, \quad 0 \leq \mu_1 \leq c_u, \tag{6}$$

$$\max_{\alpha_2, \mu_2} e_2^T \alpha_2 - \frac{1}{2}(\alpha_2^T H - \mu_2^T O)(G^T G)^{-1}(H^T \alpha_2 - O^T \mu_2) + (\epsilon - 1)e_u^T \mu_2$$

$$\text{subject to} \quad 0 \leq \alpha_2 \leq c_2, \quad 0 \leq \mu_2 \leq c_u, \tag{7}$$

where α_1, α_2, μ_1, and μ_2 are the Lagrangian coefficients, $H = [K(A, D^T) \ e_1]$, $G = [K(B, D^T) \ e_2]$, and $O = [K(U, D^T) \ e_u]$. The separating hyperplanes can be obtained from the solution of (6) and (7) by

$$[w_1^T, \ b_1]^T = -(H^T H)^{-1}(G^T \alpha_1 - O^T \mu_1),$$

and
$$[w_2^T, \ b_2]^T = (G^T G)^{-1}(H^T \alpha_2 - O^T \mu_2).$$

To avoid the possible ill conditioning when $G^T G$ or $H^T H$ are (nearly) singular, the inverse matrices $(G^T G)^{-1}$ and $(H^T H)^{-1}$ are approximately replaced by $(G^T G + \delta I)^{-1}$ and $(H^T H + \delta I)^{-1}$, where δ is a small positive scalar.

A new data point x is assigned to class $i \in \{1, -1\}$ by using the following decision rule:
$$\text{class } i = \arg\min_{i=1,2} \frac{|K(x^T, D^T)w_i + b_i|}{\|w_i\|^2}.$$

3   Improvements on Twin Bounded Support Vector machine with Universum Data

In this section, we improve $\mathfrak{U}$-TSVM to obtain two non-parallel hyperplane classifiers, named as IUTBSVM (Improvements on twin bounded support vector machine with Universum data). In this method, by introducing a slightly different in mathematical model, a different Lagrangian function is constructed so that we avoid computation of large-scale inverse matrices, that appear in classical methods.

3.1   Linear IUTBSVM

Here the data points belonging to classes $+1$ and -1 are represented by matrices A and B, respectively. The Universum data are represented by matrix U.

Linear IUTBSVM tries to find two nonparallel hyperplanes which defined as follows:

$$f_1(x) = x^T w_1 + b_1 = 0, \quad \text{and} \quad f_2(x) = x^T w_2 + b_2 = 0. \tag{8}$$

The primal problem for finding the first hyperplane can be described as follows:

$$\min_{w_1, b_1, \xi, \psi} \ \frac{1}{2}\|t\|^2 + \frac{c_2}{2}(\|w_1\|^2 + b_1^2) + c_1 e_2^T \xi + c_u e_u^T \psi$$
$$\text{subject to} \ \ Aw_1 + e_1 b_1 = t$$
$$- (Bw_1 + e_2 b_1) + \xi \geqslant e_2 \tag{9}$$
$$(Uw_1 + e_u b_1) + \psi \geqslant (-1 + \varepsilon)e_u$$
$$\xi, \psi \geqslant 0,$$

where $c_1, c_2 c_u > 0$ are parameters, e_1, e_2, e_u are vectors of ones of appropriate dimensions, and ξ and ψ are slack vectors.

The primal problem for the second hyperplane is as follows:

$$\min_{w_2, b_2, \xi, \psi} \frac{1}{2}\|t\|^2 + \frac{c_4}{2}(\|w_2\|^2 + b_2^2) + c_3 e_1^T \xi + c_u e_u^T \psi$$

$$\text{subject to } Bw_2 + e_2 b_2 = t$$

$$(Aw_2 + e_1 b_2) + \xi \geqslant e_1 \tag{10}$$

$$(Uw_2 + e_u b_2) + \psi \geqslant (1-\varepsilon)e_u$$

$$\xi, \psi \geqslant 0,$$

where $c_3, c_4, c_u > 0$ are parameters, e_1, e_2, e_u are vectors of ones of appropriate dimensions, and ξ and ψ are slack vectors.

We should note that we only added a variable t to both mathematical models. So the Lagrangian function of problem (9) reads

$$\begin{aligned}
L(\theta) &= \frac{1}{2}\|t\|^2 + \frac{c_2}{2}(\|w_1\|^2 + b_1^2) + c_1 e_2^T \xi + c_u e_u^T \psi \\
&+ \lambda^T(Aw_1 + e_1 b_1 - t) - \alpha^T(-(Bw_1 + e_2 b_1) + \xi - e_2) \\
&- \beta^T((Uw_1 + e_u b_1) + \psi + (1-\varepsilon)e_u) - p^T \xi - q^T \psi,
\end{aligned} \tag{11}$$

where $\theta = (w_1, b_1, t, \xi, \psi, \lambda, \alpha, \beta, p, q)$ are variables and $\lambda, \alpha, \beta, p, q$ are Lagrangian multipliers.

According to the necessary and sufficient KKT conditions on the Eq. (11), we have the following results:

$$\frac{\partial L}{\partial w_1} = c_2 w_1 + A^T \lambda + B^T \alpha - U^T \beta = 0 \tag{12}$$

$$\frac{\partial L}{\partial b_1} = c_2 b_1 + e_1^T \lambda + e_2^T \alpha - e_u^T \beta = 0 \tag{13}$$

$$\frac{\partial L}{\partial t} = t - \lambda = 0 \tag{14}$$

$$\frac{\partial L}{\partial \xi} = c_1 e_2 - \alpha - p = 0 \tag{15}$$

$$\frac{\partial L}{\partial \psi} = c_u e_u - \beta - q = 0 \tag{16}$$

$$\frac{\partial L}{\partial \lambda} = Aw_1 - e_1 b_1 - t = 0 \tag{17}$$

$$\frac{\partial L}{\partial \alpha} = -(-(Bw_1 - e_2 b_1) + \xi - e_2) \leqslant 0 \tag{18}$$

$$\alpha^T \frac{\partial L}{\partial \alpha} = \alpha^T(-(Bw_1 - e_2 b_1) + \xi - e_2) = 0 \tag{19}$$

$$\frac{\partial L}{\partial \beta} = -((Uw_1 - e_u b_1) + \psi + (1-\varepsilon)e_u) \leqslant 0 \tag{20}$$

$$\beta^T \frac{\partial L}{\partial \beta} = \beta^T((Uw_1 - e_u b_1) + \psi + (1-\varepsilon)e_u) = 0 \tag{21}$$

$$\frac{\partial L}{\partial p} = -\xi \leqslant 0 \tag{22}$$

$$p^T \frac{\partial L}{\partial p} = p^T \xi = 0 \tag{23}$$

$$\frac{\partial L}{\partial q} = -\psi \leqslant 0 \tag{24}$$

$$q^T \frac{\partial L}{\partial q} = q^T \psi = 0 \tag{25}$$

$$\alpha, \beta, p, q \geqslant 0 \tag{26}$$

From (12), (13), and (14), we obtain

$$w_1 = -\frac{1}{c_2}\left(A^T \lambda + B^T \alpha - U^T \beta\right) \tag{27}$$

$$b_1 = -\frac{1}{c_2}\left(e_1^T \lambda + e_2^T \alpha - e_u^T \beta\right)$$

$$t = \lambda$$

From (15), (16) and (26) we have $0 \leqslant \alpha \leqslant c_1 e_2$ and $0 \leqslant \beta \leqslant c_u e_u$. So the dual problem is stated as follows:

$$\max \ -\frac{1}{2}\begin{bmatrix} \lambda^T & \alpha^T & \beta^T \end{bmatrix} Q \begin{bmatrix} \lambda \\ \alpha \\ \beta \end{bmatrix} + c_2 \begin{bmatrix} 0 & e_2^T & (\varepsilon - 1)e_u^T \end{bmatrix} \begin{bmatrix} \lambda \\ \alpha \\ \beta \end{bmatrix}, \tag{28}$$

subject to $0 \leqslant \alpha \leqslant c_1 e_2, \ \ 0 \leqslant \beta \leqslant c_u e_u,$

where

$$Q = \begin{bmatrix} AA^T + C_2 I & AB^T & -AU^T \\ BA^T & BB^T & -BU^T \\ -UA^T & -UB^T & UU^T \end{bmatrix} + \begin{bmatrix} E & E & -E \\ E & E & -E \\ -E & -E & E \end{bmatrix},$$

and E the a matrix with all entries equal to 1. Problem (28), can be written as follows:

$$\min \ \frac{1}{2}S^T Q S + c_2 r^T S \tag{29}$$

subject to $0 \leqslant \alpha \leqslant c_1 e_2, \ \ 0 \leqslant \beta \leqslant c_u e_u.$

Here

$$Q = Q^T, \quad S = \begin{bmatrix} \lambda \\ \alpha \\ \beta \end{bmatrix}, \quad r = \begin{bmatrix} 0 \\ -e_2 \\ (1-\varepsilon)e_u \end{bmatrix},$$

Q is a symmetric positive definite matrix. Then we can solve this problem by the Matlab function "quadprog.m". By solving this problem and using the relations (27), we can find the first hyperplane.

In the similar way, we find the dual problem to problem (10) as follows:

$$\min \frac{1}{2}S^T Q S + c_4 r^T S \tag{30}$$

$$\text{subject to } 0 \leqslant \alpha \leqslant c_3 e_1, \quad 0 \leqslant \beta \leqslant c_u e_u$$

where

$$Q = \begin{bmatrix} BB^T + C_4 I & -BA^T & -BU^T \\ -AB^T & AA^T & AU^T \\ -UB^T & UA^T & UU^T \end{bmatrix} + \begin{bmatrix} E & -E & -E \\ -E & E & E \\ -E & E & E \end{bmatrix}$$

$$Q = Q^T, \quad S = \begin{bmatrix} \lambda \\ \alpha \\ \beta \end{bmatrix}, \quad r = \begin{bmatrix} 0 \\ -e_1 \\ (\varepsilon - 1)e_u \end{bmatrix}.$$

The second hyperplane is determined by the optimal solution of the dual problem and using the following expressions

$$w_2 = -\frac{1}{c_4} \left(B^T \lambda - A^T \alpha - U^T \beta \right), \tag{31}$$

$$b_2 = -\frac{1}{c_4} \left(e_2^T \lambda - e_1^T \alpha - e_u^T \beta \right).$$

It is worth pointing out that problems (29) and (30) do not face computation of the inverse matrix, in contrast to other classical methods such as TSVM and $\mathfrak{U}$-TSVM.

A new data point x is assigned to class $i \in \{1, -1\}$ by using the following decision rule:

$$\text{class } i = \arg\min_{i=1,2} \frac{|x^T w_i + b_i|}{\|w_i\|^2}.$$

3.2 Nonlinear IUTBSVM

In the classical TSVM with Universum data, we find two nonparallel hyperplanes as follows:

$$f_1(x) = K(x, D^T)w_1 + b_1 \quad \text{and} \quad f_2(x) = K(x, D^T)w_2 + b_2, \tag{32}$$

where $D = [A^T \ B^T]^T$ and K is an appropriate chosen kernel.

In this subsection, we introduce a nonlinear version of the IUTBSVM, totally different to other methods. Indeed, we do not need to consider the above kernel surfaces.

A nonlinear mapping $\Psi(\cdot)$ is used to map x into a higher dimensional feature space, i.e., $\Psi(\cdot) \colon \mathbb{R}^m \to \mathbb{R}^l$, such that

$$X = \Psi(x),$$

where $X \in \mathbb{R}^l$ and $l > m$. We call function K a kernel if there is a map $\Psi(\cdot)$ such that $K(x,y) = \langle \Psi(x), \Psi(y) \rangle$. So the primal problems in nonlinear case can be expressed as follows

$$\min \frac{1}{2}\|t\|^2 + \frac{c_2}{2}(\|w_1\|^2 + b_1^2) + c_1 e_2^T \xi + c_u e_u^T \psi$$

$$\text{subject to} \quad \Psi(A)w_1 + e_1 b_1 = t$$
$$- (\Psi(B)w_1 + e_2 b_1) + \xi \geqslant e_2 \tag{33}$$
$$(\Psi(U)w_1 + e_u b_1) + \psi \geqslant (-1 + \varepsilon)e_u$$
$$\xi, \psi \geqslant 0$$

where $c_1, c_2 > 0$, $c_u > 0$ are parameters, e_1, e_2, e_u are vectors of ones of appropriate dimensions, and ξ and ψ are slack vectors. The second primal problem is

$$\min \frac{1}{2}\|t\|^2 + \frac{c_4}{2}(\|w_2\|^2 + b_2^2) + c_3 e_1^T \xi + c_u e_u^T \psi$$

$$\text{subject to} \quad \Psi(B)w_2 + e_2 b_2 = t$$
$$(\Psi(A)w_2 + e_1 b_2) + \xi \geqslant e_1 \tag{34}$$
$$(\Psi(U)w_2 + e_u b_2) + \psi \geqslant (1 - \varepsilon)e_u$$
$$\xi, \psi \geqslant 0$$

where $c_1, c_2, c_u > 0$ are parameters, e_1, e_2, e_u are vectors of ones of appropriate dimensions, and ξ and ψ are slack vectors.

Similarly to the linear case of IUTBSVM, the dual problem to problem (33) draws

$$\min \frac{1}{2}S^T Q S + c_2 r^T S \tag{35}$$

$$\text{subject to} \quad 0 \leqslant \alpha \leqslant c_1 e_2, \quad 0 \leqslant \beta \leqslant c_u e_u$$

where

$$Q = \begin{bmatrix} K(A,A^T) + C_2 I & K(A,B^T) & -K(A,U^T) \\ K(B,A^T) & K(B,B^T) & -K(B,U^T) \\ -K(U,A^T) & -K(U,B^T) & K(U,U^T) \end{bmatrix} + \begin{bmatrix} E & E & -E \\ E & E & -E \\ -E & -E & E \end{bmatrix},$$

$$Q = Q^T, \quad S = \begin{bmatrix} \lambda \\ \alpha \\ \beta \end{bmatrix}, \quad r = \begin{bmatrix} 0 \\ -e_2 \\ (1-\varepsilon)e_u \end{bmatrix},$$

and E is the matrix with all entries equal to 1. Notice that Q is a symmetric positive definite matrix. By solving the above dual problem, the first of nonlinear hyperplane is determined as follows:

$$-\frac{1}{c_2}\left(K(x,A^T)\lambda + K(x,B^T)\alpha - K(x,U^T)\beta\right) + b_1 = 0, \tag{36}$$

where

$$b_1 = -\frac{1}{c_2}\left(e_1^T\lambda + e_2^T\alpha - e_u^T\beta\right).$$

Analogously, the dual problems of the problem (34) is:

$$\min \frac{1}{2}S^T QS + c_4 r^T S \tag{37}$$

$$\text{subject to } 0 \leqslant \alpha \leqslant c_3 e_1, \ 0 \leqslant \beta \leqslant c_u e_u$$

where

$$Q = \begin{bmatrix} K(B,B^T)+c_4 I & -K(B,A^T) & -K(B,U^T) \\ -K(A,B^T) & K(A,A^T) & K(A,U^T) \\ -K(U,B^T) & K(U,A^T) & K(U,U^T) \end{bmatrix} + \begin{bmatrix} E & -E & -E \\ -E & E & E \\ -E & E & E \end{bmatrix},$$

$$Q = Q^T, \ S = \begin{bmatrix} \lambda \\ \alpha \\ \beta \end{bmatrix}, \ r = \begin{bmatrix} 0 \\ -e_1 \\ (\varepsilon-1)e_u \end{bmatrix}.$$

By solving the above problem, the resulting nonlinear hyperplane takes the form of

$$\frac{1}{c_4}\left(K(x,B^T)\lambda - K(x,A^T)\alpha - K(x,U^T)\beta\right) + b_2 = 0, \tag{38}$$

where

$$b_2 = -\frac{1}{c_4}\left(e_2^T\lambda - e_1^T\alpha - e_u^T\beta\right).$$

Therefore we construct two decision functions as follows:

$$f_1(x) = -\frac{1}{c_2}\left(K(x,A^T)\lambda + K(x,B^T)\alpha - K(x,U^T)\beta\right) + b_1,$$

$$b_1 = -\frac{1}{c_2}\left(e_1^T\lambda + e_2^T\alpha - e_u^T\beta\right),$$

and

$$f_2(x) = -\frac{1}{c_4}\left(K(x,B^T)\lambda - K(x,A^T)\alpha - K(x,U^T)\beta\right) + b_2,$$

$$b_2 = -\frac{1}{c_4}\left(e_2^T\lambda - e_1^T\alpha - e_u^T\beta\right).$$

For any unknown input x, we assign it to the class $i \in \{1,-1\}$ by the rule

$$\text{class } i = \arg\min_{i=1,2} |f_i(x)|.$$

4   Numerical Experiments

In this section, to illustrate the validity and efficiency of the proposed method, we apply it on several UCI benchmark data sets [5] and compare our method with the $\mathfrak{U}$-SVM, TSVM, and $\mathfrak{U}$-TSVM methods. All experiments are carried out in Matlab 2019b on a PC with Intel(R) CORE(TM) i7-7700HQ CPU@2.80 GHz machine with 16 GB of RAM. For solving the dual problems of the methods, we used "quadprog.m" function in Matlab. In the experiments, we opt for the Gaussian kernel function $k(x_i, x_j) = \exp\left(\frac{-\|x_i - x_j\|^2}{\gamma^2}\right)$.

Notice that we used 5-fold cross-validation to assess the performance of the algorithms in aspect of classification accuracy.

4.1   Parameter Selection

The accuracy highly depends on the parameter values. Therefore choosing the parameters is very important for performance of classifiers. That is why we adopted the grid search method to choose the best values of the parameters for each algorithms [6].

To reduce the computational cost of the parameter selection, we set the regularization parameter values $c_1 = c_3$ and $c_2 = c_4$ in IUTBSVM method and $c_1 = c_2$ for $\mathfrak{U}$-TSVM. The optimal values for c_1, c_2, c_3 and c_4 in all methods were selected from the set $\{2^i \mid i = -8, -7, \ldots, 7, 8\}$, the parameters of the Gaussian kernel γ were selected from the set $\{2^i \mid i = -8, -7, \ldots, 7, 8\}$, parameter ϵ in $\mathfrak{U}$-TSVM, and our IUTBSVM methods was chosen from set $\{0.1, 0.2, \ldots, 0.5\}$.

4.2   Results Comparisons and discussion for UCI Data Sets

For a more detailed analysis, we compared our proposed IUTBSVM with $\mathfrak{U}$-SVM, TSVM, and $\mathfrak{U}$-TSVM on several benchmark data sets from UCI machine learning repository [5] including Australian, Pima Indian Diabetes, Heart-Statlog, German, Haberman, Hayes-Roth, House-vote84, Housing, Ionosphere, Mushrooms, Sonar, Spect, Wdbc, and Bupa.

For each data set, we randomly select the 40% of data sets from different classes to generate the Universum data.

The experimental results are displayed in Table 1. From the prospective of the classification accuracy, the proposed method IUTBSVM outperforms the others in 12 out of 14 data sets. Therefore the results demonstrated that our proposed IUTBSVM is more efficient than the other standard methods in the aspect of classification accuracy.

Table 1. Classification accuracy of 𝔘-SVM, TSVM, 𝔘-TSVM, and IUTBSVM with Gaussian kernel.

Data sets	𝔘-SVM	TSVM	𝔘-TSVM	IUTBSVM
	Acc ± std	Acc ± std	Acc ± std	Acc ± std
Australian	44.6 3 ± 0.59	65.93 ± 3.35	59.12 ± 3.84	**68.40 ± 1.88**
Pima Indian Diabetes	65.10 ± 0.23	67.44 ± 5.82	69.13 ± 3.67	**72.91 ± 2.48**
Heart-Statlog	44.44 ± .06	**85.18 ± 2.92**	72.59 ± 10.00	80.37 ± 4.26
German	31.60 ± 1.55	73.20 ± 2.63	70.30 ± 1.03	**76.10± 1.24**
Haberman	73.53 ± 0.53	75.51 ± 5.36	68.96 ± 6.02	**75.74 ± 4.36**
Hayes-Roth	50.00 ± 1.68	54.00 ± 10.83	60.85 ± 8.57	**64.47 ± 11.93**
House-vote84	84.13 ± 2.74	95.38 ± 2.61	85.06 ± 2.60	**97.01 ± 1.29**
Housing	84.85 ± 3.47	93.08 ± 0.03	89.13 ± 2.10	**94.86 ± 1.29**
Ionosphere	70.10 ± 6.74	89.45 ± 2.60	88.06 ± 6.13	**96.86 ± 1.086**
Mushrooms	95.67 ± 1.28	98.67 ± 0.31	99.21 ± 0.14	**100.00 ± 0.00**
Sonar	46.63 ± 1.00	65.45 ± 9.79	69.75 ± 5.23	**89.90 ± 3.11**
Spect	58.79 ± 0.42	70.37 ±7.12	51.39 ± 9.72	**73.38 ± 8.01**
Wdbc	86.28 ± 2.14	87.53 ± 3.42	72.71 ± 8.70	**92.61 ± 1.50**
Bupa	42.02 ± 0.00	**69.56 ± 3.39**	62.60 ± 2.78	56.52 ± 8.26

5   Conclusions

In this paper, we proposed a new inverse free 𝔘-TSVM and named this method as IUTBSVM. The main contribution of this work is introducing a different Lagrangian function to that by changes in the primal problems, leading to specific dual problems. In contrast to the other methods, we do not need to compute inverse matrices, which is always a complicated computing task. We also suggested a reduced kernel to decrease computational time of nonlinear IUTBSVM. The numerical experiments performed on several UCI benchmark data sets illustrate efficiency of the proposed method.

References

1. Arabasadi, Z., Alizadehsani, R., Roshanzamir, M., Moosaei, H., Yarifard, A.A.: Computer aided decision making for heart disease detection using hybrid neural network-genetic algorithm. Comput. Methods Programs Biomed. **141**, 19–26 (2017)
2. Bazikar, F., Ketabchi, S., Moosaei, H.: Dc programming and DCA for parametric-margin ν-support vector machine. Appl. Intell. **50**(6), 1763–1774 (2020)
3. Cai, Y.D., Ricardo, P.W., Jen, C.H., Chou, K.C.: Application of SVM to predict membrane protein types. .f Theoret. Biol. **226**(4), 373–376 (2004)
4. Cherkassky, V., Mulier, F.M.: Learning from Data: Concepts, Theory, and Methods. Wiley, New York (2007)
5. Dua, D., Graff, C.: UCI machine learning repository (2017). http://archive.ics.uci.edu/ml

6. Hsu, C.W., Chang, C.C., Lin, C.J., et al.: A practical guide to support vector classification (2003). https://www.csie.ntu.edu.tw/~cjlin/papers/guide/guide.pdf

7. Jayadeva, Khemchandani, R., Chandra, S.: Twin support vector machines for pattern classification. IEEE Trans. Patt. Anal. Mach. Intell. **29**(5), 905–910 (2007)

8. Ketabchi, S., Moosaei, H., Razzaghi, M., Pardalos, P.M.: An improvement on parametric ν -support vector algorithm for classification. Ann. Oper. Res. **276**(1–2), 155–168 (2019)

9. Liu, C.L., Hsaio, W.H., Lee, C.H., Chang, T.H., Kuo, T.H.: Semi-supervised text classification with universum learning. IEEE Trans. Cybernet. **46**(2), 462–473 (2015)

10. Mangasarian, O.L., Wild, E.W.: Multisurface proximal support vector machine classification via generalized eigenvalues. IEEE Trans. Patt. Anal. Mach. Intell. **28**(1), 69–74 (2005)

11. Moosaei, H., Ketabchi, S., Razzaghi, M., Tanveer, M.: Generalized twin support vector machines. Neural Process. Lett. **53**(2), 1545–1564 (2021)

12. Noble, W.S.: Support vector machine applications in computational biology. In: Schoelkopf, B., Tsuda, K., Vert, J.P. (eds.) Kernel Methods in Computational Biology, pp. 71–92. MIT Press, Cambridge(2004)

13. Qi, Z., Tian, Y., Shi, Y.: Twin support vector machine with universum data. Neural Netw. **36**, 112–119 (2012)

14. Richhariya, B., Tanveer, M.: EEG signal classification using universum support vector machine. Exp. Syst. Appl. **106**, 169–182 (2018)

15. Sinz, F.H., Chapelle, O., Agarwal, A., Schölkopf, B.: An analysis of inference with the universum. In: Proceedings of the 20th International Conference on Neural Information Processing Systems, NIPS 2007, pp. 1369–1376., Curran Associates Inc., Red Hook (2008)

16. Tang, L., Tian, Y., Li, W., Pardalos, P.M.: Structural improved regular simplex support vector machine for multiclass classification. Appl. Soft Compu. **91**, 106235 (2020)

17. Tang, L., Tian, Y., Li, W., Pardalos, P.M.: Valley-loss regular simplex support vector machine for robust multiclass classification. Knowl. Based Syst. **216**, 106801 (2021)

18. Tang, L., Tian, Y., Pardalos, P.M.: A novel perspective on multiclass classification: regular simplex support vector machine. Inf. Sci. **480**, 324–338 (2019)

19. Tang, L., Tian, Y., Yang, C., Pardalos, P.M.: Ramp-loss nonparallel support vector regression: robust, sparse and scalable approximation. Knowl.-Based Syst. **147**, 55–67 (2018)

20. Tong, S., Koller, D.: Support vector machine active learning with applications to text classification. J. Mach. Learn. Res. **2**(Nov), 45–66 (2001)

21. Vapnik, V.: The Nature of Statistical Learning Theory. Springer, New York (2013). https://doi.org/10.1007/978-1-4757-3264-1

22. Vapnik, V., Chervonenkis, A.: Theory of Pattern Recognition. Nauka, Moscow (1974)

23. Wang, X.Y., Wang, T., Bu, J.: Color image segmentation using pixel wise support vector machine classification. Patt. Recogn. **44**(4), 777–787 (2011)

24. Weston, J., Collobert, R., Sinz, F., Bottou, L., Vapnik, V.: Inference with the universum. In: Proceedings of the 23rd International Conference on Machine Learning, ICML 2006, pp. 1009–1016 (2006)

Hybridising Self-Organising Maps with Genetic Algorithms

Abtin Nourmohammadzadeh$^{(\boxtimes)}$ and Stefan Voß

Institute of Information Systems (IWI), University of Hamburg, Hamburg, Germany
{abtin.nourmohammadzadeh,stefan.voss}@uni-hamburg.de
https://www.bwl.uni-hamburg.de/en/iwi

Abstract. The aim of this work is to develop a hybridised approach consisting of the Self-Organising Map (SOM), which is an artificial neural network, and a Genetic Algorithm (GA) to tackle large and complex optimisation problems by decomposition. The approach is tested on the travelling salesman problem (TSP), which is a very important problem in combinatorial optimisation. The SOM clusters the nodes into a specified number of groups. Then, the resulting smaller TSPs within each cluster are solved, and finally, the solutions of the clusters are connected to build a high quality solution for the whole TSP. In the two latter steps, a GA is used. Our approach is examined on some random examples including 100 to 10000 nodes as well as 12 benchmark instances. The results with various numbers of clusters are compared to each other and also to the case of solving the TSP with all nodes and without any clustering. Furthermore, the effect of each component algorithm in the hybridised approach is evaluated by some complementary experiments in which the clustering and the optimisation algorithm are replaced by other methods.

Keywords: Clustering · Self-Organising Map (SOM) · Genetic Algorithm (GA) · Travelling Salesman Problem (TSP)

1 Introduction

Metaheuristics have been extensively used as efficient alternatives for classical optimisation methods to find appropriate solutions for complex problems in operations research. Meanwhile, the decomposition of hard problems into a number of sub-problems which can be tackled in an easier way is considered as one other methodology to find good solutions in shorter computation times. Clustering can be used as an approach to decompose problems. Artificial Neural Networks (ANNs) [2] have a strong ability in clustering. Hence, the development and application of methods which consist of metaheuristics and ANNs together can be followed to achieve the advantages of both in an optimisation process.

In this work, we devise a method which employs a neural network called Self-Organising Map (SOM) [21] to cluster or break an optimisation problem into a

© Springer Nature Switzerland AG 2021
D. E. Simos et al. (Eds.): LION 2021, LNCS 12931, pp. 265–282, 2021.
https://doi.org/10.1007/978-3-030-92121-7_22

number of easier sub-problems. Then, the sub-problems are tackled by a genetic algorithm (GA). Consequently, the partial solutions are combined together with the aid of another GA to build a complete solution for the original problem.

A problem which properly suits this solution methodology is the Travelling Salesman Problem (TSP). The TSP is a classical NP-hard combinatorial optimisation problem and is referred to as easy to describe but hard to solve [26]. It involves N cities or nodes and a salesman who seeks to find the shortest possible tour through all nodes, which includes each node exactly once.

The TSP was originally defined in 1832 [1] and is one of the most thoroughly investigated optimisation problems. Many real-world applications have been found for the TSP, for example, in vehicle routing and communication networks [24]. This increases the significance of finding a good solution for the TSP within a short time. A wide variety of solution methodologies including exact, heuristic and metaheuristic approaches have been applied to the TSP.

In [9], a very early attempt in solving the TSP is addressed, which uses a Branch and Bound algorithm on an IBM 7090 computer. It was found that the average computational time grows exponentially while the number of nodes increases. In a considerable attempt, [4] certifies an optimal TSP tour through 85900 cities. Various metaheuristics such as the Ant Colony Optimisation (ACO) [32], Simulated Annealing (SA) [14], and GAs [25] are used and also new algorithms are still being developed to deal with the TSP. In the metaheuristic optimisation, it is aimed at finding a (near) optimal solution by iterative improvement of candidate solutions, for details see [13]. Novel metaheuristics are also employed to solve the TSP such as the fixed set search (FSS) in [19] or the POPMUSIC in [30].

Our approach can be applied to the TSP as follows. The SOM divides the nodes into a number of groups and so the original TSP is decomposed to some smaller TSPs. Henceforth, it is analogous to solving a special variant of this problem known as the clustered TSP [3,16,18,23,25,28]. In the clustered TSP, there are some pre-determined clusters of the cities, and the salesman must visit the cities of each cluster consecutively. Thus, the whole problem is decomposed to some smaller TSPs, which are easier to solve. Consequently, the partial solutions corresponding to clusters are linked together to build a complete tour for the original TSP. The GA is responsible for providing good solutions within clusters and also an appropriate connection of these solutions, which leads to an optimal or sub-optimal overall solution to the entire TSP.

The remainder of this paper is organised as follows: In Sect. 2, a short overview of some related works is provided. Section 3 is devoted to explaining the solution methodologies. The numerical results and related comparisons are covered in Sect. 4. Finally, Sect. 5 presents some conclusions and recommends directions for future research.

2   Related Works

There have been several works regarding the application of clustering in the TSP. [28] uses an automatic clustering approach in which the cluster's size is controlled by a parameter. In [25], a GA for the TSP is proposed, computational results are reported on a set of Euclidean problems, and comparisons are provided with another heuristic. A novel evolutionary algorithm based on a clustering algorithm is proposed for the TSP in [23]. Furthermore, the tours are improved by a local search. Simulations on some standard test problems are made, and the results verify the efficiency of the proposed approach.

In a work similar to ours, [18] applies a Firefly Algorithm (FA) and K-means clustering to the problem. Their approach consists of the three steps of clustering the nodes, finding the optimal path in each cluster, and reconnecting the clusters, too. [12] addresses clustering the cities, then using the NEH heuristic, which provides an initial solution that is refined using a modification of the metaheuristic Multi-Restart Iterated Local Search (MRSILS).

The applications of the clustered TSP in several fields like vehicle routing, manufacturing, computer operations, examination timetabling, cytological testing, and integrated circuit testing are described in [22]. [16] uses approximation algorithms for a clustered TSP. [3] proposes the use of a memetic algorithm (MA) that combines the global search ability of a GA with a local search to refine solutions to the clustered TSP.

Different methods can be used for clustering; however, the SOM is chosen as a capable clustering tool in numerous works. [10] uses the SOM for robust clustering. In a very recent research, [20] gives chest x-ray images of COVID-19 patients to an SOM network and finds a distinct classification between sick and healthy patients. A combination of SOM and K-means to study the marketing of online games in Taiwan is the subject of [31]. The method provides a good evaluation of the market segmentation. In [29], an SOM is used for clustering the flood disaster risks.

The application of SOMs for the TSP has been observed in some papers. [8] uses a modified Kohonen self-organising map with one-dimensional neighbourhood to solve the symmetrical TSP. The solutions generated by the Kohonen network are improved using the 2-opt algorithm. [5] aims at looking for the incorporation of an efficient initialisation method and the definition of a parameter adaptation law to achieve better results and faster convergence of the SOM when it is applied to the TSP. [11] proposes novel adaptation rules for the SOM to solve the prize-collecting travelling salesman problem (PC-TSP), which aims to find a cost-efficient tour to collect prizes by visiting a subset of a given set of locations.

An overview of the related works shows that in dealing with the TSP, there is a lack of an approach which uses an SOM for the node clustering and also applies a GA to tackle the resulting sub-problems. Hence, this subject is investigated in this work.

3 Solution Methodologies

In this section, the general concepts of the SOM and the GA are explained in the first two parts and subsequently, our approach, which uses these two methods, is introduced.

3.1 Self-Organising Map

An SOM or self-organising feature map (SOFM), introduced by *Kohonen* in [21], is a type of ANNs which is trained by unsupervised learning. The SOM provides a low-dimensional and discretised representation of the input space of the training samples, called a map, and is, therefore, a method to do dimensionality reduction. The difference between SOMs and other ANNs is that SOMs apply competitive learning rather than error-correction learning (such as backpropagation with gradient descent), and it is also envisaged by their neighbourhood function usage to preserve the topological properties of the input space.

SOMs operate in two phases like most neural networks: training and mapping. Training builds the map using input examples (a competitive process, also called vector quantisation), whereas in the mapping phase an input vector is automatically classified.

The map space includes components called nodes or neurons. It is pre-defined, usually as a finite two-dimensional region where nodes are arranged in a regular hexagonal or rectangular grid. A weight vector is associated with each node, which is a position in the input space; that is, it has the same dimension as each input vector. While nodes in the map space stay fixed, training consists in moving weight vectors toward the input data (reducing a distance metric) without spoiling the topology induced from the map space. Thus, the self-organising map describes a mapping from a higher-dimensional input space to a lower-dimensional map space. Once trained, the map can classify a vector from the input space by finding the node with the closest (smallest distance metric) weight vector to the input space vector.

Figure 1 illustrates the training process of a simple SOM network. It can be seen, that a neuron of the network is affected by any single datum in the training set. However, in each iteration, the neuron is mostly influenced by the nearest datum and moved toward it. Finally, after iterative changes of the neurons' positions, the SOM network tends to lay on the data and approximate them.

3.2 Genetic Algorithm

A GA is a metaheuristic inspired by the process of natural selection. GAs belong to the larger class of evolutionary algorithms (EAs). GAs are commonly used to generate high-quality solutions to optimisation and search problems by relying on biologically inspired operators such as mutation, crossover and selection, see [17].

GAs usually start with a population of random candidate solutions and apply the genetic operators in each iteration to a proportion of the population to

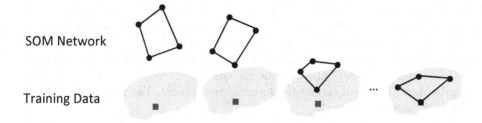

Fig. 1. Training process of the SOM

improve the solutions. The algorithm continues iteratively until a termination condition occurs.

3.3 Our Approach

Our solution approach uses both the SOM and the GA, and consists of three steps. In the first step, an SOM of a given size is applied to our data of nodes' (cities') coordinates. According to the mentioned training process, the cities' coordinates are used as the training data and they iteratively contribute to the movements of the SOM neurons or the network shape. According to the final position of the map, the cities are clustered based on the nearest SOM neuron to them. Figure 2 visualises an example of the SOM clustering of cities when once an SOM of the size 1×2 and once one of the size 2×2 is used. So, the cities of each cluster are known and the sub-TSPs are constituted. Then, the GA is applied to each sub-problem to minimise the length of the sub-tours. Another GA is used which tries to find a good order for connecting the sub-tours. The GA codes the order of visiting the cities and clusters as strings of unique numbers. After this step, a solution for the entire original TSP is found. Figure 3 depicts a solution built according to this method for a small TSP by connecting the cities of each cluster, and then, the clusters. Figure 4 shows the related encoding used in the GA.

These codes are in form of strings which include the city labels. The first element of the string is the label of the first city that the salesman visits, the second element represents the second city and it continues so up to the last element, which is the last city of the tour. The code for connecting the clusters is exactly the same and in the overall solution, the last city in the solution of one cluster is joined with the first city in the solution of the next cluster.

Our GA begins with a population ($nPop$) of random initial solutions. In each iteration, a proportion ($Cross\_rate$) of candidate solutions of the population is selected for doing crossovers according to the roulette wheel selection method [7]. One-point crossover is used. In addition, another proportion ($nMutat$) of the population is chosen randomly from the population for mutation. In mutations, two elements or genes from the encoding strings (see Fig. 4) are randomly chosen and their positions are exchanged. Our GA terminates as soon as a pre-defined maximum number of iterations ($Maxit$) is executed. Parameter tuning, i.e. determining the values of $nPop$, $Cross\_rate$, $Mutat\_rate$, $Maxit$, is separately done for each sub-problem and the cluster-connecting problem. To set the first three parameters, the Taguchi method [15] is used, whereas, for $Maxit$, experiments with increasing numbers of iterations are conducted. Whenever results are not improved, we stop and choose the last effective $Maxit$. The pseudocode of our GA is given as Algorithm 2.

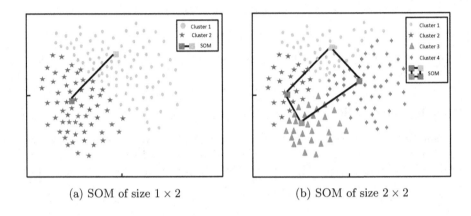

(a) SOM of size 1×2 (b) SOM of size 2×2

Fig. 2. SOM clustering of cities

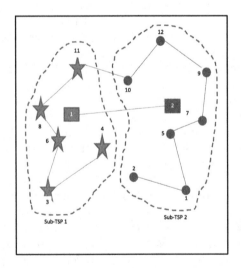

Fig. 3. An example solution

Order of visiting the cities within the clusters:

Sub- TSP1 | 4 | 3 | 6 | 8 | 11 |

Sub- TSP2 | 10 | 12 | 9 | 7 | 5 | 1 | 2 |

Order of connecting the clusters: | 1 | 2 |

Fig. 4. Encodings scheme of the GA

Algorithm 1: Our Algorithm for the TSP

Data: Nodes' coordinates
Result: A good quality feasible solution for the TSP
1 - Apply the SOM to the city coordinates.
2 - Cluster the nodes based on the final position of the SOM nodes
3 - Solve the sub-TSPs within the clusters by the GA
4 - Apply the GA to find a good order of travelling between the clusters

---

Algorithm 2: Our GA

Data: Probem input
Result: A good quality solution
1 - Set the GA parameters, $nPop$, $Cross\_rate$, $Mutat\_rate$ and $Maxit$
2 - Generate $nPop$ random candidate solutions.
3 - Evaluate the population.
4 - $It = 1$.
5 **while** $It \leq Maxit$ **do**
6   - Choose the candidates for crossover by roulette wheel selection.
7   - Do the crossovers.
8   - Choose the candidates for mutation randomly.
9   - Do the mutations.
10   - Merge the crossover and mutation results with the population.
11   - Sort the population and choose the first best $nPop$ of them as the new population.
12   - $It = It + 1$.
13 **end**
14 - Report the best solution of the population.

---

4 Computation Results

In this section, the computational results of our method are presented and it is compared to other methods. The computational experiments of this paper are executed on a computer with a Core(TM) i7 processor, 3.10 GHz CPU and 16 GB of RAM. PYTHON is used for programming the algorithms.

As the first series of experiments, TSPs with random positions of cities are generated in a $[0, 100] \times [0, 100]$ area. The number of cities or nodes are from 100 to 10000. For each problem size, 20 random instances are tested and their average is regarded.

As mentioned in the previous section, before each GA application, its parameters are tuned by the Taguchi design of experiments and trial and error method. Figure 5 shows an example for this parameter setting. In Taguchi, three levels are considered as the potential values of each parameter and a number of experiments is conducted with some designed combinations of the levels. Accordingly, a quantity called signal to noise (S/N) ratio is calculated for each level. The smaller S/N values are better in this minimisation problem. However, in the trial and error approach used to determine the maximum number of iterations, the experiments are done by growing numbers of $Maxit$, starting from 50 iterations.

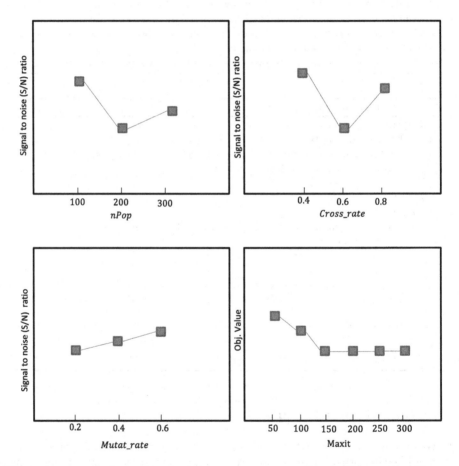

Fig. 5. An example for tuning the GA parameters: according to the results, $nPop = 100$, $Cross\_rate = 0.6$, $Mutat\_rate = 0.2$ are chosen by the Taguchi method and $Maxit = 150$ is set by the trial and error approach.

For better understanding, the results are normalised in the interval [0,1] based on the minimum of zero and a maximum equal to the worst objective value among the solutions of the first GA iteration. Our approach with different sizes of SOM map and another approach based on the method presented in [18] using the K-means classification and firefly optimisation algorithm are applied to the generated test problems. To have a better judgement between these two methods, the considered numbers of clusters in K-means, i.e. Ks, are equal to the total number of SOM nodes (clusters). In addition, the approach of considering all TSP cities at once without any clustering is used to find solutions for the same problems and the results are shown in Table 1. The average results of all methods are summarised in Tables 2, 3, 4 and 5. In the first column, the problem sizes or the numbers of TSP cities are shown. For each method, firstly, the average numbers of cities in the considered clusters are shown, then the average objective values, and finally, the execution times are tabulated. It is worth noting that the time required for setting the GA parameters and training the SOM are considered in the reported execution times.

Table 1. Results of the GA for solving the complete instances without any clustering

Size	Solving the whole problem	
	Obj. value	Time (s)
100	0.43	32.61
200	0.45	82.18
500	0.52	189.26
1000	0.56	292.40
2000	0.62	508.37
5000	0.63	1530.75
10000	0.68	3751.30

Table 2. Clustering with the SOM of size 1 × 2 and the K-means with K = 2

Size	Our approach			K-means and Firefly		
	Clusters	Obj. value	Time (s)	Clusters	Obj. value	Time (s)
100	(48, 52)	0.44	32.25	(46, 54)	0.45	33.08
200	(92, 108)	0.46	81.05	(110, 90)	0.48	82.10
500	(235, 265)	0.49	132.65	(242, 258)	0.49	147.12
1000	(462, 538)	0.49	185.47	(396, 604)	0.52	198.65
2000	(1082, 918)	0.52	310.25	(1260, 740)	0.54	318.72
5000	(2237, 2763)	0.57	562.75	(2062, 2938)	0.59	576.42
10000	(5734, 4266)	0.63	1689.82	(5842, 4158)	0.66	1720.48

Table 3. Clustering with the SOM of size 2 × 2 and K-means with K = 4

Size	Our approach			K-means and Firefly		
	Clusters	Obj. value	Time	Clusters	Obj. value	Time
100	(29, 28, 23, 20)	0.48	35.91	(28, 28, 22, 22)	0.48	37.10
200	(42, 48, 52, 58)	0.46	71.15	(44, 50, 54, 42)	0.49	76.32
500	(110, 134, 127, 129)	0.47	121.32	(121, 132, 114, 133)	0.50	132.41
1000	(225, 280, 263, 232)	0.51	158.62	(231, 292, 246, 231)	0.53	169.72
2000	(565, 441, 532, 463)	0.48	219.35	(572, 448, 582, 398)	0.51	232.46
5000	(1320, 1157, 1408, 1115)	0.54	312.86	(1292, 1163, 1452, 1093)	0.56	324.66
10000	(2608, 2572, 2380, 2440)	0.58	528.35	(2636, 2532, 2420, 2412)	0.60	549.08

Table 4. Clustering with the SOM of size 3 × 3

Size	Our approach		
	Clusters	Obj. value	Time
100	(10, 13, 9, 12, 11, 12, 11, 12, 10)	0.54	43.27
200	(17, 15, 23, 18, 31, 20, 27, 30, 19)	0.48	78.15
500	(55, 62, 65, 63, 52, 48, 55, 48, 52)	0.47	121.32
1000	(115, 120, 118, 98, 107, 111, 108, 98, 125)	0.49	148.83
2000	(230, 202, 251, 203, 238, 208, 242, 202, 224)	0.46	172.05
5000	(540, 564, 550, 576, 569, 530, 585, 528, 558)	0.48	215.88
10000	(1203, 1127, 1037, 1152, 1218, 1069, 1138, 1065, 991)	0.53	321.6

Table 5. Clustering with the K-means with K = 9

Size	K-means and Firefly		
	Clusters	Obj. value	Time
100	(9, 14, 10, 12, 14, 10, 12, 9, 10)	0.56	51.36
200	(20, 18, 25, 19, 25, 22, 28, 28, 15)	0.53	106.16
500	(58, 60, 68, 62, 50, 45, 53, 50, 54)	0.50	130.62
1000	(117, 122, 121, 103, 105, 109, 101, 102, 120)	0.51	164.37
2000	(235, 206, 246, 195, 239, 205, 245, 207, 222)	0.48	190.52
5000	(546, 572, 555, 578, 580, 541, 582, 531, 515)	0.50	232.96
10000	(1184, 1095, 949, 1168, 1225, 1056, 1195, 1092, 1036)	0.56	328.60

As it is evident, except for the smallest sizes containing 100 and 200 cities, the clustering methods can provide better solutions in shorter computational times. As the problem size grows, having more clusters, i.e. more nodes (neurons) in the SOM network and larger Ks in the K-means, increases the ability of the methods in finding good quality solutions more quickly. In comparison to the approach which uses the K-means and the Firefly algorithm, it is observed that our SOM with the complementary GA is superior in terms of both the objective value and the computation time. Figure 6 and Fig. 7 illustrate the average objective values and computational times of all methods, respectively.

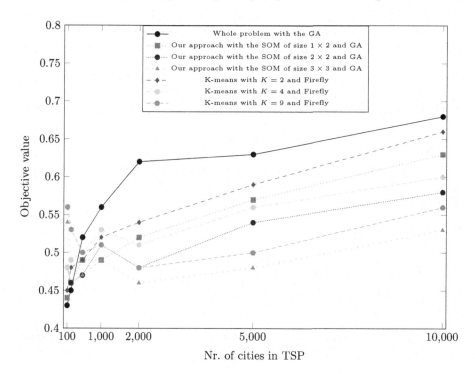

Fig. 6. Objective function comparison between the methods

Table 6. The results of applying the methods to 12 benchmark TSP instances from TSPLIB [27]

Instance	Our approach		K-means and Firefly		Whole problem	
	Obj. value	Time	Obj. value	Time	Obj. value	Time
eil51	0	16.17	0.1	18.09	0.12	15.82
berlin52	0	17.42	0.08	17.68	0.08	15.95
pr76	0.10	25.92	0.12	29.72	0.15	35.81
rat99	0.10	27.60	0.14	31.53	0.16	37.06
kroA100	0.07	36.15	0.15	42.78	0.15	48.92
lin105	0.11	38.61	0.16	44.32	0.20	57.61
pr124	0.08	38.77	0.10	46.50	0.16	55.21
pr136	0.12	43.62	0.15	46.56	0.20	57.90
pr150	0.14	44.36	0.20	46.93	0.22	62.82
rat195	0.15	47.25	0.21	51.43	0.25	63.91
kroA200	0.10	48.39	0.18	54.21	0.21	65.31
ts225	0.11	52.76	0.21	57.02	0.24	67.05

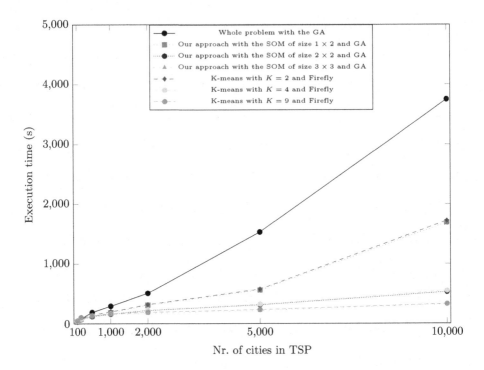

Fig. 7. Execution time comparison between the methods

Table 7. The p-values of the pairwise comparisons of the methods in terms of the objective value and the execution time based on the Friedman test with the Bergmann-Hommel post hoc procedure: a = Whole problem with the GA, b = Our approach with the SOM of size 1 × 2 and GA, c = Our approach with the SOM of size 2 × 2 and GA, d = Our approach with the SOM of size 3 × 3 and GA, e = K-means with $K = 2$, f = K-means with $K = 4$, g = K-means with $K = 9$

P-values	a	b	c	d	e	f	g
a	–	6.21×10^{-7}	2.01×10^{-8}	3.81×10^{-10}	1.72×10^{-5}	3.43×10^{-7}	6.12×10^{-8}
b	6.21×10^{-7}	–	5.52×10^{-4}	4.87×10^{-5}	2.25×10^{-5}	9.12×10^{-6}	6.44×10^{-8}
c	2.01×10^{-8}	5.52×10^{-4}	–	4.25×10^{-4}	2.18×10^{-5}	3.12×10^{-5}	7.36×10^{-7}
d	3.81×10^{-10}	4.87×10^{-5}	4.25×10^{-4}	–	6.01×10^{-3}	3.72×10^{-5}	4.12×10^{-6}
e	1.72×10^{-5}	2.25×10^{-5}	2.18×10^{-5}	6.01×10^{-3}	–	5.20×10^{-3}	1.15×10^{-4}
f	3.43×10^{-7}	9.12×10^{-6}	3.12×10^{-5}	3.72×10^{-5}	5.20×10^{-3}	–	3.17×10^{-3}
g	6.12×10^{-8}	6.44×10^{-8}	7.36×10^{-7}	4.12×10^{-6}	1.15×10^{-4}	3.17×10^{-3}	–
P-values	a	b	c	d	e	f	g
a	–	3.92×10^{-9}	8.05×10^{-10}	4.22×10^{-12}	6.52×10^{-6}	8.20×10^{-8}	2.17×10^{-11}
b	2.32×10^{-9}	–	1.84×10^{-5}	7.15×10^{-6}	3.16×10^{-7}	7.30×10^{-8}	5.12×10^{-10}
c	1.07×10^{-10}	6.82×10^{-6}	–	5.78×10^{-5}	3.62×10^{-6}	2.08×10^{-6}	9.96×10^{-8}
d	8.50×10^{-13}	6.65×10^{-7}	5.80×10^{-6}	–	7.72×10^{-4}	1.01×10^{-6}	7.42×10^{-8}
e	1.02×10^{-5}	3.21×10^{-6}	1.98×10^{-5}	5.05×10^{-4}	–	7.01×10^{-4}	6.19×10^{-4}
f	9.90×10^{-8}	3.82×10^{-7}	1.15×10^{-7}	8.42×10^{-6}	7.20×10^{-4}	–	6.65×10^{-4}
g	3.76×10^{-9}	5.91×10^{-9}	2.19×10^{-9}	3.29×10^{-8}	6.39×10^{-6}	3.08×10^{-4}	–

The next series of experiments includes the examination of our approach and the others on 12 benchmark TSP instances from the TSPLIB [27] including 51 to 225 cities. The results are given in Table 6. The first column contains the instance names, which include each a number showing the number of cities. For example $eil51$ and $ts225$ include 51 and 225 cities, respectively. The number of clusters is equal to $\lfloor \frac{Number\ of\ cities}{20} \rfloor$ or one for every 20 cities. The objective values are here normalised in [0,1] like the previous experiments but based on the minimums equal to the known optimal values. The solution of each instance with each of the methods is replicated 10 times due to the stochastic nature of the methods and the possibility of a different result in each replication. The average objective value is referred to for each instance. As it is evident, our approach outperforms the others in many instances in terms of both the objective value and the execution time.

To find out if there are statistically significant differences between the results and execution times of the three methods, a non-parametric method called the Friedman test with the Bergmann-Hommel post hoc procedure [6] is applied based on the results of all instances. The calculated p-values are presented in Table 7, which indicate that the hypothesis of the equality of objective values and computation times of the methods are rejected regarding the confidence level of 0.05.

Finally, in order to assess the contribution of each constituent algorithm, we exchange the optimisation algorithms between the two compared hybrid approaches, i.e. the SOM-GA and the K-means-Firefly. It means that the Firefly optimisation algorithm is hybridised with the SOM and the GA is combined with the K-means. The two new approaches are called SOM-Firefly and K-means-GA, respectively. Figure 8 and 9 provide a comparative illustration of the solution quality and the execution time of the two mentioned hybrid approaches along with those of the SOM-GA and the K-means-Firefly algorithm.

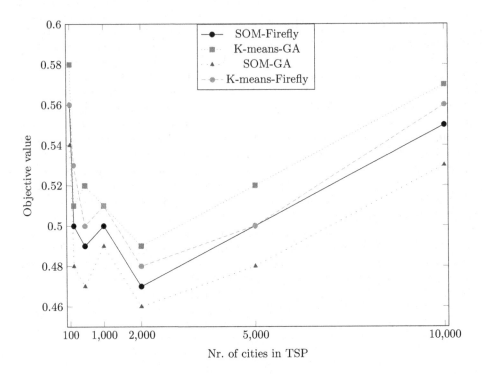

Fig. 8. Objective function comparison between all the methods

It can be observed that the solution quality drops slightly while the computation time grows if the Firefly algorithm replaces the GA as the optimisation algorithm. On the other hand, K-means-GA results are even weaker than those of K-means-Firefly in most of the cases, which shows that this combination of the component algorithms has not been successful. Nonetheless, this approach is the fastest among the four. All in all, it can be deduced from this examination that the SOM plays a more considerable role in the success of our hybridised SOM-GA optimisation methodology than the GA and both algorithms are a good match together.

The four methods are also statistically compared and the corresponding p-values according to the Friedman test with the Bergmann-Hommel post hoc procedure are given in Table 8. These p-values show significant differences between the performance of any pair of the methods.

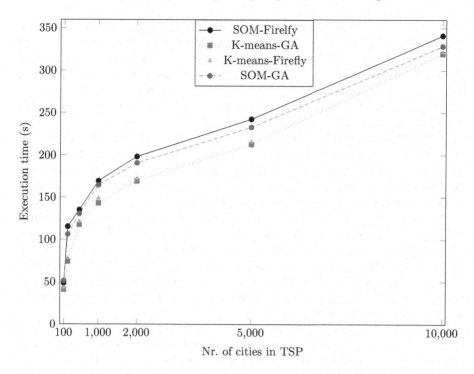

Fig. 9. Execution time comparison between all the methods

Table 8. The p-values of the pairwise comparisons of the methods based of the Friedman test with the Bergmann-Hommel post hoc procedure: a = SOM-Firefly, b = K-means-GA, c = SOM-GA, d = K-means-Firefly

P-values	a	b	c	d
a	–	2.92×10^{-4}	1.70×10^{-5}	5.36×10^{-7}
b	2.92×10^{-4}	–	3.41×10^{-4}	6.08×10^{-6}
c	1.70×10^{-5}	3.41×10^{-4}	–	2.76×10^{-6}
d	5.36×10^{-7}	6.08×10^{-6}	2.76×10^{-6}	–
P-values	a	b	c	d
a	–	1.56×10^{-6}	8.94×10^{-6}	1.98×10^{-9}
b	9.51×10^{-5}	–	5.06×10^{-5}	7.40×10^{-8}
c	5.62×10^{-6}	1.40×10^{-5}	–	4.77×10^{-7}
d	2.58×10^{-9}	3.02×10^{-7}	5.22×10^{-8}	–

5   Conclusions

In this work, an optimisation approach including a clustering phase by the SOM and two GA phases is developed. The GA is embedded in our method to solve the sub-problems and also to find a good connection of the sub-tours. This approach is applied to instances of the TSP. The results verify that this approach is more capable in comparison to solving the whole problem at once as well as another approach with clustering and metaheuristic application. For larger instances, considering more clusters provides better results in shorter elapsed times. In an additional series of experiments, the contributions of the SOM and the GA in the whole approach are measured. This shows that the clustering algorithm or the SOM has a larger effect on the success of our hybrid SOM-GA approach.

Nonetheless, we are aware of many other successful methods in dealing with the TSP in the literature. So the TSP instances are just used to evaluate the abilities of our approach in the first step. As a future extension, this solution methodology can be applied to other optimisation problems, in which clustering might be applicable. One worth topic for further research can be examining the impact of having more clusters with regard to the fact that the problems within clusters are solved independently. Another future direction in terms of the application for the TSP is to use our method for larger TSPs and TSPs with multiple salesmen. In addition, a TSP with non-deterministic cost of travelling between the cities, which is more realistic, can be investigated. Other clustering approaches can be tested as well.

As mentioned above, examining our method on other combinatorial optimisation problems is strongly recommended. This should be done with the idea in mind not to necessarily derive some new "best" method for the considered problem(s) but to gain additional insights into popular solution concepts that are more and more discussed in the realm of artificial intelligence.

References

1. Der Handlungsreisende - wie er sein soll und was er zu tun hat, um Aufträge zu erhalten und eines glücklichen Erfolgs in seinen Geschäften gewiß zu sein - von einem alten Commis-Voyageur. Springer (1832)
2. Aggarwal, C.C.: Neural Networks and Deep Learning. Springer, Cham (2018). https://doi.org/10.1007/978-3-319-94463-0
3. Alsheddy, A.: Solving the free clustered TSP using a memetic algorithm. Int. J. Adv. Comput. Sci. Appl. **8**(8), 404–408 (2017). https://doi.org/10.14569/ijacsa.2017.080852
4. Applegate, D.L., et al.: Certification of an optimal TSP tour through 85,900 cities. Oper. Res. Lett. **37**(1), 11–15 (2009). https://doi.org/10.1016/j.orl.2008.09.006
5. Bai, Y., Zhang, W., Jin, Z.: An new self-organizing maps strategy for solving the traveling salesman problem. Chaos, Solitons & Fractals **28**(4), 1082–1089 (2006). https://doi.org/10.1016/j.chaos.2005.08.114
6. Bergmann, B., Hommel, G.: Improvements of general multiple test procedures for redundant systems of hypotheses. In: Bauer, P., Hommel, G., Sonnemann, E. (eds.) Multiple Hypothesenprüfung/Multiple Hypotheses Testing. MEDINFO, vol. 70, pp. 100–115. Springer, Heidelberg (1988). https://doi.org/10.1007/978-3-642-52307-6_8
7. Blickle, T., Thiele, L.: A comparison of selection schemes used in evolutionary algorithms. Evol. Comput. **4**(4), 361–394 (1996). https://doi.org/10.1162/evco.1996.4.4.361

8. Brocki, Ł, Koržinek, D.: Kohonen self-organizing map for the traveling salesperson problem. In: Jabłoński, R., Turkowski, M., Szewczyk, R. (eds.) Recent Advances in Mechatronics, pp. 116–119. Springer, Heidelberg (1999). https://doi.org/10.1007/978-3-540-73956-2_24

9. Dantzig, G., Fulkerson, R., Johnson, S.: Solution of a large-scale traveling-salesman problem. J. Oper. Res. Soc. Am. **2**(4), 393–410 (1954). https://doi.org/10.1287/opre.2.4.393

10. D'Urso, P., Giovanni, L.D., Massari, R.: Smoothed self-organizing map for robust clustering. Inf. Sci. **512**, 381–401 (2020). https://doi.org/10.1016/j.ins.2019.06.038

11. Faigl, J., Hollinger, G.A.: Self-organizing map for the prize-collecting traveling salesman problem. In: Villmann, T., Schleif, F.-M., Kaden, M., Lange, M. (eds.) Advances in Self-Organizing Maps and Learning Vector Quantization. AISC, vol. 295, pp. 281–291. Springer, Cham (2014). https://doi.org/10.1007/978-3-319-07695-9_27

12. Fuentes, G.E.A., Gress, E.S.H., Mora, J.C.S.T., Marín, J.M.: Solution to travelling salesman problem by clusters and a modified multi-restart iterated local search metaheuristic. PLoS ONE **13**(8), e0201868 (2018). https://doi.org/10.1371/journal.pone.0201868

13. Gendreau, M., Potvin, J.Y. (eds.): Handbook of Metaheuristics. Springer, Boston (2010). https://doi.org/10.1007/978-1-4419-1665-5

14. Grabusts, P., Musatovs, J.: The application of simulated annealing method for solving travelling salesman problem. In: Proceedings of the 4th Global Virtual Conference. Publishing Society, pp. 225–229 (2016). https://doi.org/10.18638/gv.2016.4.1.732

15. Grice, J.V., Montgomery, D.C.: Design and analysis of experiments. Technometrics **42**(2), 208 (2000). https://doi.org/10.2307/1271458

16. Guttmann-Beck, N., Knaan, E., Stern, M.: Approximation algorithms for not necessarily disjoint clustered TSP. J. Graph Algorithms Appl. **22**(4), 555–575 (2018). https://doi.org/10.7155/jgaa.00478

17. Haupt, R.L., Haupt, S.E.: Practical Genetic Algorithms. Wiley, Hoboken (2003). https://doi.org/10.1002/0471671746

18. Jaradat, A., Matalkeh, B., Diabat, W.: Solving traveling salesman problem using firefly algorithm and k-means clustering. In: 2019 IEEE Jordan International Joint Conference on Electrical Engineering and Information Technology (JEEIT). IEEE, pp. 586–589 (2019). https://doi.org/10.1109/jeeit.2019.8717463

19. Jovanovic, R., Tuba, M., Voß, S.: Fixed set search applied to the traveling salesman problem. In: Blesa Aguilera, M.J., Blum, C., Gambini Santos, H., Pinacho-Davidson, P., Godoy del Campo, J. (eds.) HM 2019. LNCS, vol. 11299, pp. 63–77. Springer, Cham (2019). https://doi.org/10.1007/978-3-030-05983-5_5

20. King, B., Barve, S., Ford, A., Jha, R.: Unsupervised clustering of COVID-19 chest X-ray images with a self-organizing feature map. In: 2020 IEEE 63rd International Midwest Symposium on Circuits and Systems (MWSCAS). IEEE (2020). https://doi.org/10.1109/mwscas48704.2020.9184493

21. Kohonen, T.: Self-organized formation of topologically correct feature maps. Biol. Cybern. **43**(1), 59–69 (1982). https://doi.org/10.1007/bf00337288

22. Laporte, G., Palekar, U.: Some applications of the clustered travelling salesman problem. J. Oper. Res. Soc. **53**(9), 972–976 (2002). https://doi.org/10.1057/palgrave.jors.2601420

23. Liu, D., Wang, X., Du, J.: A clustering-based evolutionary algorithm for traveling salesman problem. In: 2009 International Conference on Computational Intelligence and Security. IEEE, pp. 118–122 (2009). https://doi.org/10.1109/cis.2009.80

24. Matai, R., Singh, S., Lal, M.: Traveling salesman problem: an overview of applications, formulations, and solution approaches. In: Davendra, D. (ed.) Traveling Salesman Problem, Theory and Applications. InTech (2010). https://doi.org/10.5772/12909

25. Potvin, J.Y., Guertin, F.: The clustered traveling salesman problem: a genetic approach. In: Osman, I.H., Kelly, J.P. (eds.) Meta-Heuristics, pp. 619–631. Springer, Boston (1996). https://doi.org/10.1007/978-1-4613-1361-8_37

26. Rego, C., Gamboa, D., Glover, F., Osterman, C.: Traveling salesman problem heuristics: leading methods, implementations and latest advances. Eur. J. Oper. Res. **211**(3), 427–441 (2011). https://doi.org/10.1016/j.ejor.2010.09.010

27. Reinelt, G.: TSPLIB—a traveling salesman problem library. ORSA J. Comput. **3**(4), 376–384 (1991). https://doi.org/10.1287/ijoc.3.4.376

28. Schneider, J.J., Bukur, T., Krause, A.: Traveling salesman problem with clustering. J. Stat. Phys. **141**(5), 767–784 (2010). https://doi.org/10.1007/s10955-010-0080-z

29. Suhriani, I.F., Mutawalli, L., Widiami, B.R.A., Chumairoh: Implementation self organizing map for cluster flood disaster risk. In: Proceedings of the International Conference on Mathematics and Islam. SCITEPRESS - Science and Technology Publications, pp. 405–409 (2018). https://doi.org/10.5220/0008522604050409

30. Taillard, É.D., Helsgaun, K.: POPMUSIC for the travelling salesman problem. Eur. J. Oper. Res. **272**(2), 420–429 (2019). https://doi.org/10.1016/j.ejor.2018.06.039

31. Yu, S., Yang, M., Wei, L., Hu, J.S., Tseng, H.W., Meen, T.H.: Combination of self-organizing map and k-means methods of clustering for online games marketing. Sens. Mater. **32**(8), 2801 (2020). https://doi.org/10.18494/sam.2020.2800

32. Yuanyuan, L., Jing, Z.: An application of ant colony optimization algorithm in TSP. In: 2012 Fifth International Conference on Intelligent Networks and Intelligent Systems. IEEE, pp. 61–64 (2012). https://doi.org/10.1109/icinis.2012.20

How to Trust Generative Probabilistic Models for Time-Series Data?

Nico Piatkowski[1]([✉]), Peter N. Posch[2], and Miguel Krause[2]

[1] ME Group, Fraunhofer IAIS, 53757 Sankt Augustin, Germany
`nico.piatkowski@iais.fraunhofer.de`
[2] Finance Group, TU Dortmund, 44227 Dortmund, Germany
`{peter.posch,miguel.krause}@tu-dortmund.de`

Abstract. Generative machine learning methods deliver unprecedented quality in the fields of computer vision and natural language processing. When comparing models for these task, the user can fast and reliably judge generated data with her bare eye—for humans, it is easy to decide whether an image or a paragraph of text is realistic. However, generative models for time series data from natural or social processes are largely unexplored, partially due to a lack of reliable and practical quality measures. In this work, measures for the evaluation of generative models for time series data are studied—in total, over 1000 models are trained and analyzed. The well-established maximum mean discrepancy (MMD) and our novel proposal: the Hausdorff discrepancy (HD) are considered for quantifying the disagreement between the sample distribution of each generated data set and the ground truth data. While MMD relies on the distance between mean-vectors in an implicit high-dimensional feature space, the proposed HD relies on intuitive and explainable geometric properties of a "typical" sample. Both discrepancies are instantiated for three underlying distance measures, namely Euclidean, dynamic time warping, and Frechét distance. The discrepancies are applied to evaluate samples from generative adversarial networks, variational autoencoders, and Markov random fields. Experiments on real-world energy prices and humidity measurements suggest, that considering a single score is insufficient for judging the quality of a generative model.

Keywords: Generative models · Time series · Deep learning · Hausdorff discrepancy · MMD

1 Introduction

Most natural and social processes are not static—they generate data over time, exhibit non-linear interdependencies between non-consecutive measurements, and gradually change their fundamental dynamics. The underlying data generating mechanisms are often largely unknown, not well understood, or not fully observable at all. In these cases, the observed dynamics can be characterized only

© Springer Nature Switzerland AG 2021
D. E. Simos et al. (Eds.): LION 2021, LNCS 12931, pp. 283–298, 2021.
https://doi.org/10.1007/978-3-030-92121-7_23

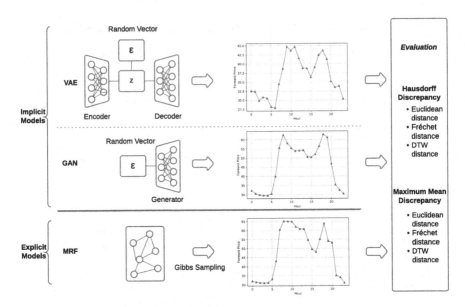

Fig. 1. Overview on the evaluation pipeline considered in this work. Time-series data is sampled from a probabilistic model and then evaluated via various measures.

indirectly, using intuitive-but-oversimplified abstractions or hard-to-understand-but-powerful non-linear blackbox models. Here, we aim at mimicking such processes, generating new data whose distribution shall be indistinguishable from the real-world measurements. Thus, allowing those methods to be used as drop-in-samplers for Monte Carlo techniques and in all situations for which samples from the underlying data distribution are required.

In this paper, we benchmark generative adversarial networks (GAN), variational autoencoders (VAE), and Markov random fields (MRF), as depicted in Fig. 1. Synthesizing new data from any of these generators we can easily simulate distributions which then can then be used in the former mentioned applications. In order to pick the best generator we assess statistical properties and measures of the artificially generated data and discuss their usefulness.

Generative models for time-series have indeed been of interest for the data science community. An overview of existing work is provided in Table 1. Here, we provide a short summary: VAE-related work on time-series data focuses mainly on data imputation [9] and anomaly detection [28], there are many studies on GANs generating synthetic time series. In particular, Esteban et al. [7] train and evaluate GANs on real-valued medical time series and measure their generation capability via Maximum Mean Discrepancy and their "Train on Synthetic, Test on Real" (TSTR) measure. Using the synthetic image generation capability of GANs Brophy et al. [3] convert EEG time series in greyscale images and use a GAN to generate synthetic greyscale images and map them back to time series. Zhang et al. [30] use GANs for smart grid data, simulating synthetic energy generation and consumption over day and night time. Chen et al. [4] use

Table 1. Overview on related work about generative models for time-series data. The column OPT indicated if a hyper-parameter optimization was conducted.

Paper	Models	Quality measure	OPT
This work	GAN, VAE, MRF	MMD, Hausdorff discrepancy	Yes
Esteban et al. [7]	RGAN, RCGAN	MMD, TSTR	No
Brophy et al. [3]	WGAN	MMD, FID	No
Zhang et al. [30]	CGAN	MMD, DTW k-means clustering	No
Wiese et al. [27]	Quant GANs	Wasserstein-1 distance, DY metric, Dependency scores	Yes
Chen et al. [4]	Wasserstein GAN, CGAN	Mode diversity, statistical resemblance	No
Yoon et al. [29]	TimeGAN	Visualization, Discriminative Score, Predictive Score (MAE)	No
Arlitt et al. [2]	HMM	Auto-correlation, marginal/conditional distributions	No
Piatkowski et al. [21]	STRF, MRF	Likelihood, NNZ ratio	No

a Wasserstein GAN and a Conditional GAN to generate realistic wind and solar power scenarios for renewable power plants and verify the statistical properties of the scenarios via mode diversity and statistical resemblance. Focusing on the generation of financial time series Wiese et al. [27] introduce Quant GANs and evaluate them via distributional metrics and dependency scores. Additionally Takahashi et al. [25] show GANs can reproduce various stylized facts of financial time series. Yoon et al. [29] introduce a time-series generation approach that combines GANs with autoregressive models. The resulting TimeGAN should preserve the temporal dynamics of time series in the generation process, in a manner that new sequences respect the original relationships between variables across time. Referring to sequence models, Kaiser and Bengio [13] present discrete autoencoding sequence models and show that sampling from the discrete latent code allows to get valid but highly diverse samples for language and translation tasks. In reinforcement learning, Metz et al. [19] show that discretizing action spaces and using sequence models can yield improvements. Referring to Markov models, Arlitt et al. [2] use a Hidden Markov model (HMM) to generate synthetic energy consumption time series data. Piatkowski et al. [21] use discrete probabilistic graphical models, like Markov random fields to predict spatio-temporal sensor states for traffic and temperature time series. Markov random fields are furthermore applied to evaluate the spatial-temporal dependency among adjacent pixels in time series of images [8].

Our contributions can be summarized as follows:

- We provide the first domain agnostic study of generative models for time-series data.

- We propose and investigate a new measure for comparing samples from two distributions based on the Hausdorff distance. In contrast to established distance metrics like the maximum mean discrepancy, our novel Hausdorff discrepancy takes the shape of the generated sample into account which gives is the perspective of assessing various statistical properties at once, including tails and multi modality.
- Our empirical comparison of three generative models and six distance measures reveals that classic explicit probabilistic models like Markov random fields still deliver the best performance in terms of reproduction of the empirical moments of the underlying distribution.

2   Generative Probabilistic Models

Probabilistic models capture the rich structures in complex data sets. These models can be divided in basically two major categories: implicit and explicit models. Explicit models give us explicit access to a likelihood function $\mathbb{P}(X)$. This allows us to compute the probability of observing specific events, analytic computation of all marginal probabilities, computing the mode, as well as sampling from the underlying distribution. Implicit models are likelihood-free and only allow us to draw samples form the estimated distribution, but cannot be used to get, e.g., pointwise evaluations of the distribution itself [20]. We do now discuss the implicit and explicit models considered in this work.

Generative Adversarial Network (GAN). The framework of generative adversarial networks [10] is a recent method for data generation. In a nutshell, two models are trained simultaneously: The generator $G : \mathcal{Z} \rightarrow \mathcal{X}$ gets as input a random vector Z from a measure $\mathbb{Q}$ over $\mathcal{Z}$, e.g. multivariate Gaussian, and shall output a new point from the data domain $\mathcal{X}$. The discriminator $D : \mathcal{X} \rightarrow [0;1]$ estimates the probability that a data point comes from the original data distribution $\mathbb{P}$. Training can be interpreted as if G and D play a two-player minimax game against each other to determine their parameters with value function $V(G;D)$.

$$\min_{G} \max_{D} V(G;D) = \mathbb{E}_{\mathbb{P}}[\log D(X)] + \mathbb{E}_{\mathbb{Q}}[\log(1 - D(G(Z)))]$$

The training procedure for G consists in maximizing the probability of D making a failure. Simultaneously, D has to maximizes the probability of assigning the correct class (generated or real data) to training examples and synthetic records from G. The random input of G comes, e.g. from a multivariate Gaussian prior $\mathbb{Q}$ with unit covariance matrix. To ensure the estimation of D near its optimal solution we alternate in an numerical implementation approach of the game between steps optimizing of D and one step of optimizing G. After several steps of training, if G and D have enough capacity, the mini game reaches a Nash equilibrium, which corresponds to the $G(Z)$ being drawn from the data domain $\mathcal{X}$. To generate new independent samples we map random input vectors from the prior $\mathbb{Q}$ with the trained generator to the data space $\mathcal{X}$ [4].

Variational Auto-Encoder (VAE). A Variational autoencoder [17,22] is a generative latent variable model which connects ideas from deep learning and Bayesian inference. Given the observable variables X and latent variables Z from a prior $\mathbb{P}(z)$, the VAE learns simultaneously a generative model $\mathbb{P}_\theta(x|z)$, parameterized by θ, and an inference model $\mathbb{Q}_\phi(z|x)$, parameterized by ϕ. The inference model, which approximates it's true posterior $\mathbb{P}_\theta(z|x) \propto \mathbb{P}(z)\mathbb{P}_\theta(x|z)$, is called encoder and encodes the data into parameters of the posterior distribution. The generative model, called decoder, learns to reconstruct samples from the posterior with some plausible surrogate. To make sampling easy, we parameterize the posterior distribution following Kingma and Welling [17] by a Gaussian with its mean and variance predicted by the encoder. Typically, encoder and decoder are parameterized by multi-layer neural networks.

To infer the marginal likelihood $\log \mathbb{P}_\theta(x)$ and train the VAE we maximize the evidence lower bound (ELBO)

$$\log \mathbb{P}_\theta(x) \geq L(\phi, \theta; x) = \mathbb{E}_{\mathbb{Q}(z|x)}[\log \mathbb{P}_\theta(x, z) - \log \mathbb{Q}_\phi(z|x)]$$
$$= -D_{\mathrm{KL}}(\mathbb{Q}_\phi(z|x)||\mathbb{P}_\theta(z)) + \mathbb{E}_{\mathbb{Q}_\phi(z|x)}[\log \mathbb{P}_\theta(x|z)]$$

The ELBO consists of the KL-divergence D_{KL}, which acts as a regularizer, encouraging the posterior $\mathbb{Q}_\phi(z|x)$ to be close to the prior $\mathbb{P}(z)$ and an expected reconstruction error [22]. As prior over the latent variables Z we choose a isotropic normal Gaussian $\mathbb{P} = \mathcal{N}(0, I)$, which is commonly used [26]. If $\mathbb{P}_\theta(x|z)$ and $\mathbb{Q}_\phi(z|x)$ are differentiable and can be computed pointwise, we can maximize the ELBO via gradient descent algorithms [18]. During the training of the VAE we use the reparametrization trick [17] to sample from posterior $\mathbb{Q}_\phi(z|x)$ to ensure that we can train the encoder and decoder as one network via backpropagation. The trained decoder can reconstruct a sensible data sample from every point in the latent space with non-zero probability under the prior and allows for the generation of new samples. Precisely, to sample from the posterior $\mathbb{Q}_\theta(z|x)$ we generate a random point z of the learned latent variable distribution by sampling $\epsilon \sim \mathcal{N}(0, I)$ and using the reparameterization trick $z = \mu_x + \Sigma_x * \epsilon$, where μ_x, Σ_x are the mean and the variance parameters of the approximate posterior, predicted by the trained encoder. Then we decode z with the trained decoder to generate new realistic time series samples [26].

Markov Random Field (MRF). Markov random fields are a class of undirected probabilistic graphical models that have been applied successfully for the generative modelling of various types of spatio-temporal data [21]. In an MRF, each time point is treated as a random variable whose dependence to the other variables is estimated explicitly via a structure learning algorithm. The resulting structure tells us which time points have the most impact on which other time-points, resulting in the conditional independence structure $G = (V, E)$. Any arbitrary time-series x of length-T can then be assigned a probability via

$$\mathbb{P}(x) = \frac{1}{Z(\theta)} \prod_{(v,u) \in E} \exp(\langle \theta, \phi_{(v,u)}(x) \rangle)$$

where $\boldsymbol{\theta} \in \mathbb{R}^d$ are the learnable model parameters and $Z(\boldsymbol{\theta})$ is the partition function.

It has been shown that Markov random fields with large cliques are equivalent to deep stochastic neural networks, so-called deep Boltzmann machines [20]. In case of discrete data, the parameters $\boldsymbol{\theta}$ of an MRF can be estimated consistently, i.e., we are asymptotically guaranteed to recover the *true* underlying probability mass function. Nevertheless, discretizing numeric data implies a subtle loss of accuracy.

3 Discrepancy

Given a ground truth sample $\mathcal{D}$ from the underlying data distribution, our main goal is to tell whether a data set $\mathcal{T}$ that was generated by a generative model is close to the distribution that underlies $\mathcal{D}$. This is of course the well-known two-sample problem, which is still an open problem in general. In the context of this paper, we are satisfied with measuring *similarity* between samples—we will not provide a definite statistical answer. However, there are indeed well known approaches for this task.

Maximum Mean Discrepancy. To test whether two samples are drawn from the same distribution, Gretton et al. [11] proposed the maximum mean discrepancy (MMD), which maps the distance between the embedding of the two samples into a reproducing kernel Hilbert space (RKHS) and determines the distance between samples in that space.

Definition 1 (Maximum Mean Discrepancy [11]). *Let $\mathcal{F}_k$ be a RKHS defined on a topological space $\mathcal{X}$ with reproducing kernel k, and $\mathbb{P}, \mathbb{Q}$ two probability measures on $\mathcal{X}$. The mean embedding of $\mathbb{P}$ in $\mathcal{F}_k$ is $\mu_k(\mathbb{P}) \in \mathcal{F}_k$ such that $\mathbb{E}_{\mathbb{P}} f(X) = \langle f, \mu_k(\mathbb{P}) \rangle_{\mathcal{F}_k}$ for all $f \in \mathcal{F}_k$. Let further $\mathcal{D}_{\mathbb{P}} = \{x_1, \ldots, x_m\}$ and $\mathcal{D}_{\mathbb{Q}} = \{y_1, \ldots, y_n\}$ denote sets of observations, sampled from $\mathbb{P}$ and $\mathbb{Q}$, respectively. The MMD and its unbiased estimate $\tilde{\mathrm{MMD}}$ are defined via*

$$\mathrm{MMD}_{\mathcal{F}_k}(\mathbb{P}, \mathbb{Q}) = \sup_{f \in \mathcal{F}_k} ||\mathbb{E}_{\mathbb{P}} f(X) - \mathbb{E}_{\mathbb{Q}} f(X)||^2_{F_k},$$

$$\tilde{\mathrm{MMD}}_{\mathcal{F}_k}(\mathbb{P}, \mathbb{Q}) = \frac{1}{|\mathcal{D}_{\mathbb{P}}|^2} \sum_{x \in \mathcal{D}_{\mathbb{P}}} \sum_{y \in \mathcal{D}_{\mathbb{P}}} k(x, y) - \frac{2}{|\mathcal{D}_{\mathbb{P}}||\mathcal{D}_{\mathbb{Q}}|} \sum_{x \in \mathcal{D}_{\mathbb{P}}} \sum_{y \in \mathcal{D}_{\mathbb{Q}}} k(x, y)$$

$$+ \frac{1}{|\mathcal{D}_{\mathbb{Q}}|^2} \sum_{x \in \mathcal{D}_{\mathbb{Q}}} \sum_{y \in \mathcal{D}_{\mathbb{Q}}} k(x, y).$$

Using a characteristic kernel, e.g. the radial basis function (RBF) kernel, MMD is a metric, which implies that $\mathrm{MMD}(\mathbb{P}, \mathbb{Q}) = 0$ if and only if $\mathbb{P} = \mathbb{Q}$ [24]. Here, we consider kernels of the form $k(x, y) = \exp(-\gamma d(x, y))$ where d is some distance function, i.e., when d is the Euclidean distance, we recover the RBF kernel.

As explained in [24], the RBF kernel is indeed characteristic. However, it is also well known that the plan euclidean distance which underlies the RBF kernel is outperformed by other measures when it comes to the comparison of time series. Thus, we will consider specialized distances for time-series data to compute the MMD, namely dynamic time warping and the Fréchet distance. While these perform well in practice, it is unknown if the induced kernel is characteristic. In this case, MMD loses its theoretically certified properties and just gives us a notion of similarity.

The Hausdorff Discrepancy. A well-known measure for the dissimilarity between two shapes is given by the Hausdorff distance. Given two compact figures in a metric space, the Hausdorff distance measures how far both shapes are from being isometric.

Definition 2 (Hausdorff distance [6]). *Let A and B be two non-empty subsets of a metric space (M, d). The Hausdorff distance between both sets is given by*

$$d_H(A, B) = \max \left\{ \sup_{x \in A} \inf_{y \in B} d(x, y), \sup_{x \in B} \inf_{y \in A} d(x, y) \right\}$$

The above distance measure has various applications in computer vision and graphics related areas. However, in data science and machine learning, the Hausdorff distance is unused—up to now. Based on $d_H(A, B)$, we define the following discrepancy between two samples from two distributions:

Definition 3 (Hausdorff Discrepancy). *Let X, Y be random variables both having state space $\mathcal{X}$ and following distributions $\mathbb{P}, \mathbb{Q}$, respectively. Let further $\mathcal{D} = \{x_1, \ldots, x_N\}, \mathcal{E} = \{y_1, \ldots, y_N\}$ be two (random) data sets of size N, containing independent samples from $\mathbb{P}$ and $\mathbb{Q}$, respectively. We define the Hausdorff discrepancy between $\mathbb{P}, \mathbb{Q}$ as*

$$\mathrm{HD}(\mathbb{P}, \mathbb{Q}) = \max \left\{ \mathbb{E}_{\mathcal{E}} \left[\max_{x \in \mathcal{D}} \min_{y \in \mathcal{E}} d(x, y) \mid \mathcal{D} \right], \mathbb{E}_{\mathcal{E}} \left[\max_{x \in \mathcal{E}} \min_{y \in \mathcal{D}} d(x, y) \mid \mathcal{D} \right] \right\}$$

Note, that max and min replace sup and inf because we assume finite data sets of known size N. The intuition is that the Hausdorff discrepancy compared the shape of the generated sample—if all data points are lying at similar positions, one could assume that the distributions which generated those sets are similar.

3.1 Distance Measures on Time-Series Data

Both, MMD and HD require another core distance which is able to measure distance between single data points. Here, we consider dynamic time warping and the Fréchet distances for that purpose.

Dynamic Time Warping. To determine the similarity between two sequences, Sankoff and Kruskal [23] introduced the dynamic time warping (DTW) to overcome the limitation of the Euclidean distance to local and global shifts in the time dimension. Given two time series X, Y of length $|X|, |Y|$ we construct a warp path

$$W = w_1, \ldots, w_K, \qquad \max(|X|, |Y|) \le k \le |X| + |Y| \qquad (1)$$

with K the length of the warp path. The kth element is defined as $w_k = (i, j)$, where i is the index from time series X and j is the index of Y. This warp path starts at $w_1 = (1, 1)$ and ends with $w_K = (|X|, |Y|)$ and can be constructed as

To find the optimal warp path we are interested in the minimum-distance warping path.

This path can be found by a dynamic programming approach [15]. We will consider the following DTW kernel for two time series X, Y:

$$k_{dtw}(x, y) = \exp(-\gamma \, \mathrm{DTW}(X, Y)). \qquad (2)$$

Fréchet Distance. The Fréchet distance is a measure of similarity of two given curves with the advantage that it takes into account the order of the points along the curves. The distance can be illustrated by a man who is walking his dog. The man walks on one curve, the dog on the other curve. Both are not allowed to go backwards but are allowed to control their speed. The Fréchet distance is than the minimal length of a leash between the man and his dog. The Fréchet distance, given two curves in $A : [a, a'] \mapsto V, B : [b, b'] \mapsto V$ in an Euclidean vector space V is defined as

$$\delta_F(A, B) = \inf_{\alpha, \beta} \max_{t \in [0,1]} d(A(\alpha(t)), B(\beta(t))) \qquad (3)$$

with a distance function d and where $\alpha : [0, 1] \mapsto [a, a'], \beta : [0, 1] \mapsto [b, b']$ range over continuous, increasing functions [1].

Similar to the DTW kernel we define the Fréchet kernel for two time series X, Y as

$$k_F(X, Y) = \exp(-\gamma \delta_F(X, Y)). \qquad (4)$$

4   Empirical Evaluation

We will now conduct a comparative study, involving all models and distance measures described above. In order to understand how good the measures work for each method, we perform a visual inspection of the synthetic data points as known from the image domain [14]. In addition, we compute the first four empirical moments. Let us now describe our experimental setting in more detail. For reproducability, all data sets and all code are available online[1].

---

[1] https://github.com/mk406/GenerativeProbabilisticModels.git.

Table 2. Hyperparameter search space for implicit probabilistic models.

Parameter	Values
Iterations	10000, 30000
Batch size	64, 128
Hidden layers units	128, 256
Hidden layer	2, 3
Droprate	0.3, 0.4
Learning rate G	0.0002, 0.0004
Learning rate D	0.002, 0.004
Discriminator iterations	2, 3

(a) GAN hyperparameters and grid values.

Parameter	Values
Epochs	500, 1000
Batch size	64, 128
Latent dimension	4, 6
Hidden layer	1, 2, 3, 4
Hidden layers units	48, 128, 256
Learning rate	0.001, 0.002, 0.004

(b) VAE hyperparameters and grid values.

4.1   Hyper-parameter Search

While artificial neural networks are universal function approximators *in general*, this is not necessary true for any single neural network. The ability to approximate a target function heavily depends on the architecture of the underlying network. Thus, we apply an extensive grid search over various hyperparameters and the network architecture to find the best model specification for the GANs and VAEs. The hyperparameters for our GAN models consist in the number of iterations of the minimax game, the batch size, the number of units of each layer of the generator and the discriminator, the number of hidden layers, the drop rate of the regularization layer of the discriminator, the number of times we train alternatively between steps of optimizing the discriminator and one optimizing step of the generator and the learning rates of the generator and discriminator. For the learning rates of the discriminator and the generator we consider the two time-scale update rule (TTUR) [12], which was shown empirically to converge to a stationary local Nash equilibrium by choosing different learning rates for generator and discriminator. Similar to Chollet [5] we choose the Leaky-ReLU activation as activation functions for the generator and the discriminator to induce sparse gradients, expect for the output layer where we use a linear activation for the generator and a sigmoid activation for the discriminator. To induce robustness we use a dropout layer in the discriminator to increase stochasticity and prevent the GAN to get stuck during the training process [5]. Both, generator and discriminator are trained using RMSProp. All hyperparameters of the generator and discriminator model are listed in Table 2a. Inspired by progressive growing [14] we choose an increasing number of units for each following hidden layer for the generator and for the discriminator vice-versa a decreasing number of units for each hidden layer. The number of units in each hidden layer are multiples of the number units hyperparameter. As layer type we choose fully connected layers for all hidden layers of the model. The hyperparameters of our VAE models consist in the numbers of epochs, batch sizes, numbers of hidden layers, numbers of units, learning rates and the numbers of units in the output layer of the encoder (input layer of the decoder), that build up the dimension of the latent space. The VAE model is trained as one neural network using the Adam optimizer [16]. For the grid search, we apply the following structure: Our encoder consists of an input layer, followed by hidden layers with an decreasing number of units and an output layer. The decoder is constructed the same way vice-versa. The number of units in each layer are multiples of the number units hyperparameter. As layer type, we choose fully connected layers for all hidden layers of the model. As activation functions we choose ReLU activation's except for the output layer of the encoder and the decoder for which we choose linear activation's. Finally, for the MRF, we apply a cth-order variant of the Chow-Liu Algorithm to estimate the underlying conditional independence structure between all points in time. The choice of c determines how many variables can interact simultaneously. This value controls how complex the functional relation between values at different time-points can be. E.g., a third order model ($c = 3$) can learn XOR-relations between values while a second order model can not. We evaluated models with $c \in \{2, 3, 4\}$. Moreover, the numeric time-series data is discretized into k bins which are defined by each time points k-quantiles. We consider $k \in \{64, 128\}$. New time-series are synthesized by performing Gibbs sampling from $\hat{\mathbb{P}}$ with 100 burn-in samples. The discrete time-series are finally de-discretized by replacing each discrete state by the mean value of its corresponding interval. Since the loss function (the negative log-likelihood) of the MRF is convex, there is no need to tune optimization related parameters like learning rate. The hyperparameter search-space is hence much smaller compared to the GAN or the VAE.

4.2 Data

Two real-world data sets are considered as ground truth. The first data set contains hourly day-ahead electricity prices of the German power market from SMARD starting from 01 January 2014 up to 30 June 2020. All 24 prices of each hour of a day are simultaneously delivered in an daily auction at the European Power Exchange (EEX). The distribution of these prices over a day exhibits different levels of autocorrelations as well as typical stylized facts, such as a morning ramp-up when productions starts and a (often less pronounced) evening ramp as people come home. The data set contains $N = 2373$ data points of length $T = 24$. The second data set contains humidity readings collected from 54 sensors deployed in the Intel Berkeley Research lab between 28 February 2004 and 05 April 2004. As the sensors measure humidity as per-second data we resample the data to hourly data for each sensor using the mean. In order to eliminate possible unrealistic outliers and defect sensors we only keep relative humidity values between 0% and 100% and removed incomplete days. The data set contains $N = 1033$ data points of length $T = 24$. All GANs, VAEs, and MRFs (in total 550 per data set) are then used to sample 10 new data sets, each containing the same number of N data points as the corresponding ground truth data set.

Table 3. Results for energy prices data set with the Maximum Mean Discrepancy (MMD) and the respective core distances. E = EUCLID, D = DTW, F = FRECHET.

	Model	MMD-E	MMD-D	MMD-F
GAN	(64, 3, 0.4, 3, 256, 10000, 0.002, 0.0002)	**0.002146**	**0.002053**	0.002232
	(64, 3, 0.4, 3, 128, 10000, 0.002, 0.0002)	0.004019	0.002237	**0.002162**
	(64, 2, 0.4, 2, 256, 30000, 0.002, 0.0002)	0.009049	0.008315	0.007140
	(64, 3, 0.3, 3, 128, 30000, 0.002, 0.0002)	0.035339	0.031543	0.020427
VAE	(1000, 6, 128, 3, 256, 0.002)	0.008521	0.007223	0.005334
	(1000, 4, 128, 4, 48, 0.002)	0.009519	0.007385	0.004971
	(1000, 6, 128, 1, 128, 0.001)	0.113087	0.091368	0.057671
	(500, 6, 128, 4, 128, 0.001)	0.245581	0.185878	0.120332
MRF	(Y128 C4)	0.019986	0.107307	0.180475
	(Y64 C2)	0.020482	0.106376	0.186436
	(Y128 C3)	0.020044	0.107198	0.180200
	(Y64 C3)	0.020556	0.108981	0.187261

Table 4. Results for energy prices data set with the Hausdorff Discrepancy (HD) and the respective core distances. E = EUCLID, D = DTW, F = FRECHET.

	Model	HD-E	HD-D	HD-F
GAN	(64, 3, 0.4, 3, 256, 10000, 0.002, 0.0002)	470.848890	205.778159	74.794100
	(64, 3, 0.4, 3, 128, 10000, 0.002, 0.0002)	475.632360	224.281684	84.998204
	(64, 2, 0.4, 2, 256, 30000, 0.002, 0.0002)	**431.590117**	214.077600	91.473012
	(64, 3, 0.3, 3, 128, 30000, 0.002, 0.0002)	451.570817	179.705832	71.061972
VAE	(1000, 6, 128, 3, 256, 0.002)	464.533760	217.399043	82.775642
	(1000, 4, 128, 4, 48, 0.002)	479.927298	171.587436	67.878177
	(1000, 6, 128, 1, 128, 0.001)	437.526727	241.730902	98.208595
	(500, 6, 128, 4, 128, 0.001)	480.036966	**136.806716**	**51.370512**
MRF	(Y128 C4)	464.181700	168.314610	74.080645
	(Y64 C2)	463.950905	173.128946	75.745309
	(Y128 C3)	463.976769	157.697151	76.692775
	(Y64 C3)	457.425072	181.018805	67.802015

To evaluate a model we compare the 10 data sets with the original data via MMD and HD measures and calculate the mean over all 10 sets.

4.3 Results

It can be seen in Tables 3, 4, 5 and 6, that both discrepancies, MMD and HD, are in favor of the neural model. However, a visual inspection of the generated series reveals, that the results of the GANs an VAEs exhibit less extreme paths

Table 5. Results for intel humidity data set with the Maximum Mean Discrepancy (MMD) and the respective core distances. E = EUCLID, D = DTW, F = FRECHET.

	Model	MMD-E	MMD-D	MMD-F
GAN	(128, 3, 0.4, 2, 256, 30000, 0.002, 0.0004)	0.005880	0.008524	0.005098
	(128, 2, 0.3, 2, 128, 10000, 0.002, 0.0004)	0.007726	0.007497	0.005851
	(128, 3, 0.4, 2, 256, 10000, 0.002, 0.0004)	0.006307	0.007784	**0.004057**
	(128, 3, 0.4, 3, 128, 10000, 0.004, 0.0004)	0.183665	0.141075	0.096112
	(64, 3, 0.4, 3, 128, 10000, 0.002, 0.0002)	0.147775	0.108733	0.065924
VAE	(1000, 4, 128, 3, 256, 0.001)	**0.004476**	0.005662	0.006548
	(1000, 4, 64, 4, 48, 0.004)	0.006513	**0.004333**	0.004907
	(1000, 6, 64, 4, 256, 0.001)	0.041662	0.027782	0.020678
	(1000, 4, 64, 4, 256, 0.001)	0.034065	0.022247	0.016933
VAE	(Y128 C4)	0.020789	0.093830	0.157937
	(Y128 C2)	0.020886	0.093393	0.158031
	(Y64 C2)	0.021723	0.094068	0.155414
	(Y64 C3)	0.022418	0.096205	0.160263

Table 6. Results for Intel humidity data set with the Hausdorff Discrepancy (HD) and the respective core distances. E = EUCLID, D = DTW, F = FRECHET.

	Model	HD-E	HD-D	HD-F
GAN	(128, 3, 0.4, 2, 256, 30000, 0.002, 0.0004)	138.288455	40.095867	15.722096
	(128, 2, 0.3, 2, 128, 10000, 0.002, 0.0004)	129.190451	42.390750	15.788611
	(128, 3, 0.4, 2, 256, 10000, 0.002, 0.0004)	127.806623	34.395313	14.353081
	(128, 3, 0.4, 3, 128, 10000, 0.004, 0.0004)	111.065010	33.471478	15.398804
	(64, 3, 0.4, 3, 128, 10000, 0.002, 0.0002)	126.163026	19.285610	10.191569
VAE	(1000, 4, 128, 3, 256, 0.001)	147.398243	25.087964	13.105644
	(1000, 4, 64, 4, 48, 0.004)	126.312667	20.525052	11.543398
	(1000, 6, 64, 4, 256, 0.001)	**105.658204**	20.046526	**8.230774**
	(1000, 4, 64, 4, 256, 0.001)	126.321157	**17.922368**	13.509255
MRF	Y128 C4	134.843024	67.664796	24.793203
	Y128 C2	136.950436	68.803114	25.629733
	Y64 C2	138.410524	67.167855	25.358533
	Y64 C3	137.430849	65.161503	25.299679

and are smoother than the MRF shapes. As the original shapes include various extreme paths, the MRF generator seems to generate more realistic samples. Comparing the first four moments of the synthetic samples in Tables 7 and 8 underpin this impression: the best statistics are generated by our MRF model.

Table 7. First four moments of the original energy price data and the generated data. The closer a method's values are to the values of the original data, the better.

	Model	Mean	Variance	Skew	Kurtosis
	Original energy prices	34.05556	219.87474	−0.53591	5.71138
GAN	(64, 3, 0.4, 3, 256, 10000, 0.002, 0.0002)	33.66485	160.34268	−0.38958	1.20340
	(64, 3, 0.4, 3, 128, 10000, 0.002, 0.0002)	33.80683	133.58552	−0.25212	1.38344
	(64, 2, 0.4, 2, 256, 30000, 0.002, 0.0002)	31.99768	125.25381	−0.22987	1.07421
	(64, 3, 0.3, 3, 128, 30000, 0.002, 0.0002)	28.66481	133.07252	−0.63094	2.39743
VAE	(1000, 6, 128, 3, 256, 0.002)	31.24921	174.50482	−0.16907	0.44592
	(1000, 4, 128, 4, 48, 0.002)	31.10025	230.43423	−0.29231	0.86731
	(1000, 6, 128, 1, 128, 0.001)	23.83706	156.17970	0.10782	−0.08631
	(500, 6, 128, 4, 128, 0.001)	16.43303	393.18478	0.14104	0.25116
MRF	(Y128 C4)	33.84105	**223.85882**	−0.53837	5.30267
	(Y64 C2)	34.02851	230.16112	−0.55635	5.09282
	(Y128 C3)	33.77426	224.10990	−0.50680	**5.27949**
	(Y64 C3)	**34.02733**	231.01562	−0.52250	4.92158

Table 8. First four moments of the original Intel humidity data set and the generated data. The closer a method's values are to the values of the original database, the better.

	Model	Mean	Variance	Skew	Kurtosis
	Original humidity data	40.17782	42.41655	−0.70570	1.42076
GAN	(128, 3, 0.4, 2, 256, 30000, 0.002, 0.0004)	41.16504	50.26162	0.19975	0.32837
	(128, 2, 0.3, 2, 128, 10000, 0.002, 0.0004)	39.86462	43.74342	0.05770	0.40597
	(128, 3, 0.4, 2, 256, 10000, 0.002, 0.0004)	39.86064	46.71354	0.06230	0.27192
	(128, 3, 0.4, 3, 128, 10000, 0.004, 0.0004)	35.05371	24.74748	0.01518	0.48845
	(64, 3, 0.4, 3, 128, 10000, 0.002, 0.0002)	35.34650	92.89869	0.45984	0.33655
VAE	(1000, 4, 128, 3, 256, 0.001)	39.90946	40.45797	−0.24321	1.05061
	(1000, 4, 64, 4, 48, 0.004)	39.40994	45.99996	−0.42118	0.86559
	(1000, 6, 64, 4, 256, 0.001)	37.45328	81.30920	−0.35601	0.42926
	(1000, 4, 64, 4, 256, 0.001)	38.06910	78.10967	−0.33326	0.44448
MRF	(Y128 C4)	**40.11159**	42.00935	**−0.71421**	1.39157
	(Y128 C2)	40.09451	**42.04563**	−0.73389	1.53284
	(Y64 C2)	40.08659	44.91162	−0.69460	**1.41515**
	(Y64 C3)	39.97168	44.30150	−0.66962	1.15169

5 Conclusion

We presented the first comparison of multiple methods for the generation of synthetic time-series. Three model classes, namely GANs, VAEs, and MRFs, have been considered on two real-world time-series data sets. For each model class, up to several hundred models have been trained. Six discrepancy measures have been used to select the best models from these sets. Looking at the standard maximum mean discrepancy, generative adversarial networks outperform variational autoencoders and Markov random fields on 4 out of 6 distance measures. However, when considering our new proposal, the Hausdorff discrepancy, this pictures changes and variational autoencoder win in 4 out of 6 cases. Finally a visual inspection of the samples shows that GANs and VAEs generate samples with less extreme values—all samples are close to the mean of the data set. Markov random field generates more realistic samples whose extreme values are similar to those of the original data. This observation is strengthened when we investigate the first four moments of the original data distribution and the generated samples of all methods: We find Markov Random fields outperform neural models, reproducing statistical properties much better than their popular but heuristic counterparts. Our findings suggest that one shall not trust a single quality measure when judging the quality of a generated data set. Even state-of-the-art measures can be misleading—they carry only limited information about the support of the learned probability measure and the major statistical properties of the underlying random variable.

Acknowledgments. Parts of this work have been funded by the Federal Ministry of Education and Research of Germany as part of the competence center for machine learning ML2R (01IS18038B). Parts of this work have been supported by the Deutsche Forschungsgemeinschaft via SFB 823.

References

1. Alt, H., Godau, M.: Computing the Fréchet distance between two polygonal curves. Int. J. Comput. Geom. Appl. **5**(01n02), 75–91 (1995)
2. Arlitt, M., Marwah, M., Bellala, G., Shah, A., Healey, J., Vandiver, B.: IoTAbench: an Internet of Things analytics benchmark. In: Proceedings of the 6th ACM/SPEC International Conference on Performance Engineering, pp. 133–144 (2015)
3. Brophy, E., Wang, Z., Ward, T.E.: Quick and easy time series generation with established image-based GANs. arXiv preprint arXiv:1902.05624 (2019)
4. Chen, Y., Wang, Y., Kirschen, D., Zhang, B.: Model-free renewable scenario generation using generative adversarial networks. IEEE Trans. Power Syst. **33**(3), 3265–3275 (2018)
5. Chollet, F.: Deep Learning with Python. Manning, Shelter Island (2018). Safari Tech Books Online
6. Dubuisson, M.P., Jain, A.K.: A modified Hausdorff distance for object matching. In: International Conference on Pattern Recognition, vol. 1, pp. 566–568. IEEE (1994)
7. Esteban, C., Hyland, S.L., Rätsch, G.: Real-valued (medical) time series generation with recurrent conditional GANs. CoRR abs/1706.02633 (2017)
8. Fischer, R., Piatkowski, N., Pelletier, C., Webb, G.I., Petitjean, F., Morik, K.: No cloud on the horizon: probabilistic gap filling in satellite image series. In: International Conference on Data Science and Advanced Analytics, pp. 546–555 (2020)

9. Fortuin, V., Baranchuk, D., Rätsch, G., Mandt, S.: GP-VAE: deep probabilistic time series imputation. In: International Conference on Artificial Intelligence and Statistics, pp. 1651–1661. PMLR (2020)

10. Goodfellow, I.J., et al.: Generative adversarial nets. In: Advances in Neural Information Processing Systems, vol. 27, pp. 2672–2680 (2014)

11. Gretton, A., Borgwardt, K.M., Rasch, M.J., Schölkopf, B., Smola, A.J.: A kernel method for the two-sample-problem. In: Advances in Neural Information Processing Systems, vol. 19, pp. 513–520 (2006)

12. Heusel, M., Ramsauer, H., Unterthiner, T., Nessler, B., Hochreiter, S.: GANs trained by a two time-scale update rule converge to a local nash equilibrium. In: Advances in Neural Information Processing Systems, vol. 30, pp. 6626–6637 (2017)

13. Kaiser, Ł., Bengio, S.: Discrete autoencoders for sequence models. arXiv preprint arXiv:1801.09797 (2018)

14. Karras, T., Aila, T., Laine, S., Lehtinen, J.: Progressive growing of GANs for improved quality, stability, and variation. In: International Conference on Learning Representations (2018)

15. Keogh, E.J., Pazzani, M.J.: Derivative dynamic time warping. In: Proceedings of the 2001 SIAM International Conference on Data Mining, pp. 1–11. SIAM (2001)

16. Kingma, D.P., Ba, J.: Adam: a method for stochastic optimization. arXiv preprint arXiv:1412.6980 (2014)

17. Kingma, D.P., Welling, M.: Auto-encoding variational bayes. arXiv preprint arXiv:1312.6114 (2013)

18. Kusner, M.J., Paige, B., Hernández-Lobato, J.M.: Grammar variational autoencoder. arXiv preprint arXiv:1703.01925 (2017)

19. Metz, L., Ibarz, J., Jaitly, N., Davidson, J.: Discrete sequential prediction of continuous actions for deep RL. arXiv preprint arXiv:1705.05035 (2017)

20. Piatkowski, N.: Hyper-parameter-free generative modelling with deep Boltzmann trees. In: Brefeld, U., Fromont, E., Hotho, A., Knobbe, A., Maathuis, M., Robardet, C. (eds.) ECML PKDD 2019. LNCS (LNAI), vol. 11907, pp. 415–431. Springer, Cham (2020). https://doi.org/10.1007/978-3-030-46147-8_25

21. Piatkowski, N., Lee, S., Morik, K.: Spatio-temporal random fields: compressible representation and distributed estimation. Mach. Learn. **93**(1), 115–139 (2013). https://doi.org/10.1007/s10994-013-5399-7

22. Rezende, D.J., Mohamed, S., Wierstra, D.: Stochastic backpropagation and approximate inference in deep generative models. In: Proceedings of the 31th International Conference on Machine Learning, pp. 1278–1286 (2014)

23. Sankoff, D., Kruskal, J.: The symmetric time-warping problem: from continuous to discrete. In: Time Warps, String Edits and Macromolecules: The Theory and Practice of Sequence Comparison, pp. 125–161. Addison Wesley (1983)

24. Sriperumbudur, B.K., Gretton, A., Fukumizu, K., Schölkopf, B., Lanckriet, G.R.: Hilbert space embeddings and metrics on probability measures. J. Mach. Learn. Res. **11**, 1517–1561 (2010)

25. Takahashi, S., Chen, Y., Tanaka-Ishii, K.: Modeling financial time-series with generative adversarial networks. Phys. A **527**, 121261 (2019)

26. Wan, Z., Zhang, Y., He, H.: Variational autoencoder based synthetic data generation for imbalanced learning. In: 2017 Symposium Series on Computational Intelligence (SSCI), pp. 1–7. IEEE (2017)

27. Wiese, M., Knobloch, R., Korn, R., Kretschmer, P.: Quant GANs: deep generation of financial time series. Quant. Finance **20**(9), 1419–1440 (2020)

28. Xu, H., et al.: Unsupervised anomaly detection via variational auto-encoder for seasonal KPIs in web applications. In: WWW Conference, pp. 187–196 (2018)

29. Yoon, J., Jarrett, D., van der Schaar, M.: Time-series generative adversarial networks. In: Advances in Neural Information Processing Systems, pp. 5508–5518 (2019)
30. Zhang, C., Kuppannagari, S.R., Kannan, R., Prasanna, V.K.: Generative adversarial network for synthetic time series data generation in smart grids. In: 2018 International Conference on Communications, Control, and Computing Technologies for Smart Grids (SmartGridComm), pp. 1–6. IEEE (2018)

Multi-channel Conflict-Free Square Grid Aggregation

Roman Plotnikov[1,2]([✉]) [ID] and Adil Erzin[1,2] [ID]

[1] Sobolev Institute of Mathematics, SB RAS, Novosibirsk 630090, Russia
{prv,adilerzin}@math.nsc.ru
[2] St. Petersburg State University, St. Petersburg 199034, Russia

Abstract. We consider minimizing the delay during an interference-conflict-free aggregation session in wireless sensor networks when the network elements can use various frequency channels for the data transmission. In general, this problem is known to be NP-hard. We focus on a particular case when sensors are positioned at the square grid nodes and have the same transmission ranges equal to their interference ranges. We propose an approximation algorithm with guaranteed estimates. The algorithm consists of two stages. At the first stage, the nodes transmit the data upwards or downwards until all data are collected at the nodes of the first row. At the second stage, all the nodes of the first row transmit the data to the sink. Also, we present a new ILP formulation of the problem on the arbitrary network topology and provide the experiment results. We use the GUROBI solver to get the optimal solutions on the small-size instances and compare the results yielded by the proposed algorithm with optimal solutions. We also compare the proposed algorithm with the known approach in a two-channel case and show how the number of channels affects the schedule length depending on the instances sizes.

Keywords: Multichannel aggregation · Square grid · Conflict-free scheduling · Heuristic algorithm · ILP formulation

1 Introduction

Wireless networks that use radio communications to transmit data have become widespread over the last 20 years. During the *convergecasting*, each network element sends a packet of aggregated data received from its child nodes and its data to the parent node once during the entire aggregation session. This requirement is dictated by the transmission process's excessive power consumption and entails building a spanning aggregation tree (AT) with arcs directed to the receiver called the base station (BS). The faster the aggregated data reaches

The Russian Science Foundation supports this research (Grant No. 19-71-10012. Project "Multi-agent systems development for automatic remote control of traffic flows in congested urban road networks").

© Springer Nature Switzerland AG 2021
D. E. Simos et al. (Eds.): LION 2021, LNCS 12931, pp. 299–314, 2021.
https://doi.org/10.1007/978-3-030-92121-7_24

BS, the better the schedule. TDMA scheduling divides time into equal-length slots, assuming that it is long enough to send or receive one data packet [4]. Minimizing the convergecast time, in this case, is equivalent to reducing the number of time slots required for all packets to reach the receiver [17].

If the only frequency channel is used for the data transmission then the problem involves two components: AT and a schedule that assigns a transmission time slot for each node so that each node transmits after all of its children in the tree have, and potentially interfering links scheduled to be sent at different time slots. The latter condition means that the TDMA schedule must be interference-free. That is, no receiving node is within the interference range of another transmitting node. There are two types of collisions in wireless networks: primary and secondary. A primary collision occurs when more than one node transmits simultaneously to the same destination. In tree-based aggregation, this happens when more than one child nodes of the same parent node send their packets during the same time slot. A secondary collision occurs when a node eavesdrops on the transmissions destined for another node. The links of the underlying communication graph cause this collision, but not in the aggregation tree. Using different frequencies for the data transmission (communication channels) allows to eliminate the conflicts and to reduce the conflict-free schedule length.

The conflict-free data aggregation problem is NP-hard [3] even if and AT is known and only one frequency channel is used [5]. Therefore, almost all existing results in literature are approximation polynomial algorithms when the network elements use one channel [1,3,11–14,16–18] or several channels [2,10,15]. In [1] a novel cross-layer approach is presented for reducing the latency in disseminating aggregated data to the BS over multi-frequency radio links. Their approach forms the AT to increase the simultaneity of transmissions and reduce buffering delay. Aggregation nodes are picked, and time slots are allocated to the individual sensors so that the readiest nodes can transmit their data without delay. The use of different radio channels allows avoiding colliding transmissions. Their approach is validated through simulation and outperforms previously published schemes.

In [12], the authors investigate the question: "How fast can information be collected from a wireless sensor network organized as a tree?" To address this, they explore and evaluate several different techniques using realistic simulation models under the many-to-one communication paradigm known as convergecasting. First, min-time scheduling on a single frequency channel is considered. Next, they combine scheduling with transmission power control to mitigate the effects of interference and show that while power control reduces the schedule length under a single frequency, scheduling transmissions using multiple frequencies is more efficient. The authors gave lower bounds on the schedule length without interference conflicts and proposed algorithms that achieve these bounds. They also evaluate the performance of various channel assignment methods and find empirically that for moderate-sized networks of about 100 nodes, multi-frequency scheduling can suffice to eliminate most of the interference. The data collection rate no longer remains limited by interference but by the routing tree's topology.

To this end, they construct degree-constrained spanning trees and capacitated minimal spanning trees and show significant improvement in scheduling performance over different deployment densities. Lastly, they evaluate the impact of varying interference and channel models on the schedule length.

In [10], the authors focus on designing a multi-channel min-latency aggregation scheduling (MC-MLAS) protocol, using a new joint approach for tree construction, channel assignment, and transmission scheduling. This paper combines orthogonal channels and partially overlapping channels to consider the total latency involved in data aggregation. Extensive simulations verify the superiority of MC-MLAS in WSNs.

In [15], the authors consider a problem of min-length scheduling for the conflict-free convergecasting in a wireless network in a case when each element of a network uses its frequency channel. This problem is equivalent to the well-known NP-hard problem of telephone broadcasting since only the conflicts between the same parent children are considered. They propose a new integer programming formulation and compare it with the known one by running the CPLEX software package. Based on numerical experiment results, they concluded that their formulation is preferable to solve the considered problem by CPLEX than the known one. The authors also propose a novel approximate algorithm based on a genetic algorithm and a local search metaheuristic. The simulation results demonstrate the high quality of the proposed algorithm compared to the best-known approaches. However, if the network has a regular structure, it is a lattice, it is solved in polynomial time. Known that in a square lattice, in each node of which information is located, the process of single-channel data aggregation is simple [9]. Moreover, in some cases, for example, when the transmission range is 1 [9] or 2 [6], one can build an optimal schedule. If the transmission range is greater than 2, then one can find a solution close to the optimal [5,7].

In [8] the two-channel min-latency aggregation scheduling on a square grid is considered. A polynomial approximation algorithm is proposed for solving this problem, and it is shown that in most cases the proposed algorithm yields a better solution than convergecasting using only one channel.

1.1 Our Contribution

In this paper, the MC-MLAS problem with a given number of channels in a square lattice is considered. We propose a new heuristic approximation algorithm and provide the precise value of the length of a schedule constructed by the algorithm depending on the instance parameters. Also, for the first time, we propose an ILP formulation of the MC-LAS. We present the simulation results where GUROBI solved the considered problem on several instances and compared our heuristic results. We compared the new algorithm with the previous one proposed for the case of two frequency channels.

The rest of the paper is organized as follows. The problem formulation is given in Sect. 2, the heuristic algorithm is described in Sect. 3, ILP formulation is presented in Sect. 4, simulation results and a posteriori analysis are provided in Sect. 5, and Sect. 6 concludes the paper.

2 Problem Formulation

We suppose that the network elements are positioned at the nodes of a $(n + 1) \times (m + 1)$ square grid. For convenience, we call the network elements the sensors, vertices, or nodes equivalently. A sink node (or BS) is located at the origin $(0,0)$. During each time slot, any sensor except the sink node can either be idle, send the data, or receive the data. We assume that each sensor has the same transmission distance $d \geq 2$ in L_1 metric. A sink node can only receive the data at any time slot. Each data transmission is performed using one of k available frequency channels (for short, further, they are referred to as channels), and each sensor can use any channel for data transmission and receiving. Besides, we suppose that the following conditions met:

- each vertex sends a message only once during the aggregation session (except the sink, which always can only receive messages);
- once a vertex sends a message, it can no longer be a destination of any transmission;
- if some vertex sends a data packet by the channel c, then during the same time slot, none of the other vertices within a receiver's interference range can use the channel c;
- a vertex cannot receive and transmit at the same time slot.

The problem consists of constructing the conflict-free min-length schedule of the data aggregation from all the vertices to the BS.

We assume that the interference range equals the transmission range. Such assumption is usually used in different WSN models. However, in many real applications, the interference range exceeds the transmission range. The proposed algorithm can be adapted to the problem with arbitrary interference range by splitting simultaneous transmissions into several conflict-free transmissions performed during different time slots.

3 Heuristic Algorithm

We call the *row* a set of sensors positioned at the grid nodes with the same ordinates. This section describes the heuristic algorithm for the approximate solution to the considered problem and performs its theoretical analysis. The algorithm consists of two stages. At the first stage, which is called a *vertical aggregation*, the nodes transmit the data upwards or downwards. By the end of this stage, all the data are sent to the nodes of the 0-th row. At the second stage, *horizontal aggregation*, all the nodes of the 0-th row transmit the data to the sink.

3.1   Vertical Aggregation

The network consists of $m + 1$ rows. Assume that $m = Md + r_v$, where $r_v \in \{0, \ldots, d - 1\}$. Let's call the distance between these rows the difference of ordinates between their elements.

Lemma 1. *If the distance between two rows equals $r \leq d$, then all elements from one row can send the data to the nodes of another row during $\lfloor (d - r)/k \rfloor + 1$ time slots.*

Proof. Note that any two vertices of the same row can simultaneously send the data vertically at a distance r only if they use different channels or the distance between them is not less than $d - r + 1$. We split all the nodes in a row into $\lceil n/(d - r + 1) \rceil$ groups in such way that each of the first $\lfloor n/(d - r + 1) \rfloor$ groups consists of $d - r + 1$ vertices that are positioned sequentially. Similarly, split each group into $\lfloor (d - r)/k \rfloor + 1$ subgroups so that each of the first $\lfloor (d - r)/k \rfloor$ subgroups of the same group consists of k vertices that are positioned sequentially. Note that for each $i \in \{1, \ldots, \lfloor (d - r)/k \rfloor + 1\}$ all vertices in the i-th subgroup of all groups can send the data vertically simultaneously without conflicts if they use the channels in the same order. Therefore, all the nodes can send the data in $\lfloor (d - r)/k \rfloor + 1$ time slots.

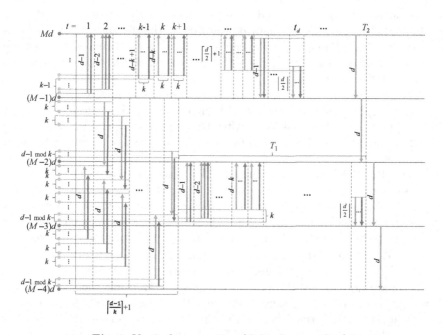

Fig. 1. Vertical aggregation (Color figure online)

Let us describe the vertical aggregation. For convenience purposes, color the rows in two colors in the following way. For each $i \in \{0, \ldots, m\}$, i-th row is colored in green if i is a multiple of d, and it is colored in blue otherwise. At first, we consider the case when $r_v = 0$, so the most remote row from the origin is green, and $M \geq 4$, so the number of rows is at least $4d + 1$. We focus on the last $4d + 1$ rows and assume that each other blue row (maybe except the first $d - 1$ blue rows) repeats the similar data transmissions as the row that lies above distance $3d$. The transmissions of the last $4d + 1$ rows are schematically presented in Fig. 1.

During the first time slots, only the blue rows transmit the data avoiding the conflicts. Each green row transmits downwards at the distance d only when all nodes above it have already transmitted their data. During the first T_1 time slots (the exact expression of this value as function of k and d is given below), the rows $(M - 1)d + 1, \ldots, Md - 1$ transmit their data to the nodes of the green rows $(M - 1)d$ and Md. Besides, during the first $\lceil (d - 1)/k \rceil + 1$ time slots all blue rows from $(M - 2)d + 1$ to $(M - 1)d - 1$ and all blue rows from $(M - 4)d + 1$ to $(M - 3)d - 1$ transmit in a following order. At the first time slot, the rows $(M - 3)d - k, \ldots, (M - 3)d - 1$ transmit simultaneously the data upwards at the distance d using different channels. At the second time slot, the rows from $(M - 3)d - 2k$ to $(M - 3)d - k - 1$ simultaneously transmit their data upwards at the distance d and the rows from $(M - 1)d - k$ to $(M - 1)d - 1$ simultaneously transmit their data downwards at the distance d. Similarly, at the third time slot, the rows from $(M - 3)d - 3k$ to $(M - 3)d - 2k - 1$ simultaneously transmit their data upwards at the distance d and the rows from $(M - 1)d - 2k$ to $(M - 1)d - k - 1$ simultaneously transmit their data downwards at the distance d. This process proceeds in a similar manner until the $(\lceil (d - 1)/k \rceil + 1)$-th time slot, when the last $d - 1 \pmod{k}$ rows (from $(M - 2)d + 1$ to $(M - 2)d + (d - 1(\bmod\ k)))$ transmit the data downwards.

After the first $\lceil (d - 1)/k \rceil + 1$ time slots, all the rows in a set $\{(M - 2)d + 1, \ldots, (M - 1)d - 1\}$ have transmitted their data as well as the rows in a set $\{(M - 4)d + 1, \ldots, (M - 3)d - 1\}$. After this, the rows from $(M - 3)d + 1$ to $(M - 2)d - 1$ transmit their data to the neighboring green rows during the next T_1 time slots.

Let us describe how $d - 1$ sequential blue rows transmit their data to the neighboring two green rows and calculate the total time of these transmissions T_1. Obviously, the longer the transmission distance, the less time a row requires for conflict-free transmission. Therefore, the rows located below the central row transmit their data upwards, and the rows located above it transmit their data downwards. According to the Lemma 1, $T_1 = 2 \sum_{r=1}^{\lceil d/2 \rceil - 1} (\lfloor r/k \rfloor + 1) + x_d(\lfloor d/(2k) \rfloor + 1)$ where $x_d = 1$ if d is even and $x_d = 0$ otherwise.

The blue rows $(M - 3)d + 1, \ldots, (M - 2)d - 1$ transmit their data during $T_2 = T_1 + \lceil (d - 1)/k \rceil + 1$ time slots. Note that if $M - 1$ is a multiple of 3, then during the first T_2 time slots all the blue rows may transmit their data if at each time slot each node below the $(M - 4)d$-th row performs the same action as the node that lies above it at the distance $3d$. In this case, after T_2 time slots only $M - 2$ time slots are required for the sequential top down transmissions by the green nodes at the distance d, and the total vertical transmission takes

$T_3 = T_2 + M - 2$ time slots. This is also the case if $M - 1 \equiv 2 \pmod 3$, because in this case the rows from 1 to $d - 1$ may perform the same actions as the rows $(M - 3)d + 1$ to $(M - 2)d - 1$ since they don't transmit below the $(M - 3)d$-th row. If $M - 1 \equiv 1 \pmod 3$, vertical aggregation may be performed in T_3 time slots if each node from 1 to $d - 1$ performs the same action as the row that lies above it at the distance $2d$. That is, during the first $\lceil (d - 1)k \rceil + 1$ time slots the nodes in these rows don't transmit, and they perform the similar vertical transmission to the neighboring green nodes at the rows 0 and d during the next T_1 time slots as the nodes in the rows $(M - 3)d + 1, \ldots, (M - 2)d - 1$.

The above reasoning is valid for the case when $M \geq 4$. Let us consider the other variants.

- $M = 1$. In this case, T_1 time slots are enough for sending the data from the blue nodes, and one more time slot is required for sending the data from d-th row to 0-th row.
- $M = 2$. In this case, the rows $1, \ldots, d - 1$ cannot transmit downwards at the distance d. The entire aggregation cannot be performed as a reduced variant of the general approach. For that reason, at first, the rows $d + 1, \ldots, 2d - 1$ transmit downwards at the distance d during the first $\lceil (d - 1)/k \rceil$ time slots, and after that during T_1 time slots the rows $1, \ldots, d - 1$ transmit to their neighboring green rows. These transmissions are the same as the transmissions made by the rows $(M - 3)d + 1, \ldots, (M - 1)d - 1$ in the general approach, and the last $2d$-th row transmits the data to the d-th row in $(T2 - 1)$-th time slot. In total, T_1 time slots are required for the vertical aggregation.
- $M = 3$. In this case, all the transmissions are performed exactly in the same way as in a general approach, and the vertical aggregation needs $T_2 + 1$ time slots.

The only case that is left to considering is one when $r_v > 0$. Except three subcases that will be mentioned below, in this case, during the first $\lceil r_v/k \rceil$ time slots the most remote r_v rows transmit downwards at the distance d while all the rows below the $(m - r_v - d)$-th row perform the same operations as in general algorithm. After that, during another T_1 time slots, the rows $(M - 1)d + 1, \ldots, Md - 1$ transmit the data to the rows $(M - 1)d$ and Md. If $\lceil r_v/k \rceil < \lceil (d-1)/k \rceil$ then the total schedule length does not increase, otherwise it increases by 1. We may conclude that $L^v_{Md+r_v} = L^v_{Md} + \lfloor \lceil r_v/k \rceil / \lceil (d - 1)/k \rceil \rfloor$, where L^v_m is the length of vertical aggregation schedule of $m + 1$ rows.

There exist three exceptions that should be considered separately. The first one is when $M = 0$. In this case, at first, the rows $1, \ldots, r_v - 1$ transmit to the rows 0 and r_v, and then, the row r_v transmits to the row 0. The total vertical aggregation takes $2 \sum_{r=1}^{\lceil r_v/2 \rceil - 1} (\lfloor (d - r_v + r)/k \rfloor + 1) + x_{r_v} (\lfloor (d - rv/2)/k \rfloor + 1) + \lfloor (d - rv)/k \rfloor + 1$ time slots, where $x_{r_v} = 1$ if r_v is even and $x_{r_v} = 0$, otherwise.

The second case is when $M = 2$. In this case, at first $\lceil r_v/k \rceil + 1$ time slots the highest r_v rows transmit at the distance d simultaneously with upward transmissions at the same distance by the rows $1, \ldots, d - 1$ in a similar way as in the general algorithm the rows $(M - 2)d + 1, \ldots, (M - 1)d - 1$ transmit the data downwards simultaneously with upward transmissions of the rows

$(M-4)d+1, \ldots, (M-3)d-1$ during the first $\lceil (d-1)/k \rceil +1$ time slots. In this case, $L^v_{Md+r_v} = L^v_{Md} + \lfloor \lceil r_v/k \rceil / \lceil (d-1)/k \rceil \rfloor + 1$.

And, finally, the third case is when $M = 1$. In this case, the schedule length is increased by $\lceil r_v/k \rceil$ time slots by the downward transmissions of the highest r_v rows at a distance d during the first time slots.

Consequently, the following lemma is true.

Lemma 2. *The vertical aggregation is performed in time $O(d^2/k + M)$.*

In a case when the number of channels is 2, and m is big, this algorithm constructs a significantly shorter schedule of vertical aggregation than the algorithm presented in [8], where the length of vertical aggregation schedule is $O(m)$.

3.2   Horizontal Aggregation

After the vertical aggregation, all the data are collected in the 0-th row. From this moment, it is required to aggregate the data to the sink node in a minimum time using only horizontal transmissions. Let us color each d-th node in green, starting from the most remote, and color other nodes, except the sink, in blue. After the node transmitted its data, it changes the color to gray.

There is no need to use more than $\lfloor d/2 \rfloor$ channels for the horizontal aggregation. In this subsection, we assume that only $k' = \min(k, \lfloor d/2 \rfloor)$ channels are used.

The main algorithm steps are illustrated in Fig. 2. During any time slot, each blue node (except the last $3d$ and, maybe, the first $n \pmod d$ nodes) repeats the node's action that lies at the distance $3d$ to the right until all blue nodes transmit. Therefore, we focus on $3d$ nodes that are the farthest from the sink. Enumerate the nodes to the right from 0 to n starting from the sink. Then for any $j \in \{0, \ldots, \lfloor n/d \rfloor\}$, the node $n - jd$ is green.

During the first $\lceil d/k' \rceil + \lceil \log_2 k' \rceil - 1$ time slots the nodes $n-3d+1, \ldots, n-2d-1$ stay idle while the next blue nodes to the right transmits in the order described below.

Let us divide the last d nodes in $\lceil d/k' \rceil$ groups and enumerate them from right to left. The last group contains $d \pmod{k'}$ elements, and each other group contains k' nodes. For each $i \in \{1, \ldots, \lceil d/k' \rceil - 1\}$, at i-th time slot, k' nodes of the $(i+1)$-th group transmit the data to the most right k' nodes using separate channels, and at $(\lceil d/k' \rceil - 1)$-th time slot the nodes of the last group transmit the data to the most right $d \pmod{k'}$ nodes. During the same time slots, the nodes $\{n-2d+1, \ldots, n-d-1\}$ perform the same actions but symmetrically about the $(n-d)$-th node. That is, at the first time slot, the nodes $n-d-k', \ldots, n-d-1$ transmit the data to the nodes $n-2d, \ldots, n-2d+k'-1$ using k' channels, at the next time slot the next to the left group of k' transmits the data to the same recipients, and so on. Notice that during the first $\lceil d/k' \rceil - 2$ time slots, any conflicts are excluded because the distance between the nodes that transmit the data to the left and the most remote $d \pmod{k'}$ nodes (which receive messages from the left) always exceeds d. Similarly, the distance between the

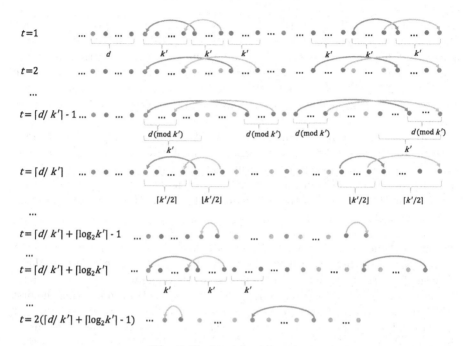

Fig. 2. Horizontal aggregation

nodes transmitting the data to the right and the nodes receiving messages from the right always exceeds d. But this is not the case for the transmissions of the $(\lceil d/k' \rceil - 1)$-th time slot when, for example, the nodes $n - d \pmod{k'} - 1, \ldots, n-1$ hear not only the nodes that transmit the data to them, but also the nodes $n - d - d \pmod{k'} - 1, \ldots, n - d - 1$ that transmit to the left. The similar situation holds for the nodes $n - 2d + 1, \ldots, n - 2d + d \pmod{k'} - 1$, that hear not only the nodes that transmit to them, but also the nodes $n - d + 1, \ldots, n - 2d + d \pmod{k'} - 1$, that transmit to the right. It is not hard to see that these transmissions can be performed without conflicts if the order of channels (from left to right) for the data transmissions to the left will be the same as the order of channels for the data transmissions to the right.

After this, the nodes $n - k', \ldots, n - 1$ transmit to the most right node in the following order. At the first time slot, the nodes $n - k', \ldots, n - \lceil k'/2 \rceil$ transmit to the right using different channels, and then similarly, at each time slot, half of the rest, not gray nodes (or the largest integer that is less than a half in a case when the number of remaining not gray nodes is odd) transmits to the right using different channels. $\lceil \log_2 k' \rceil$ time slots are enough for these transmissions. During the same time interval, the nodes $n - 2d + 1, \ldots, n - 2d + k'$ perform transmissions to the left using the different channels, which also differ from the channels used by the nodes in a group $\{n - k', \ldots, n - 1\}$. It is possible because at each time slot, not more than $\lfloor k'/2 \rfloor$ nodes of a group transmit the data. After the described transmissions, that take $T_4 = \lceil d/k' \rceil + \lceil \log_2 k' \rceil - 1$ time slots, each node of the most right $2d$ nodes is either gray or green, since each blue node of the most right $2d$ nodes has transmitted its data. But all of the nodes $\{n - 3d + 1, \ldots, n - 2d - 1\}$ are blue. During the next T_4 time slots, the nodes $\{n - 3d + 1, \ldots, n - 2d - 1\}$ perform exactly the same transmissions as the nodes $\{n - 2d + 1, \ldots, n - d - 1\}$ did starting from the first time slot. In total, after $2T_4$ time slots, all blue nodes in the most remote group of $3d$ nodes have transmitted their data.

Note that the two most right green nodes, the n-th and the $(n - d)$-th, may transmit to the $(n - 2d)$-th node during the time interval $[T_4 + 1, 2T_4]$. Although there is some freedom for choosing time slots for these transmissions. To be specific, we assign the transmission of the n-th node to the $(T4 + 1)$-th time slot and the transmission of the $(n - d)$-th node to the $2T_4$-th time slot, as it is shown in Fig. 2.

Let us denote $n = Nd + r_h$, where $N \geq 3$, $r_h \in \{0, \ldots, d - 1\}$. We assume that at any time slot each node form the r_h-th to the $(n - 3d - 1)$-th repeats the same action as the node that is located at the distance $3d$ to the right of it. Then after the first $2T_4$ time slots, all blue nodes in this range will transmit the data. It is not hard to see that the nodes $1, \ldots, r_h - 1$ can also transmit during the first $2T_4$ time slots. Indeed, since $r_h < d$, $r_h - 1$ blue nodes in a row may transmit to two neighbor nodes, 0-th and r_h-th, before $d - 1$ blue nodes in a row, using the same algorithm, i.e., in less than T_4 time slots. If N is a multiple of 3 then during the first T_4 time slots all nodes from $r_h + 1$ to $r_h + d - 1$ stay idle, and therefore the nodes $\{1, \ldots, r_h - 1\}$ may transmit during the first T_4 time slots without conflicts. If $N \equiv 1$ or 2 (mod 3) then the nodes $\{r_h + 1, \ldots, r_h + d - 1\}$ transmit during the first T_4 time slots which allows performing a conflict-free transmission by the nodes $\{1, \ldots, r_h - 1\}$ during the next T_4 time slots.

In total, all blue nodes transmit the data during $2T_4$ time slots and the most remote two green nodes. After that, another $\lceil n/d \rceil - 2$ time slots are required for sequential data transmissions of the green nodes to the sink. And, according to the explanations given above, the following lemma is proved.

Lemma 3. *If $N \geq 3$ then the length of horizontal aggregation schedule is $L_N^h = 2(\lceil d/k' \rceil + \lceil \log_2 k' \rceil) + \lceil n/d \rceil - 4 = O(d/k + \log k + N)$.*

Remark 1. Due to the lack of space we give the following horizontal aggregation schedule length for the remaining cases without explanations (however, the strategy remains the same, and the reader is able to obtain them by oneself):

- $L_0^h = \lceil r_h/k'' \rceil + \lceil \log_2 k'' \rceil$
- $L_1^h = \max(0, s_{r_h}(\lceil r_h/k'' \rceil + \lceil (d - r_h)/k' \rceil - \lceil d/k' \rceil)) + \lceil d/k' \rceil + \lceil \log_2 k' \rceil + s_{r_h}$
- $L_2^h = \lceil d/k' \rceil + \lceil \log_2 k' \rceil + \max(1, s_{r_h}(\lceil r_h/k'' \rceil - 1)) + \max(1, s_{r_h}\lceil \log_2 k'' \rceil) + s_{r_h} - 1$. Here $s_{r_h} = 1$ if $r_h > 0$ and $s_{r_h} = 0$ otherwise, and $k'' = \min(k, \lfloor r_h/2 \rfloor)$.

4   ILP Formulation

This section presents an ILP formulation of the MC-MLAS problem for solving the problem using GUROBI or CPLEX to estimate the quality of approximation algorithms. To the best of our knowledge, this is the first ILP formulation proposed for this problem.

Let n be the number of nodes in the network given as a graph $G = (V, E), V = \{v_1, v_2, \ldots, v_n\}$, where v_1 is a sink. Consider a directed graph $\boldsymbol{G} = (V, \cup \{v_0\}, A)$ which contains two oppositely directed arcs for each edge in G and additional vertex v_0 and arc (v_1, v_0). We aim to construct a min-time schedule of aggregating the data in the vertex v_0. The scheduling consists of the rooted in v_0 aggregation tree generation and assignment to each vertex a time slot and a channel to transmit the data to its recipient.

Let us denote the variables. For each $i = 0, \ldots, n$, let u_i be the number of arcs in a path in AT from v_i to v_0. For each $i, j \in \{0, \ldots, n\}$, let x_{ij} be 1 if there exists an arc (v_i, v_j) in AT and $x_{ij} = 0$, otherwise. For each $i = 0, \ldots, n$ and $t = 1, \ldots, n$, let s_{it} be 1 if v_i sends the data during t-th time slot and $s_{it} = 0$, otherwise. Let r_{it} be 1 if v_i receives the data in t-th time slot and $r_{it} = 0$, otherwise. For each $i = 0, \ldots, n$ and $p = 1, \ldots, k$, let z_{ip} be 1 if v_i sends the data using p-th channel and $z_{ip} = 0$, otherwise. And let w_{ip} be 1 if v_i receives the data using p-th channel and $w_{ip} = 0$, otherwise.

Let us introduce the parameters of the ILP model. For each $i, j = 1, \ldots, n$, let a_{ij} be 1 if v_j is within the interference range of v_i (to avoid a conflict, the time slot and the channel of v_i cannot be the same as the time slot and the channel of the vertex v_j), and $a_{ij} = 0$, otherwise.

Then, the ILP is the following.

$$\sum_{t=1}^{n} tr_{0t} \to \min; \tag{1}$$

$$\sum_{t=1}^{n} s_{it} = 1, i = 1, \ldots, n; \tag{2}$$

$$\sum_{j=0}^{n} x_{ij} = 1, i = 1, \ldots, n; \tag{3}$$

$$1 - (n+1)(1 - x_{ij}) \le u_i - u_j \le 1 + (n+1)(1 - x_{ij}) \tag{4}$$

$$s_{it} + x_{it} - r_{jt} \le 1, (i, j) \in A, t = 1, \ldots, n; \tag{5}$$

$$s_{it} + r_{it} \le 1, i = 0, \ldots, n, t = 1, \ldots, n; \tag{6}$$

$$\sum_{t'=t+1}^{n} r_{it'} + n s_{it} \le n, i = 0, \ldots, n, t = 1, \ldots, n; \tag{7}$$

$$x_{ij} + x_{lj} + s_{lt} \le 3, (i, j), (l, j) \in A, t = 1, \ldots, n; \tag{8}$$

$$\sum_{p=0}^{k} w_{ip} = 1, i = 0, \ldots, n; \tag{9}$$

$$\sum_{p=0}^{k} z_{ip} = 1, i = 0, \ldots, n; \tag{10}$$

$$w_{jp} + x_{ij} - z_{ip} \le 1, i, j = 0, \ldots, n, p = 1, \ldots, k; \tag{11}$$

$$s_{jt} + r_{it} + a_{ij} + w_{ip} + z_{jp} - x_{ji} \le 4, i, j, t = 1, \ldots, n, p = 1, \ldots, k. \tag{12}$$

In this formulation, the objective function (1) is the number of the time slots to aggregate the data in v_0. Since there is one arc ending in v_0, there is one transmission to this vertex, and $\sum_{t=1}^{n} tr_{0t}$ equals a value of the time slot of this last transmission. Constraints (2)–(3) mean that each vertex, except v_0, sends a message only once and has the only addressee. Constraints (4) require the aggregation of data via tree arcs from the leaves to the root. Constraints (5) link sending and receiving time slots: if vertex v_i sends a message to vertex v_j at time slot t then node v_j receives a message at the same time slot. Constraints (6) forbid simultaneous data transmission and receiving for each vertex. Constraints (7) state that a vertex cannot receive messages after it performed a transmission. Constraints (8) eliminate the primary conflicts: a simultaneous receiving of the data by one vertex from two different senders is forbidden. Constraints (9)–(10) define the only channel for sending and the only channel for receiving for every vertex. Constraints (11) link channels for every sender and receiver: if vertex v_i sends a message to vertex v_j, then the used by v_i channel should be the same as the receiving channel of v_j. Finally, constraints (12) provide elimination of the secondary conflicts: if vertex v_j belongs to the interference range of vertex v_i then v_i cannot use the same time slot and channel for data receiving as v_j uses for sending its message to some third vertex.

5   Simulation

We implemented the proposed algorithm using the programming language C++. We also constructed the ILP model described in Sect. 4 and used GUROBI software for its solving. Unfortunately, the model appeared to be rather large, and in time less than 3 h, GUROBI yields a guaranteed optimal solution only for small-size instances ($n, m \le 5; d, k \le 3$) in our computer x64 Intel Core i3-9100 CPU 8GB RAM. In other cases, we keep the best approximate solution found by GUROBI in 3 h. For the comparison, we also implemented the algorithm for two channels scheduling proposed in [8], which we call A_{old}. The algorithm presented in this paper we call A_{new}.

The results of the simulation for small-size instances are presented in Table 1. Both algorithms, A_{old} and A_{new}, yield optimal or near-optimal solutions. In the table, LB_1 is the lower bound that equals $\lceil \log_2(n+1)(m+1) \rceil$ (at each time slot, the number of transmitting nodes cannot exceed a half of the nodes that

Table 1. Small sizes: comparison of GUROBI, A_{old}, and A_{new}

n	m	d	k	LB_1	LB_2	G obj	A_{old}	A_{new}	G time
2	2	2	2	4	4	**4**	**4**	**4**	2.1 s
2	3	2	2	4	4	**4**	5	5	2.9 s
3	3	2	2	4	5	**5**	6	6	33.7 s
4	3	2	2	5	5	**5**	6	6	142.4 s
4	4	2	2	5	6	**6**	**6**	**6**	1200 s
5	4	2	2	5	7	**7**	**7**	**7**	486 s
5	5	2	2	6	5	7	8	9	3 h
3	3	3	2	4	5	**5**	6	6	4069 s
4	3	3	2	5	2	6	7	7	3 h
4	4	3	2	5	0	6	8	8	3 h
5	4	3	2	5	0	7	8	8	3 h
5	5	3	2	6	0	8	8	8	3 h
3	3	3	3	4	4	5	-	6	3 h
4	3	3	3	5	4	**5**	-	7	3 h
4	4	3	3	5	3	6	-	8	3 h
5	4	3	3	5	0	6	-	8	3 h
5	5	3	3	6	0	7	-	8	3 h

Table 2. Moderate sizes: comparison of A_{old} and A_{new}

$n \times m$	d	A_{old}	A_{new}					
		2 ch	2 ch	3 ch	4 ch	5 ch	7 ch	10 ch
5 × 5	3	8	8	8	8	8	8	8
	4	11	11	10	10	10	10	10
7 × 7	3	10	11	11	11	11	11	11
	4	12	13	11	11	11	11	11
	5	14	14	12	12	12	12	12
10 × 10	3	12	12	12	12	12	12	12
	4	14	13	12	12	12	12	12
	5	15	14	12	11	11	11	11
	7	21	20	18	16	16	16	16
25 × 25	3	22	22	22	22	22	22	22
	4	24	20	19	19	19	19	19
	5	24	21	19	18	18	18	18
	7	29	25	23	20	20	20	20
	10	40	35	28	25	22	21	21
50 × 50	3	38	38	38	38	38	38	38
	4	39	32	31	31	31	31	31
	5	39	31	29	28	28	28	28
	7	44	33	30	28	28	28	28
	10	52	41	34	30	27	26	25

did not communicate yet). LB_2 is the lower bound found by GUROBI. Columns "G obj" and "G time" show the objective values and running times of GUROBI. Optimums (which coincide with any of the lower bounds) are marked in bold.

In Tables 2 and 3, we present the scheduling lengths obtained by A_{old} and A_{new} for different instances. In both tables, the results obtained on the instances with different numbers of channels are given in separate columns: "2 ch", "3 ch", "4 ch", "5 ch", "7 ch", and "10 ch". Table 2 shows the results on moderate-size instances. It is seen that starting from size 10×10, the algorithm A_{new} on two channels outperforms A_{old}, and its advantage grows up with the increase of size and number of channels. However, it is worth mentioning that there exist a couple of cases when A_{old} yields the result that is slightly (by 1 time slot, to be precise) better than some of the results obtained by A_{new}: in a case $n, m = 7, d = 3$ and in a case $n, m = 7, d = 3$.

The big-size instances are presented in Table 3. Here, the gap between the results of A_{old} and A_{new} in the case of two channels significantly grows up with the increase in transmission distance and size. Also, the usage of more channels allows decreasing the schedule length significantly. This effect is also enhanced with the increase in size and transmission distance. It is also worth mention-

Table 3. Large sizes: comparison of A_{old} and A_{new}

$n \times m$	d	A_{old}	A_{new}					
		2 ch	2 ch	3 ch	4 ch	5 ch	7 ch	10 ch
100×100	3	72	72	72	72	72	72	72
	4	69	57	55	55	55	55	55
	5	69	51	49	48	48	48	48
	7	72	47	44	42	42	42	42
	10	80	51	44	40	37	36	35
	50	426	402	279	217	179	138	107
500×500	3	338	338	338	338	338	338	338
	4	319	257	255	255	255	255	255
	5	309	211	209	208	208	208	208
	7	301	161	158	156	156	156	156
	10	300	131	124	120	117	116	115
	50	630	441	311	245	205	162	129
750×750	3	504	504	504	504	504	504	504
	4	476	382	381	381	381	381	381
	5	459	311	309	308	308	308	308
	7	444	233	230	228	228	228	228
	10	437	181	174	170	167	166	165
	50	757	451	321	255	215	172	139
1000×1000	3	672	672	672	672	672	672	672
	4	632	507	505	505	505	505	505
	5	609	411	409	408	408	408	408
	7	586	304	301	299	299	298	298
	10	575	231	224	220	217	216	215
	50	885	461	331	265	225	182	149

ing that in the big-size case, the schedule length yielded by A_{new} significantly decreases with the growth of d. That happens because of the linear dependence of the schedule length on m/d and n/d.

6 Conclusion

This paper considers the problem of multi-channel conflict-free data aggregation in a square grid and proposes a new efficient approximation algorithm. A posteriori analysis confirmed that, in general, the proposed algorithm yields a significantly better solution than the previously known heuristic that can be applied for the case when only two frequencies are used. And the advantage of the new approach substantially increases with the growth of the problem size and the number of frequency channels. We also propose the first ILP formulation for the general case of the considered problem (i.e., without any restrictions on network topology) and solved several small instances using the GUROBI solver applied to the new formulation. In all these test cases, our heuristic appeared to be optimal or near-optimal.

References

1. Bagaa, M., et al.: Data aggregation scheduling algorithms in wireless sensor networks: solutions and challenges. IEE Commun. Surv. Tutor. **16**(3), 1339–1368 (2014)
2. Bagaa, M., Younis, M. and Badache, N.: Efficient data aggregation scheduling in wireless sensor networks with multi-channel links. In: MSWiM 2013, 3–8 November 2013, Barcelona, Spain (2014). https://doi.org/10.1145/2507924.2507995
3. Chen, X., Hu, X., Zhu, J.: Minimum data aggregation time problem in wireless sensor networks. In: Jia, X., Wu, J., He, Y. (eds.) MSN 2005. LNCS, vol. 3794, pp. 133–142. Springer, Heidelberg (2005). https://doi.org/10.1007/11599463_14
4. Demirkol, I., Ersoy, C., Alagoz, F.: MAC protocols for wireless sensor networks: a survey. IEEE Commun. Mag **44**, 115–121 (2006)
5. Erzin, A., Pyatkin, A.: Convergecast scheduling problem in case of given aggregation tree. The complexity status and some special cases. In: 10-th International Symposium on Communication Systems, Networks and Digital Signal Processing, Article 16, 6 p. IEEE-Xplore, Prague (2016)
6. Erzin, A.: Solution of the convergecast scheduling problem on a square unit grid when the transmission range is 2. In: Battiti, R., Kvasov, D.E., Sergeyev, Y.D. (eds.) LION 2017. LNCS, vol. 10556, pp. 50–63. Springer, Cham (2017). https://doi.org/10.1007/978-3-319-69404-7_4
7. Erzin, A., Plotnikov, R.: Conflict-free data aggregation on a square grid when transmission distance is not less than 3. In: Fernández Anta, A., Jurdzinski, T., Mosteiro, M.A., Zhang, Y. (eds.) ALGOSENSORS 2017. LNCS, vol. 10718, pp. 141–154. Springer, Cham (2017). https://doi.org/10.1007/978-3-319-72751-6_11
8. Erzin, A., Plotnikov, R.: Two-channel conflict-free square grid aggregation. In: Kotsireas, I.S., Pardalos, P.M. (eds.) LION 2020. LNCS, vol. 12096, pp. 168–183. Springer, Cham (2020). https://doi.org/10.1007/978-3-030-53552-0_18

9. Gagnon, J., Narayanan, L.: Minimum latency aggregation scheduling in wireless sensor networks. In: Gao, J., Efrat, A., Fekete, S.P., Zhang, Y. (eds.) ALGOSEN-SORS 2014. LNCS, vol. 8847, pp. 152–168. Springer, Heidelberg (2015). https://doi.org/10.1007/978-3-662-46018-4_10

10. Ghods, F., et al.: MC-MLAS: Multi-channel minimum latency aggregation scheduling in wireless sensor networks. Comput. Netw. **57**, 3812–3825 (2013)

11. Guo, L., Li, Y., Cai, Z.: Minimum-latency aggregation scheduling in wireless sensor network. J. Combinat. Optim. **31**(1), 279–310 (2014). https://doi.org/10.1007/s10878-014-9748-7

12. Incel, O.D., Ghosh, A., Krishnamachari, B., Chintalapudi, K.: Fast data collection in tree-based wireless sensor networks. IEEE Trans. Mob. Comput. **11**(1), 86–99 (2012)

13. Li, J., et al.: Approximate holistic aggregation in wireless sensor networks. ACM Trans. Sensor Netw. **13**(2), 1–24 (2017)

14. Malhotra, B., Nikolaidis, I., Nascimento, M.A.: Aggregation convergecast scheduling in wireless sensor networks. Wirel. Netw. **17**, 319–335 (2011)

15. Plotnikov, R., Erzin, A., Zalyubovskiy, V.: Convergecast with unbounded number of channels. MATEC Web Conf. **125**, 03001 (2017). https://doi.org/10.1051/matecconf/20171250

16. de Souza, E., Nikolaidis, I.: An exploration of aggregation convergecast scheduling. Ad Hoc Netw. **11**, 2391–2407 (2013)

17. Wan, P.-J. et al.: Minimum-latency aggregation scheduling in multihop wireless networks. In Proceedings of the ACM MOBIHOC, May 2009, pp. 185–194 (2009)

18. Wang, P., He, Y., Huang, L.: Near optimal scheduling of data aggregation in wireless sensor networks. Ad Hoc Netw. **11**, 1287–1296 (2013)

Optimal Sensor Placement by Distribution Based Multiobjective Evolutionary Optimization

Andrea Ponti[1], Antonio Candelieri[2(✉)], and Francesco Archetti[1]

[1] Department of Computer Science, Systems and Communication,
University of Milano-Bicocca, Milan, Italy
a.ponti5@campus.unimib.it, francesco.archetti@unimib.it
[2] Department of Economics, Management and Statistics,
University of Milano-Bicocca, Milan, Italy
antonio.candelieri@unimib.it

Abstract. The optimal sensor placement problem arises in many contexts for the identification of optimal "sensing spots", within a network, for monitoring the spread of "effects" triggered by "events". There are usually different and conflicting objectives as cost, time and reliability of the detection In this paper sensor placement (SP) (i.e., location of sensors at some nodes) for the early detection of contaminants in water distribution networks (WDNs) will be used as a running example. The best trade-off between the objectives can be defined in terms of Pareto optimality.

The evaluation of the objective functions requires the execution of a simulation model: to organize the simulation results in a computationally efficient way we propose a data structure collecting simulation outcomes for every SP which is particularly suitable for visualization of the dynamics of contaminant concentration and evolutionary optimization.

In this paper we model the sensor placement problem as a multi objective optimization problem with boolean decision variables and propose a Multi Objective Evolutionary Algorithm (MOEA) for approximating and analyzing the Pareto set.

The key element is the definition of information spaces, in which a candidate placement can be represented as a matrix or, in probabilistic terms as a histogram.

The introduction of a distance between histograms, namely the Wasserstein (WST) distance, enables to derive new genetic operators: the new algorithm MOEA/WST has been tested on a benchmark problem and a real world network. The computational metrics used are hypervolume and coverage: their values are compared with NSGA-II's in terms of the number of generations.

The experiments offer evidence of a good relative performance of MOEA/WST in particular for relatively large networks and low generation counts.

The computational analysis has been limited in this paper to WDNs but the key aspect of the method, that is the representation of feasible solutions as histograms, is suitable for problems as the detection of "fake-news" on the web where the problem is to identify a small set of blogs which catch as many cascades as early as possible and more generally Multi-objective simulation optimization problems which are also amenable to the probabilistic representation of feasible solutions as histograms.

© Springer Nature Switzerland AG 2021
D. E. Simos et al. (Eds.): LION 2021, LNCS 12931, pp. 315–332, 2021.
https://doi.org/10.1007/978-3-030-92121-7_25

Keywords: Sensor placement · Water network · Multi objective optimization ·
Evolutionary optimization · Wasserstein distance

1 Introduction

the identification of optimal "sensing spots", within a network, for monitoring the spread of "effects" triggered by "events" is referred to as the "Optimal Sensor Placement" problem and has been receiving substantial attention in the literature. Many real-world problems fit into this general framework. A wide ranging analysis of methodological as well application issues is given in [1, 2] focused respectively on sensor networks and spatial interaction models.

As an example of another problem fitting naturally in this framework one could consider the detection of "fake-news" on the web: posts are published by bloggers, are hyperlinked to other bloggers' posts and contents, and then cascade through the web. Posts are time stamped so we can observe their propagation in the blogosphere: in this setting the placement problem is to identify a small set of blogs which catch as many cascades as early as possible. In this paper sensor placement (SP) (i.e., location of sensors at some nodes) for the early detection of contaminants in water distribution networks (WDNs) will be used as a running example.

The evaluation of the objective functions requires the execution of a simulation model: to organize the simulation results in a computationally efficient way we propose a data structure collecting simulation outcomes for every SP which is particularly suitable for visualization of the dynamics of contaminant concentration and evolutionary optimization. This data structure enables the definition of information spaces, in which a candidate placement can be characterized as a matrix or, in probabilistic terms as a histogram.

In this paper we model the sensor placement problem as a multi objective optimization problem with boolean decision variables and propose a Multi Objective Evolutionary Algorithm (MOEA) for approximating and analyzing the Pareto set.

However, searching for an optimal SP is NP-hard: even for mid-size networks, efficient heuristics are required, among which evolutionary approaches are often used.

Evolutionary algorithms (EAs) have a long history in multi objective optimization, with many successful applications, also for sensor placement in WDNs.

Most of the evolutionary approaches do not use models of the objective functions and therefore cannot make predictions about elements of the search space not yet evaluated: as a consequence, a large number of function evaluations can be required.

A solution proposed in the literature to mitigate this problem is the development of problem specific operators [3–5].

Another line of research, associates with EAs probabilistic models given by Gaussian model of the objectives [6, 7]: The model is updated based on the points already evaluated in order to improve the sample efficiency of the EA.

In this paper a different distribution-based approach is taken in which the fitness of a sensor placement, computed over different simulation runs, is represented probabilistically by a discrete probability distribution.

Definition of a mapping from the search space (where each SP is represented as a binary vector) into an information space, whose elements are instances of the proposed data structure, to enable a deep analysis of the search landscape and speed up optimization.

There are many models to measure the dissimilarity or distance between two probability distributions those mostly used in machine learning are the entropy based ones like Kullback-Leibler (KL) and Jensen-Shannon (JS) whose application to gauge network dissimilarity has been first proposed in [8].

In this paper we focused on the Wasserstein (WST) distance, which is based on entirely different approach, namely optimal transport problems. The introduction of a probabilistic. distance between histograms representing sensor placements, is the key element to derive new genetic operators enabling the evolutionary algorithm to work in the information space, which are the distinguishing feature of the new algorithm MOEA/WST.

The Wasserstein distance can be traced back to the works of Gaspard Monge [9] and Lev Kantorovich [10]. Recently, also under the name of Earth Mover Distance (EMD) it has been gaining increasing importance in several fields [11–13].

Important methodological references are [14, 15] for a complete analysis of the mathematical foundations) and [16] for an up to date survey of numerical methods.

The key novelty elements of the new algorithm Multi Objective Evolutionary Algorithm are:

- The introduction of a probabilistic distance, Wasserstein distance, between sensor placements mapped into histograms in the information space.
- The formulation of Wasserstein enabled selection operators and new indicators to assess the quality of the Pareto set and to choose among its elements.
- The formulation of a crossover operators which enables the generation of "feasible by design" children from two feasible parents.

1.1 Organization of the Paper

Section 2 describes the Pareto analysis and the hypervolume indicator. Section 3 introduces the Wasserstein distance and gives the basic definitions and computational tools. Section 4 introduces the formulation of the optimization problems, the structure of the simulation model and the computation of the objective functions. Section 5 introduces the probabilistic representation of a feasible solution and the search and information space. Section 6 describes the proposed algorithm MOEA/WST focusing on the solutions proposed about evolutionary operators, termination criteria and performance indicators of the approximate Pareto set. Section 7 presents the experimental setting and the computational results. Section 8 contains concluding remarks.

2 Background Knowledge on Multiobjective Optimization: Pareto Analysis and Performance Metric

This section contains the basic definition of the solution set in multiobjective optimization through the Pareto analysis and the main indicator, the hypervolume of the quality of its approximation.

2.1 Pareto Analysis

Multiobjective optimization problem (MOP) can be stated as follows:

$$\min F(x) = (f_1(x), \ldots, f_m(x))$$

Pareto rationality is the theoretical framework to analyse multi objective optimization problems where m objective functions $f_1(x), \ldots, f_m(x)$, where $f_i(x) \;:\to\; \mathbb{R}$ are to be simultaneously optimized in the search space $\Omega \subseteq \mathbb{R}^d$. Here we use x to be compliant with the typical Pareto analysis's notation, clearly in this study x is a sensor placement s.

Let $u, v \in \mathbb{R}^m$ u is said to dominate v if and only if $u_i \geq v_i \, \forall i = 1, \ldots, n$ and $u_j > v_j$ for at least one index j.

To refer to the vector of all objectives evaluated at a location x. The goal in multi objective optimization is to identify the Pareto frontier of f(x). A point x^* is pareto optimal for the problem 2 if there is no point x such that F(x) dominate $F(x^*)$. This implies that any improvement in a Pareto optimal point in one objective leads to a deterioration in another. The set of all Pareto optimal points is the Pareto set and the set of all Pareto optimal objective vectors is the Pareto front (PF). The interest in finding locations x having the associated $F(x)$ on the Pareto frontier is clear: all of them represent efficient trade-offs between conflicting objectives and are the only ones, according to the Pareto rationality, to be considered by the decision maker.

The issue of the quality evaluation of Pareto solutions sets is the key issue in multi objective optimization. A very recent and wide survey on this issue [17] where 100 quality indicators are analyzed.

2.2 Hypervolume

To measure the progress of the optimization a natural and widely used metric is the hypervolume indicator that measures the objective space between a non-dominated set and a predefined reference vector. An example of Pareto frontier, along with the reference point to compute the hypervolume, is reported in Fig. 1.

A good approximation of the Pareto set will result into a low/high hypervolume value; thus, hypervolume is a reasonable measure for evaluating the quality of the optimization process [18].

The grey shaded area is the original hypervolume: a new point A improves the approximation to the exact Pareto front and increases the hypervolume by the blue shaded area.

The improvement of the hypervolume can also be used in the selection of the new point as in [19]. This approach has been further developed in [20] in which it is shown off to compute the gradient of the expected hypervolume improvement to speed up the selection process.

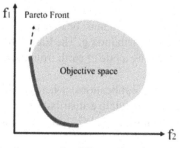
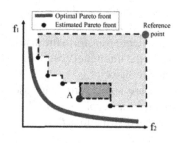

Fig. 1. An example of Pareto frontier, with the associated hypervolume, for two minimization objectives. (f_1 is the detection time and f_2 is its standard deviation)

2.3   Coverage

Given two approximations A and B of the Pareto front the C metric (Coverage) $C(A, B)$ is defined by the percentage of solutions in B that are dominated by at least one solution in A. $C(B, A)$ is the percentage of solutions in A that are dominated by at least one solution in B. $C(A, B) \neq 1 - C(B, A)$ so that both must be considered to assess. $C(A, B) = 1$ means that all solutions in B are dominated by some solution in A; $C(A, B) = 0$ implies that no solution in B is dominated by a solution in A.

3   The Wasserstein Distance – Basic Notions and Numerical Approximation

This section deals with a methodology to compute the dissimilarity or distance between two probability distributions.

Consider two discrete distributions $f(x) = \sum_{i=1}^{n} f_i \delta(x - x_i)$ and $g(x) = \sum_{i=1}^{n'} g_i \delta(y - y_i)$ where $\sum_{i=1}^{n} f_i = \sum_{i=1}^{n'} g_i = 1$ and $\delta(.)$ is the Kronecker delta.

$$\delta(x) = \begin{cases} 1 \ if \ x = 0 \\ 0 \ otherwise \end{cases}$$

The unit cost of transport between x_i and y_j is defined as the p-th power of the "ground metric" $d(x, y)$, in this case the Euclidean norm $c_{ij} = ||x_i - y_j||^p$. The transport plan γ_{ij} represent the mass transported from x_i to y_j. The WST distance between discrete distributions f and g is:

$$W_q(f, g) = \min_{\gamma_{ij} \in \Gamma} \left(\sum_{i=1}^{n} \sum_{j=1}^{n'} \gamma_{ij} |x_i - y_j|^p \right)^{\frac{1}{p}}$$

$$s.t. \sum_{j=1}^{n'} \gamma_{ij} = f_i, \sum_{i=1}^{n} \gamma_{ij} = g_j, \gamma_{ij} \geq 0$$

The constraints above ensure that the total mass transported from x_i and the total mass to y_j matches respectively f_i and g_j.

In the specific case of $p = 1$ the Wasserstein distance is also called the Earth Mover's Distance (EMD). Intuitively, $\gamma(x, y)$ indicates how much earth must be transported from x to y in order to transform the distributions f into the distribution g. The Earth Mover's distance is the minimum cost of moving and transforming a pile of earth from the shape of f into the shape of g.

There are some particular cases, very relevant in applications, where WST can be written in an explicit form. Let F and G be the cumulative distribution for one-dimensional distributions f and g in the unit interval and F^{-1} and G^{-1} be their quantile functions, then.

$$W_p(f, g) = \left(\int_0^1 \left| F^{-1}(x) - G^{-1}(x) \right|^p dx \right)^{\frac{1}{p}}.$$

In the case of one dimensional histograms the computation of WST can be performed by a simple sorting and the application of the following formula.

$$W_p(f, g) = \left(\frac{1}{n} \sum_i^n \left| x_i^* - y_i^* \right|^p \right)^{\frac{1}{p}}$$

where x_i^* and y_i^* are the sorted samples.

In this paper we use $p = 1$.

Wasserstein distances are generally well defined and provide an interpretable distance metric between distributions. Computing Wasserstein distances requires in general the solution of a constrained linear optimization problem which has, when the support of the probability distributions is multidimensional, a very large number of variables and constraints. In the general case it is shows to be equivalent to a min-flow algorithm of quadratic computational complexity [21] and, in specific cases, to linear [22, 23].

In the case of 1-dimensional histograms the computation of EMD, can be done sorting the ordered samples.

4   The Formulation of Optimal Sensor Placement

This section contains the basic optimization model for sensor placement and the simulation framework for the computation of the objective functions.

4.1   Problem Formulation

We consider a graph $G = (V, E)$ We assume a set of possible locations for placing sensors, that is $L \subseteq V$. Thus, a SP is a subset of sensor locations, with the subset's size less or equal to p depending on the available budget. An SP is represented by a binary vector $s \in \{0, 1\}^{|L|}$ whose components are $s_i = 1$ if a sensor is located at node i, $s_i = 0$ otherwise. Thus, an SP is given by the nonzero components of s.

For a water distribution network, the vertices in V represent junctions, tanks, reservoirs or consumption points, and edges in E represent pipes, pumps, and valves.

Let $A \subseteq V$ denote the set of contamination events $a \in A$ which must be detected by a sensor placement s, and d_{ai} the impact measure associated to a contamination event a detected by the ith sensor.

A probability distribution is placed over possible contamination events associated to the nodes. In the computations we assume – as usual in the literature – a uniform distribution, but in general discrete distributions are also possible. In this paper we consider as objective functions the detection time and its standard deviation.

We consider a general model of sensor placement, where d_{ai} is the "impact" of a sensor located at node I when the contaminant has been introduced at node a.

$$(P1) \begin{cases} \min f_1(s) = \sum_{a \in A} \alpha_a \sum_{i=1,...,|L|} d_{ai} x_{ai} \\ s.t. \\ \sum_{i=1,...,|L|} s_i \leq p \\ s_i \in \{0, 1\} \end{cases}$$

- α_a is the probability for the contaminant to enter the network at node a.
- d_{ai} is the "impact" for a sensor located at node i to detect the contaminant introduced at node a. In this problem d_{ai} is the detection time for a sensor located in node i od the event a.
- $x_{ai} = 1$ if $s_i = 1$ and i is the first sensor in the placement s to detect the contaminant injected at node a; 0 otherwise.

In our study we assume that all nodes have the same probability of being hit by a contamination event, that is $\alpha_a = 1/|A|$.

Therefore $f_1(s)$ is:

$$f_1(s) = \frac{1}{|A|} \sum_{a \in A} \hat{t}_a$$

where $\hat{t}_a = \sum_{i=1,...,|L|} d_{ai} x_{ai}$.

As a measure of risk, we consider the standard deviation of f_1.

$$f_2(s) = STD_{f_1}(s) = \sqrt{\frac{1}{|A|} \sum_{a \in A} (\hat{t}_a - f_1(s))^2}$$

This model can be specialized to different objective functions as:

- f_1: For each event a and sensor placement s, $\hat{t}_a$ is the minimum detection time MDT also defined as $MDT_a = \min_{i:s_i=1} d_{ai}$.
- f_2: the standard deviation of the sample average approximation of f_1.

Evaluating the above objective functions might be expensive for large scale networks because it requires to perform the hydraulic simulation for a large set of contamination events, each one associated to a different location where the contaminant is injected. Simulating the contaminant propagation allows to compute the average Minimum Detection

Time (MDT) provided by a specific SP over all the simulated events. Finally, the SP minimizing the average of the MDT is searched for. However, searching for an optimal SP is NP-hard: even for mid-size networks, efficient search methods are required, among which evolutionary approaches are often used.

4.2 Network Hydraulic Simulation

The Water Network Tool for Resilience (WNTR) [24] is a Python package used to simulate the hydraulic simulation of the net. The simulation is computationally costly as we need one execution for each contamination events. A detailed description of the data structure is given in (Ponti et al. 2021).

In this study, each simulation has been performed for 24 h, with a simulation step of 1 h. We assume $L = V$ and $A = V$ (i.e., the most computationally demanding problem configuration) with a budget constraint on the number of sensors p.

Let's denote with S^ℓ the so-called "sensor matrix", with $\ell = 1, \ldots, |L|$ an index identifying the location where the sensor is deployed at. Each row corresponds to a simulation step Δt and each column to an event-node. Each entry s_{ta}^ℓ represents the concentration of the contaminant for the event $a \in A$ at the simulation step $t = 0, .., K$, with $T_{\max} = K\Delta t$, for a sensor located in node ℓ. Without loss of generality, we assume that the contaminant is injected at the beginning of the simulation (i.e., $t = 0$).

Analogously, a "sensor placement matrix", $H^{(s)} \in \mathbb{R}^{(K+1)\times|A|}$ is defined, where every entry $h_{ta}^{(s)}$ represents the maximum concentration over those detected by the sensors in s, for the event a and at time step t.

Suppose to have a sensor placement s consisting of m sensors with associated sensor matrices $S^1, \ldots, S^m$, then $h_{ta} = \max_{j=1,\ldots,m} s_{ta}^j \forall a \in A$ (Fig. 2).

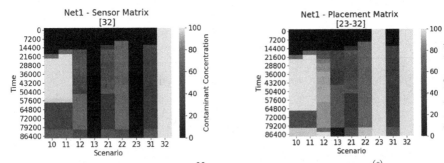

Fig. 2. The entries of the left matrix are s_{ta}^{32}. The entries of the right matrix are $h_{ta}^{(s)}$ where s denote a placement in which sensors are deployed in nodes 23 and 32.

Indeed, we can now explicit the computation of $\hat{t}_a$ in the of $f_1(s)$ and $f_2(s)$: $\hat{t}_a$ is the minimum time step at which concentration reaches or exceeds a given threshold τ for the scenario a, that is $\hat{t}_a = \min_{t=1,\ldots,K}\{h_{ta} \geq \tau\}$.

5   Distributional Representation and the Information Space

This section contains one key contribution of the paper: the representation of the candidate solutions as discrete probability distributions and the description of the structure and functionality of the information space.

5.1   Probabilistic Representation of a Solution

The information in $H^{(s)}$ about a placement can be represented as a histogram. We consider the time steps in the simulation $\Delta t_i = t_i - t_{i-1}$ where $i = 1, \ldots, k$ are equidistanced in the simulation time horizon $(0, T_{MAX})$. We consider the discrete random variable $|A_i|$ where $A_i = \{a \in A : \hat{t}_a \in \Delta t_i\}$ that is the number of events detected in Δt_i (Fig. 3).

For each placement, the matrix $H^{(s)}$ can be represented by an histogram (Fig. 4).

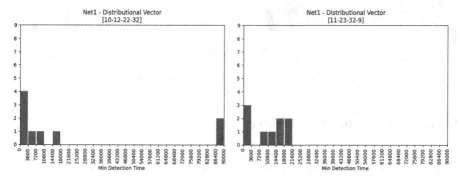

Fig. 3. The histogram of a sensor placement with $p = 4$ ($|A| = 9$).

We have added an extra bin whose value represents for any sensor placement (histograms) the number of contamination events which were undetected during the simulation (and hence the detection probability). Histogram can be scalarized amounting to a ranking of SP. The "ideal" placement is that in which $|A^{\wedge}1| = |A|$. The relation between SP and histograms is many to one: one histogram indeed can be associated to different SP.

Intuitively the larger the probability mass in lower Δt_i the better is the sensor placement, the larger the probability mass in the higher Δt the worse is sensor placement. The worst SPs are those for which no detection took place in the simulation horizon.

5.2   Search Space and Information Space

Our search space consists of all the possible SPs, given a set L of possible locations for their deployment, and resulting feasible with respect to the constraints in (P). Formally, $s \in \Omega \subseteq \{0, 1\}^{|L|}$. Beyond the computation of f_1 and f_2, the matrix $H^{(s)}$ offers a much richer representation and enables the computation of the histograms.

For the sake of simplicity, let's denote with π this computational process:

$$s \xrightarrow{\pi} H^{(s)} \Rightarrow \phi\left(H^{(s)}\right) = (f_1(s), f_2(s))$$

We use $\phi(H^{(s)})$ to stress the fact that the computation is actually performed over $H^{(s)}$ – within the "information space" – and then it generates the observation of the two objectives $(f_1(s), f_2(s))$.

In this paper the landscape is explored through histograms in the information space and their Wasserstein distance.

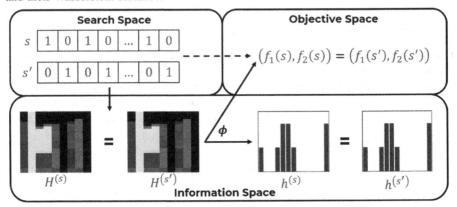

Fig. 4. Search space, information space and objectives.

6 The Algorithm MOEA/WST

This section contains the analysis of the new algorithm proposed. It is shown how all the mathematical constructs presented in the previous sections are structured in the MOEA/AST algorithm. The Sect. 6.1 offers a global view of the interplay of all algorithmic components which are described in the following Sects. 6.2 to 6.6.

6.1 General Framework

See Fig. 5.

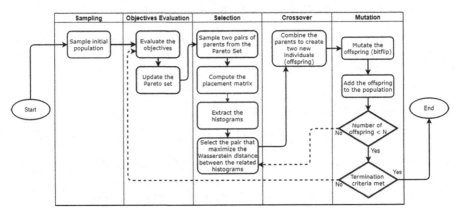

Fig. 5. The general framework of MOEA/WST.

6.2 Chromosome Encoding

In the algorithm each chromosome (individual) consists in a $|L|$-dimensional binary array that encodes a sensor placement. Each gene represents a node in which a sensor can be placed. A gene assumes value 1 if a sensor is located in the corresponding node, 0 otherwise.

6.3 Initialization

The initial population is given by 40 individuals. Our algorithm randomly samples the initial chromosomes. All the individuals in the population have to be different (sampling without replacement).

Among this population we select the non-dominated solutions (Initial Pareto set).

6.4 Selection

In order to select the pairs of parents to be mated using the crossover operation, we have introduced a problem specific selection method that takes place into the information space.

First, we randomly sample from the actual Pareto set two pairs of individuals (F_1, M_1) and (F_2, M_2). Then we choose the pair (F_i, M_i) as the parents of the new offspring, where $i = \arg\max_{i \in \{1,2\}} D(F_i, M_i)$. This favors exploration and diversification (Fig. 6).

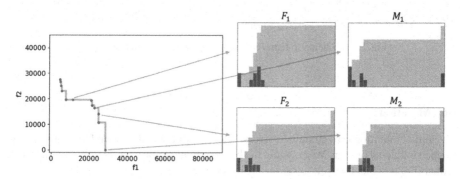

Fig. 6. Histogram representation of points of the approximate Pareto set.

In this paper we used the Wasserstein distance between the histograms corresponding to the sensor placement F_i and M_i.

If at least one individual of the pair of parents is not feasible (i.e., the placement contains more sensors than the budget p) the Constraint Violation (CV) is considered instead. Let's $c = [c_i]$ be a generic individual and p the budget, the Constraint Violation is defined as follow.

$$CV(c) = \max\left(0, \sum_i c - p\right)$$

Then we choose the pair of parents (F_i, M_i) with $i = \arg\min_{i \in \{1,2\}} (CV(F_i) + CV(M_i))$.

6.5 Crossover

The standard crossover operators applied to sensor placement might generate unfeasible children which might induce computational inefficiency in terms of function evaluations. To avoid this in MOEA/WST it has been introduced a problem specific crossover which generates two "feasible-by-design" children from two feasible parents.

Denote with $x, x' \in \Omega$ two feasible parents and with J (FatherPool) and J' (MotherPool) the two associated sets $J = \{i : x_i = 1\}$ and $J' = \{i : x'_i = 1\}$. To obtain two feasible children, c and c' are initialized as $[0, \ldots, 0]$. In turn, c and c' samples an index from J and from J', respectively, without replacement. Therefore, the new operator rules out children with more than p non-zero components.

In the following figure (Fig. 7), an example comparing the behaviour of our crossover compared to a typical 1-point crossover.

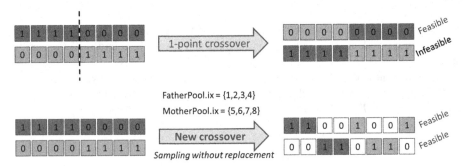

Fig. 7. A schematic representation of the new crossover operator.

6.6 Mutation

The aim of mutation is to guarantee diversification in the population and to avoid getting trapped into local optima. We consider the bitflip mutation operator, for which a mutation probability is typically used to set the "relevance" of exploration in GA.

We have been using the bitflip mutation in Pymoo (each gene has a probability of mutation of 0.1).

7 Computational Results

The search space for this example is quite limited, allowing us to solve the problem via exhaustive search. More precisely, only 561 SPs are feasible according to the constraints in (P). Thus, we exactly know the Pareto set, the Pareto frontier and the associated hypervolume.

We have used Pymoo by setting 100 generations and a population size of 40.

After 100 generations NSGA-II and MOEA/WST obtained the same results in terms of hypervolume and coverage.

There is anyway a significant difference in terms of coverage of the approximate Pareto front (APF) to the optimal one (OPF) between the two algorithms. A smaller value of $C(OPF, APF)$ means a better approximation. The index $C(APF, OPF)$ is not meaningful because no solution in the APF can dominate OPF (Fig. 8).

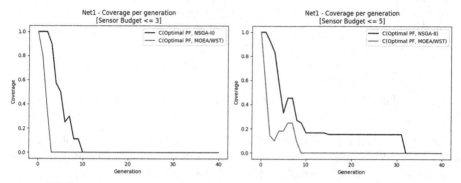

Fig. 8. Coverage between the optimal Pareto frontier and the approximate one.

7.1 Hanoi

Hanoi is a benchmark used in the literature [25].

The following Figs. (9, 10, 11, 12, 13, 14, 15 and 16) display the average value and standard deviation of hypervolume (y-axis) obtained from experiments for different values of p. In each experiment 30 replications have been performed to generate the estimation sample.

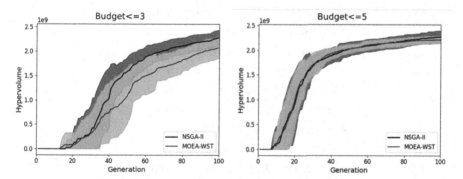

Fig. 9. Hypervolume per generation $p \leq 3$ (left) and $p \leq 5$ (right).

In terms of hypervolume MOEA/WST and NSGA-II offer a balanced performance.

The difference in values of hypervolume between MOEA/WST and NSGA-II has been tested for statistical significance for different values of p and different generations/iterations counts.

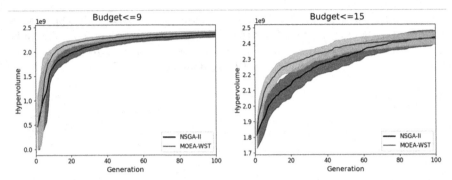

Fig. 10. Hypervolume per generation $p \leq 9$ (left) and $p \leq 15$ (right).

We run the Wilcoxon test for MOEA/WST and NSGA-II for the samples in generations 25/50/100 (each new generation requires 10 function evaluations). The null hypothesis (H0) is that the samples are from the same distribution (Table 1).

Table 1. Comparing hypervolume of MOEA/WST against NSGA-II (values are $\times 10^9$) with respect to different budgets "p" and number of generations. Statistical significance has been investigated through a Wilcoxon test (p-value is reported).

p	Generations	MOEA/WST	NSGA-II	MOEA/WST vs NSGA-II p-value
3	25	0.2188 (0.3790)	0.2190 (0.4538)	0.811
	50	1.0828 (0.5942)	**1.5883 (0.2969)**	<0.001
	75	1.7426 (0.2608)	**2.0804 (0.1797)**	<0.001
	100	2.0769 (0.2225)	**2.2808 (0.1601)**	<0.001
5	25	1.4279 (0.4120)	1.4253 (0.4577)	0.900
	50	2.0264 (0.0910)	2.0250 (0.1439)	0.423
	75	2.1552 (0.0878)	2.1579 (0.1258)	0.809
	100	2.2140 (0.0785)	2.2702 (0.1299)	0.088
9	25	**2.2084 (0.0608)**	2.0404 (0.1079)	<0.001
	50	**2.3189 (0.0521)**	2.2350 (0.0721)	<0.001
	75	**2.3625 (0.0461)**	2.3064 (0.0587)	<0.001
	100	2.3880 (0.0389)	2.3517 (0.0529)	0.006
15	25	**2.2891 (0.0560)**	2.1830 (0.0715)	<0.001
	50	**2.3827 (0.0456)**	2.3170 (0.0612)	<0.001
	75	2.4116 (0.0421)	2.4001 (0.0482)	0.432
	100	2.4317 (0.0415)	2.4364 (0.0444)	0.552

The following figures display the graphs of coverage as the function of number of generations. The graphs show that MOEA/WST improves comparatively it coverage as the number p increases.

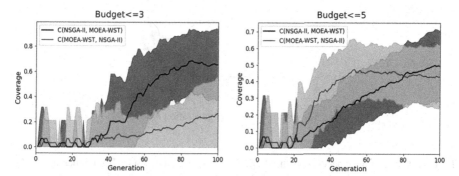

Fig. 11. Coverage per generation (left p = 3, right p = 5)

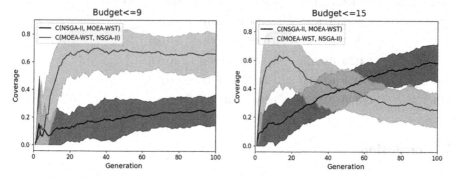

Fig. 12. Coverage per generation (left p = 9, right p = 15)

To add further elements of assessment of the comparative performance of the methods in more challenging conditions the authors have tested them on the real-life problem described in Sect. 7.4.

7.2 Neptun

Neptun is the WDN of the Romanian city of Timisoara, with an associated graph of 333 nodes and 339 edges, analyzed in the European project Icewater [8]. Therefore, each sensor placement in Neptun is a boolean vector of 333 components.

We run the Wilcoxon test for MOEA/WST and NSGA-II for the samples in generations 50, 100, 150, 200 and 250. The null hypothesis (H0) is that the samples are from the same distribution (Table 2).

In Neptun the comparative performance of MOEA/WST is quite impressive in terms of hypervolume.

It is also worth remarking that NSGA-II comes from Pymoo, a consolidated software framework, while MOEA/WST is still highly experimental.

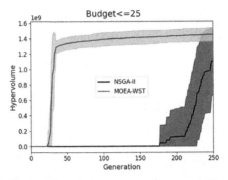

Fig. 13. Hypervolume per generation $p \leq 25$.

Table 2. Comparing hypervolume of MOEA/WST against the NSGA-II's (values are $\times 10^9$) Statistical significance has been investigated through a Wilcoxon test (p-value is reported).

Generations	MOEA/WST	NSGA-II	MOEA/WST vs NSGA-II p-value
50	**1.3377 (0.0826)**	0.0000 (0.0000)	<0.001
100	**1.3916 (0.0900)**	0.0000 (0.0000)	<0.001
150	**1.4150 (0.0915)**	0.0000 (0.0000)	<0.001
200	**1.4350 (0.0848)**	1.2232 (0.0374)	<0.001
250	**1.4530 (0.0880)**	1.1042 (0.0448)	<0.001

8 Conclusions

The key result of this paper has been the proposal of MOEA/WST, a new evolutionary algorithm on discrete structures.

Its distinguishing feature is that it works not only in the search space but in an information space which allows individuals (sensor placements) to be mapped into probability distributions whose distance is given by the WST distance between the discrete distributions associated to placements. The role of WST distance is to capture the interplay of sensor placements and to model the dependence of informational utility of placing a sensor in a location on the presence of pre-existing sensors. On two small scale benchmark networks, MOEA/WST and NSGA-II offer a balanced performance in terms of hypervolume.

On a real-world network with 333 nodes the hypervolume values of MOEA/WST are much better as a function of number of generations.

The computational analysis of MOEA/WST has been limited in this paper to WDNs but the key aspect of the method, that is the representation of feasible solutions as histograms, is suitable for any problem in which we are monitoring the spread of "effects" triggered by "events" as the detection of "fake-news" on the web: in this setting the placement problem is to identify a small set of blogs which catch as many cascades as early as possible.

Acknowledgements. This study has been partially supported by the Italian project "PERFORM-WATER 2030" – programma POR (Programma Operativo Regionale) FESR (Fondo Europeo di Sviluppo Regionale) 2014–2020, innovation call "Accordi per la Ricerca e l'Innovazione" ("Agreements for Research and Innovation") of Regione Lombardia, (DGR N. 5245/2016 - AZIONE I.1.B.1.3 – ASSE I POR FESR 2014–2020) – CUP E46D17000120009.This study has also been partially supported by the Italian project ENERGIDRICA co-financed by MIUR.We greatly acknowledge the DEMS Data Science Lab for supporting this work by providing computational resources (DEMS – Department of Economics, Management and Statistics).

References

1. Boginski, V.L., Commander, C.W., Pardalos, P.M., Ye, Y.: Sensors: Theory, Algorithms, and Applications. Springer, New York (2011). https://doi.org/10.1007/978-0-387-88619-0
2. Mallozzi, L., D'Amato, E., Pardalos, P.M.: Spatial Interaction Models. Springer, Cham (2017). https://doi.org/10.1007/978-3-319-52654-6
3. Deb, K., Myburgh, C.: A population-based fast algorithm for a billion-dimensional resource allocation problem with integer variables. Eur. J. Oper. Res. **261**, 460–474 (2017)
4. Vesikar, Y., Deb, K., Blank, J.: Reference point based NSGA-III for preferred solutions. In: 2018 IEEE Symposium Series on Computational Intelligence (SSCI), pp. 1587–1594. IEEE (2018)
5. Wang, Y., van Stein, B., Bäck, T., Emmerich, M.: A tailored NSGA-III for multi-objective flexible job shop scheduling. In: 2020 IEEE Symposium Series on Computational Intelligence (SSCI), pp. 2746–2753. IEEE (2020)
6. Knowles, J.D.: ParEGO: a hybrid algorithm with on-line landscape approximation for expensive multiobjective optimization problems. IEEE Trans Evol Comput **10**, 50–66 (2006). https://doi.org/10.1109/TEVC.2005.851274
7. Zhang, Q., Liu, W., Tsang, E.P.K., Virginas, B.: Expensive multiobjective optimization by MOEA/D with Gaussian process model. IEEE Trans Evol Comput **14**, 456–474 (2010). https://doi.org/10.1109/TEVC.2009.2033671
8. Schieber, T.A., Carpi, L., Díaz-Guilera, A., Pardalos, P.M., Masoller, C., Ravetti, M.G.: Quantification of network structural dissimilarities. Nat. Commun. **8**, 1–10 (2017)
9. Monge, G.: Mémoire sur la théorie des déblais et des remblais. Histoire de l'Académie Royale des Sciences de Paris (1781)
10. Kantorovich, L.: On the transfer of masses. In: Doklady Akademii Nauk, pp. 227–229 (1942). (in Russian)
11. Bonneel, N., Peyré, G., Cuturi, M.: Wasserstein barycentric coordinates: histogram regression using optimal transport. ACM Trans. Graph. **35**, 71–81 (2016)
12. Huang, G., Quo, C., Kusner, M.J., Sun, Y., Weinberger, K.Q., Sha, F.: Supervised word mover's distance. In: Proceedings of the 30th International Conference on Neural Information Processing Systems, pp. 4869–4877 (2016)
13. Weng, L.: From GAN to WGAN. arXiv preprint arXiv:190408994 (2019)
14. Villani, C.: Optimal Transport: Old and New. Springer, Heidelberg (2008). https://doi.org/10.1007/978-3-540-71050-9
15. Peyré, G., Cuturi, M.: Computational optimal transport: with applications to data science. Found. Trends® Mach. Learn. **11**, 355–607 (2019)
16. Peyré, G., Cuturi, M.: Computational optimal transport. Found. Trends Mach. Learn. **11**, 355–607 (2019). https://doi.org/10.1561/2200000073
17. Li, M., Yao, X.: Quality evaluation of solution sets in multiobjective optimisation: a survey. ACM Comput. Surv. (CSUR) **52**, 1–38 (2019)

18. Fleischer, M.: The measure of Pareto optima applications to multi-objective metaheuristics. In: Fonseca, C.M., Fleming, P.J., Zitzler, E., Thiele, L., Deb, K. (eds.) EMO 2003. LNCS, vol. 2632, pp. 519–533. Springer, Heidelberg (2003). https://doi.org/10.1007/3-540-36970-8_37
19. Beume, N., Naujoks, B., Emmerich, M.T.M.: SMS-EMOA: Multiobjective selection based on dominated hypervolume. Eur. J. Oper. Res. **181**, 1653–1669 (2007). https://doi.org/10.1016/j.ejor.2006.08.008
20. Yang, K., Emmerich, M., Deutz, A., Bäck, T.: Multi-objective Bayesian global optimization using expected hypervolume improvement gradient. Swarm Evol. Comput. **44**, 945–956 (2019)
21. Pele, O., Werman, M.: Fast and robust earth mover's distances. In: 2009 IEEE 12th International Conference on Computer Vision, pp. 460–467. IEEE (2009)
22. Atasu, K., Mittelholzer, T.: Linear-complexity data-parallel earth mover's distance approximations. In: Chaudhuri, K., Salakhutdinov, R. (eds.) Proceedings of the 36th International Conference on Machine Learning, ICML 2019, Long Beach, California, USA, 9–15 June 2019, pp. 364–373. PMLR (2019)
23. Shirdhonkar, S., Jacobs, D.W.: Approximate earth mover's distance in linear time. In: 2008 IEEE Conference on Computer Vision and Pattern Recognition, pp. 1–8. IEEE (2008)
24. Klise, K.A., Murray, R., Haxton, T.: An Overview of the Water Network Tool for Resilience (WNTR) (2018)
25. Vasan, A., Simonovic, S.P.: Optimization of water distribution network design using differential evolution. J. Water Res. Plan. Manage. **136**, 279–287 (2010)
26. Khuller, S., Moss, A., Naor, J.S.: The budgeted maximum coverage problem. Inf. Process. Lett. **70**, 39–45 (1999)

Multi-objective Parameter Tuning with Dynamic Compositional Surrogate Models

Dmytro Pukhkaiev[✉], Oleksandr Husak, Sebastian Götz[ID], and Uwe Aßmann

Software Technology Group, Technische Universität Dresden, Dresden, Germany
{dmytro.pukhkaiev,oleksandr.husak,sebastian.goetz1,
uwe.assmann}@tu-dresden.de

Abstract. Multi-objective parameter tuning is a highly-practical black-box optimization problem, in which the target system is expensive to evaluate. To identify well-performing solutions within the limited budget, a substitution of the target system with a surrogate model, its cheap-to-evaluate approximation, introduces immense benefits. Some surrogates may be more successful for particular objective functions, other at certain stages of optimization. Alas, most state-of-the-art approaches do not address this issue, requiring either to be selected at design time; or lack granularity, changing all models for all objective functions simultaneously. In this paper we provide an approach allowing to individually assign surrogate models to different objective functions and to dynamically combine them into multi-objective compositional surrogate models. To ensure a high prediction quality, our approach contains a model validation strategy based on the cross-validation principle. Moreover, we unite multiple compositional surrogates within a portfolio to even further increase the quality of the search process. Finally, we use the proposed validation strategy to enable a dynamic sampling plan, allowing to get high-quality solutions with even fewer evaluations. The evaluation with a WFG benchmark suite for multi-objective optimization showed that our approach outperforms existing multi-objective model-based approaches.

Keywords: Parameter tuning · Hyperparameter optimization · Multi-objective optimization · Surrogate models

1 Introduction

Parameter tuning (PT) problems are omnipresent in science and engineering and are also known as (but are not limited to): parameter configuration [3,15], hyperparameter optimization [9,17,20] and design space exploration [28,29]. The goal of PT is to find an optimal *configuration* for a certain *target system* with respect to one or several *non-functional properties*. The target system can vary in its evaluation cost: from a cheap-to-evaluate synthetic function [40] to a software system [3], whose single evaluation can span to hours or even days of continuous

© Springer Nature Switzerland AG 2021
D. E. Simos et al. (Eds.): LION 2021, LNCS 12931, pp. 333–350, 2021.
https://doi.org/10.1007/978-3-030-92121-7_26

execution. A configuration denotes a combination of *parameters*, which serve as an input to the target system, affecting its quality.

While the optimization of cheap target systems is often done with help of heuristic-based solvers [2,5,32], expensive target systems can be evaluated only a limited number of times, depending on the optimization budget, requiring a thorough choice of configurations to be evaluated. Here, surrogate-model-based approaches [1,8,18,21,23,24,28,34] come into play, where under a surrogate model[1] we understand an approximation of the target system, which is cheap to evaluate.

In this paper we consider solving multi-objective expensive PT problems, limiting the related work to model-based multi-objective optimization (MBMO). Existing MBMO approaches can be split into three major groups, differing in the type of their surrogate models: 1) multiple single-objective models for different objectives [1,8,18,24,34], 2) pure multi-objective models [10,23], able to handle multiple dimensions with a single surrogate model, and 3) scalarized models [21,28], transforming the problem into a single-objective one. Existing MBMO approaches have two major deficiencies:

P1. Coarse Granularity. A lack of a fine-granular assignment of surrogate models to objective functions prevents from using the best-suited surrogates.

P2. Static Sampling Plan. The number of configurations to be sampled randomly is decided at design time, although its optimal value is problem-dependent and cannot be determined in advance.

In this paper we aim to improve the quality of MBMO approaches with a novel approach called TutorM. It provides the following contributions:

Compositional Surrogate Models. TutorM is able to combine different surrogate models into *compositional* surrogate models, which are automatically switched at runtime, based on their accuracy. Compositional surrogates allow to achieve a prediction quality unreachable by its parts in isolation **(P1)**.

Portfolio of Surrogate Models. Our approach can group multi-objective surrogate models (both pure and compositional) into a *portfolio* and then combine their solution sets into a more diverse Pareto front approximation, further increasing the quality of the optimization **(P1)**.

Model Validation Strategy. TutorM incorporates a surrogate validation strategy, which is based on the cross-validation principle that allows to disable under-performing models at certain stages of the search process. This validation strategy enables a dynamic sampling plan, allowing to obtain high-performing configurations faster **(P2)**.

We evaluate our approach with a WFG benchmark suite for multi-objective optimization [16]. Our findings show that TutorM outperforms state-of-the-art MBMO approaches, improving the final result quality by 5 pp.

---

[1] We use the notions "surrogate model", "model" and "surrogate" interchangeably.

2   State of the Art

An intuition behind PT is to find the best performing configuration among a set of possible variants to optimize a certain black-box target system. A formal definition of a single-objective PT problem can be defined as follows:

$$x^* = \arg\min_{x \in \mathbb{S}} f(x), \qquad (1)$$

where f is a black-box target system, $x \in \mathbb{S}$ is a configuration within the search space $\mathbb{S}$ and x^* is the sought optimal configuration.

Multi-objective PT, is a multi-objective optimization (MOO) problem, which assumes finding a set of configurations $\mathbb{P}$ (Pareto front) such that a solution quality $(f_1(x), \ldots, f_k(x))$ of each configuration $x \in \mathbb{P}$ is not dominated by any other configuration within the objective space $f(x)$. A configuration $A \in \mathbb{S}$ is said to dominate another configuration $B \in \mathbb{S}$, denoted $A \succ B$ iff $f_i(A) \leq f_i(B)$ for all $i \in \{1, \ldots, k\}$ and $f_i(A) < f_i(B)$ for at least one $i \in \{1, \ldots k\}$.

Note, that the number of allowed target system evaluations is strictly limited, due to its expensive evaluation time t_f. Thus, numerous classical optimization algorithms such as: exact solvers [11] and their relaxations [35] become infeasible. Multi-objective heuristic-based solvers, often in form of evolutionary algorithms [2,5,32] are inefficient in such a setting as they also operate directly with the target system, using its output as a reference for the subsequent search.

When the number of feasible evaluations is limited, a proxy between the target system and the optimizer is necessary. The role of such a proxy is often played by a surrogate model, an approximation $\hat{f}$ of the real target system f, whose evaluation $t_{\hat{f}}$ is cheap in comparison to the target system: $t_{\hat{f}} \ll t_f$.

To simplify the understanding of multi-objective surrogate-model-based optimization, we must first consider a single-objective scenario.

2.1   Single-Objective Surrogate-Model-Based Optimization

Single-objective approaches that make use of surrogate models can be separated into three major groups: *offline*, *online* and *hybrid*. Offline approaches [3,22] train the model before the search process starts, during which only optimization of the surrogate model is performed. Online methods (sequential model-based optimization, SMBO) [9,17,31,33,36] enrich the model with new data during the search. Hybrid approaches use a surrogate model from previous optimization processes with the help of transfer learning [14,37]. Despite some minor differences between the groups, the concept of surrogate-based-optimization can be described with help of classic SMBO [19,27]. In the following, we are providing notes for the variations, when necessary. A simplified SMBO workflow is depicted in Fig. 1.

An SMBO process starts with obtaining an initial data set, where each item is a pair of a configuration and its corresponding value of the objective function. A randomized *sampling plan* (Sobol sampling, Mersenne Twister, etc.) is normally utilized for selecting configurations. The initial sampling set is treated as minimal

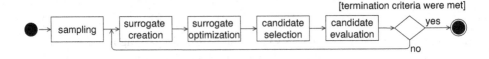

Fig. 1. Simplified SMBO/MBMO workflow

evidence about the search needed to *create* an accurate surrogate model. Note, that even some offline approaches [3] utilize the initial sampling set as input for the trained surrogate model, which then generates promising configurations.

The surrogate model can be created in a form of a frequentist [31,33] or a Bayesian [6,37] regression, a random forest [17], a Generative Adversarial Network [3], Gaussian Process Bandits [14], but is not limited to them. In some cases [9,30], several surrogate models are created. After being created, the model is used by an optimization algorithm: multi-start local search [14,17], grid search [3,31], pattern search [22], Nelder-Mead Simplex [33], random search [30].

After finding an optimal point (or several points [4,25]) using the surrogate, this configuration is evaluated with the target system and the process iterates until some termination criteria is reached, which can be: time [9,14,17], number of evaluations [9,14,17,37], performance evaluation metric [31,36]. In some cases, surrogate optimization may be again substituted with the sampling plan to randomly increase exploration [9] or based on the quality of the prediction [31].

2.2 Multi-objective Surrogate-Model-Based Optimization

Model-based multi-objective optimization (MBMO) is a generalization of SMBO to treat MOO problems. The general workflow remains unchanged, while internal strategies of *surrogate creation* and *surrogate optimization* are subjects to changes.

Surrogate Creation. MBMO has three major variants of surrogate models: *multiple single-objective models*, *pure multi-objective model* and *scalarized model*.

Utilization of multiple single-objective models is a widely used approach [1,8,18,24,34], where each objective function is approximated by a respective surrogate. It is the most straightforward solution, since multiple surrogates do not address possible correlations between the objective functions. The set of used surrogate instances is rich: Support Vector Machines (SVMs) [34], random forests [8], Gaussian Process Regressors (GPRs) [8,24], radial basis functions [1,18,24], polynomial regressions [24]. Note, that *all* listed approaches are using *the same* surrogate model for different dimensions, though in [34] the authors adapt hyperparameters of each SVM individually.

Pure multi-objective surrogate model aims at building a single surrogate model, able to handle multiple dimensions simultaneously [10,23]. Pareto-SVM [23] constructs an aggregated surrogate model by combining a One-class

SVM with a Regression SVM, where the former is used to characterize dominated, nondominated and yet unexplored configurations and the latter to map the characterized configurations onto target values. These values are then used to guide the optimization algorithm. Multivariate GPR is another viable choice for a multi-objective surrogate [10].

Scalarizing can be used as a pre-processing step that allows to use SMBO. ParEGO [21] performs scalarizing with random weights on each iteration, allowing it to move over the non-dominated front. ParEGO uses GPR as a surrogate model. Hypermapper 2.0 [28] applies multiple random forests and then scalarizes them into a single-objective acquisition function which is then optimized.

Surrogate Optimization. The optimization step is determined by the type of surrogate model(s). Scalarized surrogate models can be complemented with classical SMBO optimizations: local search [28], Nelder-Mead Simplex [21]. Multi-objective surrogate models are optimized primarily with MOEA approaches: NSGA-II [10,23], CMA-ES [23]. Multiple single-objective models are optimized with either a single optimizer: NSGA-II [18], CMA-ES [8], custom MOEA [34], local search [8]; or multiple single-objective optimization algorithms which are then aggregated into a non-dominated front: Sequential Quadratic Programming [24], local search [1]. Optimization is wrapped up in the *selection* phase where a single [21] or multiple configurations [1,8,18,23,24,26,34,39] are evaluated using the target system and are used to update the surrogate model(s).

3   Problem Definition

Presented MBMO approaches possess two major deficiencies:

P1. Coarse Granularity. Adapting the no-free-lunch theorem for optimization (NFLT) [38] to the context of surrogate models, we can assume that there exist no surrogate model superior to all other models for every possible PT problem, which is also true for different objective functions. However, most of the discussed approaches (except for [24]) use static surrogates. In [24] the authors mention that different instances of surrogates can be beneficial for various search space surfaces. They address this issue by presenting a portfolio of low-level surrogates (GPR, RBF and polynomial regression), which are used as a prediction basis on different iterations. A decision, which low-level surrogate to select, is based on the prediction of a high-level surrogate (RankBoost ranking approach [12]). However, the low-level models are not assigned individually for different objectives, but are duplicated for each objective, lacking granularity and violating the NFLT. The same rationale is applicable for the runtime performance of models, although partially addressed by [24] via dynamic switching of surrogates, the inability of changing the model for a specific objective function remains.

P2. Static Sampling Plan. Another issue of MBMO approaches is related to their sampling plans. All the aforementioned approaches have a static sampling plan, i.e., the user has to specify the size of the initial sampling set, which may vary depending on the problem, either slowing the search process down by not using an accurate model or underfitting it. For SMBO approaches, only in [31] the authors developed a dynamic sampling plan that validated the surrogate model (regression) based on the quality of its prediction and substituted the model with sampling until it was accurate again. Unfortunately, their validation strategy was domain-dependent (study of energy-efficiency); thus, leaving the question of the dynamic sampling plan's general effectiveness open.

Research Questions. The goal of this paper is to increase quality of multi-objective PT by increasing its granularity w.r.t. surrogate-objective coupling and providing a dynamic sampling strategy. To reach it, we aim to answer the following research questions:

RQ1. Does a dynamic combination of different single-objective models into a multi-objective compositional surrogate benefit the optimization quality?

RQ2. Does a combination of several models into a portfolio further enhance the optimization quality?

RQ3. Is it possible to develop a universal dynamic sampling plan? What is the effect of its utilization?

4   Dynamic Compositional Surrogate Models with TutorM

General Workflow. We start with comparing the workflows of MBMO and our approach (TutorM), whose simplified versions are depicted in Fig. 1 and 2 respectively. For the sake of simplicity, assume we are using a single multi-objective surrogate model and a two-dimensional experiment.

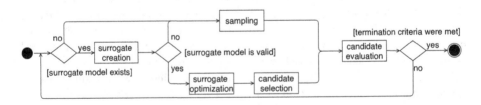

Fig. 2. Simplified TutorM workflow

From Fig. 1 and 2 we can see that the MBMO process is only partially iterative, while TutorM is purely iterative. The main difference is the *sampling strategy*: MBMO starts with a sampling plan, whose size is specified by the user.

The optimal sampling size is problem-dependent and thus, unituitive. Both being too large and too small poses problems to the optimization. The former results in underachieving solutions, when the model is already accurate, but is not used. The latter implies a start with an over-fitted surrogate model, which can follow with a wrongly selected search direction.

In our approach, we recognize that the purpose of the sampling plan and of the surrogate model is the same: namely, to provide new configurations for the evaluation. The only difference is in their internal logic: the sampling plan is experiment-independent, while surrogate-model-based selection is performed based on the current approximation of the search space being experiment-dependent. Consequently, at the experiment's beginning, when no surrogate model is available, TutorM starts sampling configurations randomly (with Sobol sequence), switches to the surrogate-model-based selection if the surrogate model withstands a *validation* check and switches back to the sampling plan usage if the model does not satisfy this check anymore (dynamic sampling plan).

Validation Strategy. The model validation strategy decides whether to use model-based configuration selection or to stick with the sampling plan. Its purpose is to show how well the surrogate extrapolates already performed measurements (variance) and how well it can evaluate yet unseen data (bias).

To discuss our validation strategy, we need to rewind our search process forward until the following state: assume, we have reached iteration i, where $0 < i \ll \infty$. Until this point, $n \geq i$ configurations have been evaluated by the target system, i.e., one or more configurations can be evaluated per iteration.

The main metric of our validation strategy is the coefficient of determination:

$$R^2(y, \hat{y}) = 1 - \frac{\sum_{i=0}^{n_{samples}-1}(y_i - \hat{y}_i)^2}{\sum_{i=0}^{n_{samples}-1}(y_i - \bar{y}_i)^2}, \tag{2}$$

where $\bar{y} = \frac{1}{n_{samples}} \sum_{i=0}^{n_{samples}-1} y_i$, $\hat{y}_i$ is a predicted value of the i-th sample, y_i is the true value of the respective sample and $n_{samples}$ is the overall number of samples. This metric may be misleading if used straightforwardly, as there is no mechanism to prevent over-fitting [28,31]. Thus, we apply a two-staged validation strategy (see Fig. 3). In the outer stage, we split n measured configurations into two sets: $0.75n$ for training and $0.25n$ testing, rounded to a larger integer for the training set. We leave the testing set intact until the *surrogate evaluation*.

In the inner stage, we need to split the $0.75n$ configurations from the training set further into inner training and testing sets (with the same train/test split, leaving $0.56n$ for inner training and $0.19n$ for inner testing). The inner training and testing sets are used for the surrogate creation. In our approach n gradually grows from 0, enabling model creation starting from 10 configurations (6 for training, 2 for inner testing, 2 for outer testing). Until this point the model is considered invalid and the sampling plan is utilized. As at the beginning of the search process n is very small, we would like to avoid a separate inner testing set. Thus, instead of a simple hold-out split we are using k-fold cross-validation.

We subdivide the outer training set $0.75n$ into k equal folds, and perform k rounds of model creation, changing the fold used for testing in each round. We compute the accuracy at each of k rounds and compare it to a user-defined threshold (in form of a hyperparameter). If the accuracy of the model does not pass the threshold for *all* k variants, we consider the model invalid and use the sampling plan.

If the model passed the cross-validation phase, we create it using the full training set of $0.75n$ configurations and evaluate with the outer training set of $0.25n$. Here, we define a second threshold to determine if the model can extrapolate the unknown data. In case of success, we use this model as a basis for subsequent *surrogate optimization* and *candidate selection* steps (see Fig. 2).

The accuracy of the model that passed the outer validation phase is also used to *rank* surrogate. But, the ranking is only possible if several *different* models are available, leading us to our main contribution: *compositional surrogate models*.

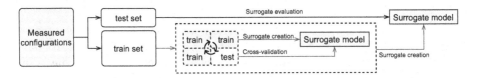

Fig. 3. Surrogate model validation strategy

Compositional Surrogate Models. The general idea is very simple and yet powerful: utilize *separate* surrogate models to approximate different objective functions in a multi-objective optimization problem, which solves the coarse-granularity problem of the existing approaches with multiple single-objective models [1,8,18,24,34].

To construct such a compositional surrogate we used the classic *composite design pattern* [13], where the compositional surrogate plays a role of a composite, while individual surrogates are leaves. Composite pattern ensures a unified treatment of SMBO surrogates and their compositions, allowing to easily switch between single- and multi-objective cases.

Utilizing compositional surrogates enables the following enhancements to the MBMO workflow:

Dynamic Surrogate Creation. Starting from this point we assume having two single-objective surrogate models and one multi-objective model (see Fig. 4).

At the start of the search process, all surrogate models are registered within TutorM, which operates with them individually. It means that during the cross-validation phase each surrogate is validated on every objective function (and on all objective functions simultaneously for multi-objective models, see Fig. 4). On different iterations, some models may or may not pass the cross-validation. Those, which passed, form m sets of usable models, where m is the number of objective functions. Combining these sets, we create s multi-objective composites, where s is the largest number of models that passed the cross-validation threshold for an objective function (in Fig. 4 three single-objective models passed the cross-validation for the objective $o1$, resulting in three compositional surrogates). Newly created compositional surrogates are then ranked depending on their accuracy. The best-ranked composite model is used for optimization and candidate selection.

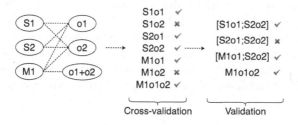

Fig. 4. Creation and validation of compositional models. $S\{X\}$ represent single-objective surrogate models, $M\{X\}$ multi-objective models, $o\{X\}$ objective function, ✓ a model passed the respective validation phase, ✗ the validation phase was not passed

Surrogate Portfolio. To find the best-ranked compositional model TutorM creates a set of composites from the models that passed the validation. Using a single, *even the most accurate*, model is wasteful for such an effort. Thus, we adapt the stacking concept of [1] to support multi-objective surrogates. We unite the created models into a *portfolio* and bind an optimization algorithm to each of them, using the models instead of the target system. Each optimizer constructs a Pareto front approximation, which are then used in the *candidate selection* stage.

Candidate Selection. During candidate selection, one or several configurations should be picked for an evaluation with the target system. In TutorM this number is exposed as a hyperparameter (default 10). If the most accurate surrogate model is used, these configurations are randomly sampled from its Pareto front approximation. In case of a surrogate portfolio, Pareto front approximations from all valid models are stacked together into a unified solution, which is more diverse than each single solution it is combined from. Afterwards, the configurations are randomly sampled from the stacked solution as in the previous case. Stacking helps to increase exploration, while keeping the exploitation at a comparable level, as all the models which are combined are valid.

The search process continues until one or several termination criteria are met, in our case it is the number of function evaluations.

5   Evaluation

We evaluate our approach with a WFG [16] benchmark suite for multi-objective optimization, implemented within `pagmo` library for massively parallel optimization [7]. All benchmark functions are continuous. We consider bi-objective problems with two-dimensional parameter spaces (if applicable). For WFG2 and 3 we consider bi-objective problems with a three-dimensional parameter space.

We consider **relative hypervolume** (r-h) as our primary metric. In general, hypervolume denotes an area dominated by a solution set, which is calculated using a static reference point. In our evaluation we compare *relative hypervolumes*, where 100.0 is a hypervolume of a baseline solution: NSGA-II with 50.000 evaluations. We also discuss the width of the non-dominated front if necessary.

Table 1 contains a description of all the approaches we use in the evaluation as well as their mean relative hypervolume over all problems. For a fair comparison we use the same implementations of the surrogates and optimization algorithms for all strategies, changing only the way they are combined. The only exception is the *scalarized-model-based* approach, where we use Hypermapper 2.0 [28].

Table 1. Evaluated approaches and their mean relative hypervolumes aggregated by all problems. GPR - Gaussian Process Regressor, MLPR - Multi-layer Perceptron Regressor, GBR - Gradient Boosting Regressor, SVR - Support Vector Regressor, RF - Random forest.

Surrogate type	Surrogate model	Optimization algorithm	Budget	r-h
No model	-	NSGA-II	50000	99.725
			1000	77.056
Scalarized model	RF	Local search	1000	75.888
Multiple single- objective models	[GPR,GPR]	MOEA/D + NSGA-II	1000	91.907
	[MLPR,MLPR]			92.205
	[GBR,GBR]			90.646
	[SVR,SVR]			84.220
Pure multi-objective model	GPR	MOEA/D + NSGA-II	1000	92.772
Compositional model	[GPR,SVR,MLPR,GBR]	MOEA/D + NSGA-II	1000	**97.652**

The measurements were performed on an Intel Core i7-8700 CPU workstation with 64 GB RAM using Fedora Server 29 running on kernel 4.18. Each experiment was repeated 5 times[2].

5.1 Results

We discuss the approaches in order of the increasing result quality (according to Table 1). Hypermapper 2.0 showed the weakest performance. The reason for that can be explained with Fig. 5, which shows final Pareto-front approximations using the WFG1 problem instance. From Fig. 5 we can see that this scalarization-based approach biased one of the objective functions ($f2$), not being able to explore the full width of the real non-dominated front.

---

[2] Evaluated strategies and benchmark setup are available online: https://github.com/Valavanca/compositional-system-for-hyperparameter-tuning/tree/LION21.

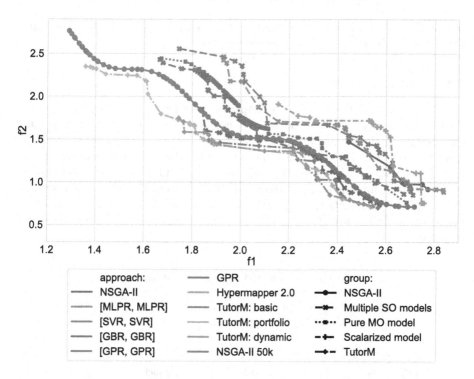

Fig. 5. Final Pareto front approximations (WFG1). $f1$ and $f2$ are the objective functions, which should be minimized.

NSGA-II shows similar final performance to Hypermapper 2.0. It was able to reach 77 r-h in 2% of the budget, meaning that the optimization process slows down immensely, bringing only 23 pp. improvement in the next 49.000 evaluations. Figure 5 shows us that NSGA-II also biased $f2$, which apparently leads to the local optima, preventing from discovering the real surface of the non-dominated front. However, we can also see that within a sufficient budget, NSGA-II is able to get out of the local optima.

Approaches with multiple single-objective models show superior results compared to both heuristic-based and scalarized approaches. The closeness to the real Pareto front depends on a type of the surrogate model, being the furthest for MLPR and closest for GBR duplications (see Fig. 5). Table 1 also shows us that using the same model in pure (GPR) and multi-model setting ([GPR,GPR]) is almost negligible. However, in Fig. 5 we can see that the multi-model GPR provided a wider Pareto front approximation, while the pure GPR moved closer to the real Pareto front, but in a narrower region.

Utilization of compositional surrogate models pays of with the best results. In terms of the final Pareto front approximation, TutorM shows a comparable solution quality to the pure GPR (advancing over the baseline with 50.000 evaluations), but is significantly wider (see Fig. 5). From Table 1 we can see that in general TutorM provides the solution quality comparable to the baseline (only 2 pp. less), while saving 98% of evaluation effort.

5.2   Runtime Behavior

To understand the reasons for the superior final performance of TutorM, we need to discuss its runtime behavior. We use WFG1 as a running example. Our paper contains three contributions. Therefore, to see the effect of each improvement we split TutorM into three variants, gradually adding enhancements: 1) basic version with compositional models only, 2) with the surrogate portfolio, 3) with dynamic sampling plan. Different configurations of TutorM are gathered in Table 2. Figure 6 depicts the runtime performance of the evaluated approaches. For a better clarity, we left only the best-performing approach among each surrogate type.

Table 2. Evaluated variants of TutorM.

Variant	Cross-validation threshold	Outer validation threshold	Model selection	Initial sampling size
TutorM: basic	$-\infty$	$-\infty$	Best	100
TutorM: portfolio	$-\infty$	$-\infty$	Stack	100
TutorM: dynamic	0.65	0.7	Stack	0

Compositional Surrogate Models. The only difference of TutorM in its basic version from classic MBMO approaches lies in having a set of models instead of a predefined surrogate model. In this version, TutorM does not use its validation strategy, meaning that *all* built models are combined into compositional surrogates. Moreover, TutorM does not use its portfolio, implying that only *the most accurate model* is used by the optimizer to guide the search.

Utilization of compositional surrogate models demonstrates a start similar to the other approaches, for which over-fitting is the most probable reason. Despite not using its validation strategy, TutorM still relies on the accuracy-metric to pick the "best-performing" model. While the number of evaluations is too small, optimization of a chosen model results in a flatter start. When the models become more accurate (at around 400 evaluations), R^2 becomes more meaningful, resulting in better overall performance of TutorM, during the subsequent iterations of the search process.

Thus, we can answer **RQ1**: utilization of compositional surrogate models improves the overall quality of obtained solutions in comparison to a static utilization of the same surrogate models. However, such benefits can be gained only in a long run, when all models become more accurate.

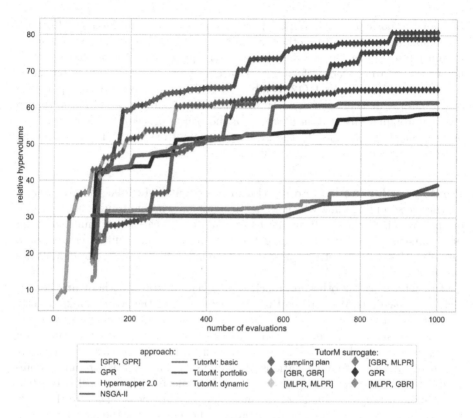

Fig. 6. Alteration of r-h for WFG1. Diamonds represent the most accurate surrogate model at the respective iteration. The respective surrogate is used by TutorM: basic, while portfolio-based versions combine predictions of multiple surrogates.

Portfolio of Compositional Surrogate Models. Here, we enhance the basic version of TutorM with a portfolio support, allowing it to combine the results obtained from the optimization of all surrogate models on each iteration.

From Fig. 6 we can see that a simultaneous utilization of all available combinations of surrogate models completely negates the problem observed for the basic version: slow start. Portfolio gives a wider set of configurations, out of which the candidates for the evaluation are selected. This results in an instant growth of r-h and the quality improvement preserves the same trend as of the basic version, for which the search accelerated with the growing accuracy of models.

Now, let us consider which surrogate was the most accurate at each iteration (colored diamonds in Fig. 6). We can see that the most accurate model is very similar for all TutorM variants throughout the search process. Nevertheless, the versions with portfolio show a much steeper growth of r-h, especially at the start of the search, when the surrogates are still unstable. In this situation a single model can easily predict an under-achieving configuration, which is the case for the basic TutorM. With a portfolio, on contrary, solution sets from *all* surrogates are combined together, meaning some configurations may reappear in solution sets of various models. These overlapping configurations have a much higher chance to be well-performing, as various extrapolations pointed on them.

Thus, we can answer **RQ2**: combination of several models into a portfolio overcomes the deficiency of basic TutorM, allowing to create more diverse Pareto front approximations at each iteration of the search process. It allows selecting more promising solutions for the evaluation right from the start of the search process, enhancing the overall quality of optimization.

Portfolio of Dynamic Compositional Surrogate Models. Finally, we add a dynamic sampling plan, which allows deciding on whether to use the surrogate or the sampling strategy from the beginning of the search. Dynamic sampling plan pursues two goals: 1) freeing the user from an unintuitive choice for a sampling size; 2) promoting the search space exploration, when the models are inaccurate.

In contrast to the other approaches, this version starts without any configurations measured. We can see that the beginning of the search looks exactly like for the approaches with static sampling: r–h grows extremely slow using the random search. However, as soon as the first model passes the validation check, solution quality raises dramatically, meaning that in case of extremely expensive target system evaluations, dynamic sampling plan is strongly recommended. During several subsequent iterations, the models are still overfitted, resulting in a frequent change of the most accurate surrogate. Despite this fact, the utilization of surrogate models are already beneficial over the random search.

When the static approaches transfer into the surrogate-based optimization phase, TutorM is already far ahead in terms of the solution quality. With the growing number of performed evaluations the validation strategy starts slowing the search down, forbidding inaccurate models to take part in the optimization process. These models, inaccurate on their own, can be beneficial when used in the portfolio, providing diversity to the combined Pareto front approximation. Thus, portfolio-based TutorM without model validation outperforms the version with dynamic sampling in the long run.

The answer to **RQ3** is the following: a dynamic sampling strategy can be extremely convenient during the beginning of the search process, which is of the utmost importance for expensive target systems. At a certain point, the validation strategy hinders the search by restricting certain models, narrowing diversity of the Pareto front approximation.

5.3   Threats to Validity

There are several threats to internal validity. The choice of internal hyperparameters in TutorM, e.g., of the values for validation thresholds can either bloat the number of compositional models in the portfolio or degrade TutorM to random search, if all the models were to be prohibited. The choice of the optimizer to work with the surrogate model and its hyperparameters can also be critical for the quality of the obtained solutions. Surrogate models themselves are vulnerable to hyperparameters. Thus, an approach that can configure parameters of surrogates *at runtime*, would have been of an immense aid to the search and is a promising direction for future work.

A threat to external validity lies in our benchmark set, which consists of well-known, but still synthetic benchmarks only. Adding an evaluation of a real-world scenario will increase the external validity.

6   Conclusion and Future Work

Parameter configuration of contemporary software systems, hyperparameter optimization of machine learning algorithms and parameter tuning of meta-heuristics in search-based software engineering are just a small subset of the use cases of multi-objective surrogate-based optimization. In this paper we have identified two major issues common for state-of-the-art approaches in this field: *coarse granularity* and *a static sampling plan.*

To tackle these issues we propose an approach that allows combining different surrogate models into compositional surrogate models, which delivers solutions of a higher quality than reachable by the individual models they are comprised of. Moreover, we enhance our approach with a portfolio of surrogate models, allowing to combine solutions from different surrogates. Such a portfolio diversifies intermediate Pareto front approximations and further improves the result quality. Finally, we present a dynamic sampling strategy that provides high-quality solutions faster. Our findings show that TutorM outperforms state-of-the-art MBMO approaches, improving the final result quality by 5 pp.

There are several promising directions worth of further investigation. On the one hand, parameter control of surrogate models can increase the quality of the obtained solutions. On the other hand, enabling dynamism in portfolios can increase performance of our approach by discarding non-scaling models.

Acknowledgments. This work was supported (in part) by the German Research Foundation (DFG) within the Collaborative Research Center SFB 912–HAEC and project HybridPPS (Project ID 418727532). It was also supported by BMBF project "Software Campus" (grant no. 01IS17044), AIF/IFL-funded project "IPS Framework" (grant no. 20961 BR/2) and SAB-funded project "PROSPER" (grant no. 100379935).

References

1. Akhtar, T., Shoemaker, C.A.: Efficient multi-objective optimization through population-based parallel surrogate search. arXiv preprint arXiv:1903.02167 (2019)
2. Azzouz, R., Bechikh, S., Ben Said, L.: Dynamic multi-objective optimization using evolutionary algorithms: a survey. In: Bechikh, S., Datta, R., Gupta, A. (eds.) Recent Advances in Evolutionary Multi-objective Optimization. ALO, vol. 20, pp. 31–70. Springer, Cham (2017). https://doi.org/10.1007/978-3-319-42978-6_2

3. Bao, L., Liu, X., Wang, F., Fang, B.: Actgan: automatic configuration tuning for software systems with generative adversarial networks. In: Proceedings of the 34th IEEE/ACM International Conference on Automated Software Engineering, ASE 2019, pp. 465–476. IEEE Press (2019)
4. Bartz-Beielstein, T., Lasarczyk, C.W.G., Preuss, M.: Sequential parameter optimization. In: 2005 IEEE Congress on Evolutionary Computation, vol. 1, pp. 773–780 (2005)
5. Bechikh, S., Elarbi, M., Ben Said, L.: Many-objective optimization using evolutionary algorithms: a survey. In: Bechikh, S., Datta, R., Gupta, A. (eds.) Recent Advances in Evolutionary Multi-objective Optimization. ALO, vol. 20, pp. 105–137. Springer, Cham (2017). https://doi.org/10.1007/978-3-319-42978-6_4
6. Bergstra, J., Yamins, D., Cox, D.D.: Hyperopt: a python library for optimizing the hyperparameters of machine learning algorithms. In: Proceedings of the 12th Python in Science Conference, pp. 13–20. Citeseer (2013)
7. Biscani, F., et al.: esa/pagmo2: pagmo 2.15.0, April 2020. https://doi.org/10.5281/zenodo.3738182
8. Bischl, B., Richter, J., Bossek, J., Horn, D., Thomas, J., Lang, M.: mlrmbo: a modular framework for model-based optimization of expensive black-box functions. arXiv preprint arXiv:1703.03373 (2017)
9. Falkner, S., Klein, A., Hutter, F.: BOHB: robust and efficient hyperparameter optimization at scale. In: Dy, J., Krause, A. (eds.) Proceedings of the 35th International Conference on Machine Learning. Proceedings of Machine Learning Research, vol. 80, pp. 1437–1446. PMLR, Stockholm, Sweden, 10–15 July 2018
10. Feng, Z., Wang, J., Ma, Y., Ma, Y.: Integrated parameter and tolerance design based on a multivariate gaussian process model. Eng. Optim. **53**(8), 1349–1368 (2021). https://doi.org/10.1080/0305215X.2020.1793976
11. Fomin, F.V., Kaski, P.: Exact exponential algorithms. Commun. ACM **56**(3), 80–88 (2013)
12. Freund, Y., Iyer, R., Schapire, R.E., Singer, Y.: An efficient boosting algorithm for combining preferences. J. Mach. Learn. Res. **4**(November), 933–969 (2003)
13. Gamma, E.: Design Patterns: Elements of Reusable Object-oriented Software. Pearson Education India, Delhi (1995)
14. Golovin, D., Solnik, B., Moitra, S., Kochanski, G., Karro, J., Sculley, D.: Google vizier: a service for black-box optimization. In: Proceedings of the 23rd ACM SIGKDD International Conference on Knowledge Discovery and Data Mining, KDD 2017, pp. 1487–1495. Association for Computing Machinery, New York (2017)
15. Horn, D., Bischl, B.: Multi-objective parameter configuration of machine learning algorithms using model-based optimization. In: 2016 IEEE Symposium Series on Computational Intelligence (SSCI), pp. 1–8 (2016)
16. Huband, S., Hingston, P., Barone, L., While, L.: A review of multiobjective test problems and a scalable test problem toolkit. IEEE Trans. Evol. Comput. **10**(5), 477–506 (2006)
17. Hutter, F., Hoos, H.H., Leyton-Brown, K.: Sequential model-based optimization for general algorithm configuration. In: Coello, C.A.C. (ed.) LION 2011. LNCS, vol. 6683, pp. 507–523. Springer, Heidelberg (2011). https://doi.org/10.1007/978-3-642-25566-3_40
18. Isaacs, A., Ray, T., Smith, W.: An evolutionary algorithm with spatially distributed surrogates for multiobjective optimization. In: Randall, M., Abbass, H.A., Wiles, J. (eds.) Progress in Artificial Life, pp. 257–268. Springer, Heidelberg (2007)
19. Jones, D.R., Schonlau, M., Welch, W.J.: Efficient global optimization of expensive black-box functions. J. Glob. Optim. **13**(4), 455–492 (1998)

20. Kaltenecker, C., Grebhahn, A., Siegmund, N., Apel, S.: The interplay of sampling and machine learning for software performance prediction. IEEE Softw. **37**(4), 58–66 (2020). https://doi.org/10.1109/MS.2020.2987024
21. Knowles, J.: Parego: a hybrid algorithm with on-line landscape approximation for expensive multiobjective optimization problems. IEEE Trans. Evol. Comput. **10**(1), 50–66 (2006)
22. Lama, P., Zhou, X.: Aroma: Automated resource allocation and configuration of mapreduce environment in the cloud. In: Proceedings of the 9th International Conference on Autonomic Computing, ICAC 2012, pp. 63–72. Association for Computing Machinery, New York (2012)
23. Loshchilov, I., Schoenauer, M., Sebag, M.: A mono surrogate for multiobjective optimization. In: Proceedings of the 12th Annual Conference on Genetic and Evolutionary Computation. pp. 471–478. GECCO 2010, Association for Computing Machinery, New York (2010)
24. Lu, X., Sun, T., Tang, K.: Evolutionary optimization with hierarchical surrogates. Swarm Evol. Comput. **47**, 21–32 (2019), special Issue on Collaborative Learning and Optimization based on Swarm and Evolutionary Computation
25. López-Ibáñez, M., Dubois-Lacoste, J., Cáceres, L.P., Birattari, M., Stützle, T.: The IRACE package: iterated racing for automatic algorithm configuration. Oper. Res. Perspect. **3**, 43–58 (2016)
26. Mlakar, M., Petelin, D., Tušar, T., Filipič, B.: GP-DEMO: differential evolution for multiobjective optimization based on gaussian process models. Euro. J. Oper. Res. **243**(2), 347–361 (2015)
27. Mockus, J., Tiesis, V., Zilinskas, A.: The application of Bayesian methods for seeking the extremum. Towards Glob. Optim. **2**(117–129), 2 (1978)
28. Nardi, L., Koeplinger, D., Olukotun, K.: Practical design space exploration. In: 2019 IEEE 27th International Symposium on Modeling, Analysis, and Simulation of Computer and Telecommunication Systems (MASCOTS), pp. 347–358 (2019)
29. Palesi, M., Givargis, T.: Multi-objective design space exploration using genetic algorithms. In: Proceedings of the Tenth International Symposium on Hardware/Software Codesign, CODES 2002, pp. 67–72. Association for Computing Machinery, New York (2002)
30. Pukhkaiev, D., Semendiak, Y., Götz, S., Aßmann, U.: Combined selection and parameter control of meta-heuristics. In: 2020 IEEE Symposium Series on Computational Intelligence (SSCI), pp. 3125–3132 (2020)
31. Pukhkaiev, D., Götz, S.: BRISE: energy-efficient benchmark reduction. In: Proceedings of the 6th International Workshop on Green and Sustainable Software, pp. 23–30. ACM (2018)
32. Qu, B., Zhu, Y., Jiao, Y., Wu, M., Suganthan, P., Liang, J.: A survey on multi-objective evolutionary algorithms for the solution of the environmental/economic dispatch problems. Swarm Evol. Comput. **38**, 1–11 (2018)
33. Trindade, Á.R., Campelo, F.: Tuning metaheuristics by sequential optimisation of regression models. Appl. Soft Comput. **85**, 105829 (2019)
34. Rosales-Pérez, A., Coello, C.A.C., Gonzalez, J.A., Reyes-Garcia, C.A., Escalante, H.J.: A hybrid surrogate-based approach for evolutionary multi-objective optimization. In: 2013 IEEE Congress on Evolutionary Computation, pp. 2548–2555 (2013)
35. Roubíček, T.: Relaxation in Optimization Theory and Variational Calculus, vol. 4. Walter de Gruyter, Berlin (2011)

36. Sarkar, A., Guo, J., Siegmund, N., Apel, S., Czarnecki, K.: Cost-efficient sampling for performance prediction of configurable systems. In: 2015 30th IEEE/ACM International Conference on Automated Software Engineering (ASE), pp. 342–352 (2015)
37. Wistuba, M., Schilling, N., Schmidt-Thieme, L.: Hyperparameter optimization machines. In: 2016 IEEE International Conference on Data Science and Advanced Analytics (DSAA), pp. 41–50 (2016)
38. Wolpert, D.H., Macready, W.G.: No free lunch theorems for optimization. IEEE Trans. Evol. Comput. 1(1), 67–82 (1997)
39. Yuan, J., Wang, K., Yu, T., Fang, M.: Reliable multi-objective optimization of high-speed WEDM process based on gaussian process regression. Int. J. Mach. Tools Manuf. 48(1), 47–60 (2008)
40. Zitzler, E., Deb, K., Thiele, L.: Comparison of multiobjective evolutionary algorithms: empirical results. Evol. Comput. 8(2), 173–195 (2000)

Corrected Formulations for the Traveling Car Renter Problem

Brenner Humberto Ojeda Rios[(✉)], Junior Cupe Casquina[(✉)],
Hossmell Hernan Velasco Añasco[(✉)], and Alfredo Paz-Valderrama[(✉)]

Escuela Profesional de Ingeniería de Sistemas, Universidad Nacional de San Agustín
de Arequipa, Arequipa, Peru
{bojeda,jcupec,hvelasco,apazv}@unsa.edu.pe

Abstract. This paper presents two corrected formulations to the mixed integer quadratically constrained programming model of the Traveling Car Renter Problem (CaRS), proposed by da Silva and Ochi (2016, An efficient hybrid algorithm for the Traveling Car Renter Problem. *Expert Systems with Applications*, 64, 132–140). In the CaRS, various vehicle types are available for rent in the cities, each one with its own rental cost; when a car is returned to the city where it was rented, an additional tax must be charged, the objective is to construct a Hamiltonian circuit that minimizes the total cost of the circuit plus the return cost of cars. We highlight the original formulation errors, propose corrections to these errors, provide an analytical validation of the corrections, and present computational experiments using a MIP solver.

Keywords: Traveling car renter problem · Mathematical programming · Optimization

1 Introduction

Goldbarg et al. [5] presented a variant of the classical Traveling Salesman Problem that models central aspects of renting a car from the customer's viewpoint. Such variant is called Traveling Car Renter Problem (CaRS). The goal is to perform the tour at the lowest possible cost. In CaRS, various vehicle types are available for rent, each one with its own characteristics and operating costs (fuel consumption, toll fees, and rental cost); additionally, there is an extra charge to be paid to return a vehicle to the city where it was rented if it was delivered in a different city [2].

The first mathematical formulation (an integer quadratic programming model) for CaRS was proposed in [3], but something seems to be missing. Da Silva and Ochi [7] confirmed those in the instance *Mauritania10e* from the CaRS library [8], where optimal values of the experiments evidencing that some constraints are missing. Consequently, da Silva and Ochi [7] proposed another mathematical formulation based on the famous Miller-Tucker-Zemlin Traveling Salesman Problem (MTZ-TSP) formulation [6], but there is evidence of it was not

© Springer Nature Switzerland AG 2021
D. E. Simos et al. (Eds.): LION 2021, LNCS 12931, pp. 351–363, 2021.
https://doi.org/10.1007/978-3-030-92121-7_27

formulated correctly, for considering wrongly the relationship among the variables $p_{i,j}^c$, w_i^c and r_i^c. All tested instances in [3] and [7] were retrieved from the CaRS library [8].

CaRS is defined as a set of different types of cars C and a graph $G(V, E)$, where V is a set of n cities (vertices) and E is a set of roads (edges) composed by pairs of cities. The cost of using a car $c \in C$ on a road (i, j) is given by D_{ij}^c. An additional cost F_{ij}^c must be paid every time a car c is rented in a city i and delivered to j, with $i \neq j$, which corresponds to the extra fee to return the car c to i. The goal is to build a Hamiltonian circuit that minimizes the total cost of the tour plus the extra fee of returning the cars.

This paper highlights the errors in the mixed integer quadratically constrained programming model of the CaRS and proposes two corrected models. The original mathematical formulation appeared on p. 135 of [7] is repeated here for convenience:

$$\min \sum_{(i,j) \in E} \sum_{c=1}^{|C|} D_{i,j}^c * x_{i,j}^c + \sum_{i \in V} \sum_{j \in V} \sum_{c=1}^{|C|} F_{i,j}^c * p_{i,j}^c \tag{1}$$

$$s.t. \quad u_i - u_j + (n-1) * x_{i,j}^c + (n-3) * x_{j,i}^c \leq n - 2$$
$$\forall i, j = 2, \ldots, n; i \neq j \quad \forall c \in C \tag{2}$$

$$a_j^c = \sum_{i=1}^{n} x_{i,j}^c \quad \forall j \in V \quad \forall c \in C \tag{3}$$

$$\sum_{c=1}^{|C|} a_j^c = 1 \quad \forall j \in V \tag{4}$$

$$d_i^c = \sum_{j=1}^{n} x_{i,j}^c \quad \forall i \in V \quad \forall c \in C \tag{5}$$

$$\sum_{c=1}^{|C|} d_i^c = 1 \quad \forall i \in V \tag{6}$$

$$r_i^c \geq a_i^c - d_i^c \quad \forall i \in V \quad \forall c \in C \tag{7}$$

$$\sum_{i \in V} r_i^c \leq 1 \quad \forall c \in C \tag{8}$$

$$w_i^c \geq d_i^c - a_i^c \quad \forall i \in V \quad \forall c \in C \tag{9}$$

$$\sum_{i \in V} w_i^c \leq 1 \quad \forall c \in C \tag{10}$$

$$\sum_{c=1}^{|c|} r_1^c = 1 \tag{11}$$

$$\sum_{c=1}^{|c|} w_1^c = 1 \tag{12}$$

$$p_{i,j}^c = w_i^c * r_j^c \quad \forall i,j = 1,\dots,n \quad \forall c \in C \tag{13}$$

Parameter $D_{i,j}^c$ represents an operational cost associated with each car type c and $arc(i,j) \in A$, $i \neq j$, while $F_{i,j}^c$ represents the return fee that must be paid by the car renter every time a car c is taken from city i and delivered in city j. The binary decision variable $x_{i,j}^c = 1$ if car renter goes from city i to j using car c. The continuous decision variable u_i represents the order in which city i is visited ($1 \leq u_i \leq n-1 \quad \forall i = 2,\dots,n$). The binary decision variable $a_j^c = 1$ if car renter arrives city j using car c. The binary decision variable $d_i^c = 1$ if car renter departs from city i using car c. The following two variables (r_i^c and w_i^c) are similar to a_j^c and d_i^c, the difference is that they are used to indicate the change of car between sub-tours. The binary decision variable $r_i^c = 1$ if car renter returns car c when visiting i. The binary decision variable $w_i^c = 1$ if car renter withdraws car c when visiting i. Finally, the binary decision variable $p_{i,j}^c = 1$ if return fee is paid for taking car c from city i to city j.

The objective function (1) adds travel costs and extra fees for returning the cars to the cities where they were rented. The constraint (2) is adapted from the Miller-Tucker-Zemlin formulation for the CaRS presented by [1]. Restrictions (3) and (4) ensure that each city is visited only once by exactly one car. Restrictions (5) and (6) ensure that each city is left only once for exactly one car. Restrictions (7) and (9) models the rental and delivery of each car. Restrictions (8) and (10) ensure that each car is rented only once. Restrictions (11) and (12) ensure that a single car must be rented and delivered to the first visited city (city 1). Finally, constraint (13) defines the rate of return that must be paid whenever a car c is rented in the city i and delivered to j, $i \neq j$.

The remainder of this paper is organized as follows. In Sect. 2, the errors in the original mixed integer quadratically constrained programming model are explained. Corrections to the model are proposed in Sect. 3 and are validated in Sect. 4.

2 Explanation of Errors in the Original Formulation

According to the version of CaRS studied in [7], any car must be rented at most once (CaRS with no repetition) and the return fees paid must be calculated considering the city 1 as the first city on the tour. The lack of restrictions on $p_{i,j}^c$ can

$$1 \xrightarrow{3} 2 \xrightarrow{3} 16 \xrightarrow{3} 3 \xrightarrow{3} 15 \xrightarrow{3} 21 \xrightarrow{3} 20 \xrightarrow{3} 26 \xrightarrow{3} 22 \xrightarrow{3} 25 \xrightarrow{3} 13 \xrightarrow{3} 18 \xrightarrow{3} 19 \xrightarrow{3} 17 \xrightarrow{2}$$
$$23 \xrightarrow{2} 9 \xrightarrow{2} 11 \xrightarrow{2} 7 \xrightarrow{2} 8 \xrightarrow{2} 5 \xrightarrow{2} 24 \xrightarrow{2} 6 \xrightarrow{2} 4 \xrightarrow{2} 12 \xrightarrow{2} 10 \xrightarrow{3} 14 \xrightarrow{3} 1$$

Fig. 1. Inconsistent solution of the mixed integer quadratically constrained programming model proposed by [7].

lead to wrong optimal value. In the instance *BrasilAM26n* from CaRS library [8], for example, suppose that Table 1 variables are all equal to 1, which corresponds to solution in the Fig. 1. Although it meets all formulation constraints, it is clearly inconsistent. Car 3 is rented twice (cities 1 and 10). Moreover, generated solution by the mathematical formulation of [7] has value equal to 201, which is contradictorily lower than the optimal value (202). Our experiments confirmed those optimal values evidencing that some constraints are missing in the original mathematical formulation. Both variables r_i^c and w_i^c are the cause of the problem, the value of $p_{i,j}^c$ depends on these ones variables, and the return fees paid depends on $p_{i,j}^c$.

In the previous example (see Fig. 1), variables r_{10}^2, r_{17}^3 and r_1^1 take the value 1, since setting all values $x_{i,j}^c$ to 1 generates unambiguously a single solution, so the cars 2 is delivered in the city 10 (r_{10}^2), whereas the car 3 is delivered twice, first in city 17 (r_{17}^3) and then in city 1 (r_1^1). Something similar happens with the variable w (w_{17}^2, w_1^1 and w_{10}^3 are equals to 1). Consequently, $p_{i,j}^c$ is erroneously calculated in the constraint (13), the return fees paid is calculated only considering rented cars in the cities 10 and 17, without considering the city 1 as the first city to rent a car in the tour. This explains why the value of the solution in Fig. 1 is 201 rather than 215. In addition, both r_i^c and w_i^c show a strange behavior ($r_1^1 = 1$ and $w_1^1 = 1$). Due to the complicated relationship between r_i^c and w_i^c in the CaRS, the constraints (7) and (9) in the original mathematical formulation are incorrect. As a result, the original mathematical formulation admits inconsistent solutions that allows to rent the same vehicle type more than once on the tour and calculate the return fees without considering the city 1 as the first city of the tour.

Table 1. Variables of da Silva and Ochi formulation [7] that creates an inconsistent solution if are set to 1

Var.	Meaning	Var.	Meaning	Var.	Meaning	Var.	Meaning
$x_{1,2}^3$	$1 \xrightarrow{3} 2$	$x_{20,26}^3$	$20 \xrightarrow{3} 26$	$x_{19,17}^3$	$19 \xrightarrow{3} 17$	$x_{8,5}^2$	$8 \xrightarrow{2} 5$
$x_{2,16}^3$	$2 \xrightarrow{3} 16$	$x_{26,22}^3$	$26 \xrightarrow{3} 22$	$x_{17,23}^2$	$17 \xrightarrow{2} 23$	$x_{5,24}^2$	$5 \xrightarrow{2} 24$
$x_{16,3}^3$	$16 \xrightarrow{3} 3$	$x_{22,25}^3$	$22 \xrightarrow{3} 25$	$x_{23,9}^2$	$23 \xrightarrow{2} 9$	$x_{24,6}^2$	$24 \xrightarrow{2} 6$
$x_{3,15}^3$	$3 \xrightarrow{3} 15$	$x_{25,13}^3$	$25 \xrightarrow{3} 13$	$x_{9,11}^2$	$9 \xrightarrow{2} 11$	$x_{6,4}^2$	$6 \xrightarrow{2} 4$
$x_{15,21}^3$	$15 \xrightarrow{3} 21$	$x_{13,18}^3$	$13 \xrightarrow{3} 18$	$x_{11,7}^2$	$11 \xrightarrow{2} 7$	$x_{4,12}^2$	$4 \xrightarrow{2} 12$
$x_{21,20}^3$	$21 \xrightarrow{3} 20$	$x_{18,19}^3$	$18 \xrightarrow{3} 19$	$x_{7,8}^2$	$7 \xrightarrow{2} 8$	$x_{12,10}^2$	$12 \xrightarrow{2} 10$
				$x_{14,1}^3$	$14 \xrightarrow{3} 1$	$x_{10,14}^3$	$10 \xrightarrow{3} 14$

3   Proposed Formulations

In this subsection, we refer to the da silva's model as the original model. Due to the bad relationship between r_i^c and w_i^c in the original model, the restrictions (7) and (9) are incorrect. As a result, the model admits inconsistent solutions that allow the same type of car to be rented more than once on tour.

3.1   First Correction Proposal - Model01

The first formulation incorporates four changes for the original model. First, the constraint (3) is replaced by (14). Second, the constraint (5) is replaced by (15). In the first two changes we include $i \neq j$ to prevent that the optimal solution takes the value zero. Third, the constraint (7) is replaced by the constraint (16). Fourth, the constraint (9) is replaced by (17).

$$a_j^c = \sum_{i=1, i \neq j}^{n} x_{i,j}^c \quad \forall j \in V \quad \forall c \in C \tag{14}$$

$$d_i^c = \sum_{j=1, j \neq i}^{n} x_{i,j}^c \quad \forall i \in V \quad \forall c \in C \tag{15}$$

$$r_i^c = a_i^c \cdot (1 - d_i^c) \quad \forall i \in V \quad \forall c \in C \tag{16}$$

$$w_i^c = d_i^c \cdot (1 - a_i^c) \quad \forall i \in V \quad \forall c \in C \tag{17}$$

Table 2. Values of the binary decision variable r_i^c in the constraint (7) of the original model and the constraint (16) of the corrected model.

		Original model	Corrected model
a_i^c	d_i^c	$r_i^c \geq a_i^c - d_i^c$	$r_i^c = a_i^c(1 - d_i^c)$
0	0	0|1	0
0	1	0|1	0
1	0	1	1
1	1	0|1	0

In Table 2, all possible combinations of the binary decision variables a_i^c and d_i^c are shown, to check the constraint (7).

1. If the salesman does not reach the city i using the car c and does not leave the city i using the same car c (i.e., if $a_i^c = 0$ and $d_i^c = 0$), by CaRS definition, the correct constraint will force the salesman to not return the car c when visiting i ($r_i^c = 0$); however, the constraint (7) allows an ambiguous behavior r_i^c (1 or 0).
2. If the salesman does not reach the city i using the car c and leaves the city i using the same car c (i.e., if $a_i^c = 0$ and $d_i^c = 1$), by definition of CaRS, the correct restriction will force the salesman to not return the car c when he visits i ($r_i^c = 0$); however, constraint (7) allows an ambiguous behavior r_i^c (1 or 0) once again.

3. If the salesman arrives in the city i using the car c and does not leave the city i using the same car c (i.e., if $a_i^c = 1$ and $d_i^c = 0$), by CaRS definition, the correct constraint will force the salesman to return the car c when visiting i ($r_i^c = 1$).
4. If the salesman arrives in the city i using the car c and leaves the city i using the same car c (i.e., if $a_i^c = 1$ and $d_i^c = 1$), by definition of CaRS, the correct restriction will force the salesman to not return the car c when visiting i ($r_i^c = 0$); however, constraint (7) allows an ambiguous behavior r_i^c (1 or 0) again.

The same reasoning applies to explain the fourth change, when the constraint (9) is replaced by the constraint (17). Table 3 is used to check the constraint (9), it shows the values of the variable w_i^c in the constraint (9) of the original model and the constraint (17) of the corrected model.

Table 3. Values of the binary decision variable w_i^c in the constraint (9) of the original model and constraint (17) of the corrected model.

		Original model	Corrected model
a_i^c	d_i^c	$w_i^c \geq d_i^c - a_i^c$	$w_i^c = d_i^c(1 - a_i^c)$
0	0	0|1	0
0	1	1	1
1	0	0|1	0
1	1	0|1	0

3.2 Second Correction Proposal - Model02

The second formulation incorporates four changes. First, the restrictions (3) and (5) are replaced by the restrictions (18) and (19), respectively. These two changes include the condition $i \neq j$ for the sums, preventing the optimal solution takes a value equal to zero. Second, the constraints (7) and (9) are replaced by the equations (20) and (21), respectively. The last two changes are based on the DFJ model proposed by [4].

$$a_j^c = \sum_{i=1, i \neq j}^{n} x_{i,j}^c \quad \forall j \in V \quad \forall c \in C \tag{18}$$

$$d_i^c = \sum_{j=1, j \neq i}^{n} x_{i,j}^c \quad \forall i \in V \quad \forall c \in C \tag{19}$$

$$r_i^c = \left(\sum_{j \in V} x_{j,i}^c \right) \left(\sum_{\substack{c' \in C, \; h \in V \\ c' \neq c}} x_{i,h}^{c'} \right) \quad \forall c \in C, \forall i \in V \tag{20}$$

$$w_i^c = \left(\sum_{j \in V} x_{i,j}^c \right) \left(\sum_{\substack{c' \in C, \, h \in V \\ c' \neq c}} x_{h,i}^c \right) \quad \forall c \in C, \forall i \in V \tag{21}$$

For didactic reasons, four new variables are used to explain the equations (20) and (21) (two variables for each equation). In the equation (20), r_i^c can be represented as the product of $A_{j,i}^c$ and $B_{i,h}^{c'}$, where:

$$A_{j,i}^c = \left(\sum_{j \in V} x_{j,i}^c \right) \tag{22}$$

$$B_{i,h}^{c'} = \left(\sum_{\substack{c' \in C, \, h \in V \\ c' \neq c}} x_{h,i}^c \right) \tag{23}$$

Since r_i^c is a binary variable, then $A_{j,i}^c$ and $B_{i,h}^{c'}$ are also binary variables. $A_{j,i}^c$, indicates whether the car c travels from city j to city i ($A_{j,i}^c = 1$) or not ($A_{j,i}^c = 0$). $B_{i,h}^{c'}$, indicates whether the car c' travels from the city i to h ($B_{i,h}^{c'} = 1$) or not ($B_{i,h}^{c'} = 0$), where the car c' is different from the car used by the salesman to arrive to the city i.

Table 4 shows the values of the variable r_i^c in the restrictions (7) and (20). It is used to check the constraint (7).

Table 4. Values of the binary decision variable r_i^c in the constraint (7) of the original model and constraint (20) of the corrected model.

Original model			Corrected model		
a_i^c	d_i^c	$r_i^c \geq a_i^c - d_i^c$	$A_{j,i}^c$	$B_{i,h}^{c'}$	$r_i^c = A_{j,i}^c \cdot B_{i,h}^{c'}$
0	0	0|1	0	1	0
0	1	0|1	0	0	0
1	0	1	1	1	1
1	1	0|1	1	0	0

1. If the salesman does not reach the city i using the car c (i.e., $A_{j,i}^c = 0$ or $a_i^c = 0$) and leaves the city i using the car c' (i.e., $B_{i,h}^{c'} = 1$ or $d_i^c = 0$), by definition of CaRS, the correct restriction will force the salesman to not return the car c when visiting i ($r_i^c = 0$). However, constraint (7) allows an ambiguous behavior for r_i^c (1 or 0).
2. If the salesman does not reach the city i using the car c (i.e., $A_{j,i}^c = 0$ or $a_i^c = 0$) and does not leave the city i using the car c' (i.e., $B_{i,h}^{c'} = 0$ or $d_i^c = 1$), by the CaRS definition, the correct constraint will force the salesman to not return the car c when visiting i ($r_i^c = 0$). However, constraint (7) allows an ambiguous behavior for r_i^c (1 or 0).

3. If the salesman arrives to city i using the car c (i.e., $A^c_{j,i} = 1$ or $a^c_i = 1$) and leaves the city i using the car c' (i.e., $B^{c'}_{i,h} = 1$ or $d^c_i = 0$), by definition of CaRS, the correct constraint forces the salesman to return the car c when visiting i ($r^c_i = 1$).

4. If the salesman arrives to city i using the car c (i.e., $A^c_{j,i} = 1$ or $a^c_i = 1$) and does not leave the city i using the car c' (i.e., $B^{c'}_{i,h} = 0$ or $d^c_i = 1$), by definition of CaRS, the correct restriction will force the salesman to not return the car c when visiting i ($r^c_i = 0$). However, constraint (7) allows an ambiguous behavior for r^c_i (1 or 0).

In the above explanation, note that $1 - B^{c'}_{i,h}$ is equivalent to d^c_i, just as $A^c_{j,i}$ is equivalent to a^c_i. On the other hand, the equation (21), w^c_i can be represented as the product of $E^c_{\cdot j}$ and $M^{c'}_{h,j}$, where:

$$E^c_{i,j} = \left(\sum_{j \in V} x^c_{i,j} \right) \tag{24}$$

$$M^{c'}_{h,j} = \left(\sum_{c' \in C, h \in V \atop c' \neq c} x^c_{h,i} \right) \tag{25}$$

$E^c_{i,j}$ indicates whether there is a car c traveling from city i to city j ($E^c_{i,j} = 1$) or not ($E^c_{i,j} = 0$). $M^{c'}_{h,j}$ indicates whether there is a car c' traveling from city h to city j ($M^{c'}_{h,j} = 1$) or not ($M^{c'}_{h,j} = 0$), where the car c' is different from the car with which the salesman arrives in the city i.

Table 5. Values of the binary decision variable w^c_i in the constraint (9) of the original model and the constraint (21) of the corrected model.

Original model			Corrected model		
a^c_i	d^c_i	$w^c_i \geq d^c_i - a^c_i$	$M^{c'}_{h,j}$	$E^c_{i,j}$	$w^c_i = M^{c'}_{h,j} \cdot E^c_{i,j}$
0	0	0|1	1	0	0
0	1	1	1	1	1
1	0	0|1	0	0	0
1	1	0|1	0	1	0

In Table 5, all possible combinations of variables for the constraints (9) and (21) are shown. The explanation for replacing the equation (9) with (21) is analogous to the previous one. Note that $a^c_i = 1 - M^{c'}_{h,j}$ and $d^c_i = E^c_{h,j}$.

4   Experiments

Model01 and Model02 were implemented on a computer with an Intel Core i5-2450M processor, 2.50 GHz x4 CPU, and 3.8 Gb of RAM in the Ubuntu operating system. Gurobi 7.0.0 was used as an optimizer (two *threads* were employed). To model CaRS, the Gurobi C++ interface was used. The processing time of the experiment was limited to 5000 s.

In order to define the lower limits for some problems in the CArSLIB library, tests were performed on 56 instances (28 Euclidean and 28 non-Euclidean) of the problem. These instances are available at http://www.dimap.ufrn.br/lae/en/projects/CaRS.php.

Tables 6 and 7 show the results of models Model01 and Model02 for non-Euclidean and Euclidean instances, respectively. The columns *Name*, n, $|C|$, and *Tempo* show, respectively, the instance name, the number of cities, the number of available cars, and the execution timeout in seconds (required by the *solver*). The columns Model01 and Model02 display, respectively, the results of the formulations presented in Sect. 3. The *Sol* column shows the cost of the optimal solution or the best entire solution found by the *solver* until the time limit is reached. The *Lim* column displays the lower bound calculated by the *solver*. The column *Gap* shows the percentage deviation derived from the values presented in *Sol* and *Lim*. The derived percentage deviation is calculated using the formula (26). The value 0.00%, in the column *Gap*, indicates that the instance was solved by the *solver* optimally with the corresponding model. The value 5000 in the *Time* column means that the *solver* has stopped due to the time limit.

$$Gap = \frac{|(Lim - Sol)|}{Sol} \times 100 \qquad (26)$$

In Table 6, Model01 stands out from other model for its speed in finding optimal solutions or closer to optimum (in most instances). For example, between the *Peru13n* and *Argelia15n* instances (with the exception of *BrasilAM26n*), Model01 finds the optimal solutions in less time than the other model. Specifically, in *Argelia15n*, Model01 finds the optimal solution in 225.06 s, while Model02 in 1648.91. With the exception of the *BrasilSP32n* instance, all models were able to solve the same instances optimally. In the *BrasilSP32n* instance, Model02 was the only model that found the optimal solution. For instances with more than 32 cities, the *solver* has reached the time limit.

In Table 7, Model02 found more solutions close to the optimum than Model01. For instances between *Bolivia10e* (inclusive) and *Russia17e* (inclusive), with the exception of *Argelia15e*, Model01 found optimal solutions in less time than the other model. In the instances *BrasilPR25e*, *BrasilMG30e* and *BrasilRS32e*, Model01 turns out to be slower than Model02. With the exception of *BrasilCO40e*, all models were able to find optimal solutions for the same instances. Model02 was the only model that found the optimal solution for *BrasilCO40e*. Model02 tends to achieve better values for the objective function than other model.

360   B. H. O. Rios et al.

Table 6. Results for non-Euclidean instances

Instance			Model01				Model02			
Name	n	|C|	Time	Sol	Lim	Gap(%)	Time	Sol	Lim	Gap(%)
Bolivia10n	10	3	**0,91**	681	681,0	**0,00**	0,94	681	681,0	**0,00**
Peru13n	13	2	**0,10**	693	693,0	**0,00**	0,23	693	693,0	**0,00**
Angola12n	12	2	**0,09**	656	656,0	**0,00**	0,47	656	656,0	**0,00**
EUA17n	17	2	**0,31**	822	822,0	**0,00**	0,9	822	822,0	**0,00**
Congo15n	15	2	**0,19**	886	886,0	**0,00**	1,06	886	886,0	**0,00**
Argentina16n	16	2	**0,32**	894	894,0	**0,00**	0,53	894	894,0	**0,00**
BrasilRJ14n	14	2	**0,10**	167	167,0	**0,00**	0,81	167	167,0	**0,00**
BrasilRN16n	16	2	**0,37**	188	188,0	**0,00**	0,52	188	188,0	**0,00**
BrasilAM26n	26	3	5,90	202	202,0	**0,00**	**5,57**	202	202,0	**0,00**
AfricaSul11n	11	3	**9,20**	714	714,0	**0,00**	11,07	714	714,0	**0,00**
Arabia14n	14	5	**111,08**	1026	1026,0	**0,00**	205,74	1026	1026,0	**0,00**
Argelia15n	15	3	**225,06**	863	863,0	**0,00**	1648,91	863	863,0	**0,00**
Cazaquistao15n	15	5	491,61	1043	1043,0	**0,00**	**309,29**	1043	1043,0	**0,00**
Australia16n	16	4	346,51	1061	1061,0	**0,00**	**289,35**	1061	1061,0	**0,00**
Canada17n	17	4	400,05	1136	1136,0	**0,00**	**159,22**	1136	1136,0	**0,00**
China17n	17	3	**2054,43**	918	918,0	**0,00**	2308,87	918	918,0	**0,00**
Russia17n	17	5	**5000,00**	1094	929,0	15,08	**5000,00**	1098	924,0	15,85
BrasilPR25n	25	3	422,16	226	226,0	**0,00**	**160,99**	226	226,0	**0,00**
BrasilSP32n	32	4	**5000,00**	257	242,0	5,84	**5000,00**	254	254,0	**0,00**
BrasilMG30n	30	4	**5000,00**	272	260	4,41	**5000,00**	271	261,0	**3,69**
BrasilRS32n	32	4	**5000,00**	269	250,0	7,06	**5000,00**	269	247,0	8,18
BrasilCO40n	40	5	**5000,00**	579	527,9	8,81	**5000,00**	586	528,0	9,90
BrasilNO45n	45	5	**5000,00**	546	507,0	7,14	**5000,00**	559	507,0	9,30
att48nA	48	3	**5000,00**	987	965,9	**2,13**	**5000,00**	987	966,0	**2,13**
BrasilNE50n	50	5	**5000,00**	613	577,9	5,71	**5000,00**	622	578,0	7,07
berlin52nA	52	3	**5000,00**	1303	1274,9	**2,15**	**5000,00**	1314	1275,0	2,97
rd100nB	100	4	**5000,00**	1365	1290,1	5,49	**5000,00**	1400	1274,0	9,00
Londrina100n	100	3	**5000,00**	1152	1135,9	1,39	**5000,00**	1148	1136,0	**1,05**

Table 8 summarizes the computational results presented in Tables 6 and 7. The rows show: the number of best solutions obtained with each model (*# Best Solutions*); the number of best processing times to achieve the best values of the objective function (*# best Times*); the number of problems solved optimally (*# Problems solved*); the minimum (*minimum gap*) percentage deviation; the medium percentage deviation (*average gap*); maximum (*maximum gap*) percentage deviation; and the average processing time (*Average time*).

For non-Euclidean instances, as shown in Table 8, Model01 achieved the best values for the objective function, best processing times and best average time. Model02 achieved an optimal solution more than the other model and the lowest

Table 7. Results for Euclidean instances

Instance			Model01				Model02			
Name	n	|C|	Time	Sol	Lim	Gap(%)	Time	Sol	Lim	Gap(%)
Bolivia10e	10	3	**0,23**	592	592	**0,00**	0,35	592	592	**0,00**
Peru13e	13	2	**0,06**	672	672	**0,00**	0,15	672	672	**0,00**
Angola12e	12	2	**0,1**	719	719	**0,00**	0,62	719	719	**0,00**
EUA17e	17	2	**0,26**	912	912	**0,00**	0,71	912	912	**0,00**
Congo15e	15	2	**0,23**	756	756	**0,00**	0,56	756	756	**0,00**
Argentina16e	16	2	**0,34**	955	955	**0,00**	0,82	955	955	**0,00**
BrasilRJ14e	14	2	**0,32**	294	294	**0,00**	0,87	294	294	**0,00**
BrasilRN16e	16	2	**0,33**	375	375	**0,00**	1,19	375	375	**0,00**
BrasilAM26e	26	3	**8,89**	467	467	**0,00**	18,27	467	467	**0,00**
AfricaSul11e	11	3	**0,12**	567	567	**0,00**	0,44	567	567	**0,00**
Arabia14e	14	5	**1,91**	851	851	**0,00**	3,26	851	851	**0,00**
Argelia15e	15	3	15,15	840	840	**0,00**	**5,26**	840	840	**0,00**
Cazaquistao15e	15	5	**3,76**	904	904	**0,00**	16,00	904	904	**0,00**
Australia16e	16	4	**46,55**	1051	1051	**0,00**	123,28	1051	1051	**0,00**
Canada17e	17	4	**26,16**	1251	1251	**0,00**	29,48	1251	1251	**0,00**
China17e	17	3	**16,44**	1003	1003	**0,00**	29,55	1003	1003	**0,00**
Russia17e	17	5	**80,46**	1061	1061	**0,00**	89,7	1061	1061	**0,00**
BrasilPR25e	25	3	162,3	508	508	**0,00**	**62,51**	508	508	**0,00**
BrasilSP32e	32	4	**5000,00**	588	553	**5,95**	**5000,00**	588	518	11,90
BrasilMG30e	30	3	**783,29**	529	529	**0,00**	2031,73	529	529	**0,00**
BrasilRS32e	32	4	2107,53	491	491	**0,00**	**729,89**	491	491	**0,00**
BrasilCO40e	40	5	5000,00	672	641	4,61	**3880,07**	668	668	**0,00**
BrasilNO45e	45	5	**5000,00**	903	756	16,27	**5000,00**	861	782	**9,18**
att48eA	48	3	**5000,00**	35645	32232	9,58	**5000,00**	35176	32084	**8,79**
BrasilNE50e	50	5	**5000,00**	756	717	5,16	**5000,00**	756	718	**5,03**
berlin52eA	52	3	**5000,00**	9185	8321	**9,41**	**5000,00**	9174	8272	9,83
rat99eB	99	5	**5000,00**	3820	2464	**35,50**	**5000,00**	4027	2434	39,56
rd100eB	100	4	**5000,00**	18136	8810	51,43	**5000,00**	18032	8814	**51,12**

value of the minimum percentage deviation of the models. In general, Model01 tends to find solutions close to the optimum in less time than Model02. If the time required by the *solver* is considered, the best results were produced by Model01.

For Euclidean instances, in Table 8, Model02 achieved better values for the objective function, greater number of problems solved in an optimized way, the lowest minimum, average, and maximum percentage deviation, and the shortest processing time average. Model01 achieved the best processing times in relation to the other model. In general, Model02 produced the best results for Euclidean instances.

Table 8. Summary of computational results

	Model01	Model02
Non-Euclidean		
#Best solutions	**24**	22
#Best times	**11**	3
#Problems solved	17	**18**
Minimum gap	1,39	**1,05**
Average gap	**2,33**	2,47
Maximum gap	15,08	**15,85**
Average time	**2109,60**	2146,65
Euclidean		
#Best solutions	23	**25**
#Best times	**17**	4
#Problems solved	20	**21**
Minimum gap	**4,61**	5,03
Average gap	4,93	**4,84**
Maximum gap	**51,43**	51,12
Average time	1544,84	**1500,96**

5 Conclusion

This study presented two corrected formulations for CaRS. The models were implemented in Gurobi. An experiment with 56 CaRS instances, divided into two classes (Euclidean and non-Euclidean), was reported. The best results for the non-Euclidean instances were obtained with Model01. On the other hand, Model02 produced the best results for Euclidean instances regarding a number of problems solved to optimality and processing time computed by the solver.

References

1. Desrochers, M., Laporte, G.: Improvements and extensions to the Miller-Tucker-Zemlin subtour elimination constraints. Oper. Res. Lett. **10**(1), 27–36 (1991)
2. Felipe, D., Goldbarg, E.F.G., Goldbarg, M.C.: Scientific algorithms for the car renter salesman problem. In: 2014 IEEE Congress on Evolutionary Computation (CEC), pp. 873–879. IEEE (2014). https://doi.org/10.1109/CEC.2014.6900556
3. Goldbarg, M.C., Goldbarg, E.F., Asconavieta, P.H., da Menezes, M.S., Luna, H.P.: A transgenetic algorithm applied to the traveling car renter problem. Expert Syst. Appl. **40**(16), 6298–6310 (2013). https://doi.org/10.1016/j.eswa.2013.05.072
4. Goldbarg, M.C., Goldbarg, E.F.G., Luna, H.P.L., Menezes, M.S., Corrales, L.: Integer programming models and linearizations for the traveling car renter problem. Optim. Lett. **12**(4), 743–761 (2017). https://doi.org/10.1007/s11590-017-1138-5

5. Goldbarg, M.C., Asconavieta, P.H., Goldbarg, E.F.G.: Memetic algorithm for the traveling car renter problem: an experimental investigation. Memetic Comput. **4**(2), 89–108 (2012). https://doi.org/10.1007/s12293-011-0070-y

6. Miller, C.E., Tucker, A.W., Zemlin, R.A.: Integer programming formulation of traveling salesman problems. J. ACM **7**(4), 326–329 (1960). https://doi.org/10.1145/321043.321046

7. da Silva, A.R.V., Ochi, L.S.: An efficient hybrid algorithm for the traveling car renter problem. Expert Syst. Appl. **64**, 132–140 (2016). https://doi.org/10.1016/j.eswa.2016.07.038

8. da Silva, P.H.A.: Cars library (2011). http://www.dimap.ufrn.br/lae/en/projects/CaRS.php

Hybrid Meta-heuristics for the Traveling Car Renter Salesman Problem

Brenner Humberto Ojeda Rios[(✉)], Junior Cupe Casquina[(✉)],
Hossmell Hernan Velasco Añasco[(✉)], and Alfredo Paz-Valderrama[(✉)]

Escuela Profesional de Ingeniería de Sistemas, Universidad Nacional de San Agustín
de Arequipa, Arequipa, Peru
{bojeda,jcupec,hvelasco,apazv}@unsa.edu.pe

Abstract. The Traveling Car Renter Problem (CaRS) is a generalization of the classic Traveling Salesman Problem (TSP), where the *tour* of visits can be broken down into contiguous paths that can be traveled with different rental cars. The objective is to determine the Hamiltonian circuit that has a final minimum cost, considering the penalty paid for each vehicle change on *tour*. The penalty is the cost of returning the car to the city where it was rented. CaRS is classified as an NP-hard problem. The research focuses on hybrid procedures that combine meta-heuristics and methods based on Linear Programming to deal with CaRS. The hybridized algorithms are scientific algorithms (ScA), variable neighborhood descent (VND), adaptive local search procedure (ALSP), and a new ALSP variant called iterative adaptive local search procedure (IALSP). The following techniques are proposed to deal with the CaRS: ScA+ALSP, ScA+IALSP, and ScA+VND+IALSP. A mixed integer programming model is proposed for the CaRS, which is used in ALSP and IALSP. Nonparametric tests are used to compare the algorithms in a set of instances in the literature.

Keywords: Traveling car renter problem · Mathematical programming · Metaheuristics

1 Introduction

Hybridization of metaheuristics with other optimization techniques for harp combinatorial optimization problems has recently gained prominence within the research field of metaheuristics. Nowadays, the focus of research has changed from being rather algorithm-oriented to being more problem-oriented; consequently, a cross-fertilization of different areas of optimization, algorithms, mathematical modeling, operational research, statistics, etc., is sought to solve problems in the best possible way [3]. This change produced a large number of powerful and efficient hybrid algorithms. Some of these were successfully applied to logistic problems that occur in the car rental industry.

The car rental business is an interesting segment due to its significant growth [23], comprising USD 27.11 billion in revenue in 2015 in the U.S. while

© Springer Nature Switzerland AG 2021
D. E. Simos et al. (Eds.): LION 2021, LNCS 12931, pp. 364–378, 2021.
https://doi.org/10.1007/978-3-030-92121-7_28

the average car rental fleet grew 5% [18]. This growth trajectory has been steady since 2010 and is forecasted to continue [19]. Concerning the global car rental market, it was valued at approximately USD 58.26 billion in 2016 and is expected to reach approximately USD 124.56 billion by 2022, growing at a CAGR (compound annual growth rate) of around 13.55% between 2017 and 2022 [21]. The car rental market is expected to witness significant growth due to the increasing tourism industry. The major driving factor for the car rental market is growth in the international tourism market and expansion of international airline services, which in turn supports the demand for car rental services [14].

Goldbarg et al. [12] presented a variant of the classic Traveling Salesman Problem (TSP), which models the central aspects of renting a car from the customer's perspective, called the Traveling Car Renter Salesman Problem (CaRS). The CaRS is a generalization of the TSP, where the path can be broken down into contiguous paths that are covered by different rented cars [12]. In CaRS, a customer intends to use rented cars to visit a certain set of cities, with the idea of minimizing the cost related to car rental. Several vehicles from different companies are available in each city, generating a wide variety of choices for renting cars along the way. The TSP is a particular case of CaRS, this happens when there is only one vehicle available to rent in the first city of the tour, and there is no possibility of changing that vehicle in other cities [25]. Several meta-heuristics have been applied to the CaRS, such as Ant colony optimization [25]; transgenetic algorithms [2,11]; memetic algorithms [12]; scientific algorithms [9]; a hybrid between an evolutionary algorithm (EA) and Adaptive Local Search Procedure (ALSP) [24]; and Iterated Adaptive Local Search Procedure (IALSP) [22]. However, for the moment, the ScA is the best simple meta-heuristic (non-hybrid) for the CaRS.

The combining of linear programming techniques and metaheuristics has shown to be a powerful tool to solve combinatorial optimization problems [20]. This approach was presented for CaRS in [24]; the employed algorithm was EA+ALSP and obtained the best-known results for this problem. In this study, we investigate the potential of the ScA and VND in a new ALSP variant called iterative adaptive local search procedure (IALSP).

The article is organized into six sections, including this one. Section 2 presents the CaRS and its mathematical formulation. Section 3 presents the solutions methods to deal with CaRS. The hybrid algorithms are proposed in Sect. 4. The analysis of the results produced in the computational experiments is presented in Sect. 5, and the final conclusions are presented in Sect. 6.

2   CaRS

Formally, the CaRS is defined as a set of different types of cars C and a graph $G(V, E)$, where V is a set of n cities (vertices) and E is a set of roads (edges) composed by pairs of cities. The cost of using a car $c \in C$ on a road (i, j) is given by D_{ij}^c. An additional cost F_{ij}^c must be paid every time a car c is rented in a city i and delivered to j, with $i \neq j$, which corresponds to the extra fee to

Table 1. Description of the parameters of the mathematical formulation for the CaRS.

Parameter	Description
$|C|$	Cardinality of set C
$F_{i,j}^c$	Fee to return car c rented in city i and delivered to city j
$D_{i,j}^c$	Cost of transporting the car c on the edge (i,j)

Table 2. Description of the variables of the mathematical formulation for the CaRS.

Var.	Description
$x_{i,j}^c$	Indicates whether the salesman is going from the city i to j with the car c ($x_{i,j}^c = 1$) or not ($x_{i,j}^c = 0$)
$p_{i,j}^c$	Indicates whether the car c is rented in i and is returned in j ($p_{i,j}^c = 1$) or not ($p_{i,j}^c = 0$)
a_j^c	Indicates whether the salesman arrives in city j using the car c ($a_j^c = 1$) or not ($a_j^c = 0$)
d_i^c	Indicates whether the salesman leaves the city i using the car c ($d_i^c = 1$) or not ($d_i^c = 0$)
r_i^c	Indicates whether the salesman delivers the car c when visiting the city i ($r_i^c = 1$) or not ($r_i^c = 0$)
w_i^c	Indicates whether the salesman rents the car c in the city i ($w_i^c = 1$) or not ($w_i^c = 0$)
u_i	Represents the order in which the city i is visited ($1 \le u_i \le n-1 \; \forall i = 2,\dots,n$)

return the car c to i. The goal is to build a Hamiltonian circuit that minimizes the total cost of the tour plus the extra fee of returning the cars.

2.1 Mathematical Formulation

We present a mathematical formulation based on the model introduced by da Silva [24], which is based on the formulation *Miller-Tucker-Zemlin Traveling Salesman Problem* (MTZ-TSP) [15]. Table 1 and 2 describe the parameters and variables used in the formulation.

$$\min \sum_{(i,j)\in E} \sum_{c=1}^{|C|} D_{i,j}^c * x_{i,j}^c + \sum_{i\in V}\sum_{j\in V}\sum_{c=1}^{|C|} F_{i,j}^c * p_{i,j}^c \qquad (1)$$

$$s.t. \quad u_i - u_j + (n-1)*x_{i,j}^c + (n-3)*x_{j,i}^c \le n-2$$
$$\forall i,j = 2,\dots,n; i \neq j \quad \forall c \in C \qquad (2)$$

$$a_j^c = \sum_{i=1,i\neq j}^{n} x_{i,j}^c \quad \forall j \in V \quad \forall c \in C \qquad (3)$$

$$\sum_{c=1}^{|C|} a_j^c = 1 \quad \forall j \in V \tag{4}$$

$$d_i^c = \sum_{j=1, j \neq i}^{n} x_{i,j}^c \quad \forall i \in V \quad \forall c \in C \tag{5}$$

$$\sum_{c=1}^{|C|} d_i^c = 1 \quad \forall i \in V \tag{6}$$

$$r_i^c = \left(\sum_{j \in V} x_{ji}^c \right) \left(\sum_{\substack{c' \in C, h \in V \\ c' \neq c}} x_{ih}^{c'} \right) \quad \forall c \in C, \forall i \in V \tag{7}$$

$$\sum_{i \in V} r_i^c \leq 1 \quad \forall c \in C \tag{8}$$

$$w_i^c = \left(\sum_{j \in V} x_{ij}^c \right) \left(\sum_{\substack{c' \in C, h \in V \\ c' \neq c}} x_{hi}^{c'} \right) \quad \forall c \in C, \forall i \in V \tag{9}$$

$$\sum_{i \in V} w_i^c \leq 1 \quad \forall c \in C \tag{10}$$

$$\sum_{c=1}^{|c|} r_1^c = 1 \tag{11}$$

$$\sum_{c=1}^{|c|} w_1^c = 1 \tag{12}$$

$$p_{i,j}^c = w_i^c * r_j^c \quad \forall i,j = 1,\ldots,n \quad \forall c \in C \tag{13}$$

The objective function (1) adds travel costs and extra fees for returning the cars to the cities where they were rented. The constraint (2) is adapted from the Miller-Tucker-Zemlin formulation for the CaRS presented by [7]. Restrictions (3) and (4) ensure that each city is visited only once by exactly one car. Restrictions (5) and (6) ensure that each city is left only once for exactly one car. Restrictions (7) and (9) models the rental and delivery of each car. Restrictions (8) and (10) ensure that each car is rented only once. Restrictions (11) and (12) ensure that a single car must be rented and delivered to the first visited city (city 1). Finally, constraint (13) defines the rate of return that must be paid whenever a car c is rented in the city i and delivered to j, $i \neq j$.

3   Solution Methods

This section presents the algorithms used in the hybridization (metaheuristics and methods based on linear programming).

3.1 The Scientific Algorithms

The Scientific Algorithms, proposed by [9], are a metaheuristic inspired in the scientific research process. This new metaheuristic introduces the concepts of *theme*, *researches*, and *literature* from a computational point of view. There are briefly explained as follows.

1. *Researches*: population of candidate solutions.
2. *Search theme*: subset of non-fixed variables.
3. *Literature*: memory or relevant information that is used to bias the search.

The interactions of the *theme*, *researches*, and *literature* contexts result in the search procedure of the scientific algorithms.

3.2 The ALSP and IALSP Algorithms

The Iterated Adaptive Local Search Procedure (IALSP) is a new variant of the ALSP algorithm presented in [24]; both algorithms are exploration mechanisms for the problem solution space. The IALSP is inspired by the idea of avoiding the use of stages such as 1-stage, 10-stages or tree strategies, since this is a very slow process to get the best strategy to solve a problem. Thus, ALSP without step strategy becomes iterative.

Let M be a mathematical formulation and A a solution given by an algorithm, both M and A for CaRS. A can be transformed into a solution for M, denoted by A'. Let S be a set of variables of the same type in A' ($S \subset A'$), which represents a unique solution in M. To know the value of a variable in S we use the notation $x_i(S)$ as in [24]. Additional constraints $x_i = x_i(S)\ \forall x_i$ in S, denoted by $C(S)$, are added to M. Now, let $\bar{S}$ be a subset of S ($\bar{S} \subset S$) and M' an extended formulation ($M' = M \cup C(\bar{S})$). Assuming that $\bar{S} = S$ and $|S| = n$, there are 2^n distinct sub-problems including $\bar{S} = \emptyset$ and $\bar{S} = S$, since a sub-problem is a subset in S. Each sub-problem $\bar{S}$ has its own complexity, optimal solution, solution space and linear relaxation. The difficulty between sub-problems is related to $|\bar{S}|$. There are two extreme cases when S is given to a MIP. First case, if $\bar{S} = \emptyset$, all the variables will be fixed and therefore the optimization will be instantaneous. Second case, if $\bar{S} = S$, no variables will be fixed and therefore we'll have the original problem.

Variables in ALSP and IALSP can be in one of the three states: fixed, test, and non-fixed. In this paper, we consider it more appropriate to refer to the collection of variables (same status) as a set instead of a group, see [13]. T is a set of test variables, S is a set of fixed variables and $\bar{S}$ is a set of non-fixed variables.

Figure 1 shows the process of changing of variables among the sets S, T, and $\bar{S}$. In the beginning, all variables are in S. Both T and $\bar{S}$ are empty. Subsequently, a subset of variables from S is chosen randomly and it's assigned to T with $|T| = size$, S is updated with the difference between S and T. Finally, there are two cases in which variables from T change of set. In the first case, if the

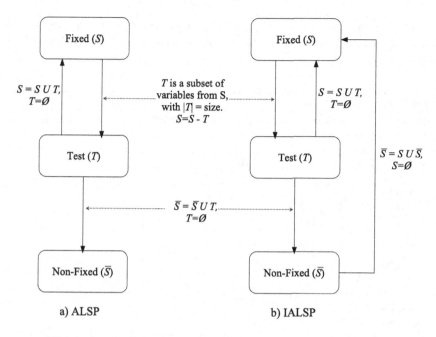

Fig. 1. Status of the variables among the ALSP and IALSP sets.

maximum gap between linear relaxation and incumbent solution values for a sub-problem M' is large or the optimization is finished by the time limit, variables from T will be added to S and T will become empty, the process is repeated again. In the second case, if optimized sub-problem finds a solution equal to or better than the incumbent solution, variables from T will be added to $\bar{S}$ and T will become empty, the process is repeated with $S = S - T$. This process continues until S becomes empty or the runtime for algorithm is exceeded. The behavior of IALSP described above is the same developed by ALSP. IALSP is different from ALSP because it widens the interaction between the sets S, T and $\bar{S}$. Thus, under certain conditions, IALSP can move all the variables of the sets $\bar{S}$ and T to S, with the aim of restarting by itself, to do this, two conditions are necessary:

- First, when the maximum time to perform a local search is exceeded. This avoids solving difficult problems early.
- Second, when the maximum gap between linear relaxation and incumbent solution values for a sub-problem is exceeded. This avoids solving a problem with similar complexity as the original problem.

3.3 VND Algorithm

The VND [16] consists of exploring the space of solutions through systematic exchanges of neighborhood structures until reaching a local minimum, which is minimal in relation to all these neighborhoods (with a strong emphasis on intensification). Its search methodology consists of a deterministic neighborhood change.

Algorithm 1. Pseudocode for VND.

Require: An initial solution (π), a runtime parameter (*prop_loctime*), and a set of
neighborhood structures (ζ).
Ensure: The best solution
1. **while** *prop_loctime* not exceeded **do**
2. $k \leftarrow 1$
3.   **while** $k \leq |\zeta|$ **do**
4. $\pi' \leftarrow$ localSearch(π,k)
5.   **if** $f(\pi') < f(\pi)$ **then**
6. $\pi \leftarrow \pi'$
7. $k \leftarrow 1$
8.   **else**
9. $k \leftarrow$ k+1
10.   **end if**
11.   **end while**
12. **end while**
13. **return** π

In this work, the VND algorithm is adapted to deal with the CaRS. The pseudocode of the Algorithm 1 presents the structure of the algorithm in question. The first step of the algorithm is the definition of a set of neighborhood structures. Let ζ be a set of three ordered neighborhood structures, consisting of N_1, N_2 and N_3. Such order aims to facilitate the search for closer neighbors and whose computational cost is lower in relation to searches in more distant neighborhoods. The number of neighborhood structures is defined as $|\zeta|$. Let π be an initial solution generated by an arbitrary algorithm. The i-th neighborhood of π ($N_i(\pi)$) contains all the solutions that can be obtained from π by applying the i-th neighborhood structure. From the initial solution π, the method explores the first neighborhood structure, N_1, through a local search. If it does not find a better solution, the algorithm starts to explore the second neighborhood structure and successively up to N_3. Whenever the method finds a better solution than the current π, the algorithm restarts the local search in the first neighborhood explored. The algorithm ends when the best solution is not improved after *prop_loctime* seconds.

The final solution should be the local optimum for all neighborhoods. As a result, the chance of achieving a global optimum is greater when using VND instead of a single neighborhood structure.

Local Searches for VND. Let $\sigma = \sigma_1, \ldots, \sigma_n$ a permutation or sequence that defines the current order of cities on the tour, where σ_i is the city in the i-th position, for $i = 1, \ldots, n$. The neighborhood of a permutation σ comprises all the sequences that can be reached by applying a local search operator. The neighborhood structures used were *independent swaps* [5], *2-opt* [6] and *3-opt* [4]. The Algorithm 2 presents the local search procedure for the VND.

Algorithm 2. Local search procedure in VND.

Require: An initial solution π and an integer k.
Ensure: The local optimum π'
  1. **if** $k = 1$ **then**
  2. $\pi' \leftarrow$ 2-opt(π)
  3. **end if**
  4. **if** $k = 2$ **then**
  5. $\pi' \leftarrow$ 3-opt(π)
  6. **end if**
  7. **if** $k = 3$ **then**
  8. $\pi' \leftarrow$ independent_swaps(π)
  9. **end if**
 10. **return** π'

2-opt and 3-opt are neighborhoods derived from exchange movements between two and between three edges belonging to a solution, respectively. *Independent swaps* are neighborhoods that can be obtained by a serie of *swaps* (exchanges of two positions). Two exchange movements are independent if $max\{i,j\} < min\{k,l\}$ or $min\{i,j\} > max\{k,l\}$. Figure 2 illustrates two independent *swaps*, where $i < j < k < l$.

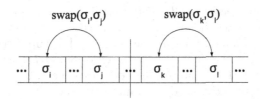

Fig. 2. Two independent *swaps*.

The initial solution π consists of two vectors: a vector of numbers representing a sequence of cities on the tour and a vector of letters representing the salesman's sequence of cars. VND applies six improvement operators (local searches) to the vector of cities. Such operators are $\alpha_1, \alpha_2, \alpha_3, \alpha_4, \alpha_5$ and α_6. Neighborhood structures incorporate improvement operators as follows: *2-opt* incorporates α_1, *3-opt* incorporates operators from α_2 to α_5 (used in sequence), and *independent swaps* incorporates the α_6 operator. Figures 3, 4, 5, 6, 7 and 8 illustrate examples of the application of the operators for a solution of CaRS. The stopping criterion for each operator was limited by the *prop_oprtime* time.

Operators from α_2 to α_5 consider 4 cities c_1, c_2, c_3 and c_4, and two adjacent intervals l_1 and l_2, where l_1 is the interval between c_1 and c_2, and l_2 the interval between c_3 and c_4.

The operator α_1 reverses the direction of cities in an interval in a *tour*. This procedure considers a city pair c_1 and c_2, so that the path between these cities is reversed (note that this path is not necessarily crossed by a single car). After the cost of the new *tour* is verified, if an improvement is found, then the solution is updated. Figure 3 shows an example of this operator reversing a path between cities 2 and 7.

Fig. 3. Example of operator α_1.

The operator α_2 exchanges two adjacent intervals l_1 and l_2 in a *tour*. After the cost of the new *tour* is verified, if an improvement is found, the solution is updated. Figure 4 shows an example of this operator changing the interval between cities 8 and 7 with the interval between cities 5 and 2.

Fig. 4. Example of operator α_2.

The operator α_3 selects two adjacent intervals l_1 and l_2, reverses the direction of l_2, and then exchanges l_1 and l_2. After the cost of the new *tour* is verified, if an improvement is found, the solution is updated. Figure 5 shows an example of this operator. First the interval between cities 8 and 7 is inverted, then this interval is changed with the interval between cities 5 and 2.

The operator α_4 selects two adjacent intervals l_1 and l_2, reverses the direction of l_1, and then exchanges l_1 and l_2. After the cost of the new *tour* is verified, if an improvement is found, the solution is updated. Figure 6 shows an example of this operator. First, the interval between cities 5 and 2 is reversed, then this interval is changed with the interval between cities 8 and 7.

The operator α_5 selects two adjacent intervals l_1 and l_2, reverses the sense of both intervals, and then exchanges l_1 and l_2. After the cost of the new *tour* is verified, if an improvement is found, the solution is updated. Figure 7 shows

Fig. 5. Example of operator α_3.

Fig. 6. Example of operator α_4.

an example of this operator. The intervals between 8 and 7, and 5 and 2 are reversed, so both intervals are switched.

Fig. 7. Example of operator α_5.

The operator α_6 randomly selects four cities c_1, c_2, c_3 and c_4, such that c_1 is visited before c_2, c_2 visited before c_3 and c_3 visited before c_4. After two independent *swaps* are applied, if an improvement is found, the solution is updated. Figure 8 shows an example of this operator making independent *swaps*. Cities 4, 2, 7 and 3 are selected. The first movement of exchange occurs between cities 4 and 2, and the second between cities 7 and 3.

Fig. 8. Example of operator α_6.

4   Proposed Hybrid Algorithms

The heuristic algorithms used in hybridization with approaches based on linear programming are scientific algorithms (ScA), an evolutionary algorithm (EA) [24], and an adaptation of the variable neighborhood descent (VND). The algorithms are based on linear programming and use a mathematical formulation internally are ALSP and IALSP.

The hybridizations investigated in this work were ScA+ALSP, ScA+IALSP, and ScA+VND+IALSP. ScA was chosen because it is the best simple (non-hybrid) metaheuristic to deal with CaRS. Each hybrid algorithm calls a MIP *solver*, which makes use of a mathematical formulation. Currently, the EA+IALSP is the state-of-the-art algorithm for the problem.

ScA+ALSP and ScA+IALSP are sequential hybrid algorithms. Thus, each proposed approach was a sequential execution of a meta-heuristic that is performed before an approach based on linear programming. These algorithms have two phases in their execution, a construction phase followed by an improvement phase. Sca was the algorithm used in the construction phase, and ALSP or IALSP was used in the improvement phase. On the other hand, ScA+VND+IALSP combines three phases: construction, heuristic improvement, and exact improvement. The Algorithm 3 presents the pseudocode for ScA+VND+IALSP.

Algorithm 3. ScA+VND+IALSP for CaRS

---
1. Generate the initial solution using ScA
2. **while** total time not exceeded **do**
3.   Improve solution using VND
4.   Improve solution using IALSP
5. **end while**
6. Inform better solution found

---

5 Computational Experiments

In order to validate the proposed algorithms, these were executed on an Intel Core i5 PC, 4Gb of RAM, on Ubuntu 16.04.2 LTS and using the C++ language. Gurobi 7.0.0 was used as MIP *solver*. The algorithms were implemented with Model01[1]. To evaluate the performance of these algorithms, computational experiments were carried out on 17 non-Euclidean instances in the CArSLIB (See Footnote 1) [25]. We consider a diverse set of instances that vary between 14 and 300 vertices. Instances with more than 100 vertices (large instances) are the most challenging in CaRS. All algorithms were parameterized with the *irace* (See Footnote 1) tool. After the parameterization was performed, 30 independent executions of each hybrid algorithm were performed for each instance.

The algorithms' general performance was analysed with the averages of the values obtained by them in each instance. The Friedman test [10] was used to verify significant differences between the solutions obtained by the algorithms. This test was performed using the XLSTAT software [1]. For the interpretation of the test, we use H_0 (null hypothesis, samples come from the same population) and H_a (alternative hypothesis, samples do not come from the same population). The level of significance was $\alpha = 0.05$. If the calculated p-value is less than the significance level $\alpha = 0.05$, we must reject the null hypothesis H_0 and accept the alternative hypothesis H_a.

If the Friedman test rejects the null hypothesis, then we must execute the Nemenyi *post-hoc* procedure [17] to find concrete pairs that produce differences. If Nemenyi is executed, the correction of Bonferroni [8] must be used because there are multiple comparison in k groups.

---

[1] The detailed experimental package of the algorithms, parameterization and comparison between the models for the CaRS is available at https://repositorio.ufrn.br/jspui/handle/123456789/24822.

Table 3. Results of the algorithms for non-Euclidean instances of the CaRS

Instance Name	n	|C|	Time	ScA Avg.	Best	EA+ALSP Avg.	Best	ScA+ALSP Avg.	Best	ScA+IALSP Avg.	Best	ScA+VND+IALSP Avg.	Best
BrasilRJ14n	14	2	0,54	167,00	**167**	168,17	**167**	167,00	**167**	167,00	**167**	167,00	**167**
BrasilRN16n	16	2	0,54	188,00	**188**	192,80	**188**	188,00	**188**	188,00	188	188,00	**188**
BrasilPR25n	25	3	9,54	226,03	**226**	228,2	**226**	226,00	**226**	226,00	**226**	226,00	**226**
BrasilAM26n	26	3	3,50	202,87	**202**	209,47	**202**	202,13	**202**	202,07	**202**	202,00	**202**
BrasilMG30n	30	4	19,17	273,03	**271**	277,83	**271**	272,97	**271**	272,27	**271**	271,1	**271**
BrasilRS32n	32	4	23,94	269,47	**269**	276,50	**269**	269,00	**269**	269,03	**269**	269,00	**269**
BrasilSP32n	32	4	23,40	257,37	**254**	262,47	**254**	256,40	**254**	255,87	**254**	255,21	**254**
BrasilCO40n	40	5	50,40	578,77	575	584,87	**574**	574,97	**574**	574,22	**574**	574,11	**574**
BrasilNO45n	45	5	62,37	553,80	547	558,23	547	548,33	540	548,10	541	545,63	**539**
att48nA	48	3	80,37	995,60	989	997,07	988	989,03	987	988,26	**987**	988,16	**987**
rd100nB	100	4	750,60	1390,70	1372	1377,30	1362	1374,53	**1358**	1368,10	**1358**	1365,30	1359
ch130n	130	5	1080,00	1681,43	1673	1654,33	1638	1650,30	1638	1649,23	1638	1646,53	**1637**
kroB150n	150	3	1080,00	2951,07	2939	2861,57	2852	2860,40	2850	2858,47	2848	2857,67	**2844**
d198n	198	4	2160,00	3161,50	3149	3065,73	3054	3069,23	3053	3067,23	**3052**	3063,30	3056
Aracaju200n	200	3	2160,00	1916,90	1909	1850,60	**1840**	1854,03	1847	1852,23	1844	1848,73	**1840**
Teresina200n	200	5	2160,00	1389,57	1379	1365,89	1355	1363,16	1352	1358,47	1351	1356,07	**1349**
Curitiba300n	300	5	3240,00	2157,73	2140	2124,60	2113	2120,37	2111	2115,87	2108	2101,20	**2091**

Table 4. Multiple comparisons using the Nemenyi procedure for non-Euclidean instânces.

Algorithm	Frequency	Sum of *ranks*	Avg. of *ranks*	Groups		
ScA+VND+IALSP	17	21,500	1,265	A		
ScA+IALSP	17	38,000	2,235	A	B	
ScA+ALSP	17	49,500	2,912		B	C
ScA	17	72,000	4,235			C
EA+ALSP	17	74,000	4,353			C

Table 5. *p*-values between pairs of algorithms for non-Euclidean instânces

	ScA	EA+ALSP	ScA+ALSP	ScA+IALSP	ScA+VND+IALSP
ScA	1	1,000	0,105	**0,002**	**0,000**
EA+ALSP	1,000	1	0,060	**0,001**	**0,000**
ScA+ALSP	0,105	0,060	1	0,723	**0,020**
ScA+IALSP	**0,002**	**0,001**	0,723	1	0,380
ScA+VND+IALSP	**0,000**	**0,000**	**0,020**	0,380	1

After carrying out the experiments, Friedman's test found significant differences between the solutions obtained by the ScA, EA+ALSP, ScA+ALSP, ScA+IALSP, and ScA+VND+IALSP algorithms, with a *p*-value less than 0.0001. Since the calculated *p*-value is less than the significance level 0.05, we must reject the null hypothesis H_0 of equality of means and accept the alternative hypothesis H_a. The risk of rejecting the null hypothesis H_0, while it is true is less than 0.01%.

Table 3 shows the results of the ScA, EA+ALSP, ScA+ALSP, ScA+IALSP and ScA+VND+IALSP. The best solutions are in boldface. The columns related with the instances show: the instance name, *Name*; the number of cities (vertices) in the graph, n; the number of cars, $|C|$; and the maximum processing time, in seconds, given for each algorithm. Columns *avg.* and *best* show the average and the best solution found by the algorithms in the 30 independent executions, respectively.

Table 4 presents the results of the Nemenyi procedure (two-tail test). Four columns are shown, in addition to the column with the name of the algorithms. These columns are the number of averages for each instance for each algorithm (*frequency*), mean of *ranks*, sum of *ranks*, and group of the results of the algorithms. In this procedure, 3 groups of algorithms are identified (A, B and C). Those algorithms with the best performance belong to group A, with the worst performance to group C and the rest to group B. Algorithms that do not have statistical differences between them show the same letters. The results shows that ScA+VND+IALSP and ScA+IALSP are different to the ScA, EA+ALSP, and ScA+ALSP algorithms, because they correspond to different groups.

A Table 5 presents the p-values between pairs of algorithms. Significant differences are shown between the ScA+VND+IALSP algorithm, and the ScA, EA+ALSP and ScA+ALSP algorithms, given that the its p-values are less than the significance level of 0.05. The same table shows significant differences between the ScA+IALSP algorithm, and the ScA and EA+ALSP algorithms.

The results show a clear superiority of ScA+IALSP and ScA+VND+IALSP algorithms for the state of the art (ScA and EA+ALSP), finding equal or better solutions in all instances.

6   Conclusion

This work presented hybrid approaches of meta-heuristics with methods based on linear programming for the CaRS. A mathematical formulation was presented. Three algorithms have also been proposed for the CaRS, two of which are sequential hybridizations and the other is a more complex hybridization, involving VND as part of *feedback* of the algorithm. The sequential hybrid algorithms were ScA+ALSP and ScA+IALSP. The last hybrid algorithm implemented was ScA+VND+IALSP. In terms of solution quality of the hybrid algorithms, the best results were obtained by ScA+VND+IALSP and ScA+IALSP, followed by ScA+ALSP. The experiments demonstrated the efficiency of ScA+VND+IALSP, which obtained better solutions at the same time than state-of-the-art algorithms. In this sense, the results are considered very promising, so it was possible to show the relevance of developing hybrid approaches to solve CaRS. The hybrid algorithms ScA+VND+IALSP and ScA+IALSP obtained the expected results taking advantage of the power of local searches of the VND and IALSP algorithms; these obtained better results among all the investigated algorithms. The proposed algorithms are very promising due to the fact that, in almost all instances tested, significant improvements have been achieved in relation to the state of the art.

References

1. Addinsoft: Data analysis and statistical solution for Microsoft excel, Paris, France (2017)
2. Asconavieta, P.H., Goldbarg, M.C., Goldbarg, E.F.: Evolutionary algorithm for the car renter salesman. In: 2011 IEEE Congress on Evolutionary Computation (CEC), pp. 593–600. IEEE (2011)
3. Blum, C., Raidl, G.R.: Hybrid Metaheuristics: Powerful Tools for Optimization. Springer, Cham (2016). https://doi.org/10.1007/978-3-319-30883-8
4. Christofides, N., Eilon, S.: An algorithm for the vehicle-dispatching problem. J. Oper. Res. Soc. **20**, 309–318 (1969). https://doi.org/10.1057/jors.1969.75
5. Congram, R.K., Potts, C.N., van de Velde, S.L.: An iterated dynasearch algorithm for the single-machine total weighted tardiness scheduling problem. INFORMS J. Comput. **14**(1), 52–67 (2002)
6. Croes, G.A.: A method for solving traveling-salesman problems. Oper. Res. **6**(6), 791–812 (1958)
7. Desrochers, M., Laporte, G.: Improvements and extensions to the Miller-Tucker-Zemlin subtour elimination constraints. Oper. Res. Lett. **10**(1), 27–36 (1991)
8. Dunn, O.J.: Multiple comparisons among means. J. Am. Stat. Assoc. **56**(293), 52–64 (1961)
9. Felipe, D., Goldbarg, E.F.G., Goldbarg, M.C.: Scientific algorithms for the car renter salesman problem. In: 2014 IEEE Congress on Evolutionary Computation (CEC), pp. 873–879. IEEE (2014)
10. Friedman, M.: The use of ranks to avoid the assumption of normality implicit in the analysis of variance. J. Am. Stat. Assoc. **32**(200), 675–701 (1937)
11. Goldbarg, M.C., Goldbarg, E.F., Asconavieta, P.H., Menezes, M.S., Luna, H.P.: A transgenetic algorithm applied to the traveling car renter problem. Expert Syst. Appl. **40**(16), 6298–6310 (2013). https://doi.org/10.1016/j.eswa.2013.05.072
12. Goldbarg, M.C., Asconavieta, P.H., Goldbarg, E.F.G.: Memetic algorithm for the traveling car renter problem: an experimental investigation. Memetic Comput. **4**(2), 89–108 (2011). https://doi.org/10.1007/s12293-011-0070-y
13. Magnus, W., Karrass, A., Solitar, D.: Combinatorial Group Theory: Presentations of Groups in Terms of Generators and Relations. Courier Corporation (2004)
14. Martyshenko, N.S., Vinichuk, O.Y.: Determining the prospects for car rental market in Primorsky krai (Russia). Int. Rev. Manag. Mark. **6**(2), 213–218 (2016)
15. Miller, C.E., Tucker, A.W., Zemlin, R.A.: Integer programming formulation of traveling salesman problems. J. ACM **7**(4), 326–329 (1960). https://doi.org/10.1145/321043.321046
16. Mladenović, N., Hansen, P.: Variable neighborhood search. Comput. Oper. Res. **24**(11), 1097–1100 (1997)
17. Nemenyi, P.: Distribution-free multiple comparisons. In: International Biometric Society, Washington, DC, vol. 18, p. 263 (1962)
18. Auto Rental News: U.S. car rental revenue and fleet size comparisons (from 2005–2015) (2017). http://www.autorentalnews.com/content/research-statistics.aspx. Accessed 07 July 2017

19. Oliveira, B.B., Carravilla, M.A., Oliveira, J.F.: Fleet and revenue management in car rental companies: a literature review and an integrated conceptual framework. Omega **71**, 11–26 (2016)
20. Raidl, G.R., Puchinger, J.: Combining (integer) linear programming techniques and metaheuristics for combinatorial optimization. In: Blum, C., Aguilera, M.J.B., Roli, A., Sampels, M. (eds.) Hybrid Metaheuristics. SCI, vol. 114, pp. 31–62. Springer, Heidelberg (2008). https://doi.org/10.1007/978-3-540-78295-7_2
21. ZION Market Research: Car rental market by car type (luxury cars, executive cars, economy cars, SUV cars and MUV cars) for local usage, airport transport, outstation and others: global industry perspective, comprehensive analysis, size, share, growth, segment, trends and forecast, 2016–2022 (2017). https://www.zionmarketresearch.com/report/car-rental-market. Accessed 07 July 2017
22. Rios, B.H.O., Goldbarg, E.F.G., Goldbarg, M.C.: A hybrid metaheuristic for the traveling car renter salesman problem. In: 2017 Brazilian Conference on Intelligent Systems (BRACIS). IEEE, October 2017. https://doi.org/10.1109/bracis.2017.20
23. Seay, S., Narsing, A.: Transitioning to a lean paradigm: a model for sustainability in the leasing and rental industries. Acad. Strateg. Manag. J. **12**(1), 113 (2013)
24. da Silva, A.R.V., Ochi, L.S.: An efficient hybrid algorithm for the traveling car renter problem. Expert Syst. Appl. **64**, 132–140 (2016). https://doi.org/10.1016/j.eswa.2016.07.038
25. da Silva, P.H.A.: O problema do caixeiro viajante alugador: um estudo algorítmico. Ph.D. thesis, Universidade Federal do Rio Grande do Norte (2011)

HybridTuner: Tuning with Hybrid Derivative-Free Optimization Initialization Strategies

Benjamin Sauk[1] and Nikolaos V. Sahinidis[2(✉)]

[1] Department of Chemical Engineering, Carnegie Mellon University,
Pittsburgh, PA 15213, USA
[2] H. Milton Stewart School of Industrial and Systems Engineering and School
of Chemical and Biomolecular Engineering, Georgia Institute of Technology,
Atlanta, GA 30332, USA
nikos@gatech.edu
https://sahinidis.coe.gatech.edu

Abstract. To utilize the full potential of advanced computer architectures, algorithms often need to be tuned to the architecture being used. We propose two hybrid derivative-free optimization (DFO) methods to maximize the performance of an algorithm after evaluating a small number of possible algorithmic configurations. Our autotuner (a) reduces the execution time of dense matrix multiplication by a factor of $1.4\times$ compared to state-of-the-art autotuners, (b) identifies high-quality tuning parameters within only 5% of the computational effort required by other autotuners and (c) can be applied to any computer architecture.

Keywords: Autotuners · Derivative-free optimization · GPU computing

1 Introduction

As the landscape of high performance computing evolves, the need to design software that is optimal on a variety of computer architectures has grown. To meet this need, algorithms are designed with tunable parameters to allow for performance portability. Tuning is the problem of selecting a set of parameters that maximize the performance of an algorithm [39]. Identifying an optimal set

---

This work was conducted as part of the Institute for the Design of Advanced Energy Systems (IDAES) with funding from the Office of Fossil Energy, Cross-Cutting Research, U.S. Department of Energy. This work used the Extreme Science and Engineering Discovery Environment (XSEDE), which is supported by National Science Foundation grant number ACI-1548562. Specifically, it used the Bridges system, which is supported by NSF award number ACI-1445606, at the Pittsburgh Supercomputing Center (PSC). We also gratefully acknowledge the support of the NVIDIA Corporation with the donation of the NVIDIA Tesla K40 GPU used for this research.

© Springer Nature Switzerland AG 2021
D. E. Simos et al. (Eds.): LION 2021, LNCS 12931, pp. 379–393, 2021.
https://doi.org/10.1007/978-3-030-92121-7_29

of parameters in this space is challenging as it requires solving a multi-extremal optimization problem in the absence of an explicit algebraic function to relate input tuning parameters to an output performance metric.

To overcome the challenges in this problem, parameter tuning has been solved with derivative-free optimization techniques [35]. As solution strategies in this field involve querying a black-box, the number of simulations required to identify optimal solutions to a problem increases with the number of variables. In practice, tuning problems are challenging in cases with more than thirty variables, though some methods have been observed to find reasonable solutions for problems with hundreds of tuning variables [3]. As there is no way to validate whether optimal parameters have been found for problems with a large search space, many users determine tuning parameters with heuristics, often producing inefficient algorithms.

Autotuning aims to automate the tuning process, where a programmer defines lower and upper bounds for all parameters and a search strategy selects the best set of parameters. Autotuners have been employed for compiler optimization, and machine learning applications. In machine learning, where hyperparameters affect classification accuracy, Bayesian Optimization is commonly utilized for hyperparameter tuning [7]. Applications of autotuning have found widespread usage in computing with graphics processing units (GPUs) [25,36], optimizing dense linear algebra kernels [13,25,29,38], and compiler optimization [3,4,40], among other problems.

In this work, we investigate the benefits of new autotuning strategies based on hybridizing derivative-free optimization algorithms. In contrast to modeling the tuning space with a highly sophisticated neural network, or developing a model after weeks of training, hybrid derivative-free optimization algorithms enable near-optimal tuning parameters on any system in less time than more exhaustive methods. The primary contributions of this paper are as follows:

1. We propose an algorithmic framework to combine local and global DFO approaches for parameter tuning. Two methods, Bandit DFO and Hybrid DFO, are introduced that identify the best or near-optimal solutions for all problems that are considered in this work.
2. We demonstrate that, by combining local and global DFO strategies, it is possible to improve the performance of dense matrix multiplication by a factor of 1.4× compared to optimal parameters identified by OpenTuner [3], ActiveHarmony [39] and Bayesian Optimization [7].
3. We share our implementation with the community facilitating the development and use of hybrid autotuning algorithms. We provide the proposed Bandit DFO and Hybrid DFO as free and open-source software available at https://github.com/bsauk/HybridTuner. The proposed methods can be used with any derivative-free optimization solvers. We include several open-source DFO solvers in our implementation.

In Sect. 2, we review related literature, including the field of autotuning, the literature on derivative-free optimization algorithms, and hybrid tuning algorithms closely related to this work. In Sect. 3, we propose hybrid DFO algorithms and describe advantages of hybrid methods over other approaches. In Sect. 4, we present a computational comparison between autotuners and our proposed hybrid methods to measure the performance of the proposed method for tuning matrix multiplication. We provide conclusions in Sect. 5.

2   Literature Review

2.1   Autotuners

Algorithmic parameter tuning or autotuning has been studied over the last two decades [39]. ActiveHarmony is one of the first autotuners developed [39]. It uses a Nelder-Mead simplex search strategy to suggest points to evaluate that may be potentially optimal. Audet and Orban were the first to utilize derivative-free optimization in the autotuning space when they tuned a trust-region algorithm with four tuning parameters with the MADS algorithm [6]. Later, Audet et al. developed the Optimization of Algorithms (OPAL) framework and used it to tune an algorithm called DFO with nine tuning parameters [5]. PetaBricks generates several algorithms from a list of algorithmic options, and automates code generation from provided inputs [2]. Algorithms tuned by the PetaBricks compiler are divided into sub-problems that are tuned independently of each other using a genetic algorithm. OpenTuner is an autotuner that combines several local search strategies to determine high quality solutions [3], by solving the multi-armed bandit with sliding window, area under the curve credit assignment problem (AUC Bandit Meta Technique) [32]. This technique balances exploiting strategies that have performed well in recent iterations with the potential benefit of having other strategies explore the search space. Hutter et al. explored using a local DFO solver to tune the CPLEX algorithm [21], and then investigated using a global DFO solver to tune SAT solvers [20].

Other autotuning approaches hide the tuning process from users. These autotuners are designed for domain-specific languages where each autotuner is developed for one specific application, and optimizes an algorithm given assumptions that must hold true for the particular application of interest. Automatically Tuned Linear Algebra Software (ATLAS) [44], is a software that was developed by Whaley et al. to automate the empirical tuning and optimization of basic linear algebra subroutines (BLAS). ATLAS optimizes the performance of BLAS on any computer, regardless of architecture, without requiring an extensive knowledge of the system or the underlying linear algebra involved. For BLAS and LAPACK libraries, algorithmic developers have developed autotuning approaches for their algorithms based on heuristics and experimental observations to allow for performance portability [25]. The Optimized Sparse Kernel Interface (OSKI) [43] improves the performance of sparse matrix kernels. OSKI uses heuristics to tune sparse algorithms automatically without any input.

In the machine learning literature, hyperparameter tuning refers to techniques to select parameters for neural networks, such as learning rate, momentum, and categorical decisions, such as the type of activation function to select. Some parameters have been observed to affect the accuracy of classification models produced by neural networks or support vector machines. As accuracy is affected by the selection of hyperparameters, considerable effort has been invested in hyperparameter optimization. Tuning these algorithms is challenging and is typically addressed with random search, grid search, or with Bayesian Optimization using Gaussian Process models [8]. Bayesian Optimization is widely regarded as the most efficient way to perform hyperparameter tuning [7,8,37]. Other stochastic strategies, such as the Covariance Matrix Adaption Evolution Strategy, have been observed to outperform Bayesian Optimization solvers as the iteration budget increases [27]. While the focus of this work is on online parameter tuning, there has also been some work developing models offline with the F-race [9] and irace algorithms [26]. These methods rely on sampling the search space to identify statistically significant parameters and determining an optimal set of parameters, and they have been applied to tuning ant colony optimization algorithms.

2.2   Derivative-Free Optimization Algorithms

DFO algorithms are divided into groups depending on how they search for an optimal solution. One such distinction is between direct methods, such as the Nelder-Mead simplex [30], that search over a certain pattern and model-based methods, such as the trust region method [34], that rely on a model to guide the search. DFO solvers are also classified by whether they search for local or global solutions, and whether they are deterministic or employ stochastic elements. For a more detailed discussion of DFO solvers, readers are referred to [35]. In this work, several DFO solvers that are investigated are listed in Table 1, and expanded upon in the remainder of this section.

Table 1. DFO solvers considered

Solver	Type
HOPSPACK [33]	local, direct
SID-PSM [12]	local, direct
SNOBFIT [22]	global, model
DAKOTA MESH ADAPTIVE SEARCH (MADS) [1]	local, direct
DAKOTA SOGA [1]	global, stochastic
TOMLAB/glcDirect [19]	global, deterministic
TOMLAB/glcFast [19]	global, deterministic
TOMLAB/glcSolve [19]	global, deterministic

2.3   Existing Hybrid Tuning Algorithms

Hybrid algorithms combine simple solvers that may otherwise be unable to escape from a local optima. Hybrid search algorithms have been demonstrated to be effective at identifying optimal tuning parameters for different computational algorithms [3]. In several cases, local DFO solvers have been combined with global strategies to improve their accuracy with fewer function evaluations [14,42].

One line of research involving deterministic hybrid tuning algorithms combines the DIRECT algorithm with implicit filtering [11], pattern search techniques [17], or with surrogate models [18]. In the work of Hemker et al. [18], in every iteration, instead of evaluating a point at the center of a box, as is done in DIRECT, the point to be evaluated is determined by minimizing a surrogate over the area of the domain that is being considered. The authors test this implementation on a problem with 17 variables and bound constraints, but found that the hybrid implementation was unable to determine an optimal solution after 400 function evaluations.

Griffin et al. [17] developed a hybrid framework that combines DIRECT and APPSPACK [16]. APPSPACK is a predecessor of HOPSPACK and is also a generating set search DFO solver. The work of Griffin et al. focuses on solving small problems with two, three, or four variables, and describes how the framework scales with an increasing number of CPU cores. The authors utilize the asynchronous nature of the DIRECT algorithm, and parallelize the search process. Their algorithm does not use DIRECT to initialize the starting point of the APPSPACK algorithm, but instead interleaves the execution of the two algorithms throughout the entire search.

OpenTuner is a hybrid strategy that is initialized with a random starting point, and then explores the search space with several local search strategies [3]. The multi-armed bandit problem is maximized after every iteration to select a solver to locate the next trial point. OpenTuner has been demonstrated to perform well for parameter tuning on high-dimensional problems, such as tuning GCC, while it identifies optimal solutions for smaller problems within a fraction of the iterations required by exhaustive search.

3   Proposed Hybrid Tuning Algorithms

Our proposed methodology derives from the observation that certain global DFO strategies identify high quality solutions quickly, but then struggle to escape from a local minima to obtain a globally optimal solution. We propose two hybrid methodologies, Bandit DFO and Hybrid DFO. We create Bandit DFO by combining a global DFO solver with several local DFO solvers that are selected by solving the multi-armed bandit function. In Hybrid DFO, we initialize a local DFO strategy with the solution found by a global DFO strategy that is executed for a small fraction of the computational budget.

The methods we propose here borrow ideas from OpenTuner, but significantly improve tuning performance by combining local with global DFO solvers. Additionally, merely implementing a global DFO solver in the OpenTuner framework would be insufficient to obtain near-optimal solutions as quickly as the proposed Bandit DFO method. A key improvement in our implementation is the frequency at which we solve the multi-armed bandit problem. We identified that solving the bandit problem less frequently results in faster convergence to optimal solutions, when including model-based DFO solvers.

3.1   Multi-armed Bandit Technique

The multi-armed bandit problem with a sliding window, area under the curve credit assignment problem is a technique developed for programmatic autotuning [3,32]. This problem is inspired by attempting to maximize profit when gambling on numerous slot machines [15]. If multiple machines are available, initially, one has to sample each of the machines to discover the expected profit from each game. Then, there is a balance between exploiting machines that have worked well with exploring options that have not been evaluated recently.

This framework translates well into a hybrid autotuner. In hybrid algorithms, there is a trade-off between exploiting solvers that are currently performing well, and exploring the potential of other solvers with unknown or previously poor performance. Different techniques have been proposed in the literature to balance the trade-off between these objectives, such as simulated annealing [28] and particle swarm optimization [24].

The multi-armed bandit problem is solved at every iteration in the Open-Tuner framework to select which local solver to use at the current iteration. This approach also includes a sliding window to bias the current selection towards solvers that have performed well recently, while ignoring results outside of the current time window. The following optimization problem is solved to determine which technique (t) to select:

$$\underset{t}{\operatorname{argmax}} \frac{2}{N_t(N_t + 1)} \sum_{i=1}^{N_t} iV_{ti} + C\sqrt{\frac{2 \log W}{H_t}}$$

Here, N_t is the number of times that technique t has been used in the current time window, and V_{ti} is 1 if the use of technique t in time period i resulted in discovering a better solution. The length of the sliding window is defined as W. H_t is the frequency of using technique t in the current time window. Finally, the non-negative parameter C controls the exploration and exploitation trade-off. Large values of this parameter put more emphasis on exploration; smaller values on exploitation. Traditionally, this maximization problem is solved at every iteration to determine the solver that is expected to perform the best in the next iteration.

While solving the bandit problem at every iteration may be appropriate when using local direct DFO methods, in practice, we observed that model-based DFO strategies require several iterations before suggesting a better candidate solution. To allow model-based DFO methods to find better trial points, we introduced another hyperparameter, n, into the algorithm to control the frequency of solving the multi-armed bandit problem. This modification is critical to the performance of our algorithm, and can be set to one to recover the original formulation. Figure 1 outlines our multi-armed bandit function implementation. During initialization, it is possible to select a starting point either manually, or by using a DFO solver for a small number of function evaluations. In this work, we initialize Bandit DFO with the DIRECT search method. The hyperparameters in this implementation are:

- C to control the exploration and exploitation trade-off;
- Max_{Evals} is the tuning function limit;
- W is the number of iterations to consider in the sliding window; and
- n is the number of iterations between solving the multi-armed bandit problem.

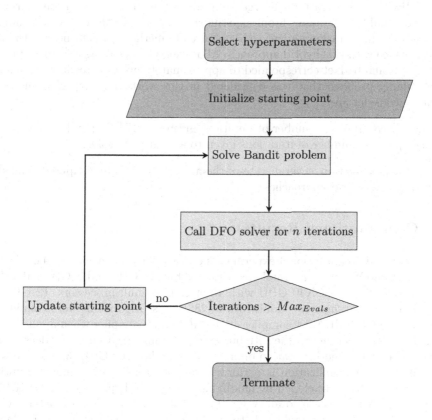

Fig. 1. Algorithmic framework of proposed Bandit DFO method

3.2 Initialization Strategy

The proposed Hybrid DFO algorithm is outlined in Fig. 2. It has previously been observed that the DIRECT algorithm locates near-optimal solutions with a small number of iterations but is unable to converge to a globally optimal solution. We improve the performance of DIRECT by initializing a local DFO solver with the solution returned from DIRECT after a small number of iterations. DIRECT has been combined with local optimizers before, but its application was limited to small problems [16,23]. To extend this idea to large-scale problems, we investigated what percentage of the experimental budget to allocate to global search to identify a near-optimal solution.

We initialize the HOPSPACK and SID-PSM algorithms with a starting point identified from TOMLAB glcDirect. While HOPSPACK and SID-PSM are local direct search DFO solvers, they have been observed experimentally to identify the best measured solutions in previous tuning experiments [36]. We have observed that initializing these solvers with an intermediate solution obtained by DIRECT, leads to faster convergence to near-optimal solutions in comparison to other similar techniques. From experimentation, we identified that assigning 5%–10% of the iteration budget to a global DFO method leads to near-optimal performance with this hybrid approach. Allocating global solvers 5%–10% of the computational budget corresponded to approximately two or three iterations for each tuning problem that was considered in this work. The hyperparameters that we consider are:

- n_{Global} controls the number of iterations given to DIRECT; and
- n_{Local} is the number of iterations given to a local DFO solver.

In the next section, several state-of-the-art autotuners are compared against our proposed hybrid approaches.

4   Computational Results

We conducted computational experiments on two different machines. The first, running CentOS7, with an Intel Xeon E5-1630 at 3.7 GHz and 8 GB of RAM, with a NVIDIA Tesla K40 GPU with 15 streaming multiprocessors, 12 GB of RAM, and a peak memory bandwidth of 288 GB/s. Algorithms were compiled with the NVCC CUDA 9.1 compiler or the GCC 4.8.5 compiler when applicable. For the other experiments, the Pittsburgh Supercomputing Center, Bridges, was used to perform experimentation on a NVIDIA Tesla P100 GPU [31,41]. Dense matrix-matrix multiplication is performed on each GPU with a matrix multiplication algorithm developed by modifying example code provided by NVIDIA in the CUDA 9.1 release to allow for the inclusion of tunable parameters. We created the modified example code to create a tuning space that considers algorithmic options and NVCC compiler optimizations. The parameter space for this problem is 3.4×10^{11} unique combinations. The GCC examples are codes from the PolyBench 4.2.1 benchmark suite [45].

Given the size of the parameter space and that the objective function is a black-box, we are unable to enumerate all parameter combinations to determine an optimal set of tuning parameters for each problem. Instead, we compared solvers by their performance relative to the best solution found in each experiment. Solvers that identify parameters that produce the highest observable performance in the shortest number of function evaluations are regarded as the best autotuner in each experiment. In all of the figures provided below, Bandit refers to our Bandit DFO algorithm, Hybrid refers to the Hybrid DFO algorithm, and Bayesian refers to solving the problem with Bayesian Optimization.

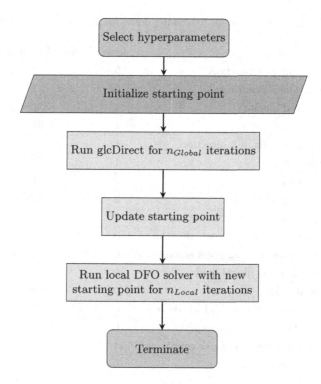

Fig. 2. Algorithmic framework of proposed Hybrid DFO method

4.1 Matrix Multiplication on the Tesla K40

In the matrix multiplication experiment, we addressed the problem of tuning 17 parameters. These parameters are listed in Table 2, along with their corresponding lower and upper bounds. Parameters consist of categorical choices and integer decisions. Categorical choices are represented as 0 or 1 binary decisions such as whether to use GPU shared memory. Integer choices include methods for optimizing spatial locality through tiling. GPU specific parameters, such as the number of threads in a thread block, are adjusted by varying an integer value, while maintaining hardware constraints. The inner loop of matrix multiplication is unrolled based on the loop unrolling parameter. The remainder of the parameters are NVCC compiler optimizations. We considered the parameters that the authors of [10] identified as NVCC parameters that could be tuned to outperform the -O2 or -O3 compiler flags. All parameters investigated in this experiment are integer variables. As several DFO solvers only operate on continuous variables, and may attempt to evaluate fractional trial points, we rounded values to the nearest integer before passing them to the matrix multiplication kernel.

Table 2. Algorithmic options for tuning dense matrix-matrix multiplication

Tunable parameter	Lower bound	Upper bound
Store transpose of matrix A	0	1
Store transpose of matrix B	0	1
Use of shared memory	0	1
Block size	1	63
Number of threads in x direction	1	32
Loop unrolling	1	256
No-align-double	0	1
Relocatable-device-code	0	1
Single-precision denormals support	0	1
Single-precision floating-point division	0	1
Single-precision floating-point square root	0	1
Cache modifier on global load	0	2
Optimization level	0	3
Fusion of multiplication and addition	0	1
Allow expensive optimizations	0	1
Maximum amount of register count	24	63
Preserve resolved relocations	0	1

We report results only for square matrices of size 10000 by 10000. Not shown here are results from additional experiments that we conducted with 2000 by 2000 and 6000 by 6000 matrices, for which we observed similar trends to those of the figures shown below on both the Tesla K40 and the Tesla P100 GPUs. The tuning objective was to maximize performance of the matrix multiplication kernel. Performance was measured in gigaflops (GFLOPs) and was calculated as the number of operations required to perform dense matrix multiplication divided by the amount of time required to perform all of the operations. All solvers were allowed to call the matrix-matrix multiplication kernel 1000 times, and the best performance obtained in the experiment was reported against the iteration number in the figures below. The same random starting point was given to all solvers, if an initial starting point was accepted.

As shown in Fig. 3 for multiplication of 10000 by 10000 matrices, we observed that the autotuners have different performance profiles. ActiveHarmony terminated before finding a solution above 200 GFLOPs, less than half of the peak performance obtained by our Bandit DFO. OpenTuner identified a solution with a performance of 350 GFLOPs after 300 function evaluations, using 10 times more function evaluations than the Hybrid DFO solvers to obtain a similar result. The Bayesian strategy has a similar performance to Bandit DFO for the first 400 iterations. However, the algorithm was terminated by the operating system on our machine because it ran out of memory after 500 iterations.

The proposed hybrid initialization technique discovered a solution with a performance over 300 GFLOPs within the first 20 iterations. Hybrid DFO never escaped from the locally optimal solution initially obtained by the TOMLAB solvers and terminates after 400 iterations. Both Hybrid DFO and Bandit DFO use a DIRECT algorithm initially. However, Bandit DFO explored a different search direction than Hybrid DFO, leading to a worse performance for the first 600 iterations. After 550 function evaluations, the SID-PSM sover in Bandit DFO escaped from the previous local optima and improved to over 400 GFLOPs. Bandit DFO converged to a parameter set that outperformed the best solution obtained by the other autotuners by more than 80 GFLOPs after 600 iterations.

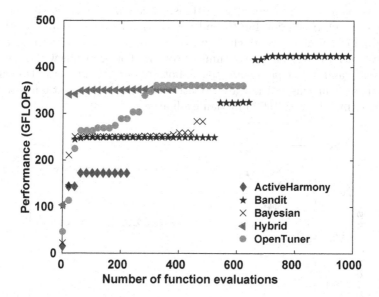

Fig. 3. Comparison of the performance of different autotuners tuning dense matrix multiplication for 10000×10000 matrices on the Tesla K40 GPU

4.2 Matrix Multiplication on the Tesla P100

We also performed the same set of matrix multiplication experiments on another GPU, the Tesla P100. Figure 4 displays the results for the multiplication of square 10000 by 10000 matrices. ActiveHarmony and OpenTuner obtained suboptimal solutions and terminated after 400 and 700 function evaluations respectively. Unlike on the Tesla K40, Bayesian Optimization was able to be run for 1000 iterations on the Pittsburgh Supercomputing Center Bridges. While we do not report all of the results here, we note that the Bayesian strategy had the same performance profile on all three of the matrix sizes that we experimented on, converging to a performance around 1600 GFLOPs. In each case, the Bayesian approach arrives at a slightly different set of optimal parameters, even though the performance was similar.

Bandit DFO was the first to achieve a performance over 2000 GFLOPs, and then outperformed all of the other solvers after 550 function evaluations. The use of multiple DFO solvers combined with the DIRECT strategy performed well in both obtaining near-optimal solutions, and converging to the best solution observed in our experiments. The second-best performing strategy was our Hybrid DFO algorithm. This algorithm was the fastest solver to find a solution with a performance over 2100 GFLOPs. Our main results on the P100 GPU align with the results on the K40 GPU. Parameters obtained by using our hybrid tuning algorithms are superior to those obtained with other autotuners, either in terms of performance, or in the number of function evaluations required to identify the best known solution. Near-optimal solutions are obtained within the first 200 iterations with the proposed algorithms, while OpenTuner, ActiveHarmony, and Bayesian Optimization fail to find near-optimal solutions after 1000 iterations. From experiments conducted here, our proposed hybrid methods are the best solvers to use for this type of tuning problem. For problems with fewer than 20 variables, and a vast parameter space, both of our hybrid methods identified near-optimal solutions quickly and identified better solutions than any other strategy within the first 1000 function evaluations.

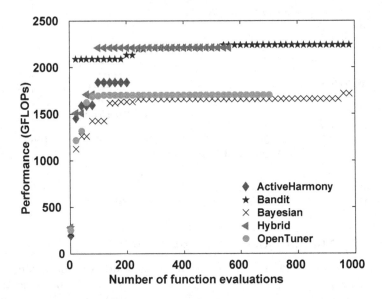

Fig. 4. Comparison of the performance of different autotuners tuning dense matrix multiplication for 10000 × 10000 matrices on the Tesla P100 GPU

5   Conclusions

This paper investigates hybrid tuning algorithms for parameter tuning. While previous approaches rely on heuristics, or local direct search derivative-free optimization algorithms, we propose hybridizing global DFO algorithms with local methods. We propose two hybrid methodologies, Bandit DFO and Hybrid DFO, that combine different DFO strategies to improve the rate at which tuners converge to an optimal solution.

We demonstrate that the two proposed hybrid algorithms outperform three state-of-the-art autotuners, ActiveHarmony, OpenTuner, and Bayesian Optimization with Gaussian Process models. Bandit DFO reduces the execution time of dense matrix multiplication by a factor of 1.4× compared to algorithms generated by other autotuners on a problem with a parameter space of 3.4×10^{11} combinations. In additional results, that we were unable to present due to space limitations, we found that the proposed algorithms have advantages over other autotuners for tuning the GCC compiler.

By combining global DFO strategies with local strategies, our hybrid algorithms identify the best observed parameters for the tuning applications that we present here. The proposed hybrid algorithms are generic and can tune problems of various sizes. To facilitate the development and use of autotuning software, we provided an open-source implementation of our Bandit DFO and Hybrid DFO algorithms for parametric autotuning.

References

1. Adams, B.M., et al.: DAKOTA, A Multilevel Parallel Object-Oriented Framework for Design Optimization, Parameter Estimation, Uncertainty Quantification, and Sensitivity Analysis: Version 6.5 User's Manual. Sandia National Laboratories, Albuquerque/Livermore (2016). https://dakota.sandia.gov/
2. Ansel, J., et al.: PetaBricks: a language and compiler for algorithmic choice. In: Proceedings of the 30th ACM SIGPLAN Conference on Programming Language Design and Implementation, pp. 38–49. Association for Computing Machinery, New York (2009)
3. Ansel, J., et al.: OpenTuner: an extensible framework for program autotuning. In: Proceedings of the 23rd International Conference on Parallel Architectures and Compilation, pp. 303–316. Association for Computing Machinery, New York (2014)
4. Ashouri, A., Mariani, G., Palermo, G., Park, E., Cavazos, J., Silvano, C.: COBAYN: compiler autotuning framework using Bayesian networks. ACM Trans. Archit. Code Optim. (TACO) **13**, 1–26 (2016)
5. Audet, C., Dang, C.-K., Orban, D.: Algorithmic parameter optimization of the DFO method with the OPAL framework. In: Suda, R., Naono, K., Teranishi, K., Cavazos, J. (eds.) Software Automatic Tuning, pp. 255–274. Springer, New York (2011). https://doi.org/10.1007/978-1-4419-6935-4_15
6. Audet, C., Orban, D.: Finding optimal algorithmic parameters using derivative-free optimization. Soc. Ind. Appl. Math. **17**, 642–664 (2006)
7. Balandat, M., et al.: BoTorch: programmable Bayesian optimization in PyTorch, pp. 1–20. arXiv preprint arXiv:1910.06403 (2019)
8. Bergstra, J., Bardenet, R., Bengio, Y., Kégl, B.: Algorithms for hyper-parameter optimization. In: Shawe-Taylor, J., Zemel, R.S., Bartlett, P.L., Pereira, F., Weinberger, K.Q. (eds.) Proceedings of the 24th International Conference on Neural Information Processing Systems, pp. 2546–2554. Curran Associates Inc., Red Hook (2011)

9. Birattari, M., Yuan, Z., Balaprakash, P., Stützle, T.: F-Race and iterated F-Race: an overview. In: Bartz-Beielstein, T., Chiarandini, M., Paquete, L., Preuss, M. (eds.) Experimental Methods for the Analysis of Optimization Algorithms, pp. 311–336. Springer, Heidelberg (2010). https://doi.org/10.1007/978-3-642-02538-9_13

10. Bruel, P., Gonzalez, M., Goldman, A.: Autotuning GPU compiler parameter using OpenTuner. In: XXII Symposium of Systems of High Performance Computing, Bangalore, India, pp. 1–12. IEEE (2015)

11. Carter, R., Gablonsky, J., Patrick, A., Kelley, C., Eslinger, O.: Algorithms for noisy problems in gas transmission pipeline optimization. Optim. Eng. **2**, 139–157 (2001). https://doi.org/10.1023/A:1013123110266

12. Custódio, A.L., Vicente, L.N.: SID-PSM: a pattern search method guided by simplex derivatives for use in derivative-free optimization. Departamento de Matemática, Universidade de Coimbra, Coimbra, Portugal (2008)

13. Davidson, A., Owens, J.: Toward techniques for auto-tuning GPU algorithms. In: Jónasson, K. (ed.) PARA 2010. LNCS, vol. 7134, pp. 110–119. Springer, Heidelberg (2012). https://doi.org/10.1007/978-3-642-28145-7_11

14. Fan, S.S., Zahara, E.: A hybrid simplex search and particle swarm optimization for unconstrained optimization. Eur. J. Oper. Res. **181**, 527–548 (2007)

15. Fialho, A., Da Costa, L., Schoenauer, M., Sebag, M.: Analyzing bandit-based adaptive operator selection mechanisms. Ann. Math. Artif. Intell. **60**, 25–64 (2010). https://doi.org/10.1007/s10472-010-9213-y

16. Gray, G.A., Kolda, T.G.: Algorithm 856: APPSPACK 4.0: parallel pattern search for derivative-free optimization. ACM Trans. Math. Softw. **32**, 485–507 (2006)

17. Griffin, J.D., Kolda, T.G.: Asynchronous parallel hybrid optimization combining DIRECT and GSS. Optim. Methods Softw. **25**, 797–817 (2010)

18. Hemker, T., Werner, C.: DIRECT using local search on surrogates. Pac. J. Optim. **7**, 443–466 (2011)

19. Holmström, K., Göran, A.O., Edvall, M.M.: User's Guide for TOMLAB 7. Tomlab Optimization. http://tomopt.com

20. Hutter, F., Hoos, H.H., Leyton-Brown, K.: Sequential model-based optimization for general algorithm configuration. In: Coello, C.A.C. (ed.) LION 2011. LNCS, vol. 6683, pp. 507–523. Springer, Heidelberg (2011). https://doi.org/10.1007/978-3-642-25566-3_40

21. Hutter, F., Hoos, H.H., Leyton-Brown, K., Stützle, T.: ParamILS: an antomatic algorithm configuration framework. J. Artif. Intell. Res. **36**, 267–306 (2009)

22. Huyer, W., Neumaier, A.: SNOBFIT-stable noisy optimization by branch and fit. ACM Trans. Math. Softw. **35**, 1–25 (2008)

23. Jones, D.R.: The DIRECT global optimization algorithm. In: Floudas, C.A., Pardalos, P.M. (eds.) Encyclopedia of Optimization, vol. 1, pp. 431–440. Kluwer Academic Publishers, Boston (2001)

24. Kennedy, J., Eberhart, R.: Particle swarm optimization. In: Proceedings of the IEEE International Conference on Neural Networks, Piscataway, NJ, USA, pp. 1942–1948 (1995)

25. Li, Y., Dongarra, J., Tomov, S.: A note on auto-tuning GEMM for GPUs. In: Allen, G., Nabrzyski, J., Seidel, E., van Albada, G.D., Dongarra, J., Sloot, P.M.A. (eds.) ICCS 2009. LNCS, vol. 5544, pp. 884–892. Springer, Heidelberg (2009). https://doi.org/10.1007/978-3-642-01970-8_89

26. López-Ibáñez, M., Dubois-Lacoste, J., Cáceres, L., Birattari, M., Stützle, T.: The irace package: iterated racing for automatic algorithm configuration. Oper. Res. Perspect. **3**, 43–58 (2016)

27. Loshchilov, I., Hutter, F.: CMA-ES for hyperparameter optimization of deep neural networks, pp. 1–15. arXiv preprint arXiv:1604.07269 (2016)
28. Metropolis, N., Rosenbluth, A.W., Rosenbluth, M.N., Teller, A.H., Teller, E.: Equation of state calculations by fast computing machines. J. Chem. Phys. **21**, 1087–1092 (1953)
29. Nath, R., Tomov, S., Dongarra, J.: An improved MAGMA GEMM for Fermi graphics processing units. Int. J. High Perform. Comput. Appl. **24**, 511–515 (2010)
30. Nelder, J.A., Mead, R.: A simplex method for function minimization. Comput. J. **7**, 308–313 (1965)
31. Nystrom, N., Levine, M., Roskies, R., Scott, J.: Bridges: a uniquely flexible HPC resource for new communities and data analytics. In: Proceedings of the 2015 XSEDE Conference: Scientific Advancements Enabled by Enhanced Cyberinfrastructure, pp. 1–8. Association for Computing Machinery, New York (2015)
32. Pacula, M., Ansel, J., Amarasinghe, S., O'Reilly, U.-M., et al.: Hyperparameter tuning in bandit-based adaptive operator selection. In: Di Chio, C. (ed.) EvoApplications 2012. LNCS, vol. 7248, pp. 73–82. Springer, Heidelberg (2012). https://doi.org/10.1007/978-3-642-29178-4_8
33. Plantenga, T.D.: HOPSPACK 2.0 user manual. Technical report SAND2009-6265, Sandia National Laboratories, Albuquerque, NM and Livermore, CA (2009). https://software.sandia.gov/trac/hopspack/
34. Powell, M.J.D.: UOBYQA: unconstrained optimization BY quadratic approximation. Math. Program. **92**, 555–582 (2002). https://doi.org/10.1007/s101070100290
35. Rios, L.M., Sahinidis, N.V.: Derivative-free optimization: a review of algorithms and comparison of software implementations. J. Glob. Optim. **56**, 1247–1293 (2013). https://doi.org/10.1007/s10898-012-9951-y
36. Sauk, B., Ploskas, N., Sahinidis, N.V.: GPU parameter tuning for tall and skinny dense linear least squares problems. Optim. Methods Softw. **35**, 638–660 (2020)
37. Snoek, J., Larochelle, H., Adams, R.P.: Practical Bayesian optimization of machine learning algorithms. In: Pereira, F., Burges, C.J.C., Bottou, L., Weinberger, K.Q. (eds.) Proceedings of the 25th International Conference on Neural Information Processing Systems, pp. 2951–2959. Curran Associates Inc., Red Hook (2012)
38. Tan, G., Li, L., Triechle, S., Phillips, E., Bao, Y., Sun, N.: Fast implementation of DGEMM on Fermi GPU. In: Proceedings of 2011 International Conference for High Performance Computing, Networking, Storage and Analysis, pp. 35–46. Association for Computing Machinery, New York (2011)
39. Ţăpuş, C., Chung, I., Hollingsworth, J.: Active harmony: towards automated performance tuning. In: Proceedings of the ACM/IEEE Conference on Supercomputing, pp. 1–11. IEEE Computer Society Press, Washington, DC (2002)
40. Tartara, M., Reghizzi, S.: Continuous learning of compiler heuristics. ACM Trans. Archit. Code Optim. (TACO) **9**, 1–25 (2013)
41. Towns, J., et al.: XSEDE: accelerating scientific discovery. Comput. Sci. Eng. **16**, 62–74 (2014)
42. Vaz, A.I.F., Vicente, L.N.: A particle swarm pattern search method for bound constrained global optimization. J. Glob. Optim. **39**, 197–219 (2007). https://doi.org/10.1007/s10898-007-9133-5
43. Vuduc, R., Demmel, J., Yelick, K.: OSKI: a library of automatically tuned sparse matrix kernels. J. Phys: Conf. Ser. **16**, 521–530 (2005)
44. Whaley, R., Petitet, A., Dongarra, J.: Automated empirical optimizations of software and the ATLAS project. Parallel Comput. **27**, 3–35 (2001)
45. Yuki, T., Pouchet, L.N.: PolyBench/C 4.2.1. https://www.cs.colostate.edu/~pouchet/software/polybench/polybench-fortran.html

Sensitivity Analysis on Constraints of Combinatorial Optimization Problems

Julian Schulte$^{(\boxtimes)}$ and Volker Nissen

Ilmenau University of Technology, Ilmenau, Germany
`julian.schulte@tu-ilmenau.de`

Abstract. Combinatorial optimization problems in practice are subject to a variety of constraints, such as resource limitations or organizational regulations. Since these model parameters can have a major impact, for example, on the performance of a scheduling system, it is crucial to know how changes in the constraints affect the optimal solution value. The question of how changes in input parameters of an optimization model, such as right-hand side values of constraints, affect the output of the model is the main concern of sensitivity analysis. Although well established in the domain of linear programming, the literature on combinatorial optimization lacks universal sensitivity analysis approaches which are applicable to practical problems. In this paper, a general approach is proposed which allows to identify how the optimal solution of a combinatorial optimization problem is affected when model parameters, such as constraints, are changed. Using evolutionary bilevel optimization in combination with data mining and visualization techniques, the suggested concept of bilevel innovization allows to find trade-offs among constraints and objective function value. Additionally, it enables decision-makers to gain insights into the overall model behavior under changing framework conditions. The concept of bilevel innovization as a tool for sensitivity analysis is illustrated, without loss of generality, by the example of the generalized assignment problem.

Keywords: Sensitivity analysis · Combinatorial optimization · Evolutionary bilevel optimization · Data mining · Generalized assignment

1 Introduction

In practical applications, combinatorial optimization problems are generally subject to a variety of constraints, such as resource and time limitations or organizational and legal regulations [21]. In personnel scheduling, these constraints may concern, for example, the size and structure of a company's workforce, staffing policies or working time regulations. In vehicle routing, as another example, vehicle capacities or fleet size may be relevant as well as due dates or time window regulations. Since these model parameters can have a major impact, for instance, on the performance of a scheduling system, it is crucial to know how changes in the constraints affect the optimal solution value. This is especially relevant in

© Springer Nature Switzerland AG 2021
D. E. Simos et al. (Eds.): LION 2021, LNCS 12931, pp. 394–408, 2021.
https://doi.org/10.1007/978-3-030-92121-7_30

(strategic) planning situations, where a company has to determine this kind of framework conditions rather than solely optimizing the combinatorial problem.

The analysis of the behavior of an optimization model and the answering of questions such as "Given a specific change of a parameter, what is the new optimal solution value?" or "What is the new optimal solution value when several parameters change simultaneously?" are the main focus of sensitivity analysis (SA) [12]. Although SA is well established in the field of linear optimization and an important part of the model building as well as decision-making process [13], the literature on combinatorial optimization lacks universal sensitivity analysis approaches which are applicable to practical problems. Hall and Posner [12] performed an extensive study of sensitivity analysis approaches for scheduling problems and provide suggestions for SA that can be applied to a variety of scheduling problems. However, they indicate various issues which remain open. From these issues (especially issues I1, I2, I9), we identified three main requirements that should at least be met for a SA approach to be useful in practice:

1. *Problem independence*: The sensitivity analysis should neither require a detailed analysis of the mathematical model nor rely on certain model or problem characteristics.
2. *Algorithm independence*: The sensitivity analysis should not rely on algorithm properties, such as duality or an algorithmic analysis of a heuristic. Thereby, a solution method can be chosen that is most suitable not for SA but to solve the problem at hand.
3. *Simultaneous independent changes of multiple parameters*: The sensitivity analysis should identify how simultaneous changes of multiple parameters affect the optimal solution. To get full insight into the model behavior, it is furthermore important that these changes can occur independently of each other, in contrast to, e.g., parametric programming approaches.

To the best of our knowledge, in the literature on combinatorial optimization there exists no approach that meets these requirements (see [1,11,15] for recent works and reviews on SA in combinatorial optimization). In the general operational research literature, however, the approach of global sensitivity analysis has been identified that fulfills the above criteria. The idea of global SA was introduced into the domain of operations research in order to overcome deficits of classical, duality-based sensitivity analysis approaches [29]. In general, global SA is not limited to optimization problems but relevant in all disciplines that deal with mathematical modeling, such as economics or engineering [3,25]. In contrast to local SA methods, which perform analyses around one point of interest in the input parameter space, e.g., by varying one parameter at a time, global SA methods perform simultaneous variations of input parameters that cover the entire parameter space, e.g., by sampling data points via Monte Carlo simulation and subsequently applying statistical methods [3,24].

Although fulfilling the mentioned requirements, the general application of global SA in combinatorial optimization is limited by its focus on uncertainty in the model input parameters and the fundamental assumption that probability distributions can be assigned to each input parameter [3]. Therefore, if uncertainty is the primary concern of the problem to be analyzed and there is information about the probability distribution of the model inputs (e.g., constraint values), global SA could be used to get a better model and problem understanding. However, if the primary interest of the sensitivity analysis is to identify efficient trade-offs regarding e.g., resource usage and objective function value, a different approach is needed.

The identification of optimal constraint configurations, such as the number of vehicles or shifts, does not necessarily have to be part of a sensitivity analysis, but can also be considered an optimization problem itself. Examples are the problems of location routing [23] or shift design [8]. While these approaches allow to find trade-offs regarding several types of constraints and the optimal solution value, they provide only limited information about how a model responds to certain input parameter changes and therefore do not allow to answer questions related to sensitivity analysis.

In order to overcome this limitation and provide decision-makers with deeper insights into the overall model behavior, Schulte et al. [26] proposed the concept of bilevel innovization (BLI). The concept was demonstrated in the context of a personnel scheduling problem which was solved using a Genetic Algorithm (GA). Here, the scheduling problem was analyzed regarding how variations in the number and qualification of employees as well as the use of different shifts influence the objective function value.

Due to its fulfillment of the aforementioned requirements and the successful application to a problem of practical size, in this paper we propose the concept of bilevel innovization as general approach for sensitivity analysis on constraints in combinatorial optimization (see Sect. 2 for more information on BLI). To illustrate the general applicability, the concept is demonstrated by the example of the generalized assignment problem. The problem was chosen because of its easy comprehensibility and high practical relevance due to its transferability to numerous fields of application, such as workforce planning, project management, vehicle routing and production planning [20, 22].

The remainder of this paper is structured as follows: In Sect. 2, the general concept of bilevel innovization is presented. In Sect. 3 and 4, the single steps of the BLI process are demonstrated in detail by the example of the generalized assignment problem. Section 3 focuses on bilevel optimization, whereas Sect. 4 illustrates the application of visualization and data mining methods within the BLI process to gain insights into model behavior. Finally, conclusions and suggestions for further research are presented in Sect. 5.

2   Bilevel Innovization

In the following, we briefly describe the general idea of bilevel innovization as well as the underlying concepts. The main assumption behind using bilevel innovization for sensitivity analysis on constraints in combinatorial optimization is that there exist trade-offs between the optimal solution value and the constraint values (e.g., resource limitations). This kind of (strategic) problem can be formulated as a hierarchical optimization problem (bilevel problem), where the lower-level problem optimizes its objective based on the parameters determined by the upper-level optimization problem. The upper-level, in turn, optimizes its objectives under consideration of the lower-level results [6,18]. In the context of BLI and SA, the upper-level problem determines the constraints of the lower-level problem, which is the actual problem under consideration for SA.

A popular approach to solve bilevel problems is using an Evolutionary Algorithm (EA) at the upper-level and any kind of optimization algorithm at the lower-level, resulting in a nested EA [27,28]. Since the upper-level problem faces at least two conflicting objectives, i.e., lower-level and upper-level objective, a multi-objective optimization problem has to be solved. Due to the usage of an EA at the upper-level, an evolutionary multi-objective algorithm is applied. Evolutionary multi-objective optimization (EMO) supports the decision-making process by providing a set of Pareto optimal solutions. The final solution to be selected by the decision-maker will, therefore, be a trade-off among the considered objectives (see [4,10] for more detailed information on EMO).

In general, the bilevel innovization process (see Fig. 1) can be divided into two parts: data generation and data analysis. The first part serves the purpose of generating the data set for the subsequent analysis by solving the investigated bilevel optimization problem. The data analysis part is based on the visual analytics process [14] and aims to gain insights into model behavior and to extract rules that support decision-making regarding possible modifications of parameters (e.g., resource allocation). Visual analytics integrates model-based and visual analysis methods and consequently combines the strengths of machine and human capabilities. Here, the user is in a constant loop of data processing (e.g., feature creation), visualization (e.g., scatter or box plots) and model building (e.g., regression or decision trees) in order to gain knowledge of the problem.

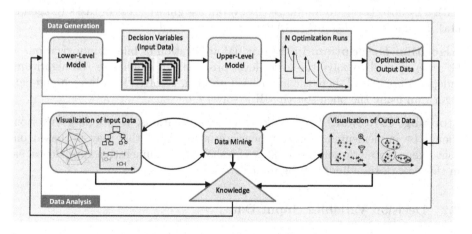

Fig. 1. Bilevel innovization process [26]

3 Data Generation

3.1 Lower-Level Model

Starting point is the actual problem to be analyzed and the lower-level optimization model, respectively. The model behavior is represented by the lower-level objective function value. The optimization algorithm here is freely selectable depending on the considered problem. Details regarding problem instances and parameter settings are described in Sect. 3.4. Due to the bilevel notation we use y instead of x as decision vector.

Problem Description: The generalized assignment problem (GAP) is concerned with the optimal assignment of n tasks to m agents and can be formulated as follows [20, 22]:

$$\min_{y} \sum_{i=1}^{m} \sum_{j=1}^{n} c_{ij} y_{ij} \tag{1a}$$

$$\text{subject to:} \quad \sum_{i=1}^{m} y_{ij} = 1 \qquad j = 1, ..., n \tag{1b}$$

$$\sum_{j=1}^{n} r_{ij} y_{ij} \le b_i \qquad i = 1, ..., m \tag{1c}$$

$$y_{ij} \in \{0, 1\} \qquad \forall i, j \tag{1d}$$

where c_{ij} is the cost and r_{ij} the capacity requirement of assigning task j to agent i. Each task j has to be assigned to one agent i (1b) and one agent can be assigned multiple tasks while not exceeding the agent's capacity b_i (1c). Whether a task j is assigned to agent i is indicated by the binary decision variable y_{ij} (1d). The objective is to minimize the overall assignment cost of tasks to agents (1a). For more details on the GAP we refer to [20, 22].

Algorithm Description: In the original application of bilevel innovization, a GA was applied to solve the lower-level problem. To emphasize the algorithm independence of BLI, here the lower-level problem is solved by a branch-and-cut algorithm using the open-source MILP solver Cbc[1].

Stopping Criterion: To provide efficient evaluations, a stopping criterion should be implemented. For metaheuristic algorithms, this may be based on convergence rate or a solution quality threshold. For enumerative approaches, as in this case, a time limit can be used.

3.2 Decision Variables (Input Data)

The next step is to determine the parameters of the lower-level model to be investigated (e.g., resource constraints). These will serve as decision variables at

---

[1] https://github.com/coin-or/Cbc.

the upper-level EMO problem. The decision variables may, for instance, change the right-hand side value of a particular constraint (in the following referred to as constraint value) or cause constraints to be added or removed.

Variable Selection: For the example of the generalized assignment problem, the issue to be analyzed is how variations in the resource constraints (Eq. 1c) affect the overall assignment cost (Eq. 1a). Therefore, the right-hand side value b_i of each resource constraint is used as one of the decision variables.

Variable Translation to Objective Value: Furthermore, it has to be determined how the selected decision variables are to be represented at the upper-level problem. One approach is to consider each decision variable as an individual objective, which is then integrated into the upper-level problem. The second approach is to use scalarization methods to aggregate multiple decision variables to a single value (for more information on scalarization see [10]).

Since there is no indication that we are facing constraints of different kinds, for the considered GAP, we decided on scalarization using linear weighting to convert the m resource constraint values into one single objective value. To reflect their initial importance, the weights are set to the normalized resource capacities of the initial data of the selected problem instance (see Sect. 3.4).

Variable Bounds: To limit the search space, the upper and lower bounds of the decision variables should be set to a reasonable value, e.g., the area on which the sensitivity analysis is focused. To get a general overview of the considered GAP, we set the upper bound of each constraint to the maximum value needed to assign all tasks to one agent and the lower bound to zero.

3.3   Upper-Level Model

Now, the resulting upper-level problem has to be modeled. For solving the upper-level problem, an evolutionary multi-objective algorithm is used. In the context of sensitivity analysis, it is assumed that at least two conflicting objectives are optimized within the bilevel optimization problem, one of which should be the lower-level objective function value.

Bilevel Problem: Bilevel optimization problems can generally be formulated as follows [27,28]:

$$\min_{x,y} \quad F(x,y) \tag{2a}$$

$$\text{subject to:} \quad G(x,y) \leq 0 \tag{2b}$$

$$y \in \operatorname*{argmin}_{y}\{f(x,y) : g(x,y) \leq 0\} \tag{2c}$$

where x and y are the vectors of decision variables determined by the upper- and lower-level problem, respectively. Moreover, $F(x,y)$ and $f(x,y)$ are the objective functions and $G(x,y)$ and $g(x,y)$ the constraints of the upper- and lower-level problem. For the sensitivity analysis, x represents the vector of constraint values. For each x, the GAP $f(x,y)$ will be optimized yielding the vector of assigned tasks y. Therefore, the upper-level objective function $F(x, y)$ is dependent on the scalarized constraint values (upper-level decision) as well as the cost of the assigned tasks at the lower level, which in turn is influenced by the determined value of each constraint.

Problem Description: The upper-level problem corresponds to the trade-off decision that is analyzed within the sensitivity analysis. For the GAP, the trade-off regarding cost and constraint values is of interest. As mentioned in Sect. 3.1, the upper-level objective is created by linear weighting of the m agent capacity constraints (3b), with the weights w_i representing the relative importance of the constraint values B_i of the initial lower-level problem data set (3c). The value range of x_i is limited by predefined upper (UB) and lower (LB) bounds (3d). The objective of the upper-level problem is to minimize the scalarized constraint values (3a) subject to constraints (3b)–(3d) and the total assignment cost resulting from the lower-level problem decision (3e).

Algorithm Description: Following the taxonomy given by Talbi [28], the algorithm applied to solve the sensitivity analysis problem is a nested constructing approach with a metaheuristic at the upper-level and an exact optimization algorithm at the lower-level. In this type of bilevel model, an upper-level metaheuristic calls a lower-level MILP solver during its fitness assessment. In doing so, the upper-level heuristic determines the decision vector x (here the values of the different constraints) as input for the lower-level algorithm, which in turn determines the decision vector y (optimized task assignment). Both decision vectors are subsequently used to solve the bilevel problem at the upper-level.

$$\min_{x,y} \quad F(W_x, y) \tag{3a}$$

$$\text{subject to:} \quad W_x = \sum_{i=1}^{m} w_i x_i \tag{3b}$$

$$w_i = \frac{B_i}{\sum_{\iota=1}^{m} B_\iota} \quad i = 1, ..., m \tag{3c}$$

$$UB \geq x_i \geq LB \quad i = 1, ..., m \tag{3d}$$

$$y \in \underset{y}{\arg\min}\{(1a) - (1d)\} \tag{3e}$$

At the upper-level, the NSGA-II [7] is used to solve the multi-objective problem of minimizing constraint values and assignment cost. The chromosome of the upper-level individuals is denoted by the one-dimensional integer vector $\mathbf{x} = (x_1, ..., x_m)$ with each value representing one constraint of the GAP. For reproduction, one-point, uniform and intermediate crossover are applied, randomly selected for each reproduction process. Furthermore, random integer as well as random walk mutation are used (see [16] for more details).

Stopping Criterion: The primary aim of solving the upper-level problem is not to find the best possible solutions to the optimization problem but to generate data for the subsequent analysis. For this purpose, a stopping criterion (see [17]) measures convergence based on how many non-dominated individuals of the current iteration dominate those of the previous iteration.

3.4   N Optimization Runs

For the actual data generation, a number of independent runs of the upper-level algorithm have to be conducted in order to obtain as many different solutions as possible for the subsequent analysis. The number of runs depends, among others, on the problem structure, the number and characteristics of the decision variables at the upper-level and the required computation time. Since BLI is to be considered an iterative process, one may start with only a few runs and gradually increase until satisfactory data is created.

Problem Instances: In this study, we use the benchmark data sets for the GAP presented in [5]. For the demonstration, the first instance of the set "gapa" with $m = 5$ agents and $n = 100$ tasks was chosen. The selection was arbitrary, however, experiments on other instances showed similar behavior.

Parameter Settings: For the upper-level GA, a population size of 50, a threshold of 5% (as recommended in [17]), and 10 restarts are chosen, with each restart having a random initial population. The lower bound of the decision variables is set to zero, the upper bound to the maximum of all constraint values required to assign all tasks to one agent. If the stopping criterion is not activated, a generation number of 100 is selected. The mutation rate is set to $1/v$, with v being the number of genes of the encoded upper-level individual. At the lower-level, a MILP solver with a time limit of 10 s is used. Experiments on the actual benchmark instances with larger time limits did not yield results where the improvements justified the considerable larger computation times.

Optimization Output Data: The last step of the data generation stage is to prepare the obtained data sets, i.e., evaluated individuals of the different optimization runs, in a suitable manner for the following visual analytics task. All values associated with each evaluated individual, such as objective or decision variable values, are henceforth referred to as features.

Data Collection: One key aspect of bilevel innovization is not only to analyze the final Pareto-optimal solution set but to collect all individuals that are evaluated at the upper-level algorithm. In order to perform a comprehensive analysis, it is further recommended to not only store the objective function value of the lower-level algorithm but to collect all data that may be useful for the analysis. Possible additional features could be constraint usage, constraint slack or, when using a heuristic at the lower-level, constraint violations.

Feature Creation: In addition to the features obtained during optimization, new features can be created based on the available data. This can either be done in advance or during the analysis based on new insights.

For the example of the GAP, we introduce two additional features for all evaluated upper-level individuals $s \in S$. The first feature (see Eq. (4)) represents the overall *capacity* of all agents and can be considered as decoding of the upper-level objective W_x (3b). The second feature (see Eq. (5)) serves as efficiency indicator and measures the euclidean distance from each solution $s \in S$ to the nearest Pareto-optimal solution $p \in P$, with f_d representing each upper-level objective in a d-dimensional objective space, here constraint values and cost (see [9] for more information on this efficiency measure).

$$capacity_s = \sum_{i=1}^{m} b_i \qquad i = 1, ..., m \qquad (4)$$

$$efficiency\_distance_s = \min_p \sqrt{\sum_{f=1}^{d} (f_{ds} - f_{dp})^2} \qquad \forall p \in P \qquad (5)$$

4   Data Analysis

In the context of bilevel innovization, each data record is at least composed of the objective values of an evaluated individual at the upper-level problem (output data) and the corresponding decision variables (input data). As is common in visual analytics, data analysis can be considered an iterative process, and one can step back to create new features or generate additional data.

4.1   Visualization of Output Data

The first step is to visualize the output data. Depending on the spread of the data, an area of interest for deeper analysis can be selected manually (e.g., range around the Pareto-optimal front). Subsequently, the filtered data can be visualized again to explore shape and distribution of the objective values. This may lead to first insights regarding the model's behavior.

Within the 10 optimization runs of the upper-level algorithm, 19,787 solutions have been evaluated. For 747 of these individuals, no feasible solution was found, i.e., the constraints were set too tight or the time limit was too short. Furthermore, for each solution, the gap between optimal solution value of the linear relaxation (regarding (1d)) and the actual solution was measured. For all feasible solutions this gap is on average 0.46%, which indicates that the lower-level stopping criterion was set to a reasonable value.

Area of Interest Selection: The left plot in Fig. 2 shows all 19,040 feasible as well as the 652 Pareto-optimal solutions (in the following referred to as efficient solutions) of the different optimization runs. To narrow the analysis and to focus on efficient trade-offs, we further filtered the data by determining cut-off points at each dimension of the objective space. The capacity cut-off was chosen by first selecting all solutions with minimum cost value and subsequently selecting the smallest capacity value within the remaining solutions. The same procedure was applied for the cost cut-off point. These (edge) cut-off points especially aim at removing solutions with the smallest possible cost value but different capacity values along the y-Axis. The capacity and cost cut-off values were set to 1,558 and 3,209, respectively. By the application of the capacity cut-off 6,987 solutions were removed, while the cost cut-off removed 25 solutions. As a result, 12,028 solutions remain in the area of interest for subsequent analysis (see Fig. 2 right).

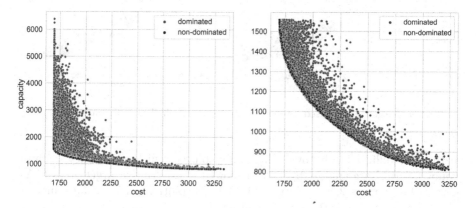

Fig. 2. Feasible objective space (left) and selected area of interest (right)

What we can already identify from the output data visualization is that, while maintaining feasible solutions, the minimum cost of 1,693 can be achieved with the maximum required capacity of 1,558. Further, a total capacity of at least 810 must met to maintain feasible solutions at maximum cost of 3,340.

Target Area Identification: Prior to a more detailed examination of the input data, further target areas within the selected area of interest have to be identi- fied. This could either be done manually [19] or by the application of a clustering algorithm to support the identification of suitable cut-off points [26]. Another option is the use of specific measures, such as the efficiency of a solution mea- sured by its distance to Pareto-optimal solutions [9]. Figure 3 exemplary shows target areas based on clustering (left) and on solution efficiency (right). For the clustering, a k-means clustering algorithm was used with k = 4 and capacity and cost as feature variables. The number of clusters was selected based on the number of desired target areas. For the target areas based on efficiency, the pre- viously introduced feature *efficiency_distance* was used. Here, the distance was normalized to a 0–1 scale and subsequently the solutions were partitioned in 10 target area groups.

The clustered target areas allow an analysis in respect of the different objec- tives, e.g., low-cost and high-cost solutions. The efficiency-based target areas can be used to examine what distinguishes efficient from inefficient solutions. In the further course of analysis, we focus on the trade-off between cost and capacity. Therefore, we chose the clustered target areas as basis. Furthermore, we focus

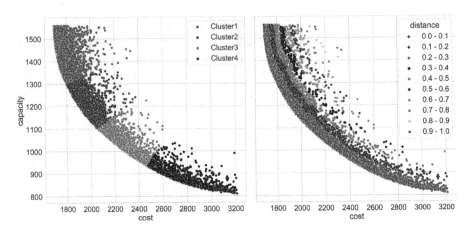

Fig. 3. Target areas based on clustering (left) and efficiency distance (right)

on the most efficient solutions. Thus, only solutions in the efficiency distance group "0.0–0.1" are analyzed. Figure 4 shows the resulting 5 target areas (TA), where TA1 contains 2,425 solutions, TA2 3,012 solutions, TA3 2,859 solutions, TA4 1623 solutions and TA5 2,109 solutions.

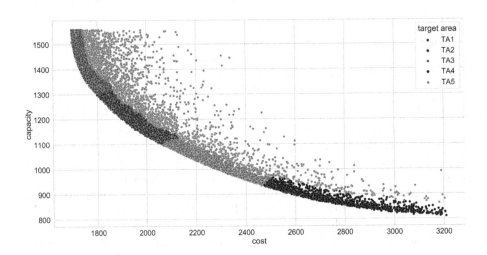

Fig. 4. Selected target areas

Target areas with low cost are further referred to as high performance areas. Consequently, TA1 is the highest and TA4 is the lowest performing area. In terms of efficiency, TA1 to TA4 are considered equivalent, with TA5 being the only inefficient area.

4.2   Data Mining and Visualization of Input Data

The next step is to further analyze the target areas by applying data mining methods, such as clustering, classification or regression. The objectives here could be either to obtain more precise rules, as to which input parameter configurations lead to which target areas, or to gain deeper insights into the individual target areas. Each objective aims to uncover interesting patterns and identify relationships between the decision variables and the model output in order to gain a better understanding of the behavior of the analyzed model and support decision-making. This process is supported by the visualization of the input data (or decision variables). Suitable visualization techniques are, among others, scatter plots, parallel coordinate plots, radar charts, pie charts or box plots.

In this study, we focus on gaining deeper insights into the structure of TA1. Thereby, we apply a k-means clustering algorithm to the input data, with the constraints b1–b5 as feature variables. The number of clusters k = 2 was chosen based on a silhouette score analysis. The aim of clustering is to identify solution groups within the target area with correlated constraint values. Figure 5 shows the resulting clusters C1 (1,424 solutions) and C2 (1,001 solutions) with respect to the output data (left) and input data (right). The parallel coordinate plot (right) visualizes the average value of each capacity constraint within the two clusters.

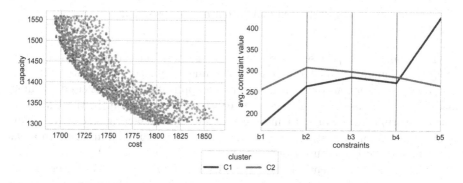

Fig. 5. Visualization of output data (left) and input data (right) of clustered target area TA1

The left plot in Fig. 5 shows that both clusters are distributed across the entire target area, with C1 being slightly more dense on the top left and C2 on the bottom right. However, not much conclusion can be drawn from this plot. The right plot, with the visualization of the input data (on which the clustering was aimed), shows a much clearer picture. Here we identify two solution groups, which are primarily differing in the use of b1 and b5. To support decision-making, the clusters are visualized once again with regard to the output data, however, this time with aggregated values. In order to be comparable, the values of the corresponding variables are first normalized to a 0–1 scale.

The resulting box plots (see Fig. 6) show the normalized values of the two upper-level objectives as well as the efficiency distance measure. While not distinguishable in the first place, the aggregated values reveal that both clusters have a different focus. Therefore, if the decision-maker is interested in low cost, cluster C1 should further be investigated. On the other hand, if the interest tends towards low capacity, further investigations should focus on C2. If there is no particular interest, focus should be on C1 due to it's higher efficiency.

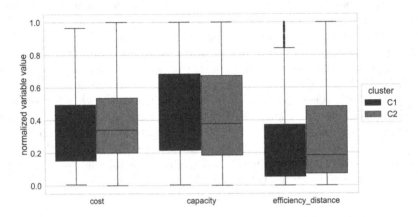

Fig. 6. Normalized output data distribution

To obtain more precise rules that support decision-making regarding the design or configuration of the constraint values (e.g., allocation of additional resources), further analysis can focus on the application of supervised learning algorithms, such as decision trees or linear/logistic regression. In the EMO literature, various kinds of data mining methods for extracting knowledge from multi-objective optimization data are described, depending on application area, problem and data structure (for a comprehensive review, see [2]).

Each of the steps mentioned in Sects. 4.1 to 4.2 may lead to a better understanding of the problem and model, respectively. This knowledge can in turn be used to restart the data generation process at any step, such as to adjust the lower-level model, add or remove decision variables, pick a new algorithm at the upper level or conduct more optimization runs.

5   Conclusions

In this paper, a problem and algorithm independent approach for sensitivity analysis on constraints of combinatorial optimization problems was proposed. The approach was demonstrated by a sensitivity analysis of the well-known and practically important generalized assignment problem. By applying bilevel innovization it was shown that while the objective value (i.e., cost) of the problem can generally be decreased by increasing the capacity constraints, higher performing solutions with respect to the cost objective and efficiency are primarily driven by an increase of constraint b5. Furthermore, the application of a clustering algorithm on the selected target area revealed two groups of solutions, each with a different focus on one of the two upper-level objectives. The insights, however, are limited in that they only apply to the selected target area. For further analysis, the individual target areas could be analyzed in more detail or supervised learning algorithms can be applied to obtain more precise rules that support decision-making. In the course of the sensitivity analysis, evolutionary bilevel optimization in combination with visualization techniques and data mining methods was used. Thereby it was possible to determine how changes in the input parameters of a model (agent capacity) affect its outputs (cost). Insights into the model behavior were gained and recommendations could be given that support the decision-making process regarding design or configuration of constraints, such as the allocation and mix of resources. By considering not only Pareto-optimal but all solutions evaluated during optimization, a broad data base was obtained to identify what distinguishes different solution groups.

Further research will investigate the use of bilevel innovization for different types of problems, both benchmark problems as well as real world problems. For example, it should be examined how problems with numerous constraints can be analyzed by applying dimensionality reduction techniques, such as principal component analysis. Furthermore, it is of interest how different types of upper- and lower-level algorithms affect the data analysis results.

References

1. Al-Maliky, F., Hifi, M., Mhalla, H.: Sensitivity analysis of the setup knapsack problem to perturbation of arbitrary profits or weights. Int. Trans. Oper. Res. **25**(2), 637–666 (2018)
2. Bandaru, S., Ng, A.H., Deb, K.: Data mining methods for knowledge discovery in multi-objective optimization: part a - survey. Expert Syst. Appl. **70**, 139–159 (2017)
3. Borgonovo, E., Plischke, E.: Sensitivity analysis: a review of recent advances. Eur. J. Oper. Res. **248**(3), 869–887 (2016)
4. Branke, J., Deb, K., Miettinen, K., Słowiński, R. (eds.): Multiobjective Optimization. LNCS, vol. 5252. Springer, Heidelberg (2008). https://doi.org/10.1007/978-3-540-88908-3
5. Chu, P.C., Beasley, J.E.: A genetic algorithm for the generalised assignment problem. Comput. Oper. Res. **24**(1), 17–23 (1997)
6. Colson, B., Marcotte, P., Savard, G.: An overview of bilevel optimization. Ann. Oper. Res. **153**(1), 235–256 (2007). https://doi.org/10.1007/s10479-007-0176-2
7. Deb, K., Pratap, A., Agarwal, S.: A fast and elitist multiobjective genetic algorithm: NSGA-II. IEEE Trans. Evol. Comput. **6**(2), 182–197 (2002)
8. Di Gaspero, L., Gärtner, J., Kortsarz, G., Musliu, N., Schaerf, A.: The minimum shift design problem. Ann. Oper. Res. **155**(1), 79–105 (2007). https://doi.org/10.1007/s10479-007-0221-1
9. Dudas, C., Ng, A.H., Pehrsson, L.: Integration of data mining and multi-objective optimisation for decision support in production systems development. Int. J. Comput. Integr. Manuf. **27**(9), 824–839 (2014)

10. Emmerich, M.T.M., Deutz, A.H.: A tutorial on multiobjective optimization: fundamentals and evolutionary methods. Nat. Comput. **17**(3), 585–609 (2018). https://doi.org/10.1007/s11047-018-9685-y

11. Fernández-Baca, D., Venkatachalam, B.: Sensitivity analysis in combinatorial optimization. In: Gonzalez, T.F. (ed.) Handbook of Approximation Algorithms and Metaheuristics, 2nd edn, pp. 455–472. Chapman and Hall/CRC (2018)

12. Hall, N.G., Posner, M.E.: Sensitivity analysis for scheduling problems. J. Sched. **7**(1), 49–83 (2004). https://doi.org/10.1023/B:JOSH.0000013055.31639.f6

13. Hillier, F.S., Lieberman, G.J.: Introduction to Operations Research, 10th edn. McGraw-Hill Education, New York (2015). internat. student ed. edn

14. Keim, D.: Mastering the Information Age: Solving Problems with Visual Analytics. Eurographics Association, Goslar (2010)

15. Kimbrough, S.O., Kuo, A., Lau, H.C.: On decision support for deliberating with constraints in constrained optimization models. In: Pelikan, M., Branke, J. (eds.) Proceedings of the 12th Annual Conference Companion on Genetic and Evolutionary Computation - GECCO 2010, p. 1833. ACM Press, New York (2010)

16. Luke, S.: Essentials of Metaheuristics: A Set of Undergraduate Lecture Notes, 2nd edn. lulu.com, Morrisville (2013). Online version 2.0 edn

17. Marti, L., Garcia, J., Berlanga, A., Molina, J.M.: An approach to stopping criteria for multi-objective optimization evolutionary algorithms: the MGBM criterion. In: 2009 IEEE Congress on Evolutionary Computation, pp. 1263–1270. IEEE (2009)

18. Migdalas, A., Pardalos, P.M., Värbrand, P. (eds.): Multilevel Optimization: Algorithms and Applications. Nonconvex Optimization and Its Applications, vol. 20. Springer, Boston (1998). https://doi.org/10.1007/978-1-4613-0307-7

19. Nojima, Y., Tanigaki, Y., Ishibuchi, H.: Multiobjective data mining from solutions by evolutionary multiobjective optimization. In: Bosman, P.A.N. (ed.) Proceedings of the Genetic and Evolutionary Computation Conference - GECCO 2017, pp. 617–624. ACM Press, New York (2017)

20. Öncan, T.: A survey of the generalized assignment problem and its applications. INFOR: Inf. Syst. Oper. Res. **45**(3), 123–141 (2007)

21. Pardalos, P.M., Du, D.-Z., Graham, R.L. (eds.): Handbook of Combinatorial Optimization. LNCS, Springer, New York (2013). https://doi.org/10.1007/978-1-4419-7997-1

22. Pentico, D.W.: Assignment problems: a golden anniversary survey. Eur. J. Oper. Res. **176**(2), 774–793 (2007)

23. Prodhon, C., Prins, C.: A survey of recent research on location-routing problems. Eur. J. Oper. Res. **238**(1), 1–17 (2014)

24. Saltelli, A.: Global Sensitivity Analysis: The Primer. Wiley, Chichester (2008)

25. Saltelli, A., Annoni, P.: How to avoid a perfunctory sensitivity analysis. Environ. Model. Softw. **25**(12), 1508–1517 (2010)

26. Schulte, J., Feldkamp, N., Bergmann, S., Nissen, V.: Knowledge discovery in scheduling systems using evolutionary bilevel optimization and visual analytics. In: Deb, K., et al. (eds.) EMO 2019. LNCS, vol. 11411, pp. 439–450. Springer, Cham (2019). https://doi.org/10.1007/978-3-030-12598-1_35

27. Sinha, A., Malo, P., Deb, K.: A review on bilevel optimization: from classical to evolutionary approaches and applications. IEEE Trans. Evol. Comput. **22**(2), 276–295 (2018)

28. Talbi, E.G.: A taxonomy of metaheuristics for bi-level optimization. In: Talbi, E.G., Brotcorne, L. (eds.) Metaheuristics for Bi-Level Optimization. SCI, vol. 482, pp. 1–39. Springer, Heidelberg (2013). https://doi.org/10.1007/978-3-642-37838-6_1

29. Wagner, H.M.: Global sensitivity analysis. Oper. Res. **43**(6), 948–969 (1995)

Author Index

Printed in the United States
by Baker & Taylor Publisher Services